2022 IEEE 9th Workshop on Wide Bandgap Power Devices & Applications (WiPDA 2022)

Redondo Beach, California, USA
7 – 9 November 2022

IEEE Catalog Number: CFP22WBP-POD
ISBN: 978-1-6654-8901-0

**Copyright © 2022 by the Institute of Electrical and Electronics Engineers, Inc.
All Rights Reserved**

Copyright and Reprint Permissions: Abstracting is permitted with credit to the source. Libraries are permitted to photocopy beyond the limit of U.S. copyright law for private use of patrons those articles in this volume that carry a code at the bottom of the first page, provided the per-copy fee indicated in the code is paid through Copyright Clearance Center, 222 Rosewood Drive, Danvers, MA 01923.

For other copying, reprint or republication permission, write to IEEE Copyrights Manager, IEEE Service Center, 445 Hoes Lane, Piscataway, NJ 08854. All rights reserved.

*** This is a print representation of what appears in the IEEE Digital Library. Some format issues inherent in the e-media version may also appear in this print version.

IEEE Catalog Number: CFP22WBP-POD
ISBN (Print-On-Demand):   978-1-6654-8901-0
ISBN (Online):   978-1-6654-8900-3
ISSN: 2641-8274

Additional Copies of This Publication Are Available From:

Curran Associates, Inc
57 Morehouse Lane
Red Hook, NY 12571 USA
Phone: (845) 758-0400
Fax: (845) 758-2633
E-mail: curran@proceedings.com
Web: www.proceedings.com

TABLE OF CONTENTS

An Academic's Perspective on SiC Power Devices: Retrospection and Prognostication..................... 1
  B. Jayant Baliga

A New Cell Topology for 4H-SiC Planar Power MOSFETs and Comparison with Hexagonal and
Octagonal Cell Topologies 7
  Shengnan Zhu, Tianshi Liu, Arash Salemi, Michael Jin, Marvin H. White, David Sheridan,
  Anant K. Agarwal

Exploring Optimum Designs for 1.2kV 4H-SiC JBS Diode Integrated MOSFETs (JBSFETs) 11
  Stephen A. Mancini, Seung Yup Jang, Dongyoung Kim, Woongje Sung

Design Optimization and Surge Current Capability of 4H-SiC Lateral Deep P$^+$ JBS Diode on Thin
RESURF Layer.. 17
  Atsushi Shimbori, Alex Q. Huang

Characterization of Near Conduction Band SiC/SiO$_2$ Interface Traps in Commercial 4H-SiC Power
MOSFETs... 22
  Hema Lata Rao Maddi, Suvendu Nayak, Vishank Talesara, Yibo Xu, Wu Lu, Anant K.
  Agarwal

Reverse Breakdown Time of Wide Bandgap Diodes 26
  Jack Flicker, Emily Schrock, Robert Kaplar

Symmetrical V_{TH}/R_{ON} Drifts Due to Negative/Positive Gate Stress in p-GaN Power HEMTs.......... 31
  Nicolò Zagni, Marcello Cioni, Maria Eloisa Castagna, Maurizio Moschetti, Ferdinando
  Iucolano, Giovanni Verzellesi, Alessandro Chini

Peak Channel Temperature Determination for an AlGaN/GaN HEMT with Raman Thermography
and MTTF Extraction for Long Term Reliability... 35
  Cristina Miccoli, Leonardo Gervasi, Viviana Cerantonio, James Pomeroy, Martin Kuball,
  Ferdinando Iucolano

High Temperature Robustness of Enhancement-Mode p-GaN-Gated AlGaN/GaN HEMT
Technology... 40
  Mengyang Yuan, Qingyun Xie, John Niroula, Mohamed Fadil Isamotu, Nitul S. Rajput,
  Nadim Chowdhury, Tomás Palacios

Effects of Oxide Electric Field Stress on the Gate Oxide Reliability of Commercial SiC Power
MOSFETs... 45
  Limeng Shi, Tianshi Liu, Shengnan Zhu, Jiashu Qian, Michael Jin, Hema Lata Rao Maddi,
  Marvin H. White, Anant K. Agarwal

A Comparison of Ion Implantation at Room Temperature and Heated Ion Implantation on the Body
Diode Degradation of Commercial 3.3 kV 4H-SiC Power MOSFETs 49
  Jiashu Qian, Tianshi Liu, Jake Soto, Mowafak M. Al-Jassim, Robert Stahlbush, Nadeemullah
  Mahadik, Limeng Shi, Michael Jin, Anant K. Agarwal

A Comparison of Short-Circuit Failure Mechanisms of 1.2 kV 4H-SiC MOSFETs and JBSFETs... 54
  Dongyoung Kim, Skylar Deboer, Seung Yup Jang, Adam J. Morgan, Woongje Sung

Failure Rate Calculation Due to Neutron Flux with SiC MOSFETs and Schottky Diodes.............. 58
  Dennis Meyer, Xuning Zhang, Reenu Garg, Bruce Odekirk, Steve Chenetz, Ehab Tarmoom,
  Kevin Speer

Scaling of EPC's 100 V Enhancement-Mode Power Transistors..64
Gordon Stecklein, Jordan Green, Christopher Wong, Joe Cao, Bob Beach

Advancement in Integration for GaN Power ICs..68
Tom Ribarich, Stephen Oliver, Marco Giandalia, Llew Vaughan-Edmunds

Threshold Voltage Behavior and Short-Circuit Capability of p-Gate GaN HEMTs Depending on
Drain- And Gate-Voltage Stress..73
Thorsten Oeder, Martin Pfost

Series Compensation for DC Microgrid Stabilization Utilizing T-Type Modular Dc Circuit Breaker..............77
Faisal Alsaif, Yue Zhang, Nihanth Adina, Khalid Alkhalid, Jin Wang

3kV 6.7mΩ·cm² 4H-SiC BJT with an Effective Junction Termination Extension (JTE)..................................82
Xixi Luo, Alex Q. Huang

Thermal Design and Experimental Evaluation of a 1kV, 500A T-Type Modular DC Circuit Breaker..............86
*Baljit Riar, Jeffrey Ewanchuk, Hailing Wu, Yue Zhang, Xiao Li, Dihao Ma, Rob Borjas, Jin
Wang*

Source Turn-Off (STO) MOSFET: A New Driving Architecture for Smart SiC Module................................91
Zhicheng Guo, Alex Q. Huang

A 5 to 50 V, −25 to 225 °C, 0.065%/°C GaN MIS-HEMT Monolithic Compact 2T Voltage
Reference..95
Ziqian Li, Yi Shen, Ang Li, Wen Liu

Embedding Solutions for Vertical SiC and GaN Power Devices 99
*Hoang Linh Bach, Anqi Huang, Yue Teng, Hubert Rauh, Andreas Schletz, Michael P. M.
Jank, Martin März*

Three-Level ANPC Inverter Common-Mode Voltage Analytical Characterization 105
Yang Huang, Xin Xia, Hua Kevin Bai, Fanning Jin, Xiaodong Shi, Bing Cheng

Bidirectional High Voltage Conversion Ratio High-Frequency DC/DC Converter with Low
Number of Components 111
Pedram Chavoshipour Heris, Zahra Saadatizadeh, Rahul Biswash, Alan Mantooth

Short Circuit Fault Induced Failure of SiC MOSFETs in DC Solid-State Circuit Breakers 116
Shuyan Zhao, Reza Kheirollahi, Hua Zhang, Fei Lu

Area-Efficient High-Voltage (HV) Lateral MOSFETs for Discrete Device Development and Power
IC Integration 122
Sundar Babu Isukapati, Seung Yup Jang, Woongje Sung

A New Layout Method for Junction Field Effect Transistors (JFETs) on 4H-SiC that Provides a
Significant Reduction in On-Resistance...127
Justin Lynch, Nick Yun, Seung Yup Jang, Adam J. Morgan, Woongje Sung

Busbar Design and Optimization for High Power Three-Phase Inverter with WBG Device 132
Yuxuan Wu, Mustafeez Ul-Hassan, Fang Luo

Analysis of a Switching Event and Its Impact on Gate Drive in Gallium-Nitride Based Bi-
Directional Switches..138
Mustafeez Ul-Hassan, Yuxuan Wu, Fang Luo

In-Situ Ultrafast Sensing Techniques for Prognostics and Protection of SiC Devices 142
Ali Parsa Sirat, Chondon Roy, Daniel Evans, James Gafford, Babak Parkhideh

A Flux Balancing Strategy for 10-KV SiC-Based Dual-Active-Bridge Converter 148
Zihan Gao, Pengfei Yao, Haiguo Li, Fred Wang

Short-Circuit Ruggedness and Partial Discharge Evaluation of a 3.3 kV SiC MOSFET Power
Module 154
*Ke Wang, Yizhou Cong, Pengyu Fu, Xiao Li, Qianyi Cheng, Boxue Hu, Jin Wang, Ashish
Kumar, Kraig Olejniczak, Daniel Pelletier, Zach Cole, Amol Deshpande, Amit Goyal*

Advantages of SiC-Based Devices on the Design of Dual-Active Bridge DC/DC Converter for DC
Faults 159
Shrivatsal Sharma, Yos Prabowo, Subhransu Satpathy, Subhashish Bhattacharya

Active Gate Driving of Cascoded SiC JFETs... 164
Arijit Sengupta, Sima Azizi Aghdam, Mohammed Agamy

High Frequency High Power Integrated Transformer Design for Resonant Converters with SiC
Devices 170
Tianlong Yuan, Feng Jin, Zheqing Li, Qiang Li

A Highly Integrated Sensorless Field Oriented Control BLDC / PMSM Inverter with 99%
Efficiency Enabled by an All-in-One System Integrated Full SiC Intelligent Power Module
(sIPM®)... 176
*Fu-Jen Hsu, Cheng-Tyng Yen, Hsiang-Ting Hung, Guan-Wei Lin, Chih-Feng Huang, Lung-
Sheng Lin, I-Chi Lin, Chih-Fang Huang, Ta-Yung Yang*

Noise Analysis of Current Sensor for Medium Voltage Power Converter Enabled by Silicon-
Carbide MOSFETs 180
*Morten Rahr Nielsen, Mathias Kirkeby, Hongbo Zhao, Dipen Narendra Dalal, Michael
Møller Bech, Stig Munk-Nielsen*

GaN Power Converter Applied to Electrocaloric Heat Pump Prototype and Carnot Cycle 186
*Stefan Moench, Richard Reiner, Kareem Mansour, Michael Basler, Patrick Waltereit,
Rüdiger Quay, Kilian Bartholomé*

Design Considerations of a GaN-Based Three-Level Traction Inverter for Electric Vehicles 192
Subhransu Satpathy, Partha Pratim Das, Subhashish Bhattacharya, Victor Veliadis

Novel High-Voltage-Gain High-Frequency Nonisolated Three-Port DC-DC Converter with Zero
Input Current Ripple and Soft Switching Capability.. 198
Zahra Saadatizadeh, Pedram Chavoshipour Heris, Alan Mantooth

Comparison of Thermally Optimized SMD Packages for 100 V GaN HEMTs in 300 kHz Buck
Converter High Current Applications.. 204
Dominik Koch, Ankit Sharma, Till Huesgen, Ingmar Kallfass

Design of High Power Converter with Single Low R_{on} Discrete SiC Device 209
Zibo Chen, Chen Chen, Qingyun Huang, Alex Q. Huang

Comparative Investigation of Current-Source Inverters Using SiC Discrete Devices and Power
Modules 215
Feida Chen, Sangwhee Lee, Thomas M. Jahns, Bulent Sarlioglu

A Medium-Voltage Transformer with Integrated Leakage Inductance for 10 kV SiC-Based Dual-Active-Bridge Converter 221
Zihan Gao, Haiguo Li, Fred Wang

Development of a 250°C 15kV Supercascode Switch Using SiC JFET Technology 227
David E. Sanabria, Randy Appert, Steven G. E. Pronko, Joshua Major, Douglas Devoto,
Karen Heinselman, Jane M. Lehr, Nicolas Gonzalez, David S. Ginley

Next Generation of GaN Single-Board High Power Stages 233
Rahil Samani, Juncheng Lu, Ignacio Galiano Zurbriggen

100V GaN for Highly Efficient 1kW Motor Drive Applications 238
Asantha Kempitiya, Hrach Amirkhanian, Srikanth Yerra, Kapil Kelkar

Compact Three-Level GaN Power Module Suitable for Active-Neutral-Point-Clamped (ANPC)
Three-Level Converter 242
Ziwei Liang, Liyan Zhu, Hua Bai, Yue Sun

Design of High Current, High Power Density GaN Based Motor Drive for All Electric Aircraft
Application 247
Waqar A. Khan, Armin Ebrahimian, Hosseini S. S. Iman, Nathan Weise

Design of Three-Level Flying Capacitor Totem Pole PFC in USB Type-C Power Delivery for
Aircraft Applications 254
Tianyu Zhao, Rolando Burgos, Bo Wen, Andrew McLean, Rodrigo Fernández Mattos

Author Index

An Academic's Perspective on SiC Power Devices: Retrospection and Prognostication

B. Jayant Baliga

Progress Energy Distinguished University Professor of Electrical Engineering

North Carolina State University

Raleigh, NC, USA

bjbaliga@ncsu.edu

Abstract—The role of an academic in the area of SiC power devices can be categorized as: (1) Fundamental Theoretical Predictions; (2) Fundamental Scientific Knowledge; (3) Fundamental Device Structures; (4) Device Structural Innovations; (5) Device Physics and Analysis; and (6) Technology Feasibility Studies. This paper provides examples of these activities during my past 40 years of effort in this field as a retrospective. New prospective device ideas that will have a strong impact on power electronics in the future are then discussed.

Keywords—*BFOM, BG-MOSFET, BiDFET, Channel length, Gate oxide thickness, HF-FOMM, Impact ionization coefficients, JBSFET, JBS Rectifier, MRBT, OCTFET, SG-MOSFET, Shielded Power MOSFET, Silicon Carbide, Specific on-resistance*

I. INTRODUCTION

In my personal opinion, an academic has the prime responsibility to demonstrate exemplary scholarship for emulation by students at the university. This was indeed true for my own advisor, Prof. Sorab K. Ghandhi, under whose tutelage I earned by Masters and Doctorate degrees at Rensselaer Polytechnic Institute. It was his support that allowed my transition from India to the United States to embark up on a career devoted to doing research leading to many innovations that have benefitted society. This was the stated objective in my graduate school application, which although naïve at that time, actually came to fruition.

The academic's role in the context of SiC power devices can be defined by the following categories:
(a) Fundamental Theoretical Predictions: The formulation of new directions for making quantum leaps in power device performance;
(b) Fundamental Scientific Knowledge: The addition of basic knowledge that is required to advance the understanding of power device performance;
(c) Fundamental Device Structures: The creation of structures that break barriers to the realization of power devices;
(d) Device Structural Innovations: The proposing and demonstration of novel device structures that enhance the on-state, switching, and safe-operating-area of power devices;
(e) Device Physics and Analysis: The generation of an improved understanding of the operation of power device structures with analytical and numerical modelling;
(f) Technology Feasibility Studies: The exploration of structural parameters beyond the boundaries used by the commercial vendors of power devices.
Examples in each of these categories are provided in this article as a retrospective view over my four-decades of activity on power devices. This is followed by a prognostication section where new device ideas are described that could have a strong impact on power electronics in the future. Many other excellent examples are available in the literature that are not discussed here.

II. FUNDAMENTAL THEORETICAL PREDICTIONS

A. Baliga's Figure-of-Merit (BFOM)

After joining the GE Research Center in 1974, I found that the power device community was content with making devices from silicon. I asked the question: *Would there be a benefit by replacing Si with another semiconductor material? How can this benefit be quantified to justify an effort to make this transition attractive?* No one seems to have taken the initiative at that time to ponder up on this. I realized that a key parameter dictating the performance of power devices was the voltage drop across it when carrying current. It was also recognized that unipolar devices were preferable to bipolar devices due to superior switching performance.

The on-state voltage drop of a high voltage unipolar device was limited by the resistance of the drift region required to support the voltage. In 1979, I was able to derive a relationship between the resistance of the drift region of a unipolar power device and the basic material properties [1,2]:

$$R_{on,sp} = \frac{4\,BV^2}{q\,\mu_n\,E_C^3}$$

where BV is the breakdown voltage, q is the electron charge, μ_n is the electron mobility, and E_C is the critical electric field for breakdown. GE allowed my publication of this powerful relationship after a few years of embargo. During this time, I supervised a team of 10 people for development of the first power devices from Gallium Arsenide [3]. They were the first wide band gap semiconductor based power devices that became products.

The denominator of the above equation is now commonly referred to as *Baliga's Figure-of-Merit* (BFOM) for power devices. It can be written as:

$$BFOM = q\,\mu_n E_C^3 = \frac{4\,BV^2}{R_{on,sp}}$$

with units of W/cm². The ratio of the BFOM for any semiconductor to that for Si is indicative of the improvement in the on-state resistance and hence performance. It should be done using parameters applicable to the same breakdown voltage that is dictated by the device application. Pertinent to this article, the improvement predicted by this equation for 4H-SiC over Si is a factor of 1000. This enormous theoretically predicted value allowed justification of many decades of effort to manufacture high quality large diameter 4H-SiC wafers and device structures that can display this performance. A paper published in 2014 documents the $R_{on,sp}$ of 4H-SiC planar-gate power MOSFETs fabricated with ratings of 900 V to 15,000 V [4]. The $R_{on,sp}$ is close to drift region resistance for devices with ratings above 3000 V with values 1000-times smaller than the ideal Si case. The

978-1-6654-8901-0/22 $31.00 © 2022 IEEE

prophecy made using the BFOM had come true after three decades of effort by the university and industrial community with considerable funding from governmental agencies.

The BFOM is now often used to justify effort on ultra-wide band gap semiconductors. In addition, the computed value for BFOM using the measured BV and $R_{on,sp}$ is now often used to demonstrate progress in the literature. This term has even begun to show up in the titles of published papers!

B. Benefits to Power Electronic Systems

In order to create a sustained effort to develop and utilize SiC power devices, it was also important to show the benefits that can be accrued to the power electronics community. By the 1990s, the Si IGBT had become widely accepted as the most suitable device for power electronics operating at voltages above 200 V [5]. However, its switching losses were known to be an impediment to operation at higher frequencies. In 1994, I was invited to contribute an article for a special issue of the Proceeding of the IEEE on Variable Frequency Motor Drives [6]. I took this opportunity to quantify the reduction in power losses that could be achieved with various emerging Si power devices and project the performance achievable with SiC power devices. The projected benefits of replacing the Si IGBT and its Si PiN anti-parallel diode with the SiC power MOSFET and Schottky antiparallel diode that

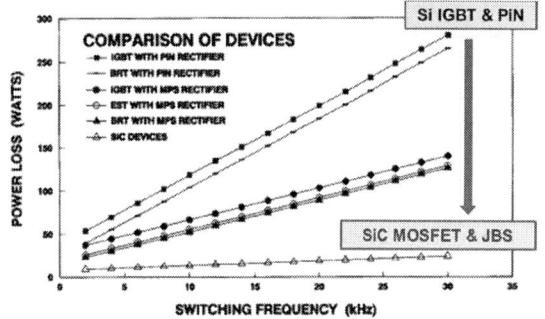

Fig. 1. Projected reduction in power losses achieved by replacing the Si IGBT and PiN antiparallel diode with emerging Si devices and projected benefits of using unipolar SiC power devices published in 1994 [6].

was published in this paper is reproduced in Fig. 1. A 10-fold reduction in power losses was projected by me encouraging pursuit of this technology for industrial and transportation applications. After a decade of effort, a paper was published with a figure exactly like that shown in Fig. 1 with experimental demonstration of the projected performance [7].

III. FUNDAMENTAL SCIENTIFIC KNOWLEDGE

SiC material was in its infancy when I proposed the BFOM in 1979. It was not possible to project the performance of the benefits of SiC due to the many poly-types with poorly quantified material properties. With the availability of some data by 1989, it was possible to project a reduction in drift region on-resistance by a factor of 100 for the 6H-SiC poly-type [2]. However, the measurements of impact ionization coefficients that determine E_C were not satisfactory due to use of optical excitation. The measurement area was large intersecting regions in the material with defects and the diode edges with enhanced electric fields that distort the data.

I assigned the measurement of the impact ionization coefficients for 6H-SiC and 4H-SiC to my graduate student Rohini Raghunathan in 1994. I procured an electron-beam microscope with beam blanking capability to make the measurement area very small (< 10 μm diameter) to avoid defects. In addition, I recognized that the measurements would be limited by enhanced electric fields at the diode edges. To solve this problem, I proposed an argon implanted edge termination for the Schottky diode test structures with which another student, Dev Alok, was able to obtain nearly ideal breakdown voltage [8]. Using these diodes and a carefully crafted extraction procedure, we were able to obtain high quality data on impact ionization coefficients as a function of electric field [9]. This data was also obtained as a function of temperature by installing a heated stage inside the microscope. Using this impact ionization data, the E_C for 4H-SiC increased significantly resulting in a BFOM of 1000-times larger than Si. In addition, we were able to show that the impact ionization coefficients are much larger at defect sites in the 4H-SiC material. These results provided even greater motivation for the creation of SiC power devices.

IV. FUNDAMENTAL DEVICE STRUCTURES

It was initially assumed that the unipolar devices made from Si could be easily replicated in SiC material. However, such clones of Si structures did not perform well making it necessary to make fundamental changes to the device structure to extract the projected performance benefits.

A. Junction Barrier controlled Schottky (JBS) Rectifier

The first high voltage (400 V) SiC Schottky rectifiers were simple structures fabricated using dots of platinum evaporated on to the 6H-SiC drift layer after surface preparation [10]. They had excellent on-state characteristics with a voltage drop of 1.1 V at 100 A/cm². They displayed no reverse recovery current when compared with large value observed for Si PiN diodes. This provided confirmation that SiC Schottky diodes would make superior anti-parallel diodes for Si IGBTs. However, the leakage current of the diode increased by 5-orders of magnitude with increasing reverse bias voltage. This is due to the large Schottky barrier lowering and tunneling current contributions [5].

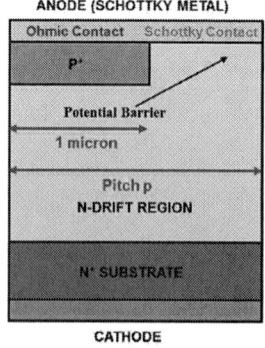

Fig. 2. The JBS Diode concept.

In 1984, I proposed the use of a P-N junction in parallel with the Schottky contact to improve the performance of Si Schottky diodes [11]. This concept, called the Junction Barrier controlled Schottky (JBS) rectifier illustrated in Fig. 2, was shown to mitigate the increase in leakage current due to the

978-1-6654-8901-0/22 $31.00 © 2022 IEEE

Schottky barrier lowering phenomenon. Although effective for Si devices with use of small area for the P$^+$ regions [12], the JBS concept has proven to be indispensable for development of 4H-SiC Schottky diodes. The first 4H-SiC JBS diodes were reported in 1998 [13,14]. The leakage current can be reduced by 4 orders of magnitude using the JBS diode structure by proper adjustment of the cell pitch p with minimal increase in the on-state voltage drop [15].

B. Shielded Planar-Gate Power MOSFET

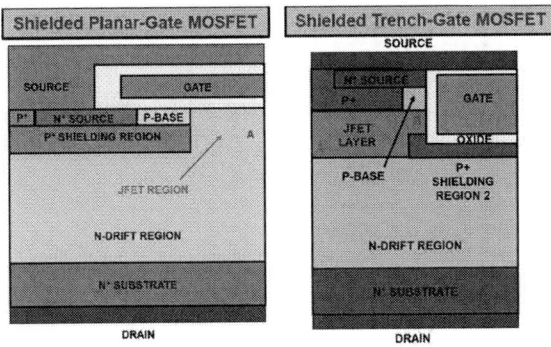

Fig. 3. Shielded SiC Power MOSFET structures.

The planar-gate 4H-SiC power MOSFETs were commercialized first. This required overcoming several major obstacles that prevented the cloning of Si device technology. The first practical problem was the lack of significant diffusion of dopants in SiC. This problem was overcome by performing staggered ion implantation of the P-base and N$^+$ source regions in place of the DMOS process [16,17]. The second physics based problem was enhanced reach-through of the depletion region due to the high electric field at the P-base/N-drift junction. This leads to very long channel lengths that make the R$_{on,sp}$ prohibitively large [15]. The reach-through problem was overcome using the P$^+$ shielding region shown in Fig. 3 [18]. The third physics based problem was development of very high (> 3 MV/cm) electric fields in the gate oxide leading to poor reliability. This problem is also overcome using the P$^+$ shielding region. It produces a potential barrier in the JFET region at location A that screens the gate oxide and brings down the electric field in the oxide. The first shielded SiC power MOSFETs were experimentally demonstrated in 1997-1998 [19,20].The shielded planar-gate 4H-SiC power MOSFET structure is now used in all commercial products.

The trench-gate 4H-SiC power MOSFET are prone to the same problems discussed above. One solution for this problem is the structure shown in Fig. 3 with a P$^+$ shielding region located at the bottom of the trench [21]. Another P$^+$ shielding region is located next to the P-base region in Fig. 3 to suppress reach-through breakdown [22]. This approach has produced working devices [23]. The commercialized trench-gate products utilize deep P$^+$ regions to bring down the electric field in the gate oxide [24,25]. An alternative approach makes use of a thick oxide at the bottom of the trench [26]. A comparison of the R$_{on,sp}$ and Q$_{gd,sp}$ for these trench-gate structures has been published [27].

V. DEVICE STRUCTURAL INNOVATIONS

Replacing the Si IGBT with a SiC power MOSFET to reduce power losses in inverters is handicapped by the larger cost of this nascent technology. One approach to offset the larger SiC chip cost is by reducing the overall cost of the power electronics by increasing the operating frequency. This reduces the cost of the passive elements and filters. SiC power device innovations are needed to reduce switching power losses rather than just the R$_{on,sp}$. This has been achieved by several structures discussed here. The high-frequency figures-of-merit HF-FOM[R$_{on}$*C$_{gd}$] and HF-FOM[R$_{on}$*Q$_{gd}$] are good metrics for gauging the overall improvement in performance achieved by using these innovations.

A. The JBSFET

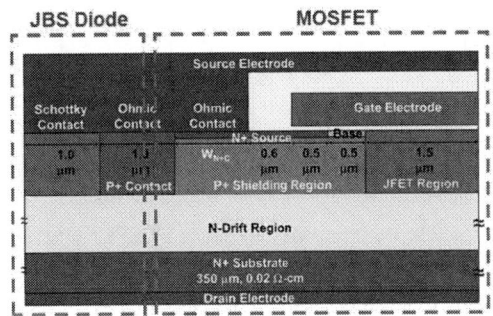

Fig. 4. The SiC Power JBSFET structure.

In principle the IGBT and its anti-parallel PiN diode can be replaced with a single SiC power MOSFET by using its body diode for current flow in the third quadrant. However, this has been shown to lead to a bipolar degradation problem [28]. In addition, the reverse recovery losses from the P-N junction body diode have been found to be significant at elevated temperature [29]. These problems can be overcome by using the JBSFET structure shown in Fig. 4 where a JBS diode is integrated within the power MOSFET cell [30,31]. Current flow in the third quadrant occurs via the JBS diode suppressing the body diode. The cell cross-section shown in Fig. 4 can be utilized with any gate electrode topology, such as the hexagonal and octagonal cases, to obtain good performance [32]. This idea has been implemented on devices with ratings of 600, 1200, 1700 and even 3300 V.

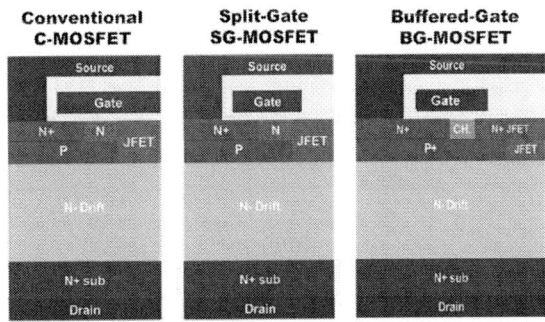

Fig. 5. The SiC Power MOSFET structural innovations.

B. The SG-MOSFET and BG-MOSFET

The dominant component of the turn-on and turn-off switching power losses for SiC power MOSFETs occurs during the drain voltage transient. The drain voltage transition time is determined by the gate-drain capacitance (C$_{gd}$) and gate charge (Q$_{gd}$) [5]. Two innovative structures that reduce these parameters are the split-gate (SG) and buffered-gate (BG) designs shown in Fig. 5 [33]. The HF-FOM of the 4H-

978-1-6654-8901-0/22 $31.00 © 2022 IEEE

SiC power MOSFET can be improved by a factor 1.7-times by utilizing the SG structure [34] while an improvement by a factor of 4-times has been experimentally demonstrated with the BG structure [35].

C. The OCTFET

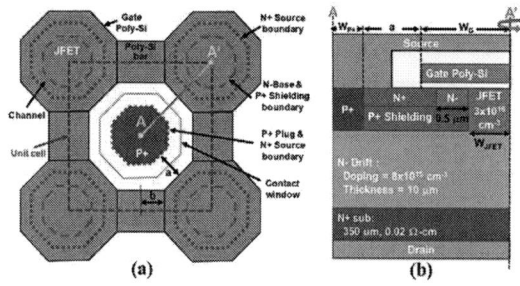

Fig. 6. The Octagonal cell SiC power MOSFET structure.

An additional structural innovation to improve the HF-FOM is by change the gate electrode layout to the Octagonal shape as illustrated in Fig. 6 [36]. The cross-section along line A-A' for this design is identical to that for the Linear cell MOSFET shown in Fig. 3 making the processing for them identical. An improvement in the HF-FOM by a factor of 1.5-2.0 times has been experimentally observed with this idea [37].

VI. DEVICE PHYSICS AND ANALYSIS

A traditional role taken by academics is in performing analyses by numerical simulations and analytical modelling that improve our understanding of the physics of SiC devices. This has of course been done throughout my 30 year academic career.

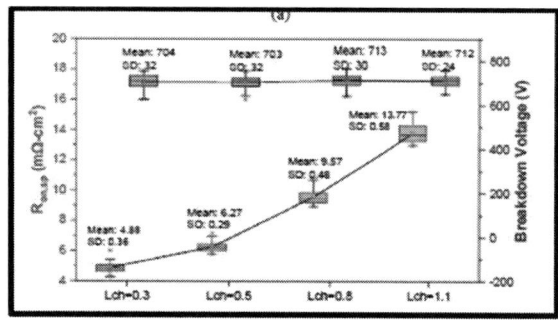

Fig. 7. Impact of reducing channel length for 4H-SiC power MOSFETs.

One recent example was demonstration of the benefits of reducing the channel length on 600 V 4H-SiC power MOSFETs. $R_{on,sp}$ was shown to monotonically improve by reducing the channel length to 0.3 µm without loss of blocking voltage capability due to short-channel effects as shown in Fig. 7 [38].

Another example was demonstration of the importance of including the ion implant straggle of the P^+ shielding region in the analysis and optimization of the 1.2 kV rated 4H-SiC planar-gate power MOSFET structure [39]. The implant straggle was shown to increase channel length and narrow the JFET width. This enlarges the optimum JFET width required

to obtain the lowest $R_{on,sp}$. It also enables improvement of the HF-FOM by a factor of 2.5-times and allows manufacturing short-channel (0.3 µm length) devices with a non-self-aligned 0.3 µm technology.

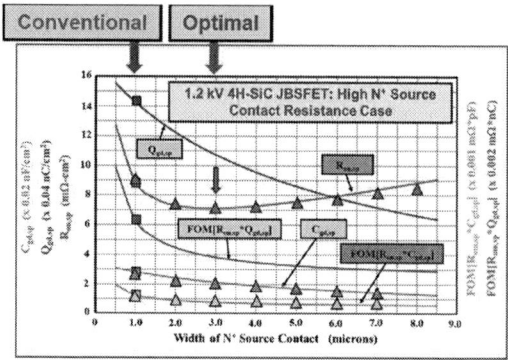

Fig. 8. Optimization of N^+ source contact width in the SiC power JBSFET structure.

A third example is the optimization of the width of the contact to the N^+ source region for the SiC JBSFETs [40]. As shown in Fig. 8, analytical modelling demonstrates that the $R_{on,sp}$ can be reduced by a factor of 1.25x by increasing the contact width from the conventional 1 µm to an optimal value of 3.0 µm, while simultaneously improving the HF-FOM by a factor of 1.65-times.

VII. TECHNOLOGY FEASIBILITY STUDIES

Another role for academic research is to explore new process technologies that could enhance SiC power MOSFET performance that are not being pursued by the commercial vendors. This type of effort was undertaken at NCSU during the past 5 years by examining the benefits of reducing the gate oxide thickness in half from the 500 Å value typically used by the industry to build product.

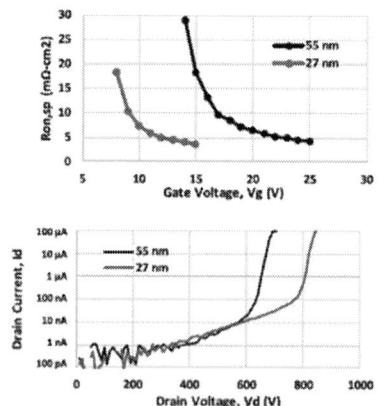

Fig. 9. Performance of 600 V rated SiC power MOSFETs with thinner gate oxide.

A systematic study was performed using numerical simulations and the measured quantification of the performance of 4H-SiC planar-gate power MOSFETs manufactured in a commercial foundry. The research showed that significant improvements are obtained with reduction of gate oxide thickness in half for 4H-SiC power MOSFETs with ratings of 600 V, 1200 V, and 1700 V [41,42,43]. The reduced gate oxide thickness allows these devices to be operated using the widely available inexpensive 15 V Si IGBT gate drive

978-1-6654-8901-0/22 $31.00 © 2022 IEEE

circuits. As an example, the $R_{on,sp}$ and breakdown characteristics for the 600 V rated 4H-SiC planar-gate power MOSFETs are reproduced in Fig. 9. The breakdown voltage is larger with the thinner oxide with oxide electric field within reliability limits. A 5-fold improvement in the $R_{on,sp}$ and HF-FOM was observed in this case. In addition, it was experimentally shown that the short-circuit withstand capability is improved with the thinner gate oxide [44]. These results may encourage the industry to move towards thinner gate oxide based products.

VIII. PROGNOSTICATION

It has been said that '*Making predictions is difficult especially about the future!*' A safer approach taken here is to describe some new device innovations that are likely to have a transformative influence on power electronics. Until my invention and commercialization of the Si IGBT, power electronics favored the current-source-inverter (CSI) due to availability of thyristors with bi-directional voltage blocking capability. The Si IGBT revolutionized power electronics by making voltage-source-inverters the dominant topology. However, this topology requires a bulky and unreliable DC bus electrolytic capacitor with limited temperature withstand capability. Reverting to the CSI topology has not been feasible due to lack of a fast switching MOS-gated reverse blocking switch. In addition, a true bi-directional blocking switch is needed to construct matrix converters with improved power density. New SiC based device innovations now provide the desired solutions to enable these circuit topologies.

A. The MBRT

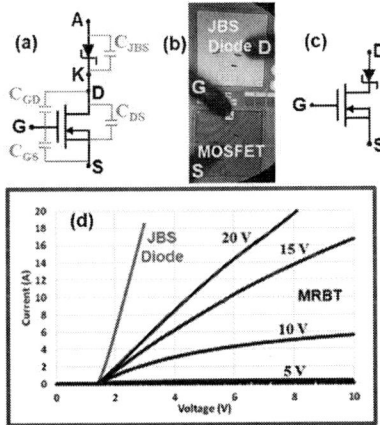

Fig. 10. The SiC Monolithic Reverse Blocking Transistor (MBRT).

A three-terminal 1.2 kV rated SiC monolithic reverse blocking transistor (MRBT) was recently created by integration of a JBS diode in series with the drain of the SiC power MOSFET [45] as illustrated in Fig. 10(a). The cathode of the JBS diode and the drain of the MOSFET share the same N^+ substrate allowing convenient co-fabrication in a commercial foundry. The fabricated MRBTs could support 1650 V in the first and third quadrants. They have diode-like on-state characteristics as shown in Fig. 10(d) (blue line) with MOS-gate controlled output characteristics. The internal unipolar devices provide the desired fast switching, low power loss behavior needed for CSI circuits.

B. The BiDFET

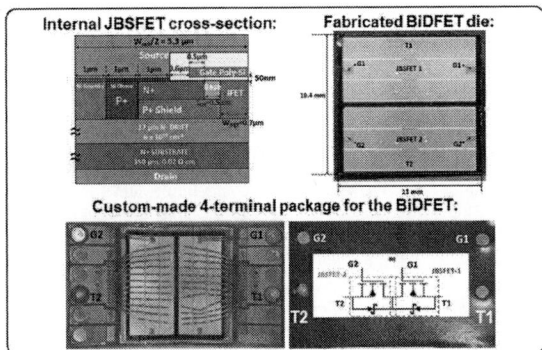

Fig. 11. The SiC Monolithic Bi-Directional Field Effect Transistor (BiDFET).

A 1.2 kV rated four-terminal SiC monolithic bi-directional field effect transistor (BiDFET) was recently created by integration of two JBSFETs in tandem configuration [46]. JBSFETs were designed with a large active area to achieve a BiDFET on-resistance of 50 mΩ with a 1 cm x 1 cm die size as shown in Fig. 11 [47]. The JBSFET cell cross-section is shown in the figure. The die was packaged in a custom designed 4-terminal compact package shown in the figure. The devices could support 1400 V with on-state resistance of 50 mΩ in both quadrants. MOS-gate controlled output characteristics with good FBSOA were observed in both quadrants. Excellent switching characteristics for the device were also experimentally confirmed making the BiDFET an ideal component for matrix converters [48]. The devices have been used to put together a 2.3 kW matrix converter with 95 % efficiency for residential photovoltaic use [49].

IX. CONCLUSIONS

The formulation of the Baliga's Figure-of-Merit in 1979 allowed prognostication of an enormous reduction in the specific on-resistance by replacing Si with SiC material. However, the available Si structures were inadequate to deliver the desired performance enhancement. Fundamental changes in the device structures allowed commercialization of 4H-SiC JBS diodes and shielded power MOSFETs after 3 decades of effort on materials growth and device processing.

Structural enhancements to improve the high frequency operation of power MOSFETs have been recently demonstrated that will allow power electronic circuits to operate at higher frequency. This should reduce the cost and size of the SiC based power electronics to below that achievable using Si technology. SiC power device technology is therefore poised to make a huge impact on electric vehicles and renewable energy generation. The resulting future benefits to society should rival that already derived using Si IGBTs during the past 3 decades.

REFERENCES

[1] B. J. Baliga, "Semiconductors for High Voltage Vertical Channel Field Effect Transistors", *J. Applied Physics*, Vol. 53, pp. 1759-1764, 1982.

[2] B. J. Baliga, "Power Semiconductor Device Figure of Merit for High Frequency Applications," *IEEE Electron Device Letters*, Vol. EDL-10, pp. 455-457, 1989.

[3] B. J. Baliga, et al. "Gallium Arsenide Sc;hottky Power Rectifiers", *IEEE Trans. Electron Devices*, Vol. ED-32, pp. 1130-1134, 1985.

[4] J. W. Palmour, "Silicon Carbide Power Device Development for Industrial Markets", *IEEE Int. Electron Devices Meeting*, pp. 1.1.1-1.1.8, 2014.

[5] B. J. Baliga, "Fundamentals of Power Semiconductor Devices", *Springer Science*, Second Edition, Chambourg, Switzerland, 2019.

[6] B. J. Baliga, "Power Semiconductor Devices for Variable-Frequency Drives", (Invited Paper in Special Issue on 'Power Electronics and Motion Control'), *Proceedings of the IEEE*, Vol. 82, pp. 1112-1122, 1994.

[7] J. Fabre, et al., "Characterization and Implementation of Dual-SiC MOSFET Modules for Future Use in Traction Converters", *IEEE Trans. Power Electronics*, Vol. 30, pp. 4079-4090, 2015.

[8] D. Alok, B. J. Baliga, and P. K. McLarty, "A Simple Edge Termination for Silicon Carbide Devices with Nearly Ideal Breakdown Voltage", *IEEE Electron Device Letters*, Vol. 15, pp. 394-395, 1994.

[9] R. Raghunathan and B. J. Baliga, "Measurement of Electron and Hole Impact Ionization Coefficients for SiC", *IEEE Int. Symp. On Power Semiconductor Devices and ICs"*, Paper O3.2, pp. 173-176, Weimar, Germany, May 1997.

[10] M. Bhatnagar, P. M. McLarty, and B. J. Baliga, "Silicon-Carbide High-Voltage (400 V) Schottky Barrier Diodes", *IEEE Electron Device Letters*, Vol. 13, pp. 501-503, 1992.

[11] B. J. Baliga, "The Pinch Rectifier: A Low-Forward-Drop High-Speed Power Diode", *IEEE Electron Device Letters*, Vol. EDL-5, pp. 194-196, 1984.

[12] M. Mehrotra and B. J. Baliga, "Very Low Forward Drop JBS Rectifiers Fabricated Using Submicron Technology", *IEEE Trans. Electron Devices*, Vol. 40, pp. 2131-2132, 1993.

[13] R. Held, N. Kaminski, and E. Nieman, "SiC Merged p-n/Schottky Rectifiers for High Voltage Applications", *Material Science Forum*, Vol. 264-268, pp. 1057-1060, 1998.

[14] F. Dahlquist, "Junction Barrier Schottky Diodes in 4H-SiC and 6H-SiC", *Material Science Forum*, Vol. 264-268, pp. 1061-1064, 1998.

[15] B. J. Baliga, "Gallium Nitride and Silicon Carbide Power Devices", *World Scientific Press*, 2017.

[16] B. J. Baliga and M. Bhatnagar, "Method of Fabricating Silicon Carbide Field Effect Transistor", *U. S. Patent 5,322,802*, Issued June 21, 1994.

[17] J. N. Shenoy, J. A. Cooper, and M. R. Melloch, "High Voltage Double-Implanted Power MOSFETs in 6H-SiC", *IEEE Electron Device Letters*, Vol. 18, pp. 93-95, 1997.

[18] B. J. Baliga, "Silicon Carbide Semiconductor Devices having Buried Silicon Carbide Conduction Barrier Layers Therein", *U. S. Patent 5,543,637*, Issued August 6, 1996.

[19] P. M. Shenoy and B. J. Baliga, "The Planar 6H-SiC ACCUFET: A New High-Voltage Power MOSFET Structure", IEEE Electron Device Letters, Vol. 18, pp. 589-591, 1997.

[20] R. Chilukuri, P. M. Shenoy, and B. J. Baliga, " Compairison of 6H-Sic and 4H-Sic High Voltage Planar ACCUFETs", *IEEE Int. Symp. On Power Semiconductor Devices and ICs*, pp. 115-118, 1998.

[21] B. J. Baliga, "Silicon Carbide Switching Device with Rectifying Gate", *U. S. Patent 5,396,085*, Issued March 7, 1995.

[22] B. J. Baliga, "Silicon Carbide Power Devices", *World Scientific Press*, Singapore, 2005. .

[23] J. Tan, J. A. Cooper, and M. R. Melloch, "High-Voltage Accumulation-Layer UMOSFET's in 4H-SiC", IEEE Electron Device Letters, Vol. 19, pp. 487-489, 1998.

[24] T. Nakamura, et al, "Novel developments towards increased SiC power device and module efficiency", *IEEE Energytech*, pp. 1–6. 2012.

[25] D. Peters, et al., "The new CoolSiC trench MOSFET technology for low gate oxide stress and high performance," *Power Conversion and Intelligent Motion Europe*, pp. 1–7, 2017.

[26] H. Takaya, et al., "A 4H-SiC Trench MOSFET with Thick Bottom Oxide for Improving Characteristics", *IEEE International Symposium on Power Semiconductor Devices and ICs*, pp. 43-46, 2013

[27] A. Agarwal, K. Han, and B. J. Baliga., "Analysis of 1.2 kV 4H-SiC Trench-Gate MOSFETs with Thick Trench Bottom Oxide", *IEEE Workshop on Wide Band Gap Power Devices and Applications*, pp. 125-129, 2018.

[28] A. Agarwal, et al., "A New Degradation Mechanism in High-Voltage SiC Power MOSFETs", *IEEE Electron Device Letters*, Vol. 28, pp. 587-589, 2007.

[29] B. Hull, et al, "Switching Performance Comparison of 1200 V and 1700 V SiC Optimized Half Bridge Power Modules with SiC Antiparallel Schottky Diodes versus MOSFET Intrinsic Body Diodes", *IEEE Applied Power Electronics Conf.*, pp. 2297-2304, 2017..

[30] W. Sung and B. J. Baliga, "Monolithically Integrated 4H-SiC MOSFET and JBS Diode (JBSFET) Using a Single Ohmic/Schottky Process Scheme", *IEEE Electron Device Letters*, Vol. 37, pp. 1605-1608, 2016.

[31] W. Sung and B. J. Baliga, "On Developing One-Chip Integration of 1.2 kV SiC MOSFET and JBS Diode (JBSFET)", *IEEE Tran. Industrial Electronics*, Vol. 64, pp. 8206-8212, 2017.

[32] K. Han, A. Agarwal, and B. J. Baliga, "Comparison of New Octagonal Cell Topology for 1.2 kV 4H-SiC JBSFETs with Linear and Hexagonal Topologies: Analysis and Experimental Results", *IEEE Workshop on Wide Bandgap Power Devices and Applications*, pp. 159-162, 2019.

[33] B. J. Baliga, "Power MOSFETs with Superior High Frequency Figure-of-Merit", *U. S. Patent 10,355,132*, Issued July 16, 2019.

[34] K. Han, W. Sung and B. J. Baliga, "Split-Gate 1.2-kV 4H-SiC MOSFET: Analysis and Experimental Validation", *IEEE Electron Device Letters*, Vol. 38, pp. 1437-1440, 2017.

[35] K. Han, W. Sung and B. J. Baliga, "A Novel 1.2 kV 4H-SiC Buffered-Gate (BG) MOSFET: Analysis and Experimental Results", *IEEE Electron Device Letters*, Vol. 39, pp. 248-251, 2018. .

[36] B. J. Baliga, "Power MOSFET and JBSFET cell topologies with Superior High Frequency Figure-of-Merit", *U. S. Patent 11,276,779*, Issued March 15, 2022.

[37] K. Han and B. J. Baliga, "The 1.2-kV 4H-SiC OCTFET: A New Cell Topology with improved High-Frequency Figures-of-Merit," *IEEE Electron Device Letters*, Vol. 40, pp. 299-302, 2019.

[38] A. Agarwal and B. J. Baliga, "Impact of Channel Length on Characteristics of 600 V 4H-SiC Inversion-Channel Planar MOSFETs", *IEEE European Solid-State Device Research Conference (ESSDERC)*, Paper A3L-G1, pp. 78-81, 2019.

[39] A. Agarwal and B. J. Baliga, "Implant Straggle Impact on 1.2 kV SiC Power MOSFET Static and Dynamic Parameters", *IEEE Journal of Electron Devices Society*, Vol. 10, pp. 245-255, March 2022.

[40] A. Agarwal and B. J. Baliga, "Optimization of Linear Cell 4H-SiC Power JBSFETs: Impact of N$^+$ Source Contact Resistance", *Power Electronic Devices and Components*, Paper 100008, pp. 1-11, 2022.

[41] A. Agarwal, K. Han, and B. J. Baliga, "600 V 4H-SiC MOSFETs Fabricated in Commercial Foundry With Reduced Gate Oxide Thickness of 27 nm to Achieve IGBT-Compatible Gate Drive of 15 V", *IEEE Electron Device Letters*, Vol. 40, pp. 1792-1796, 2019.

[42] A. Agarwal, K. Han, and B. J. Baliga, "Impact of Gate Oxide Thickness on Electrical Characteristics of 1200 V 4H-SiC Planar-Gate Power MOSFETs", *IEEE Device Research Conf.*, pp. 237-238, 2019.

[43] A. Agarwal and B. J. Baliga, "Performance Enhancement of 2.3 kV 4H-SiC Planar-Gate MOSFETs Using Reduced Gate Oxide Thickness", *IEEE Tran. Electron Devices*, Vol. 40, pp. 5029-5033, 2022.

[44] A. Agarwal, et al., "Switching and Short-Circuit Performance of 27 nm Gate Oxide, 650 V SiC Planar-Gate MOSFETs with 10 to 15 V Gate Drive Voltage", *IEEE Int. Symp. On Power Semiconductor Devices and ICs*, pp. 250-253, 2020.

[45] A. Kanale, et al, "Monolithic Reverse Blocking 1.2 kV 4H-SiC Power Transistor: A Novel, Single-Chip, Three-Terminal Device for Current Source Inverter Applications", *IEEE Tran. Power Electronics*, Vol. 37, pp. 10112-10116, 2022. .

[46] B. J. Baliga, "Monolithically Integrated AC Switch having JBSFETs therein with Commonly-Connected Drain and Cathode Electrodes", *U. S. Patent 10,804,393*, Issued October 13, 2020.

[47] K. Han, et al., "Monolithic 4-Terminal 1.2 kV/20 A 4H-SiC Bi-Directional Field Effect Transistor (BiDFET) with Integrated JBS Diodes", *IEEE Int. Symp. On Power Semiconductor Devices and ICs*, pp. 242-245, 2020.

[48] A. Kanale, et al., "Switching Characteristics of a 1.2 kV, 50 mΩ SiC Monolithic Bidirectional Field Effect Transistor (BiDFET) with Integrated JBS Diodes", *IEEE Applied Power Electronics Conf.*, pp. 1267-1274, 2021.

[49] S. Shah, et al., "Optimized AC/DC Dual Active Bridge Converter using Monolithic SiC Bidirectional FET (BiDFET) for Solar PV Applications", *IEEE Energy Conversion Congress and Exhibition*, pp. 568-575, 2022.

978-1-6654-8901-0/22 $31.00 © 2022 IEEE

A New Cell Topology for 4H-SiC Planar Power MOSFETs and Comparison with Hexagonal and Octagonal Cell Topologies

Shengnan Zhu*, Tianshi Liu*, Arash Salemi†, Michael Jin*, Marvin H. White*, David Sheridan†, and Anant K. Agarwal*

*Department of Electrical and Computer Engineering, The Ohio State University, OH, USA (zhu.2670@osu.edu)
†Alpha and Omega Semiconductor, Sunnyvale, CA, USA

Abstract—A new Dodecagonal (polygon with twelve sides, short for Dod) cell topology is designed for 4H-SiC planar power MOSFETs. The Dod cell is used in the layout design of the 650 V SiC MOSFETs. The nominal Hexagonal (Hex) cell and a recently published Octagonal (Oct) cell are also used on the 650 V SiC MOSFET layout designs for comparisons. The devices are fabricated, diced, and packaged for static and dynamic characterizations. Experimental results show that the Hex-cell MOSFET has the lowest specific ON-resistance ($R_{on,sp}$) and is suitable for high-power applications. The Dod and Oct-cell MOSFETs have much smaller gate-drain capacitance than Hex-cell MOSFETs, making them good candidates for high-frequency applications. The new Dod cell is designed with optimized channel density to have reduced $R_{on,sp}$ compared to the Oct cell.

Index Terms—SiC planar power MOSFET, Cell topology, High-frequency switching, High-frequency figure of merit (HF-FOM), Dodocagonal cell, Octagonal cell.

I. INTRODUCTION

Due to the high operation temperature, fast switching speed, and low switching loss, silicon carbide (SiC) power MOSFETs are gaining increasing attention in the semiconductor industry and electric vehicles (EVs) market [1], [2]. The JFET region and cell topology designs are studied to improve the static and dynamic performances of SiC MOSFETs [3]–[5]. The Hexagonal (Hex) cell topology is a conventional cell topology that is used in SiC power MOSFETs. The devices with Hex cell topology usually have low specific ON-resistance ($R_{on,sp}$) due to the high channel density. Recently, an octagonal (Oct) cell topology was used on 1.2 kV SiC planar power MOSFETs to improve the high-frequency figures-of-merit (HF-FOM) compared with traditional Linear cells [3]. The Oct cell is designed with minimized JFET region area to achieve a low gate-drain capacitance (C_{gd}). The C_{gd}, which is known as the Miller capacitance, essentially determines the drain voltage slew rate during the switching transient [6]. Thus, the Oct-cell devices have a high switching speed and are suitable for high-frequency applications [7]. However, the Oct cell achieves low JFET density with a low channel density, resulting in a high $R_{on,sp}$.

This work reports a new cell topology called dodecagonal (polygon with twelve sides, short for Dod) cell. The Dod cell is designed to achieve a low C_{gd} with a low JFET density. The Dod cell has a 1.6× higher channel density than the Oct

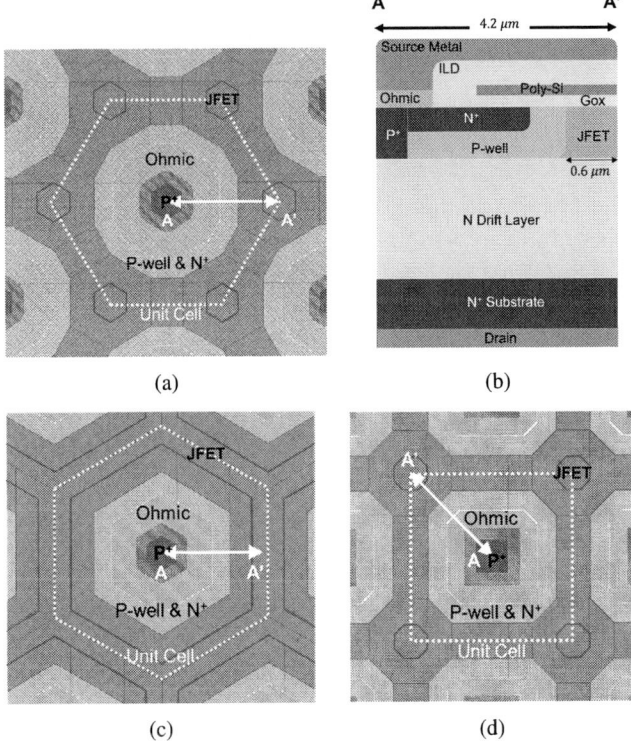

Fig. 1: (a) Layout of the Dod cell topology, (b) cross-sectional view along AA' of the designed 650 V SiC MOSFETs, (c) Layout of the Hex cell topology, and (d) Layout of the Oct cell topology.

cell, hence lower $R_{on,sp}$. The Dod cell has been used for 650 V SiC planar power MOSFETs for the first time. The Hex and Oct cells are also used on the layout designs of the 650 V SiC MOSFETs for comparisons. The devices were fabricated, packaged, and characterized.

II. DEVICE DESIGN AND FABRICATION

The Dod cell topology is shown in Fig. 1a. Six hexagonal JFET regions are located at the corners of the hexagonal unit cell. Poly-Si gate (hexagonal shapes connected by rectangles) is on the top of the JFET regions. A P^+ dodecagon (twelve-sided polygon) is located in the center of the unit cell with a

978-1-6654-8901-0/22 $31.00 © 2022 IEEE

TABLE I: Design parameters for the 650 V SiC MOSFETs

Cell topology	Dod	Hex	Oct
Half Cell pitch [μm]	4.2	4.2	4.2
Active area [mm^2]	0.64	0.64	0.64
Channel density [μm^{-1}]	0.181	0.408	0.113
JFET density [unitless]	0.055	0.265	0.034

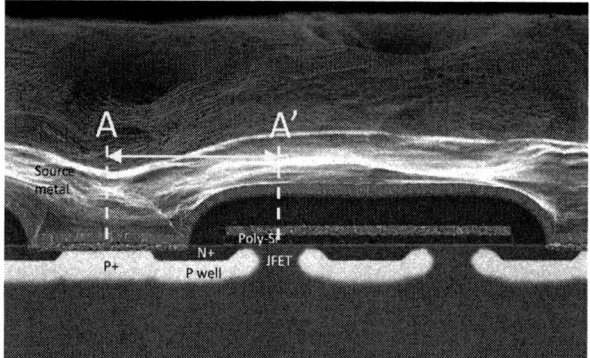

Fig. 2: Cross-sectional SEM image of the fabricated 650 V Dod-cell SiC power MOSFET. The AA' is correspond to the cross-sectional view in Fig. 1b.

dodecagonal Ohimc contact on top. The cross-sectional view of the designed 650 V SiC MOSFETs is shown in Fig. 1b. The same cross section is used for a conventional Hex cell layout (Fig. 1c) and the Oct cell layout (Fig. 1d). Design parameters for the MOSFETs with different layouts are listed in Table I. All devices have the same edge termination and die size as described in [5]. The 650 V MOSFETs are fabricated on a 6-inch SiC wafer by a state-of-art commercial SiC foundry. The cross-sectional SEM image of the fabricated Dod-cell MOSFET is shown in Fig. 2. Devices are packaged into open-cavity TO-247 packages for characterizations.

III. EXPERIMENTAL RESULTS

A. Threshold Voltage and Breakdown Voltage

The transfer and blocking characteristics of the 650 V SiC power MOSFETs are measured using a Keysight B1506A power semiconductor analyzer and the results are shown in Fig. 3. The threshold voltage (V_{th}) and the breakdown voltage (BV) of the devices are extracted and listed in Table II. Minimal V_{th} difference is observed among the MOSFETs. The devices under test (DUTs) have similar BV of $\sim$780 V. The drain leakage currents are less than 10 nA up to 650 V for all the DUTs.

B. Specific ON-resistance

Fig. 4 shows the output characteristics (measured using Keysight B1506A) of the 650 V SiC MOSFETs with different cell topologies. A 3$\times$ higher drain current (I_D) at drain voltage (V_D) of 1.5 V is observed for Hex-cell MOSFET compared to Dod-cell MOSFET, which is contributed by the higher channel density and higher JFET density of Hex cell topology.

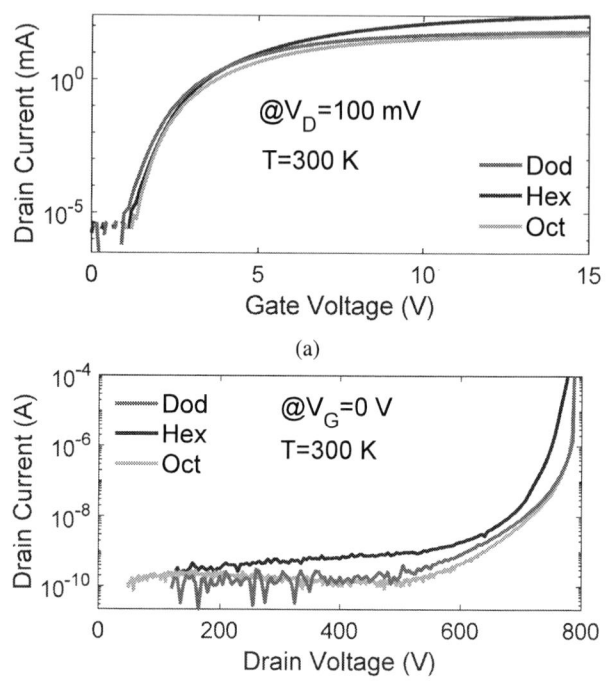

Fig. 3: (a) Transfer, (b) blocking characteristics of the 650 V SiC power MOSFETs.

TABLE II: Experimental results for the 650 V SiC MOSFETs

Cell topology	Dod	Hex	Oct
$V_{th}@I_D = 1mA$ [V]	3.38	3.38	3.65
$BV@I_D = 100\mu A$ [V]	787.2	776.7	787.3
$R_{on,sp}@V_D = 1.5V$ [$m\Omega \cdot cm^2$]	8.34	2.54	15.35
$C_{gd}@V_D = 400V$ [pF]	1.63	8.05	1.59
$HF - FOM(C_{gd} \times R_{on})$ [$\Omega \times pF$]	2124	3195	3814
dV/dt_{on} [V/ns]	11.50	*	5.71
$E_{loss,on}$ [μJ]	21.02	*	31.04
dV/dt_{off} [V/ns]	10.51	*	7.29
$E_{loss,off}$ [μJ]	4.86	*	5.72
$E_{loss,total}$ [μJ]	25.88	*	36.76
SCWT [μs]	>15	4	>15

*Not tested due to a different current rating.

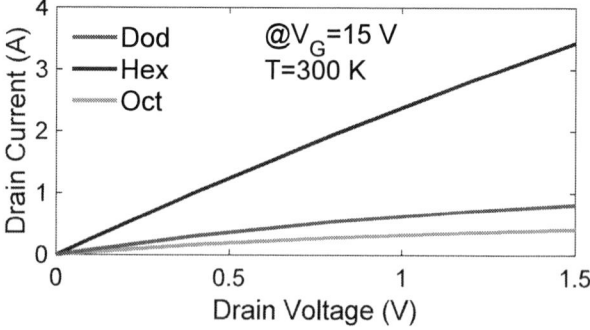

Fig. 4: Output characteristics of the 650 V SiC power MOSFETs.

978-1-6654-8901-0/22 $31.00 © 2022 IEEE

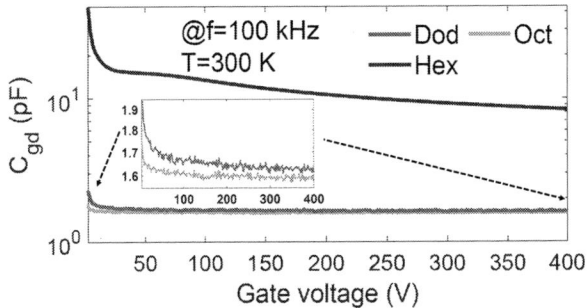

Fig. 5: Gate-drain capacitance of the 650 V SiC power MOSFETs.

The channel density is defined as the total channel width in a unit cell divided by the unit cell area. A higher channel density indicates more conduction current and lower channel resistance. The JFET density is the JFET region area in a unit cell divided by the unit cell area. A higher JFET density implies a lower JFET region resistance. The specific ON-resistances ($R_{on,sp}$) for the 650 V SiC MOSFETs are extracted and listed in Table II. The Dod-cell MOSFET produces a $1.8\times$ smaller $R_{on,sp}$ compared to the Oct-cell MOSFET.

C. Gate-drain Capacitance

The gate-drain capacitances (C_{gd}) of the fabricated 650 V SiC power MOSFETs are measured using a Keysight B1505A power semiconductor analyzer. The results are shown in Fig. 5. The Dod and Oct-cell MOSFETs have roughly one order of magnitude smaller C_{gd} than the Hex-cell MOSFET. The C_{gd} of a SiC planar power MOSFET is strongly related to the JFET region design [5]. The Dod and Oct cell topologies are designed with minimized JFET regions (low JFET densities), contributing to small C_{gd}. The C_{gd} at $V_D = 400V$ is extracted for all DUTs in Table II. Dod and Oct-cell MOSFET achieves $5\times$ smaller C_{gd} compared to Hex-cell MOSFET. The HF-FOMs of the 650 V SiC MOSFETs are also calculated. The Dod-cell MOSFET obtains the lowest HF-FOM as listed in Table. II, indicating a better high-frequency switching performance.

D. Switching Performance

It has been demonstrated that the Oct-cell devices have a faster switching performance and lower switching energy loss than the Hex-cell devices [7]. Thus the switching performance of the Dod-cell and Oct-cell MOSFETs are compared in this section. The Double-pulse tests are conducted on the Dod-cell and Oct-cell 650 V SiC power MOSFETs under a gate voltage (V_G) of 20 V, $V_D = 400V$, and $I_D = 2A$. The turn-on and turn-off waveforms are shown in Fig. 6.

The dv/dt and switching losses during turn-on and turn-off transients are extracted for all DUTs and listed in Table II. During the turn-on procedure, the Dod-cell obtains $2\times$ higher turn-on dv/dt and $1.5\times$ lower switching loss than the fabricated Oct-cell MOSFET. During the turn-off transient, the Dod-cell MOSFET shows a $1.4\times$ higher dv/dt and a lower switching loss than the measured Oct-cell MOSFET. The total switching

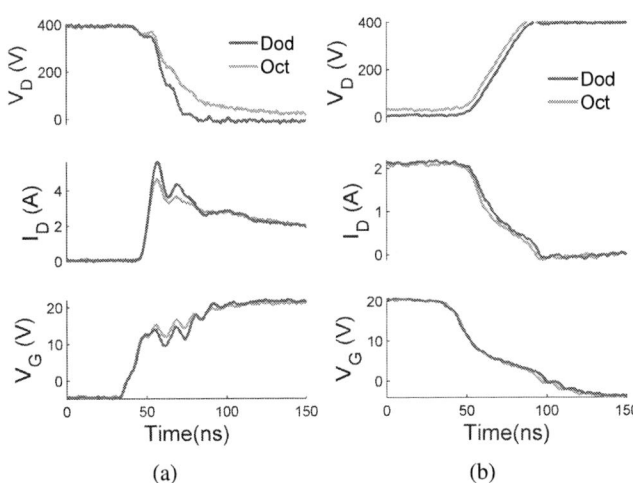

Fig. 6: (a) Turn-on, (b) turn-off waveform for the 650 V SiC power MOSFETs.

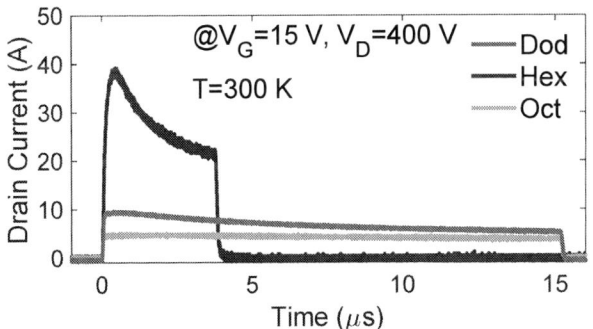

Fig. 7: Drain current waveform during short-circuit test of the 650 V SiC power MOSFETs.

losses are also calculated. Lower switching loss energy is observed for the Dod-cell MOSFET than the Oct-cell MOSFET.

E. Short-circuit Performance

The short-circuit measurements are conducted on the fabricated 650 V SiC power MOSFETs at $V_G = 15V$ and $V_D = 400V$. The Drain current waveforms are shown in Fig. 7. For the Hex-cell MOSFET, the waveform refers to that last test before device failure. For the Dod and Oct-cell MOSFETs, the devices survived for up to $15\mu s$ and the waveforms are from the last test when the time frame equals $15\mu s$. Due to the lower ON-resistance (R_{on}), the Hex-cell MOSFET shows a $4\times$ higher peak current than the Dod-cell MOSFET during the short-circuit events. The Oct-cell MOSFET shows the lowest peak current, corresponding to its highest R_{on} among all devices. The lowest R_{on} also results in the lowest short-circuit withstand time (SCWT) for Hex-cell MOSFET [8]. The measured SCWT for Hex-cell MOSFET is $4\mu s$. For the Dod and Oct-cell MOSFETs, the SCWTs are more than $15\mu s$.

IV. CONCLUSION

A new Dod cell topology is used on 650 V SiC planar power MOSFET. The conventional Hex cell and the recently

978-1-6654-8901-0/22 $31.00 © 2022 IEEE

published Oct cell are also used in the device layout design. The devices are fabricated and packaged for static and dynamic characterizations. The experimental results demonstrate that the Hex-cell MOSFET has the best static performance with the lowest $R_{on,sp}$ and can be used in high-power applications. The Dod and Oct cells are suitable for high-frequency applications due to their small C_{gd} and low HF-FOMs. Compared to the Oct cell, the Dod cell has a lower $R_{on,sp}$ due to optimized topology design that has a higher channel density.

ACKNOWLEDGMENTS

This work is supported by the Block Gift Grant from II-VI Foundation. The authors would like to thank the Power Electronics Teams at Ford Motor Company's Research and Innovation Centers for helpful discussions.

REFERENCES

[1] T. Kimoto, "Material science and device physics in SiC technology for high-voltage power devices," *Japanese Journal of Applied Physics*, vol. 54, no. 4, p. 040103, 2015.

[2] H. L. R. Maddi, S. Yu, S. Zhu, T. Liu, L. Shi, M. Kang, D. Xing, S. Nayak, M. H. White, and A. K. Agarwal, "The road to a robust and affordable SiC power MOSFET technology," *Energies*, vol. 14, no. 24, p. 8283, 2021.

[3] K. Han and B. Baliga, "The 1.2-kV 4H-SiC OCTFET: A new cell topology with improved high-frequency figures-of-merit," *IEEE Electron Device Letters*, vol. 40, no. 2, pp. 299–302, 2018.

[4] T. Liu, S. Zhu, A. Salemi, D. Sheridan, M. H. White, and A. K. Agarwal, "JFET Region Design Trade-Offs of 650 V 4H-SiC Planar Power MOSFETs," *Solid State Electronics Letters*, vol. 3, pp. 53–58, 2021.

[5] S. Zhu, T. Liu, J. Fan, H. L. R. Maddi, M. H. White, and A. K. Agarwal, "Effects of JFET Region Design and Gate Oxide Thickness on the Static and Dynamic Performance of 650 V SiC Planar Power MOSFETs," *Materials*, vol. 15, no. 17, p. 5995, 2022.

[6] Z. Chen, "Characterization and modeling of high-switching-speed behavior of SiC active devices," Ph.D. dissertation, Virginia Tech, 2009.

[7] K. J. Han, A. Kanale, B. J. Baliga, and S. Bhattacharya, "Demonstration of Superior Static, Dynamic, and Short-Circuit Performance of 1.2 kV 4H-SiC Split-Gate Octagonal Cell MOSFETs Compared with Linear, Square, and Hexagonal Topologies," in *Materials Science Forum*, vol. 1004. Trans Tech Publ, 2020, pp. 783–788.

[8] D. Xing, B. Hu, S. Yu, Y. Zhang, T. Liu, A. Salemi, M. Kang, J. Wang, and A. Agarwal, "Current saturation characteristics and single-pulse short-circuit tests of commercial SiC MOSFETs," in *2019 IEEE Energy Conversion Congress and Exposition (ECCE)*. IEEE, 2019, pp. 6179–6183.

Exploring Optimum Designs for 1.2kV 4H-SiC JBS Diode Integrated MOSFETs (JBSFETs)

Stephen A. Mancini, Seung Yup Jang, Dongyoung Kim, Woongje Sung
State University of New York Polytechnic Institute
College of Nanoscale Science and Engineering
Albany NY 12203 USA
(e-mail: mancins@sunyploy.edu)

Abstract- **Several different designs of 1.2kV-rated 4H-SiC JBS diode integrated MOSFETs (JBSFETs) have been successfully fabricated to increase the 3rd quadrant device performance while maintaining the electrical characteristics on both the forward and blocking modes of operation when compared to its MOSFET counterpart. Incorporating the Schottky area in an efficient manner to improve overall device performance has been a critical path in the development and adoption of 4H-SiC JBSFETs rather than MOSFETs co-packaged with JBS Diodes. Device design, fabrication, and electrical performances are discussed in this paper and as a result, an optimum JBSFET design is proposed.**

Index Terms- **4H-Silicon Carbide (SiC), Junction Barrier Schottky Diode, MOSFET, JBSFET, 3rd Quadrant, Design Approach, Leakage Current, Breakdown Voltage, Room Temperature Implantation.**

I. Introduction

4H-SiC offers superior benefits over Silicon due to its wide bandgap and thus a small impact ionization coefficient while simultaneously having an electron mobility similar to that of silicon, which makes it the material of choice for high voltage power devices. The wide bandgap and high critical electric field in 4H-SiC allow a thin and heavily doped epitaxial drift layer, resulting in a device that has lower resistance when compared with traditional silicon devices at a voltage rating greater than or equal to 600V [1]. However, the nearly 3 times greater bandgap of 4H-SiC brings about large built-in potential across a PN junction, causing greater overall voltage drops in bipolar devices such as PiN diodes and the PN body diode in a MOSFET structure (3rd quadrant operation). For this reason, 4H-SiC MOSFETs have traditionally been paired with an external Junction Barrier Schottky (JBS) diode, therefore bypassing the bipolar body diode during the 3rd quadrant operation thanks to the low voltage drop from the JBS diode conduction [2]. This external JBS Diode not only reduces the overall voltage drop across the device, but also has the added benefit of reducing the impacts of Basal Plane Dislocations (BPD)-related voltage degradation such as higher device resistance and increase leakage current [3-8]. Therefore, this parallel connection would result in greater device performances due

to a reduction in power loss and an increase in overall device longevity.

Despite the benefits, the external chip-to-chip connection of the MOSFET and JBS Diode results in a high wafer area consumption as the two separate devices both require their own active area and edge termination regions doubling (at least) the area needed. Due to this reason, there has been a push to develop the integration of a JBS Diode within the MOSFET cell structure, and by doing such can reduce the total wafer area [2,9]. Previous JBSFET cell structures either contain a large cell pitch to accommodate the integration of the Schottky area [2], or do not provide enough Schottky contacts within the active area as seen in Fig. 1 [10]. This causes a tradeoff between forward conduction and improved 3rd quadrant operation. Although it has been shown that Schottky contacts can be increased without negatively impacting forward conduction effectively overcoming this trade off, the resulting cell pitch remains large [11]. This results in a higher specific on resistance ($R_{on,sp}$) when compared to its MOSFET counterpart. [11]. In this study, various JBSFET cell designs and architecture were proposed and fabricated to reduce the cell pitch. The design goal is to match the forward conduction and blocking characteristics of the proposed JBSFETs with Nominal MOSFET structure while maintaining a large Schottky current during the 3rd quadrant mode of operation. As a result, an optimum JBSFET design is then proposed.

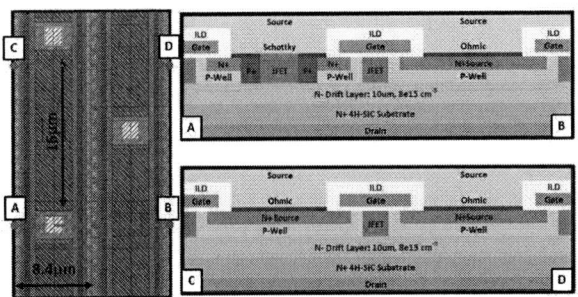

Figure 1. Top and cross-sectional views of the conventional JBSFET structure. An orthogonal layout was utilized for the Schottky contact regions indicated by the two cross sectional areas, as opposed to the traditional stripe pattern. Key device regions are labeled in each cross-section view.

978-1-6654-8901-0/22 $31.00 © 2022 IEEE

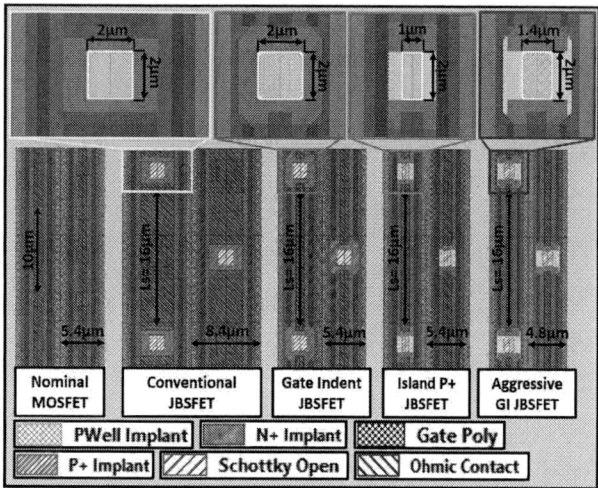

Figure 2. Top views of the fabricated Nominal MOSFET and various JBSFET cell designs utilizing the L_s of 16μm spacing between the Schottky opening design used in this study. The Schottky designs can be observed in further detail highlighted by a yellow, green, blue, and purple outlines for the Conventional, Gate Indent (GI), Island P+, and Aggressive GI JBSFET designs respectively. Borders of the gate poly regions are indicated with dotted red lines.

Table 1. Overall device design comparison between the various JBSFET cell layouts along with the Nominal MOSFET design as a comparison. Devices vary in cell pitch with 5.4μm being equivalent to that of the nominal MOSFET design, the width of the Schottky opening in the X-Direction, spacing between Schottky openings (L_s), whether the Schottky region is fully or partially enclosed by a P+ implantation region, and if the channel region is interrupted to allow for a more aggressive design.

Device Design	Cell Pitch [μm]	Schottky Open [μm]	Ls [μm]	Surrounded P+ Regions	Channel Interruption
Nominal MOSFET	5.4	N/A	N/A	N/A	0%
Conventional JBSFET	8.4	2	16, 8, 4	Fully Enclosed	0%
Island P+ JBSFET	5.4	1	16	Top + Bottom	0%
Gate Indent JBSFET	5.4	2	16, 8, 4	Fully Enclosed	20%
Aggressive GI JBSFET	4.8	1.4	16	Top + Bottom	20%

II. Device and Edge Termination Design

A. Active Area Design

The top views of the various JBSFET structures and their respective Schottky openings utilized in this study are compared in Fig. 2. For the conventional JBSFET structure, the Schottky openings is fully surrounded by P+ regions to provide a higher shielding from the electric fields and therefore suppress leakage current, throughout the structure. The channels of this device are uninterrupted as to not hinder the current flow; however, the cell pitch remains large at 8.4μm. The Gate Indent (GI) JBSFET still utilizes the fully surrounded P+ Schottky regions while maintaining the same cell pitch as the Nominal MOSFET. However, in order to make this accommodation the gate poly layer was indented

as to not violate any design criteria. By doing such, part of the N+ was cut near the Schottky opening, causing a discontinuation in the source region. The combination of the gate poly indentation along with the cut N+ region partially interrupts control over the channel . The Island P+ design does not indent the gate poly; however, the cell pitch is still reduced to that of the Nominal MOSFET by reducing the Schottky opening within the X-Direction to 1μm. For this, the P-well region of the MOSFET was used to properly shield the device in the X-direction rather than using fully surrounded P+ regions as seen in previous designs. There are still P+ regions used in shielding on the top and bottom of the Schottky opening, as the dimension in the Y-Direction is still 2μm. Finally, the Aggressive Gate Indent structure, integrated the gate indent design into the P+ Island design along with utilizing more aggressive design rules than any other structure. As a result the cell pitch was further reduced to 4.8μm. For the Conventional JBSFET, and GI structures, the spacing between the Schottky openings (L_s) was varied from 16μm, 8μm, and 4μm, to allow for greater conduction in the 3rd quadrant of operation [11]. The remining Island P+ and Aggressive GI structures only had an L_s of 16μm for this study. All Designs can be summarized in Table 1.

Optimal device dimensions used in the MOSFETs and JBSFETs were determined through 2-D forward conduction and blocking electrical simulations. Dimensions for the JFET width and channel length that resulted in low $R_{on,sp}$, leakage current, and high breakdown voltage were determined to be 1.2μm and 0.5μm, respectively. Due to the low background doping of $8 \times 10^{15} \mathrm{cm}^{-3}$ within the epitaxial drift layer, doping within the JFET region was increased to $5 \times 10^{16} \mathrm{cm}^{-3}$ to ensure the optimal dimensions previously mentioned. The optimal Schottky width (W_s) to ensure sufficient current density through a low forward voltage drop and low leakage current determined by a low maximum electric field was determined to be approximately 2μm. This optimized Schottky opening width was then integrated into the Conventional JBSFET and GI JBSFET structure.

B. Edge Termination Design

To provide near ideal breakdown voltage of the devices while simultaneously minimizing $R_{on,sp}$ an n-epitaxial drift layer was designed with a thickness of 10μm and a nitrogen doping concentration of $8 \times 10^{15} \mathrm{cm}^{-3}$. To further ensure this near ideal breakdown voltage, a "hybrid edge termination" structure was implemented [12]. This hybrid edge termination structure combines the strengths of both the Ring Assisted Junction Termination Extension (RA-JTE) and the Multiple Floating Zone Junction Termination Extension (MFZ-JTE) structures. This provides a wide range of JTE dose easing device fabrication while simultaneously ensuring high breakdown voltages [12].

III. Device Fabrication

For this study, 1.2kV-rated 4H-SiC devices including MOSFETs, JBSFETs, PiN and JBS Diodes were fabricated on the same wafer and thus were all subjected to the same process flow and design rules. A heavily doped N+ buffer layer followed by a 10μm thick, $8 \times 10^{15} cm^{-3}$ n-doped epitaxial layer was grown on 4H-SiC substrates was used. Aluminum and Nitrogen were used during the ion implantation process and formation of the P+/p-well/JTE and JFET/N+ regions, respectively. For the p-well implantation, a profile was designed as such to create an accumulation mode channel. This creates a channel with higher mobility and thus reduces the channel resistance and overall, $R_{on,sp}$ of the device [13]. All implantation steps for both Aluminum and Nitrogen ions were performed at room temperature of 25°C. After the implantation steps, a 1650°C activation anneal was performed for 10 minutes utilizing a carbon capping layer to reduce lattice damage originating from carbon vacancies [14].

Following the implantation and activation annealing steps, a 50nm thick gate oxide layer was formed using thermal oxidation followed by the deposition and patterning of the polysilicon gate. After the polysilicon gate formation, an interlayer dielectric was uniformly deposited, and then opened to allow the formation of an ohmic on the source region. Nickel was then deposited, and the silicide was pre-annealed at 750°C for 2 minutes utilizing a rapid thermal anneal (RTA) process to partially form the frontside ohmic contacts. Residual nickel that did not undergo the silicidation process was then stripped. Nickel was then deposited, on the backside of the wafers and all the nickel metal was annealed at 1000°C for 2 minutes using an RTA process. This annealing process forms both the front and backside ohmic contacts of the devices. The Schottky spacings for the various JBSFETs were then opened using a reactive ion etch process to ensure tighter dimension control of the designs. Titanium was then deposited to form the Schottky contacts on the n-epitaxial drift layer [15]. Following the titanium deposition, a thin layer of titanium nitride followed by a 4μm thick aluminum metal layer were deposited and patterned to form the final source and gate contacts for each of the various devices. Finally, a nitride followed by a thick polyimide layer were deposited and patterned to passivate the frontside of the wafer.

IV. Results

A. Forward Conduction Output Characteristics

Fig. 3 shows the typical forward conduction characteristics of the various 16μm spacing JBSFET structures, along with the Nominal MOSFET for a comparison. For the uninterrupted channel structures, the forward $R_{on,sp}$ can be attributed to the overall cell pitch of the structure. The average $R_{on,sp}$ for the nominal MOSFET and Island P+ JBSFET design was determined to be 4.13mΩ·cm² and 4.17mΩ·cm² respectively; both of which having a cell pitch of 5.4μm. Ron,sp for the structures was determined at a gate bias of $V_{gs} = 20V$, and a drain-source current of $I_{ds} = 0.1A$. The conventional JBSFET design with a cell pitch of 8.4μm, or ~35% large than the nominal MOSFET, had an average

$R_{on,sp}$ of 5.11 mΩ·cm² or an increase of ~23% compared to the MOSFET counterpart. However, this correlation between cell pitch and $R_{on,sp}$ can not be observed to the same effect within the channel interruption JBSFET structures. For the GI JBSFET with $L_s = 16μm$, the average Ron,sp was determined to be 4.37mΩ·cm²; nearly a 7% increase when compared to the MOSFET counterpart despite the two structures possessing the same cell pitch. The same can be

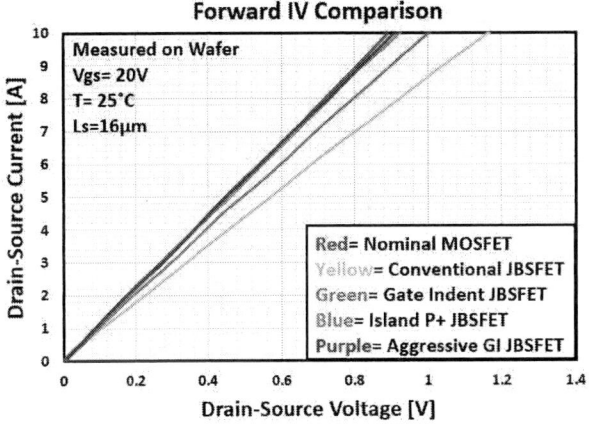

Figure 3. Typical output IV characteristics measured from the various 16μm spacing JBSFETs along with the nominal MOSFET as a comparison. All curves are representative and were measured on wafer at 25°C at a gate bias of $V_{gs} = 20V$. It should be noted that these Ron,sp measurements were performed on-wafer and would be greatly improved after packaging into TO-247s (10 – 20% improvement from our previous studies).

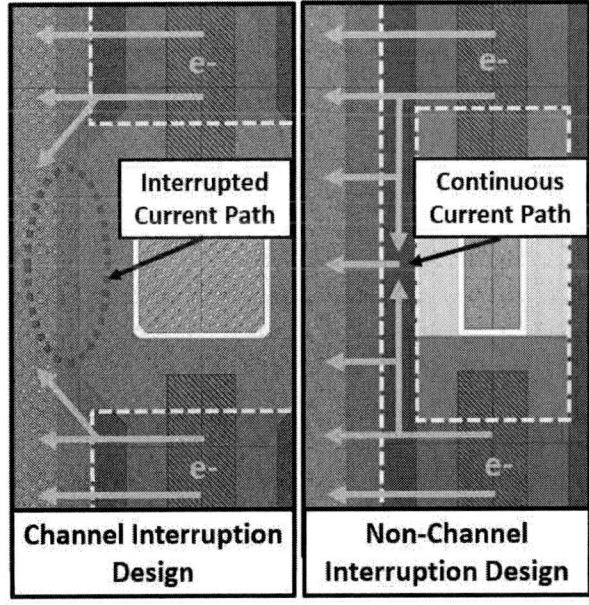

Figure 4. Schematic representation of the electron current path being interrupted for both the Channel Interruption Design used in the Gate Indent JBSFETs (left) and the Non-Channel Interruption Designs used in the Island P+ and the conventional JBSFETs (Right). Electron current path can be seen in orange, flowing from the N+ Ohmic contacts to the JFET regions of each device. Boundaries of the N+ source regions are outlined with dotted yellow lines.

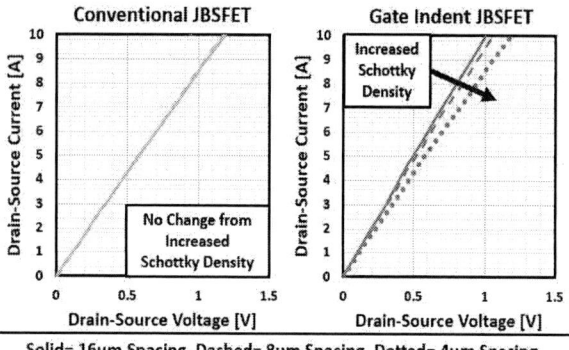

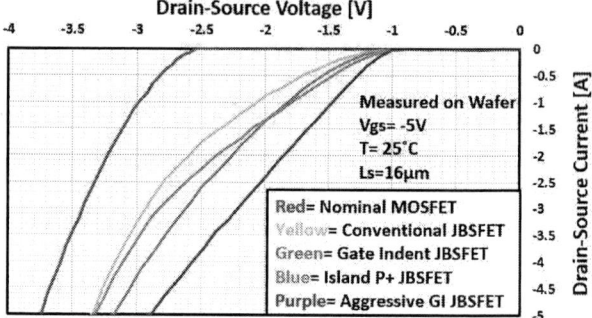

Figure 5. Typical output IV characteristics measured from the Conventional JBSFET (left) and the Gate Indent JBSFET (right) with different spacing between the Schottky contacts. Solid, Dashed, and dotted lines represent the JBSFETs with 16μm, 8μm, and 4μm between the Schottky contacts respectively. Negligible change can be observed within the Conventional structure however Ron,sp increases for the Gate Indentation structures as Schottky density increases. All curves are representative and were measured on wafer at 25˚C at a gate bias of V_{gs}=20V.

Figure 6 Typical 3rd quadrant IV characteristics of the various JBSFET designs along with the Nominal MOSFET structure. All curves are representative and were measured on wafer at 25˚C using the 16μm spacing between the Schottky opening design. A gate bias of V_{gs}=-5V was applied to minimize any current flowing through the channels.

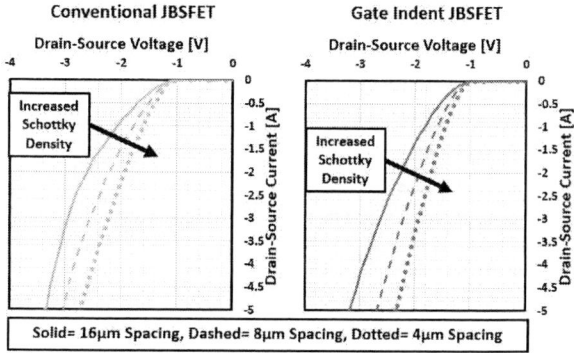

Figure 7. Typical 3rd quadrant measured from the Conventional JBSFET (left) and the Gate Indent JBSFET (right) with different spacing between the Schottky contacts. Solid, Dashed, and dotted lines represent the JBSFETs with 16μm, 8μm, and 4μm between the Schottky contacts respectively. Unlike the forward conduction behaviors, the increased Schottky density resulted in an overall increase in 3rd quadrant output. All curves are representative, measured on wafer at 25˚C and a gate bias of V_{gs}=-5V was applied.

observed within the Aggressive GI structure as well, with the structure providing an Ron,sp of 4.11mΩ·cm². This mimics the forward conduction performance of both the Island P+ JBSFET design and the nominal MOSFET, despite having a 12.5% reduction in cell pitch. The poorer performances of the gate indentation structures can be attributed to an interruption in the electron current path as observed in Fig. 4. Since the N+ source region is continuous for the uninterrupted channel designs, current can move throughout the source structure and utilize the entirety of the channel region for forward conduction. However, the N+ source regions within the Gate Indented structures are discontinuous and thus prevents to utilization of the entire channel regions. This effect is more prominent as the Schottky frequency is increased, thereby further reducing the channel current paths, to accommodate a greater Schottky area. As the spacing between Schottky contacts decreased from 16μm, 8μm, and 4μm, the average $R_{on,sp}$ of the Gate Indent JBSFET measured was 4.37mΩ·cm², 4.83mΩ·cm², and 4.93mΩ·cm², respectively, whereas the $R_{on,sp}$ of the conventional JBSFET (non-interrupted design) remained nearly consistent at 4.98mΩ·cm², 5.02mΩ·cm², and 4.97mΩ·cm² respectively as seen in Fig. 5.

B. 3rd Quadrant Output Characteristics

The typical 3rd quadrant output characteristics can be observed in Fig. 5. The source-drain voltage (V_{sd}) at 5A for the Nominal MOSFET, and various 16μm spacing JBSFET designs (Conventional, Gate Indent, P+, and Aggressive GI) were 3.59V, 3.38V, 3.01V, 3.23V, and 2.85V respectively. Although the Gate Indent and aggressive Gate Indent JBSFET designs resulted in poor forward conduction characteristics, the same cannot be said for the 3rd quadrant mode of operation. The smaller cell pitch combined with a sufficient Schottky opening allows for a higher unipolar current. However, this current flow is still lower than that of the JBS diode used in MOSFET co-packaging, which fabricated under the same conditions has a forward voltage

drop (V_f) of 2.19V at 5A. Since the 3rd Quadrant conduction is proportional to the Schottky opening area, decreasing L_s from 16μm, 8μm, and 4μm, resulted in an overall increase in 3rd quadrant performance for both channel interruption and non-channel interruption designs unlike the forward conduction behaviors. As L_s decrease to 16μm, 8μm, and 4μm for the two structures, the V_{sd} at 5A are 3.38V, 3.00V and 2.65V for the conventional JBSFET and 3.01V, 2.85V, and 2.50V for the Gate Indent JBSFET, respectively as observed in Fig. 7.

C. Forward Blocking Output Characteristics

Forward blocking behaviors between the various JBSFET designs along with the Nominal MOSFET can be observed in Fig. 8. All JBSFET device types show a near ideal breakdown voltage of 1600V by utilizing the hybrid edge termination as previously discussed. The breakdown voltages were determined when the device had reached a

978-1-6654-8901-0/22 $31.00 © 2022 IEEE

drain-source current (I_{ds}) reading of 1 mA. Although all the JBSFET designs share breakdown voltages same as the Nominal MOSFET counterpart, the leakage currents between these designs can vary quite considerably. JBSFET designs ranging from highest to lowest leakage currents were the GI, Aggressive GI, Conventional, and Island P+ JBSFETs respectively. The GI JBSFET and the Island P+ JBSFET have the highest and lowest leakage currents respectively, which can be attributed to the electric fields at the Schottky opening as observed in Fig 9. However, the same trend cannot be observed in the Conventional and Aggressive GI structures. The higher leakage seen in the Aggressive GI compared to the conventional design can be attributed to an increase in Schottky area due to a smaller cell pitch whereas the Conventional structure has the smallest Schottky area due to the widest cell pitch. Moreover, the aggressive design rules would be attributed to the higher leakage current observed in the Aggressive GI. From observing the forward blocking capabilities of the various JBSFET designs, it can be determined that the Island P+ design is optimal due to its ideal breakdown voltage and suppression of leakage current at the 1.2kV blocking rating.

Table 2. Overall device electrical performances between the various JBSFET structures and nominal MOSFET structures. Average $R_{on,sp}$, Vsd at -5A, and leakage current at the 1200V blocking conditions are shown as the Schottky spacings (L_s) are varied for the JBSFET structures. Values in black represents actual data points, whereas values in grey represent predicted data based off their channel interruption or non-channel interruption designs.

Device Type	Ls [μm]	Ron,sp [mΩ*cm²]	Vsd (@-5A, Vgs=-5V) [V]	Leakage Current (@1200V) [A]
Nom. MOSFET	N/A	4.13	3.60	9.6x10⁻⁸
Conventional JBSFET	16	4.98	3.38	4.6x10⁻⁸
	8	5.02	3.00	6.9x10⁻⁸
	4	4.97	2.65	3.1x10⁻⁸
Gate Indent JBSFET	16	4.37	3.01	1.1x10⁻⁶
	8	4.83	2.85	9.8x10⁻⁷
	4	4.93	2.49	1.2x10⁻⁶
Island P+ JBSFET	16	4.17	3.22	1.9x10⁻⁸
	8	4.20	2.86	N/A
	4	4.16	2.52	N/A
Aggressive GI JBSFET	16	4.11	2.84	4.6x10⁻⁷
	8	4.54	2.69	N/A
	4	4.64	2.35	N/A

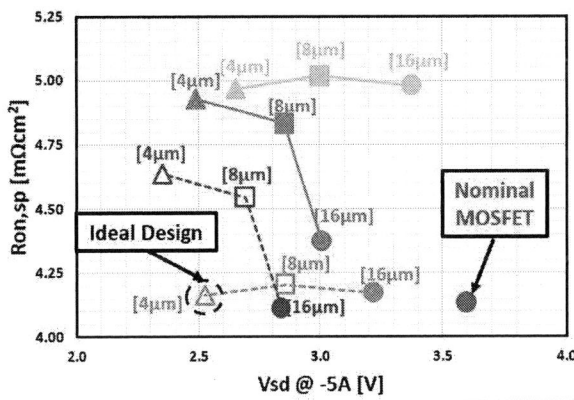

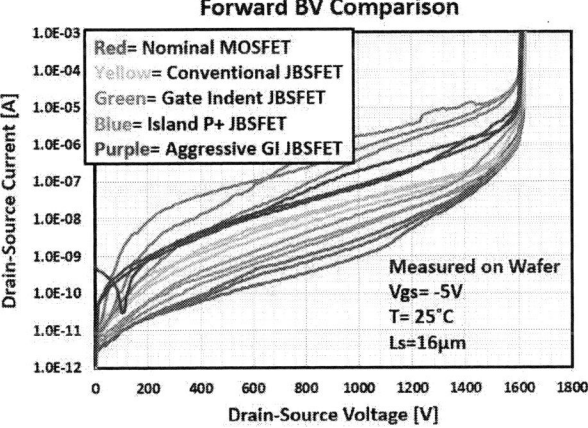

Figure 8. Forward blocking characteristics of the various 16μm spacing JBSFET designs along with the nominal MOSFET. All leakage curves were performed at V_{gs}= -5V to minimize any leakage from the channel, on wafer, and at 25°C.

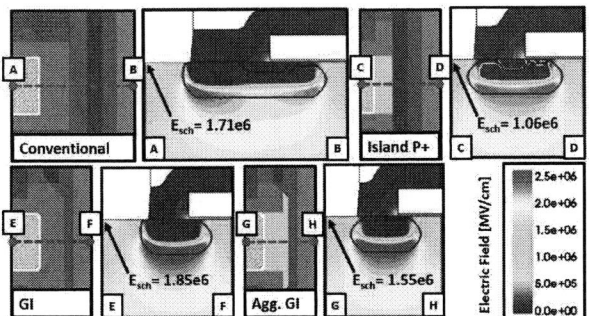

Figure 9. Simulated electric field distributions of the various JBSFET structures at a forward blocking voltage of 1200V and a gate bias of V_{gs}=-5V. Higher electric fields in the middle of the Schottky opening can be observed in the Gate Indent design, yielding larger leakage currents when compared to the other JBSFET designs.

Figure 10. Trade off comparison curves using the average $R_{on,sp}$ values and V_{sd} at -5A for various JBSFETs as the spacing between the Schottky contacts (L_s) decreases along with the nominal MOSFET as a comparison. Actual data (solid filled) were obtained for all the nominal and GI designs however predicted data (hollow filled) was used for the remained JBSFET designs, based on their channel interruptions (see Table 1). Circle, Square and Triangle data points represents the 16μm, 8μm, and 4μm spacing between the Schottky opening designs respectively. The ideal design based on this study is highlighted.

D. Device Trade-Offs Comparisons

An overall comparison between the various JBSFET and MOSFET structures can be observed in Table 2. and Fig. 10 where the spacing between the Schottky openings were varied. As mentioned in the previous sections, the uninterrupted channel design seen in the conventional JBSFET structure resulted in the average $R_{on,sp}$ values remaining constant while simultaneously increasing the 3rd quadrant performance as the Schottky frequency increased. This trend can therefore be predicted and extrapolated for the Island P+ JBSFET design; maintaining forward conduction and increased 3rd quadrant output as L_s is reduced. However,

978-1-6654-8901-0/22 $31.00 © 2022 IEEE   15

this trend is not observed in the channel interruption designs (GI designs); Increasing the Schottky frequency interrupts a greater amount of the channel area, causing larger on-resistance during the output characteristics (1st Quadrant). Subsequently this trend can be extrapolated to the aggressive gate indent design. increasing 3rd quadrant at the cost of forward conduction. Therefore, to achieve the same cell pitch (and thus similar $R_{on,sp}$) as the MOSFET, the Island P+ JBSFET is an optimal design, which can provide sufficient Schottky current by scaling the Schottky opening frequency without negatively impacting forward conduction, while simultaneously maintaining a well suppressed leakage current during blocking mode of operation.

V. Conclusions

Several 1.2kV JBS diode integrated MOSFETs (JBSFETs) utilizing various device architectures have been successfully fabricated. Previous JBSFET structures have resulted in high Ron,sp values when compared to the nominal MOSFET counterpart, primarily attributed to the larger cell pitch to accommodate the JBS diode integration. However, in this study two methods (Gate Indent, and Island P+ Design) have been examined to alleviate this issue and reduce the cell pitch of the JBSFET to that of the MOSFET structure. Of the two methods, the Gate Indent approach did not match forward conduction of the nominal MOSFET and lacked the capabilities scaling the Schottky frequency. However, the Island P+ design was shown to be optimal design approach due to comparable $R_{on,sp}$ and leakage current when compared to that of the Nominal MOSFET, and Although the fabricated 16μm spacing structure does not provide adequate Schottky current, the uninterrupted channel design allows for an increase in Schottky frequency without negatively affecting other device performances. Therefore, the 4μm spacing Island P+ JBSFET would offer a near ideal design providing sufficient Schottky current while matching the static performances of the Nominal MOSFET.

References

[1] B Jayant Baliga, *Fundamentals of power semiconductor devices.* Cham, Switzerland: Springer, 2019.

[2] W. Sung and B. J. Baliga, "On Developing One-Chip Integration of 1.2 kV SiC MOSFET and JBS Diode (JBSFET)," *IEEE Transactions on Industrial Electronics*, vol. 64, no. 10, pp. 8206–8212, Oct. 2017, doi: 10.1109/tie.2017.2696515.

[3] A. Agarwal, H. Fatima, S.Haney, S.-H. Ryu, "A New Degradation Mechanism in High-Voltage SiC Power Mosfets". *IEEE Electron Device Letters vol. 28, no. 7, pp.587-589, 2007.* doi.: 10.1109/LED.2007.897861

[4] K. Konishi, R. Fujita, Y. Mori, and A. Shima, "Investigation of Forward Voltage Degradation due to Process-Induced Defects in 4H-SiC MOSFET," *Materials Science Forum*, vol. 924, pp. 365–368, Jun. 2018, doi: 10.4028/www.scientific.net/msf.924.365.

[5] K. Konishi, R. Fujita, Y. Mori, and A. Shima, "Inducing defects in 3.3 kV SiC MOSFETs by annealing after ion implantation and evaluating their effect on bipolar degradation of the MOSFETs," *Semiconductor Science and Technology*, vol. 33, no. 12, p. 125014, Nov. 2018, doi: 10.1088/1361-6641/aae814.

[6] K. Konishi, R. Fujita, and A. Shima, "Modeling and Evaluation of Stacking Fault Expansion Velocity in Body Diodes of 3.3 kV SiC MOSFET," *Journal of Electronic Materials*, vol. 48, no. 3, pp. 1704–1713, Jan. 2019, doi: 10.1007/s11664-018-06901-0.

[7] R. E. Stahlbush, K. N. A. Mahakik, A. J. Lelis, and R. Green, "Effects of Basal Plane Dislocations on SiC Power Device Reliability," *2018 IEEE International Electron Devices Meeting (IEDM)*, Dec. 2018, doi: 10.1109/iedm.2018.8614623.

[8] S. A. Mancini, S. Y. Jang, D. Kim and W. Sung, "Static Performance and Reliability of 4H-SiC Diodes with P+ Regions Formed by Various Profiles and Temperatures," *2022 IEEE International Reliability Physics Symposium (IRPS)*, 2022, pp. P62-1-P62-6, doi: 10.1109/IRPS48227.2022.9764538.

[9] W. Sung and B. J. Baliga, "Monolithically Integrated 4H-SiC MOSFET and JBS Diode (JBSFET) Using a Single Ohmic/Schottky Process Scheme," *IEEE Electron Device Letters*, vol. 37, no. 12, pp. 1605–1608, Dec. 2016, doi: 10.1109/led.2016.2618720.

[10] N. Yun, J. Lynch, and W. Sung, "Area-Efficient, 600V 4H-SiC JBS Diode-Integrated MOSFETs (JBSFETs) for Power Converter Applications," IEEE Journal of Emerging and Selected Topics in Power Electronics, vol. 8, no. 1, pp. 16–23, Mar. 2020, doi: 10.1109/jestpe.2019.2947284.

[11] S. A. Mancini, S. Y. Jang, D. Kim and W. Sung, "Increased 3rd Quadrant Current Handling Capability of 1.2kV 4H-SiC JBS Diode-Integrated MOSFETs (JBSFETs) with Minimal Impact on the Forward Conduction and Blocking Performances," *2021 IEEE 8th Workshop on Wide Bandgap Power Devices and Applications (WiPDA)*, 2021, pp. 101-106, doi: 10.1109/WiPDA49284.2021.9645152.

[12] W. Sung and B. J. Baliga, "A Near Ideal Edge Termination Technique for 4500V 4H-SiC Devices: The Hybrid Junction Termination Extension," *IEEE Electron Device Letters*, vol. 37, no. 12, pp. 1609–1612, Dec. 2016, doi: 10.1109/led.2016.2623423.

[13] W. Sung, K. Han, and B. J. Baliga, "A comparative study of channel designs for SiC MOSFETs: Accumulation mode channel vs. inversion mode channel," *2017 29th International Symposium on Power Semiconductor Devices and IC's (ISPSD)*, May 2017, doi: 10.23919/ispsd.2017.7988996.

[14] M. K. Linnarsson, H. M. Ayedh, A. Hallén, L. Vines, and B. G. Svensson, "Surface Erosion of Ion-Implanted 4H-SiC during Annealing with Carbon Cap," *Materials Science Forum*, vol. 924, pp. 373–376, Jun. 2018, doi: 10.4028/www.scientific.net/msf.924.373.

[15] N. Yun *et al.*, "On the Development of 1700V SiC JBS Diodes in a 6-Inch Foundry," *Materials Science Forum*, vol. 963, pp. 558–561, Jul. 2019, doi: 10.4028/www.scientific.net/msf.963.558.

Design Optimization and Surge Current Capability of 4H-SiC Lateral Deep P⁺ JBS Diode on Thin RESURF Layer

Atsushi Shimbori and Alex Q. Huang

Semiconductor Power Electronics Center, The University of Texas at Austin, Texas 78758, USA

ashimbo@utexas.edu, aqhuang@utexas.edu/ Phone: (919) 208-3114

Abstract— In this paper, we explore the optimization of "saddle" type finger layout for lateral SiC RESURF (Reduced Surface Electric Field) diodes where concave/convex finger ends combined with mesa etch isolation achieve over 82% of the inner cell breakdown voltage. Furthermore, added deep P⁺ JBS pattern near the anode Schottky contact significantly improves robustness against field crowding when cell pitch/finger width is reduced to densify the chip. Improved surge current capability compared with pure Schottky structure is investigated through mix-mode TCAD simulation. The proposed design methodologies offer guidance for future efforts toward realizing a monolithic SiC high voltage integrated circuit.

Keywords—RESURF, 4H-SiC, JBS diode, Edge Termination, 3D TCAD simulation, SiC HVIC, and Surge Current Capability

I. INTRODUCTION

4H-SiC Schottky diodes offer fast switching and low forward drops compared to Si PIN diodes. This is also the same when the device is designed laterally for easier monolithic integration. However, its limited ruggedness against current overloads is a concern for high power applications. The proposed deep P⁺ JBS structure offers both improved surge current capability and robustness against field crowding associated with cell pitch reduction within the "saddle" type finger layout. The activation of bipolar conduction within these JBS pattern is strongly dependent on the device structure and layout design; thus, further investigation is necessary through three-dimensional (3D) TCAD simulation.

II. INNER CELL STRUCTURE AND EPI WAFER DESIGN

Standard finger-type layout is utilized for these proposed lateral power devices and the inner cell structures are initially optimized through two-dimensional (2D) TCAD simulation. The cross section of the proposed 4H-SiC lateral RESURF deep P⁺ JBS diode cell structure is shown in Fig. 1(a). The cross section of this cell is drawn along the x-y axis, while the finger runs in the z axis direction which will eventually be terminated by a mesa etch isolation with a concave anode pad pattern as shown in Fig. 1(b) cathode finger end structure. On the other hand, the anode finger will eventually end in the inner chip direction with a convex shape. These repeating multiple fingers form a "saddle" shape layout structure, mitigating the electric field crowding associated with both the edge termination of the active area and the formation of the inner cathode pads. While the high voltage is supported by the n-type/p-type RESURF layers grown onto a 4H-SiC n-type substrate by chemical vapor deposition (CVD) with parameters optimized through 2D numerical simulations, additional care to control the electric

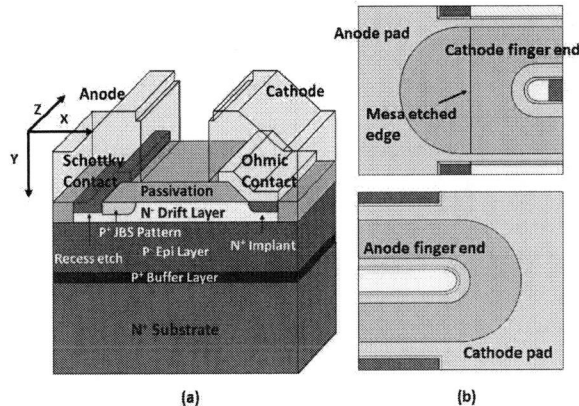

Figure 1. (a) The cell structure of the proposed lateral deep P⁺ JBS diode (b) Top view diagram showing cathode and anode finger end structure

field distribution at the finger ends through 3D simulations is necessary to avoid premature breakdown [1]. Fig. 2 shows the results of the 2D simulation of the blocking voltage and the specific on-resistance as a function of the RESURF layer charge of PIN unit cell diode. The peak blocking voltage is obtained at a RESURF charge of $8.5 \times 10^{12} cm^{-2}$. Excessive charge instantly leads to breakdown voltage reduction; thus, epi wafer specification was picked slightly below the optimum value at the peak (circled in doted red line). Because of this, the surface electric field cut of a RESURF device shows a balanced parabolic shape and significantly reduces the peak electric field at the anode junction through 2-dimensional depletion.

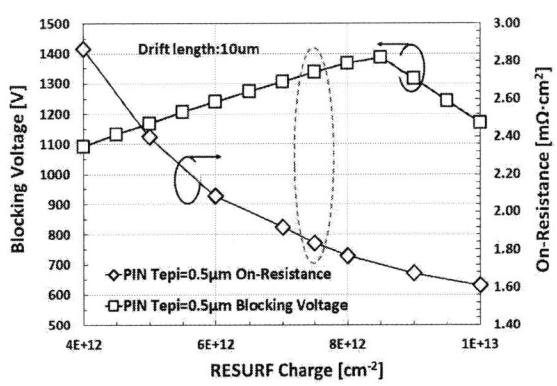

Figure 2. Blocking voltage and on-resistance of a lateral RESURF diode cell as a function of the RESURF layer charge

978-1-6654-8901-0/22 $31.00 © 2022 IEEE

The obtained epi wafer design parameters are summarized in Table I. P⁻ epi layer thickness and doping concentration were set to 10μm and 3×10^{15} cm⁻³, respectively, to assure >600 V vertical blocking voltage with additional consideration of the 3D electric field crowding associated with actual fabrication.

TABLE I Epi Wafer Design Parameters

Parameters	Values
N⁻ epi drift concentration	1.5×10^{17} cm⁻³
P⁻ epi drift concentration	3×10^{15} cm⁻³
P⁺ buffer concentration	5×10^{17} cm⁻³
N⁻ epi thickness	0.5 μm
P⁻ epi thickness	10 μm
P⁺ buffer thickness	1 μm

While RESURF effect alone is sufficient in improving 2D device performance through increased doping concentration and improved blocking voltage, the still high electric field at the anode contacts will induce premature breakdown when we reduce the bend radius to lay greater numbers of fingers in a given chip area, as shown in the 3D simulations (Fig. 3).

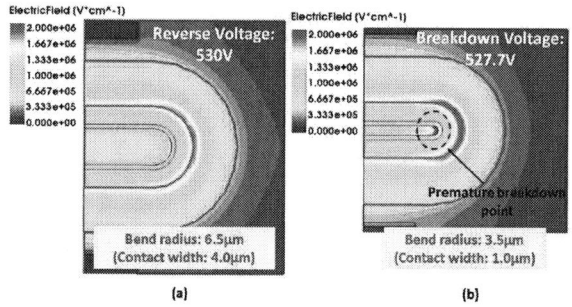

Figure 3. Top view 2D electric field distribution of pure Schottky (a) wide anode finger end (contact width:4μm) (b) fine anode finger end (contact width:1μm) at similar reverse voltage condition

In order to improve robustness against this field crowding, placing a p⁺ pattern deeper than the recess etched Schottky contact (insert of Fig.4) provides further field shielding, protecting the anode junction from premature breakdown during cell pitch reduction. The surface electric field cut comparison illustrated in Figure 4 plots indicates the effectiveness of the deep P⁺ JBS pattern, significantly lowering the electric field peak at the anode Schottky contact. The validity of this design approach can also be confirmed through the impact ionization distribution plot shown in Fig. 5. These deep P⁺ JBS patterns at the finger ends act as guard rings, mitigating the field crowding until the breakdown takes place at the originally designed vertical n-p junction as indicated in the circled black dotted line of Fig. 5(b). Through these guard-ring assisted finger end structures, the deep P⁺ JBS structure can sustain its original blocking capability with minimum penalty in forward

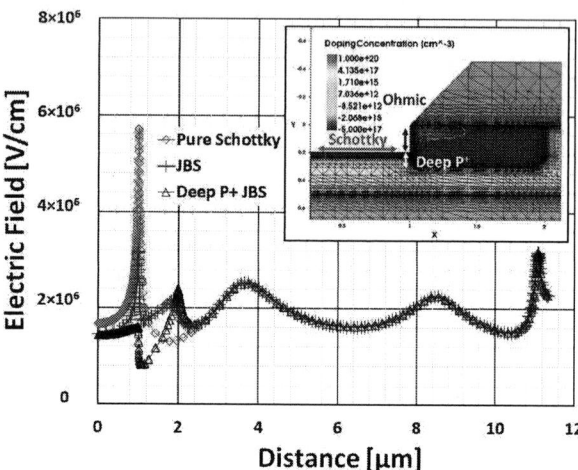

Figure 4. Surface electric field cut comparison of cuts made precisely below the anode Schottky contact. The insert of Figure 4 illustrates the cross section of deep P⁺ JBS structure

conduction, while the pure Schottky structure loses more than 50% of its blocking capability when fingers are packed tight (Fig. 6 contact width=1μm). The 3D breakdown voltage simulation results of both optimized cathode and anode finger ends show over 82% of the 2D simulation of inner cell (*BV*=1338V), indicating the effectiveness of the proposed layout design (Fig. 7).

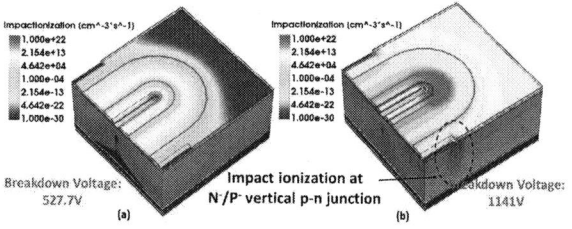

Figure 5. Impact ionization distribution of (a) fine anode finger without P⁺ guard end (b) fine anode finger with P⁺ guard end at respective breakdown voltage

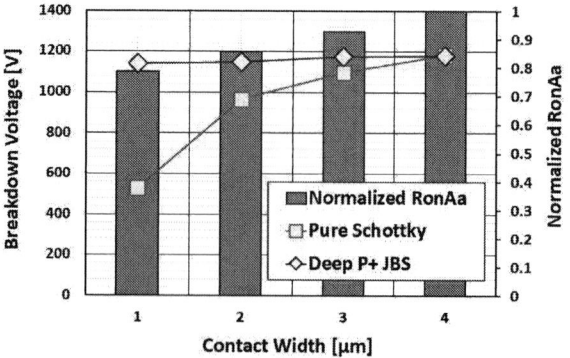

Figure 6. Breakdown voltage dependence of SBD and deep P⁺ JBS diode against contact width and normalized RonAa representing performance improvement through cell pitch reduction

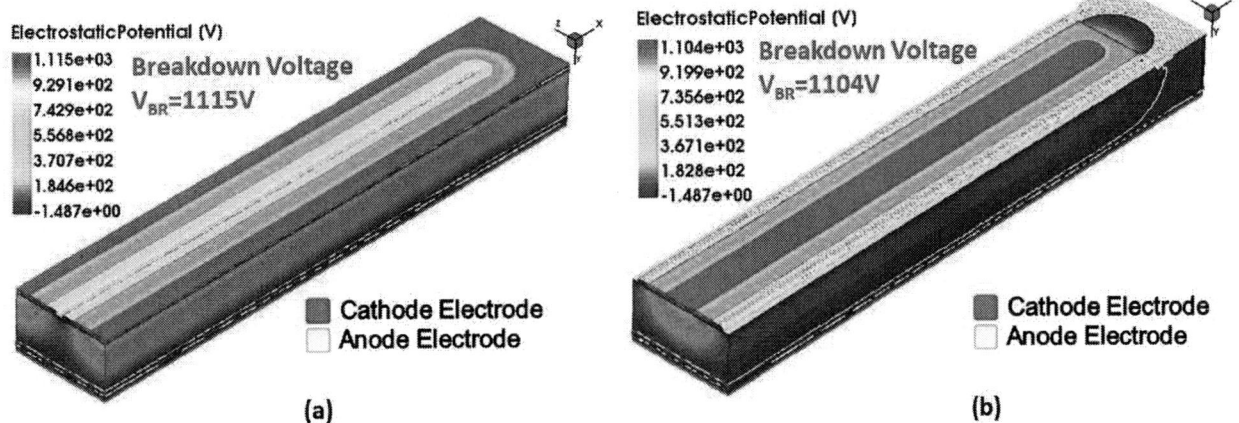

Figure 7. (a) Anode finger and (b) Cathode finger in 100μm length with colored contacts indicating the 2 different electrodes (Pink: Cathode and Yellow: Anode)

III. CELL PITCH REDUCTION AND PERFORMANCE IMPROVEMENT

Reduction of cell pitch involving smaller bend radius will improve the forward performance of lateral power device. While the wider contact width (4μm) fingers can spread slightly more current at the finger ends, if the fingers are extended significantly longer (>200μm) than their cell pitch, both wider (4μm) and finer (1μm) contact width fingers will eventually conduct a nearly equivalent amount of current with negligible on-resistance difference. This is because, assuming that specific contact resistance is kept low during fabrication ($\rho_C < 10^{-5}$ $\Omega \cdot cm^2$), the majority of the current conduction at the electrodes will occur at the edge of the metal contacts. While the current flowing through the semiconductor (drift region) is uniform, this

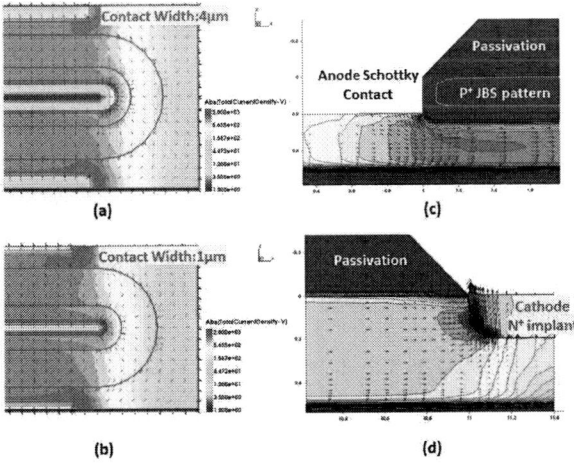

Figure 8. Top view of anode finger ends showing the total current density distribution (a) 4μm contact width (b) 1μm contact width Total current density vector plot of (c) anode electrode and (d) cathode electrode

phenomenon, known as "current crowding", occurs at the edge of the contact as shown in the vector plot of the total current density distribution (Fig. 8) [2]. This allows to cut down on the excessive contact width (shown in Figure 8(a) as light and dark blue regions with low current density) without compromising the device's current conduction capability. The on-resistance difference of 200μm in length, 4μm contact width finger and 1μm contact width finger is less than ~0.2%, and the performance improvement (RonAa reduction) will solely follow the ratio of the cell pitch reduction as shown in the bar graph of Figure 6. In addition to the forward performance improvement through cell pitch reduction, the deep P+ JBS structure will also be able to sustain its original high blocking capability coming from the RESURF effect while the pure Schottky structure is limited from its still high electric field at the anode contacts. 1μm contact width is the physical resolution limit of our cleanroom facilities, and this will be sufficient in dealing with current crowding at the contact edges.

IV. SURGE CURRENT CAPABILITY

Current surge simulation was conducted using mixed-mode TCAD simulation (Synopsys) utilizing an electro-thermal model of the diode tested in circuit shown in Fig. 9 (a) and (b). 50 Hz sinusoidal waveform (T=10ms) was set to flow in the forward direction of the diode. The thermal boundary conditions, including thermal resistivity, were specified in SDevice "Thermode", where the "Substrate" electrode is set to emulate a 300K, backside cooling configuration. Estimation of surface thermal resistance [$cm^2 \cdot K/W$] is done, assuming the dies are attached to a slightly bigger copper base plate with a thickness of 1mm [3]. When high-current short pulse is induced to a pure Schottky structure, the mobility of the carrier degrades greatly with increased device temperature, causing the forward *I-V* to lie down and the forward voltage drop to increase. This creates an uncontrollable positive feedback or "thermal runway", leading to instant catastrophic device failure as shown in Fig. 11(a). Failure criteria for the surge current simulation is the

978-1-6654-8901-0/22 $31.00 © 2022 IEEE

melting temperature of aluminum (T=933.5K or 660.35°C) [4] and the pure Schottky structure reaches this destructive point at the initial 2ms of the surge event. On the other hand, the P⁺ JBS pattern not only provides field shielding, but when the current reaches higher densities, carrier injection through the equally spaced p⁺ layer occurs (Fig. 9 (c)). At 6 ms, the injection of minority carriers (holes) into the n-type top RESURF layer is at maximum and clearly visible in the hole current density plot shown in Fig. 9 (d).

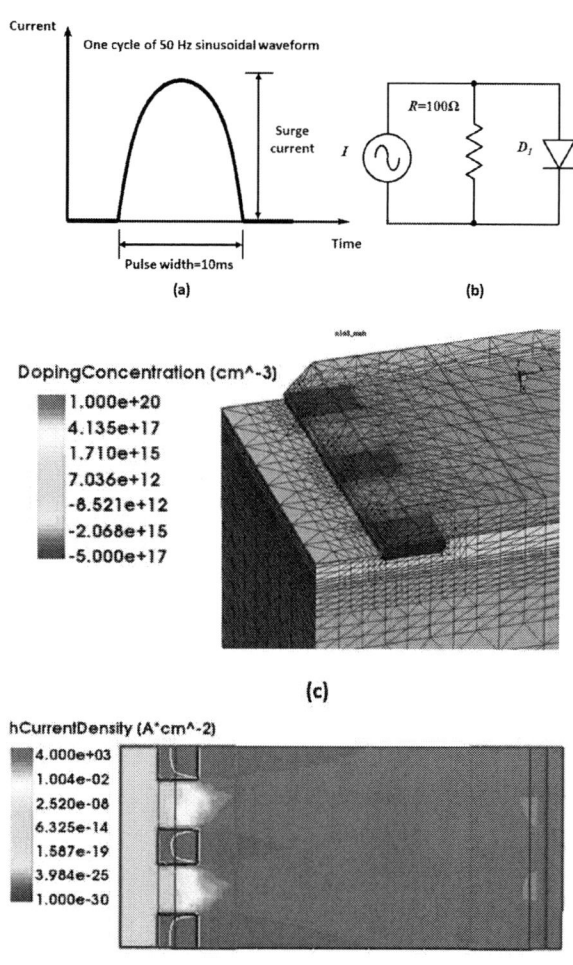

(a)

(b)

(c)

(d)

Figure 9. (a) Forward surge current waveform (b) Schematic of current surge test circuit for mixed-mode TCAD simulation (c) 3D deep P⁺ JBS meshed structure (d) Hole current density plot showing minority carrier injection during surge current event at 6ms (1ms past peak surge current)

The smaller the P⁺ JBS regions that make contact and occupy the anode electrode, the higher is the current density necessary to activate the minority carrier (hole) injection. The deep P⁺ JBS structure with 0.3μm depth requires less current density for the activation of its p-region, deriving from larger surface area of JBS pattern occupying the anode electrode, and the rise of the device temperature at initial 2ms of surge event is significantly lower than that of the structure with 0.2μm depth (shown in Fig.

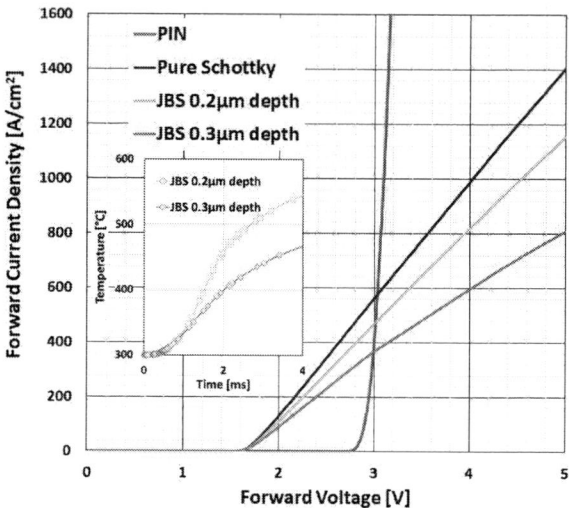

Figure 10. Forward *I-V* comparison of lateral PIN, Schottky, and deep P⁺ JBS diode with different p-region depth (0.2μm and 0.3μm). The temperature rise difference is shown according to p-region depth variation simulated as 150μm substrate thickness

10). This structural difference will eventually lead to approximately ~80°C peak temperature difference (JBS 0.3μm depth: 477°C peak temperature, JBS 0.2μm depth: 557°C peak temperature), making the former design more robust against current surge event. Contrastingly, this robustness is also in trade-off relationship with the forward performance of the designed device as shown in Figure 10. Both deep P⁺ JBS diodes will suffer consequences of higher on-resistance compared to a pure Schottky structure in return for more robustness against surge currents and field crowding at the anode contacts. The designer can further optimize the depth of the p-region JBS pattern in combination with the recess etch to attain the desired robustness/forward performance trade-off relationship specific to the application needs. The simulated deep P⁺ JBS diode withstands sixteen times the rated current (500A/cm²). Additionally, two different chip thicknesses (350μm and 150μm) were simulated for deep P⁺ JBS structure (Fig. 11). The results indicate reduced thermal resistance from thinner semiconductor material will further improve surge current robustness. Similar surge waveforms and temperature transients are also obtained with other TCAD simulator (Silvaco Atlas) indicating validness of our model and simulation [5].

V. CONCLUSION

The optimized "saddle" type finger layout of 4H-SiC lateral deep P⁺ JBS diode where concave/convex finger ends combined with mesa etch isolation achieved over 82% of the inner cell breakdown voltage. The added deep P⁺ JBS pattern significantly improves robustness against surge currents and field crowding. The simulated deep P⁺ JBS diode withstands sixteen times the rated current during surge event and this can be further optimized through the depth of the p-region JBS pattern in combination with the recess etch to attain the desired robustness/forward performance trade-offs and can be further improved through reduced thermal resistance from thinner semiconductor material.

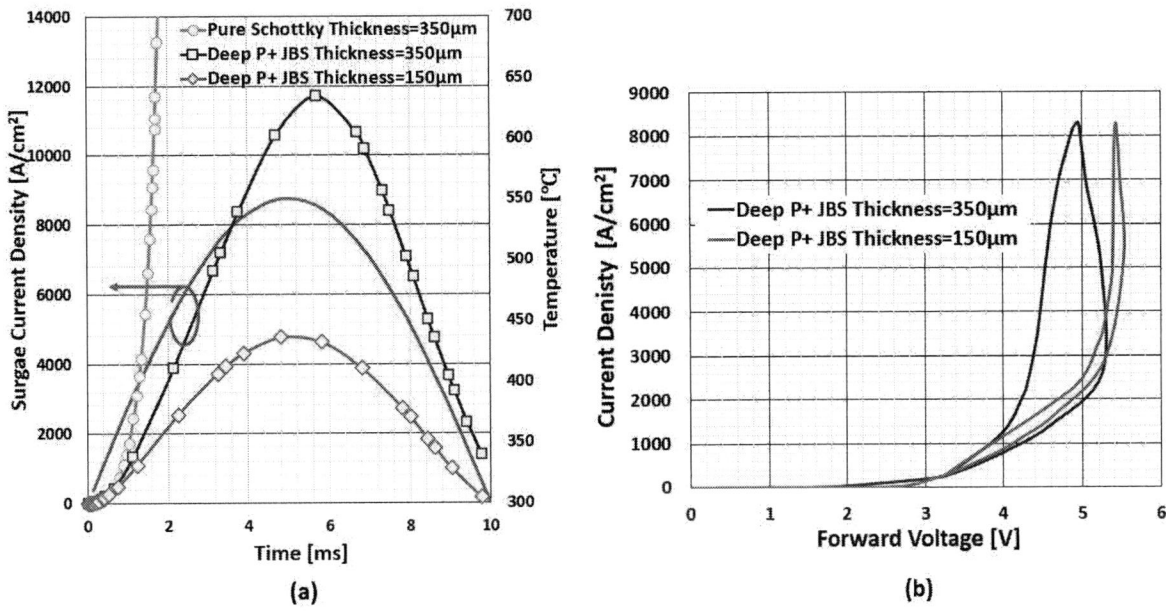

Figure 11. (a) Temperature transient of lateral deep P⁺ JBS diode during a current surge simulation (b) *I-V* characteristics of current surge event

REFERENCES

[1] Atsushi S., Alex Q.H., Design methodologies and fabrication of 4H-SiC lateral Schottky barrier diode on thin RESURF layer featured, *Appl. Phys. Lett.* 120 (12) (2022) 122103.

[2] H. Yu et al., "A Simplified Method for (Circular) Transmission Line Model Simulation and Ultralow Contact Resistivity Extraction," in IEEE Electron Device Letters, vol. 35, no. 9, pp. 957-959, Sept. 2014, doi: 10.1109/LED.2014.2340821.

[3] Buttay, Cyril & Wong, Hiu-Yung & Wang, Boyan & Xiao, Ming & Dimarino, Christina (2020). Surge Current Capability of Ultra-Wide-Bandgap Ga2O3 Schottky Diodes. Microelectronics Reliability. 114. 10.1016/j.microrel.2020.113743.

[4] Fichtner, Susanne & Lutz, J. & Basler, Thomas & Rupp, Roland & Gerlach, Rolf. (2014). Electro-Thermal Simulations and Experimental Results on the Surge Current Capability of 1200 V SiC MPS Diodes.

[5] W. Zhong et al., "Electro-thermal Analysis of 1.2kV-100A SiC JBS Diodes Under Surge Current Stress," 2019 16th China International Forum on Solid State Lighting & 2019 International Forum on Wide Bandgap Semiconductors China (SSLChina: IFWS), 2019, pp. 22-25, doi: 10.1109/SSLChinaIFWS49075.2019.9019788.

Characterization of Near Conduction Band SiC/SiO$_2$ Interface Traps in Commercial 4H-SiC Power MOSFETs

Hema Lata Rao Maddi[*]
Department of Electrical & Computer Engineering
The Ohio State University
Columbus, OH, USA
(Global Foundries since October 10, 2022)
Email:mhlrao2019@gmail.com

SuvenduNayak, Vishank Talesara, Yibo Xu, Wu Lu, and
Anant K. Agarwal
Department of Electrical & Computer Engineering
The Ohio State University
Columbus, OH, USA

Abstract—**It is well known that the high density of interface traps (D$_{it}$) near the conduction band (CB) edge limits the net inversion layer charge in the conduction band of a 4H-SiC/SiO$_2$ MOSFET in strong inversion. Measurements at cryogenic temperatures are necessary to study the presence of interface trap density (D$_{it}$) up to the CB edge. In our study, we extracted the threshold voltage (V$_T$) from cryogenic (10 K) to high temperature (500 K) and calculated the V$_T$ shift which is a measure of interface trapped negative charge (ΔQ$_{it}$). This shift in negative charge is used to estimate a relative magnitude of D$_{it}$ near the CB edge. The higher the value of ΔV$_T$, the higher the D$_{it}$ for a given sample.**

Keywords—4H-SiC, Power MOSFETs, Interface traps, cryogenics, Threshold voltage shift

I. Introduction

There have been several studies on planar gate MOSFETs analyzing their characteristics at both room and high temperatures, but very few have investigated their performance at cryogenic temperatures [1, 2]. The electron trapping at the semiconductor-oxide interface at room temperature is shown in Fig.1. This effect is responsible for the increase in the inversion channel resistance of 4H-SiC MOSFETs. The reduced inversion layer electron density in combination with enhanced scattering via charged traps at the interface, charged NIOTs, and fixed positive charge in the oxide (Q$_F$) near the interface result in low field-effect electron mobility. The industry standard post oxidation anneal in Nitric Oxide has led to the reduction of interface state density (D$_{it}$) near the conduction band edge which has made commercial MOSFETs possible. However, there is a wide variation of D$_{it}$ values in MOSFETs from different vendors due to different processes and p-well doping. The motivation for this work is to determine D$_{it}$ values near the conduction band (CB) edge for devices from several vendors. Since D$_{it}$ rises towards the CB edge, the interface traps near or at the CB edge are most important in determining the field-effect electron mobility. Most room temperature techniques such as subthreshold characteristics [3] can determine D$_{it}$ up to 0.1 eV below the edge of the conduction band. The use of cryogenic temperature allows us to determine D$_{it}$ up to the edge of the conduction band [2,4]. Thus, by measuring the change in V$_T$ (ΔV$_T$) from high temperature to cryogenic temperature, one can get a good estimate of the relative magnitudes of D$_{it}$ near the CB edge in different samples. It is crucial to determine a relative estimation of D$_{it}$ near the conduction band edge for commercial power MOSFETs from different vendors because it determines a large part of the specific on-resistance.

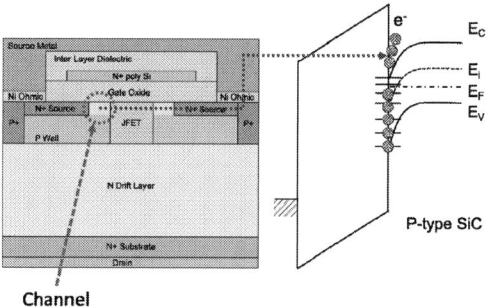

Fig. 1. Cross-sectional view of SiC MOSFET and energy band diagram at room temperature showing electron trapping at the interface.

II. Method for Estimation of D$_{IT}$

In order to estimate the trap density at the SiO$_2$/SiC interface, we used threshold voltage shifts at different temperatures. The temperature is varied from 500 K to 10 K using liquid helium as a cooling medium. D$_{it}$ closer to the conduction energy level (E$_C$) can be determined at low temperatures by monitoring the subthreshold slope deterioration in the drain current versus gate voltage (I$_D$-V$_{GS}$) characteristics of MOSFETs [3]. Alternatively, it may also be assessed from threshold voltage shift caused by temperature changes. To explain this, two energy band diagrams are illustrated in Fig. 2. Both diagrams are drawn at the onset of strong inversion (at V$_G$ = V$_T$) such that the surface potential is 2ϕ_F. Since ϕ_F increases at a lower temperature (Fig. 3), the Fermi level at the interface moves closer to the conduction band edge. The interface states, between the Fermi levels at room temperature and low temperature ($\Delta\phi_F$) are filled at low temperature and give rise to an additional negative charge, ΔQ$_{it}$.

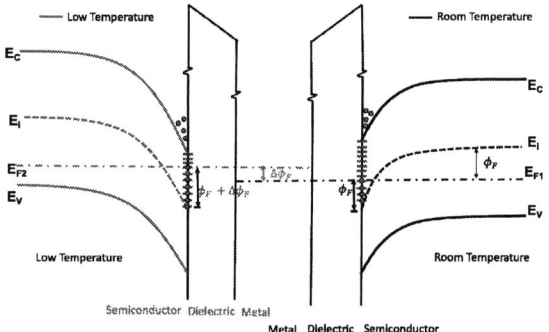

Fig. 2. Energy band diagram at ϕ_S=2ϕ_F showing fermi potential difference at low and high temperatures.

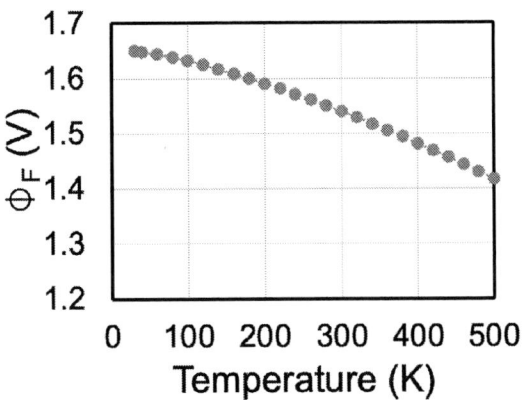

Fig. 3. Graph showing the variation of ϕ_F with temperature.

In other words, with the decrease in temperature, the Fermi-level (E_F) moves closer to the valence band (E_V) in the p-type bulk semiconductor. Therefore, at $\phi_S=2\phi_F$ additional band bending is required to induce sufficient electron concentration at the surface. Conversely, at high temperatures, E_F moves away from the valence band and therefore band banding required is less. As the temperature increases, electron concentration in the conduction band increases due to the detrapping of the carriers from the interface states above E_F. This, results in fewer negative charges at the interface, leading to a decrease in threshold voltage. The surface potential and the Fermi level can be related as

$$E_{CS} - E_F = E_g/2 - q(\phi_S - \phi_F) \qquad (1)$$

where, E_{CS} is the conduction band level at the surface, E_F is the Fermi level, E_g is the energy bandgap, ϕ_S is the surface potential, ϕ_F is the Fermi potential and q is the elementary charge.

The threshold voltage equation is expressed as [5]

$$V_T = \phi_{MS} + 2\phi_F + sqrt\ (4\epsilon_s q N_a \phi_F)/C_{ox} - Q_{it}/\ C_{ox} - Q_F/\ C_{ox} \qquad (2)$$

where, ϕ_{MS} is the metal-semiconductor work function difference, N_a is the acceptor concentration, C_{ox} is the oxide capacitance per unit area, Q_F is the fixed oxide charge per unit area and Q_{it} is the interface trapped charge (negative) per unit area.

Threshold voltage shift from cryogenic temperature (10 K) to high temperature (500 K), ΔV_T, can be essentially represented as (neglecting changes in ϕ_F with temperature),

$$\Delta V_T \sim \Delta Q_{it}/C_{ox} \qquad (3)$$

III. EXPERIMENTAL SETUP

An experimental setup to analyze the static characterization of SiC power MOSFETs from different vendors can be seen in Fig. 4. A lakeshore TTP-4 cryogenic probe station, a liquid helium dewar, a lakeshore (mention model) temperature controller, and a Keysight B1506 parametric analyzer comprise the experimental setup. The device under test is attached to the probe station temperature stage, whose temperature is adjustable from 10K to 500K.

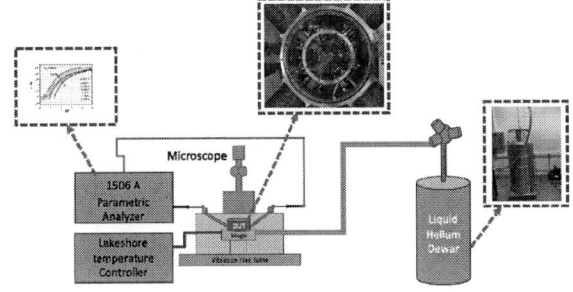

Fig. 4. Schematic diagram showing the experimental setup for cryogenic measurements.

The stage temperature was lowered using liquid helium, which was controlled by a PID feedback loop for optimal stabilization within 0.1 K. All the tests were conducted in a vacuum (10^6 mbar at 300K) to avoid condensation. The three terminals of power MOSFET (TO-247 package) were connected to the Keysight B1506 analyzer to perform the static characterization of the device. The 4H-SiC D-MOSFETs from different vendors involved in this study are tabulated in Table 1. It is to be noted that all the commercial MOSFETs from different vendors are packaged. A review of studies [6,7] evaluating SiO_2/SiC interfaces shows that N_2O/NO anneal methods are proven to be most effective for the Si face of 4H-SiC which has been widely used in industries. In many vendors' processes, NO/N_2O annealing is employed to reduce the interface state density without significantly affecting the oxide reliability [8].

TABLE I. VOLTAGE, CURRENT RATINGS, AND TYPICAL ON-RESISTANCE OF DIFFERENT VENDORS

Vendors	Type	Voltage Rating	Current rating	Typ. R_{ds-on}
C	Planar MOSFET	1200 V	12 A	500 mΩ
E	Planar MOSFET	1200 V	36 A	80 mΩ
K	Planar MOSFET	1200 V	7 A	350 mΩ

IV. RESULTS AND DISCUSSIONS

Table 1 shows the type of commercial MOSFET, voltage, and current rating of three vendors, named C, E, and K. Fig. 5 shows the I_D-V_{GS} characteristics of vendor E with a drain voltage of 100 mV in a temperature range of 10 K -500 K. It is to be noted that the I_D-V_{GS} characteristics for vendor C and K show similar behavior. The threshold voltage is measured using the linear extrapolation method [9] for each vendor. The threshold voltage shift as a function of temperature for commercial MOSFETs from three vendors is depicted in Fig. 6.

An approximate value of oxide thickness is extracted by using voltage-to-ramp breakdown measurements at room temperature. Fig.7 represents the voltage-to-ramp breakdown measurements for each vendor. With an oxide breakdown field of 11 MV/cm [10, 11], DUT oxide thickness extracted from vendors C, E, and K, are 472, 435, and 356 Å, respectively. A p-type base doping concentration, $N_a = 2 \times 10^{17}$ cm^{-3} has been assumed for all samples.

978-1-6654-8901-0/22 $31.00 © 2022 IEEE

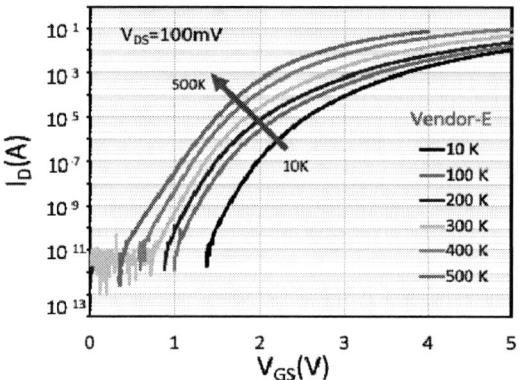

Fig. 5. I_D-V_{GS} characteristics of Vendor E in a temperature range of 10K - 500 K.

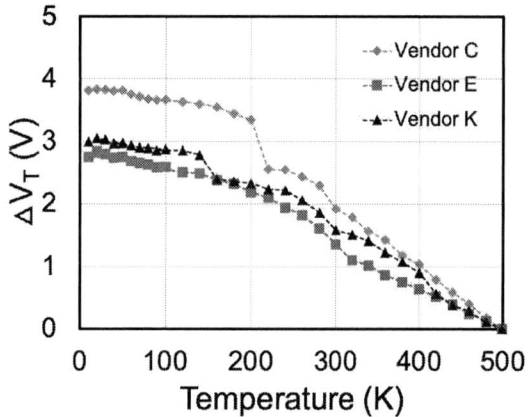

Fig. 6. Temperature-dependent threshold voltage shift for vendors C, E, and K.

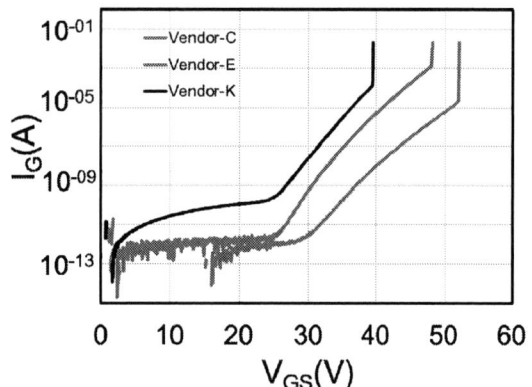

Fig. 7. Voltage ramp-to-breakdown measurements for vendors C, E, and K.

The p-well doping is generally not known in commercial devices. However, as Fig. 8 shows, $N_a = 2 \times 10^{17}$ /cm³ seems to be a reasonable value for a $V_T \sim 4.83$ V assuming Q_F and Q_{it} terms cancel each other in (2). The value of Q_F and Q_{it} is assumed as 5×10^{12}/cm³. In practice, Q_{it} term dominates over Q_F and can potentially add another 1 V to V_T.

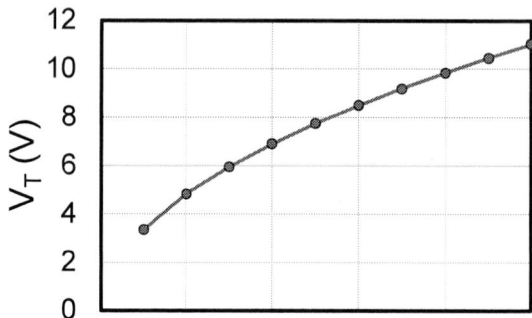

Fig. 8. Graph showing the variation of V_T with p-base doping concentration (N_a). Gate oxide thickness of 40 nm was considered.

If we closely observe the threshold voltage shift at low temperature in Fig. 6, one can easily predict that the device from Vendor C has the highest density of states (D_{it}) near the CB edge as compared to other vendors. The relative magnitude of D_{it} is also plotted in Fig. 9. The higher D_{it} would give rise to lower field-effect (FE) mobility in devices from vendor C and therefore higher MOS channel resistance.

To justify the above statement, two-dimensional simulations were performed using Silvaco TCAD and the dependence of FE-mobility with the interface traps is investigated. Channel mobility of a SiC device is lower due to the scattering by ionized interface states present when the device is in the on state [12]. This scattering is called coulomb scattering, and it is significant in SiC with a higher density of trap states. ALT.CVT (modified Lambordi model) is used as the mobility model in the simulation. This model includes surface phonon scattering, surface roughness, and coulomb scattering.

The acceptor-type interface trap is defined throughout the band gap of SiC as shown in Fig. 10. The resultant field-effect mobility with varied acceptor trap density is given in Fig. 11. I_D-V_{GS} characteristics with two different acceptor trap density is plotted in Fig. 12 to see the effect of FE-mobility. The decrease in trap concentration will shift the threshold voltage to the negative side and mobility will be increased, which is also evident from the mobility plot.

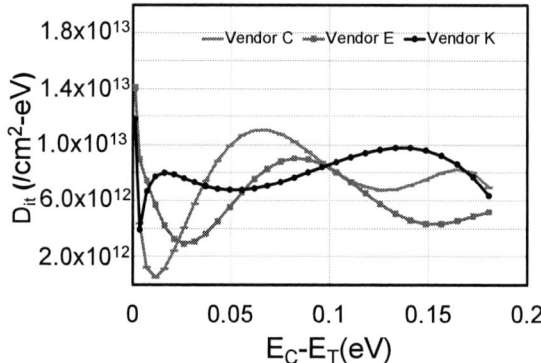

Fig. 9. Graph showing the extracted D_{it} for a temperature range of 10 K to 500 K.

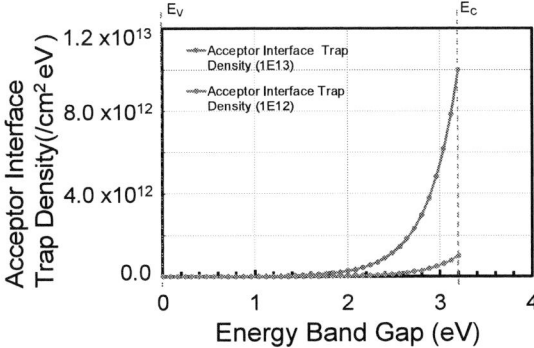

Fig. 10. Profile of acceptor interface trap density within the bandgap.

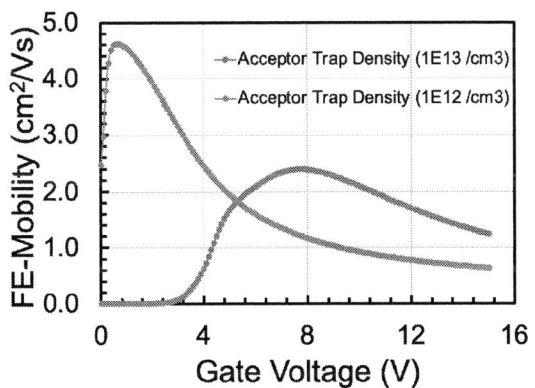

Fig. 11. Variation of FE-mobility with two different acceptor trap density profiles.

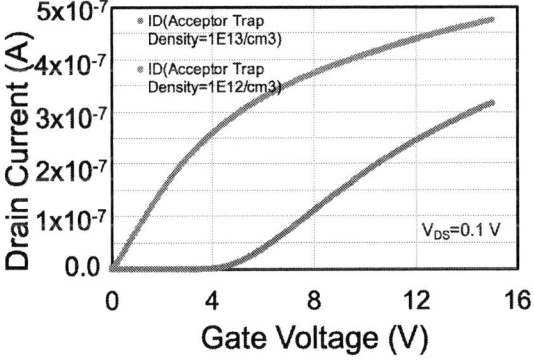

Fig. 12. I_D-V_{GS} characteristics (transfer characteristics) with two different acceptor interface trap density profiles.

V. CONCLUSIONS

A simple method of extraction of interface state density in commercial MOSFETs is presented. The exponential increase in D_{it} towards the conduction band edge is visible as the temperature is lowered from 500 K to 10 K. Commercial MOSFETs can easily be compared in terms of

threshold voltage shift which in turn is a measure of interface state density. As observed, the D_{it} value for vendor "C" is higher than vendor "E" and vendor "K" which can also be predicted from a higher threshold voltage variation (ΔV_T =3.82 V) in vendor "C" as compared to 2.75 V in vendor "E" and 2.99 V in vendor "K" for the entire range of measured temperatures.

ACKNOWLEDGMENT

We are grateful for the funding from the Vehicle Technology Office at the US Department of Energy, Washington DC, and the II-VI Foundation.

REFERENCES

[1] H. Yoshioka K. Hirata, "Characterization of SiO₂/SiC interface states and channel mobility from MOSFET characteristics including variable-range hopping at cryogenic temperature," AIP Adv., vol.8, p.045217, Apr. 2018.

[2] T. Kobayashi, S. Nakazawa, T. Okuda, J. Suda, and T. Kimoto, "Interface state density of SiO₂/p-type 4H-SiC (0001), (11 2̄ 0), (1 1̄00) metal-oxide-semiconductor structures characterized by low-temperature subthreshold slopes," Appl. Phys. Lett., vol.108(15), p.152108, Apr. 2016.

[3] S. Yu, M. H. White, A. K., and Agarwal, "Experimental Determination of Interface Trap Density and Fixed Positive Oxide Charge in Commercial 4H-SiC Power MOSFETs," IEEE Access, vol. 9, pp.149118-149124, Nov. 2021.

[4] H. Yoshioka, J. Senzaki, A. Shimozato, Y. Tanaka, and H. Okumura, "N-channel field-effect mobility inversely proportional to the interface state density at the conduction band edges of SiO₂/4H-SiC interfaces," AIP Advances, vol.5(1), p.017109, Jan. 2015.

[5] S. M. Sze, K. K. Ng, "Physics of Semiconductor Devices,", Wiley, Third edition, Chapter 6, p. 315, 2007

[6] T. Kimoto, Y. Kanzaki, M. Noborio, H. Kawano, and H. Matsunami, "Interface properties of metal oxide semiconductor structures on 4H-SiC (0001) and (1120) formed by N₂O oxidation,'" Jpn. J. Appl. Phys., vol. 44, no. 3, pp. 1213-1218, Mar. 2005.

[7] G. Y. Chung, C. C. Tin, J. R. Williams, K. McDonald, R. K. Chanana, R. A. Weller, S. T. Pantelides, L. C. Feldman, O. W. Holland, M. K. Das, and J. W. Palmour, "Improved inversion channel mobility for 4H-SiC MOSFETs following high temperature anneals in nitric oxide," IEEE Electron Device Lett., vol. 22, no. 4, pp. 176-178, Apr. 2001.

[8] H.-F. Li, S. Dimitrijev, H. B. Harrison, and D. Sweatman, "Interfacial characteristics of N₂O and NO nitrided SiO₂ grown on SiC by rapid thermal processing," Appl. Phys. Lett., vol. 70, no. 15, pp. 2028-2030, Apr. 1997.

[9] A. Ortiz-Conde, F. J. García-Sánchez, J. Muci, A. T. Barrios, J. J. Liou, and C.-S. Ho, "Revisiting MOSFET threshold voltage extraction methods," Microelectron. Rel., vol. 53, no. 1, pp. 90-104, Jan. 2013.

[10] U. Schwalke, M. Poelzl, T. Sekinger, and M. Kerber, "Ultra-thick gate oxides: Charge generation and its impact on reliability," Microelectron. Reliab., vol. 41, no. 7, pp. 1007 1010, Jul. 2001.

[11] G. Liu, A. C. Ahyi, Y. Xu, T. Issacs-Smith, Y. K. Sharma, J. R. Williams, L. C. Feldman, and S. Dhar, "Enhanced inversion mobility on 4H-SiC (112̄0) using phosphorus and nitrogen interface passivation," IEEE Electron Device Lett., vol. 34, no. 2, pp. 181 183, Jan. 2013.

[12] S. Dhar, S. Haney, L. Cheng, S.-R. Ryu, A. K. Agarwal, L. C. Yu, and K. P. Cheung, "Inversion layer carrier concentration and mobility in 4H-SiC metal-oxide-semiconductor field-effect transistors," J. Appl. Phys., vol. 108, no. 5, Sep. 2010.

978-1-6654-8901-0/22 $31.00 © 2022 IEEE

Reverse Breakdown Time of Wide Bandgap Diodes

Jack Flicker
Power Electronics and Energy Conversion Systems
Sandia National Laboratories
Albuquerque, NM USA
jdflick@sandia.gov

Emily Schrock
Directed Energy Missions
Sandia National Laboratories
Albuquerque, NM USA
eschroc@sandia.gov

Robert Kaplar
Semiconductor Material and Device Sciences
Sandia National Laboratories
Albuquerque, NM USA
rjkapla@sandia.gov

Abstract—In order to evaluate the time evolution of avalanche breakdown in wide and ultra-wide bandgap devices, we have developed a cable pulser experimental setup that can evaluate the time-evolution of the terminating impedance for a semiconductor device with a time resolution of 130 ps. We have utilized this pulser setup to evaluate the time-to-breakdown of vertical Gallium Nitride and Silicon Carbide diodes for possible use as protection elements in the electrical grid against fast transient voltage pulses (such as those induced by an electromagnetic pulse event). We have found that the Gallium Nitride device demonstrated faster dynamics compared to the Silicon Carbide device, achieving 90% conduction within 1.37 ns compared to the SiC device response time of 2.98 ns. While the Gallium Nitride device did not demonstrate significant dependence of breakdown time with applied voltage, the Silicon Carbide device breakdown time was strongly dependent on applied voltage, ranging from a value of 2.97 ns at 1.33 kV to 0.78 ns at 2.6 kV. The fast response time (< 5 ns) of both the Gallium Nitride and Silicon Carbide devices indicate that both materials systems could meet the stringent response time requirements and may be appropriate for implementation as protection elements against electromagnetic pulse transients.

Keywords—impact ionization, avalanche breakdown, EMP protection, SiC, GaN, WBG, UWBG

I. INTRODUCTION

One of the primary advantages of wide and ultra-wide bandgap (U/WBG) devices is their increased breakdown electric field compared to conventional semiconductor materials [1]. In vertical devices, this increased voltage hold-off allows for thinner active regions, which results in lower cost and faster switching dynamics. The importance of breakdown electric field for semiconductor device can be seen in typical figures of merit (FOMs) that are used to compare performance across materials systems. For example, the Baliga Figure of Merit (BFOM) [2] scales as the cube of the breakdown electric field, the Huang FOM (HCAFOM) [3] scales as the breakdown electric field squared, and the Johnson FOM (JFOM) [4] scale linearly with the breakdown electric field. Although it is important to note that these FOMs are approximations that do not consider the dependence of breakdown electric field on doping, temperature, and a variety of other more complicated factors [5], they are still utilized extensively to compare the appropriateness of possible materials systems in a given application.

The intrinsic process which limits voltage hold-off in semiconductor devices is impact ionization or avalanche breakdown [6]. Due to the importance of understanding avalanche breakdown, significant effort has been expended in the literature on elucidating the process using a variety of analytical and phenomenological models [7-10] and measuring breakdown electric field and/or impact ionization coefficients for U/WBG materials [5]. These measurements typically are carried out under static or pseudo-static conditions to evaluate the magnitude of breakdown electric field.

U/WBG devices with high static voltage hold-off and fast carrier dynamics are ideal for utilization as protection elements against high-voltage, fast transient wavefronts, for example for high voltage electromagnetic pulse (EMP) protection of the electrical grid. The fastest protection elements on the grid are typically metal oxide varistors (MOVs) composed of packed granular power columns of zinc oxide (ZnO) powder that are designed as lightning surge arresters (LSAs) and tested to ~1 μs response times [11].

EMP is a complicated multi-step process that is typically split into three regimes; a fast risetime wave (<10 ns) known as E1, a medium risetime wave (~1 μs) known as E2, and a slow wave (~minutes to hours) known as E3 [12]. While conventional LSAs may offer protection against E2 dynamics, modeling/testing has shown that they are too slow to arrest the E1 component of the EMP wavefront [13, 14] with charge-migration rates at grain boundary interfaces limiting response times to tens to hundreds of nanoseconds [15, 16].

Due to their small size, high breakdown electric field, and fast dynamic response U/WBG diodes may be ideal protection devices on the electrical grid to protect against E1-type transients. These diodes can be placed in the power system to provide coordinated protection at fast time scales (~ns) before traditional LSAs can actuate (~1 μs). These diodes would utilize avalanche breakdown, so that any applied voltage greater than the breakdown voltage (V_{br}) would induce conduction and shunt current to ground. The highly non-linear conduction would quickly clamp voltages at the diode breakdown voltage until traditional LSAs can react (~1 μs).

Utilization of U/WBG diodes as protection devices depends not only on the static breakdown electric field, but also on the dynamics of the avalanche process, which determines the timescale of voltage clamping. While significant effort has been made to understand the static or pseudo-static characteristics of the avalanche mechanism in U/WBG device, much less effort has been made to evaluate the dynamics of this process. In this work, we describe a test setup to evaluate

---

The authors are thankful for the support from ULTRA, an Energy Frontier Research Center funded by the U.S. Department of Energy (DOE), Office of Science, Basic Energy Sciences (BES), under Award No. DE-SC0021230. The work is also supported, in part, by ARPA-E's OPEN+ Kilovolt Devices Cohort directed by Dr.Isik Kizilyalli.

978-1-6654-8901-0/22 $31.00 © 2022 IEEE

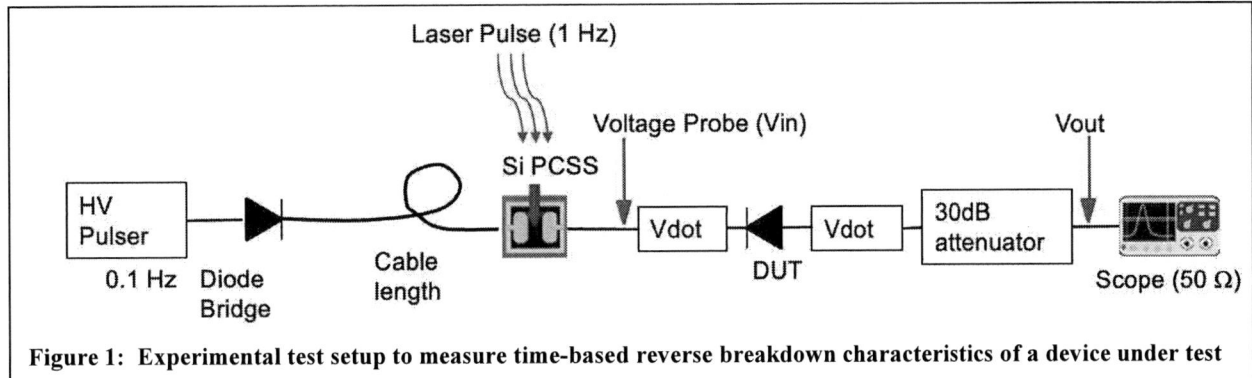

Figure 1: Experimental test setup to measure time-based reverse breakdown characteristics of a device under test

the time dynamics of diode avalanche breakdown and describe results on vertical SiC Schottky and GaN pin diodes.

II. REVERSE BREAKDOWN TESTING SETUP

In previous work [17] we have described a novel pulser measurement capability to evaluate ultra-fast reverse recovery time of WBG diodes based on previous work on light emitting diodes [18]. Briefly, this reverse recovery measurement is a subset of time domain reflectometry (TDR) measurements where an input pulse waveform interacts with a reflected waveform from the cable termination interface (i.e., the load). The nature of this interaction can give information regarding both the time-of-flight and load impedance based on the reflection coefficient, Γ. An open circuit results in $\Gamma=1$ with constructive interference between the incident and reflected wave, resulting in a measured voltage doubling of the incident wave. A short circuit results in $\Gamma=-1$ with destructive interference with the incident wave. A matched impedance (typically 50 Ω) results in $\Gamma=0$ with no reflected wave. By evaluating the evolution of the incident waveform over time (and thus the value of $\Gamma(t)$), it is possible to track the evolution of load impedance over time with very high time resolution.

This pulser setup was previously utilized to measure the reverse recovery time of vertical Gallium Nitride (v-GaN) diodes with resolution ~100ps. To measure reverse recovery, the device under test (DUT) was forward biased by a small voltage in the conducting regime ($\Gamma\approx-1$). A high voltage incident pulse was applied to force the diode into a blocking state ($\Gamma\approx1$). Capacitive (V-dot) probes [19] were utilized to evaluate the time-evolution of the incident pulse to extract the evolution of Γ over time.

To change the measurement from the previous time-based reverse recovery measurements (conduction-to-blocking) to time-based reverse breakdown (blocking-to-breakdown conduction), this pulser experimental setup has been slightly altered. Figure 1 shows a diagram schematic of the experimental setup.

The main components of the system are a pulse generating circuit, the device under test, and diagnostic components. The pulse generating system is composed of a high voltage pulse charging circuit (capable of pulse voltages up to 3 kV), a diode bridge and cable length for charge storage (to give control over pulse length), a Silicon (Si) photoconductive semiconductor switch (PCSS) which controls voltage application to the DUT, and a triggering laser. Details of these components can be found in [17].

The DUT is electrically connected to a copper strip line on a standard printed circuit board (PCB) FR4 substrate. This PCB is sandwiched between two aluminum outer conductor pieces to form a two-port, high-bandwidth, 50 Ω enclosure that is connected to the cable pulser via N-type connectors. Since the diode does not need to be biased prior to measuring (as in [17]), instead of terminating with a current viewing resistor (CVR) and biasing voltage, the cable pulser is terminated with an impedance-matched (50 Ω) 4 GHz oscilloscope (Tektronix TDS7404). Diagnostic voltage probes are present both before and after the DUT, allowing evaluation of both the reflected and conducted pulses on either side of the DUT.

In this setup, the pulse generation circuit generates a fast rise-time reverse-biasing pulse which is applied to the diode with a 130 ps rise-time via the transmission line (10-90% threshold under open circuit). When the pulse reaches the diode being tested, it forces the diode from a non-biased blocking state to a reverse-bias blocking state. This reverse bias blocking state has a very high impedance, and the pulser will be terminated by the DUT with a reflection coefficient that is nearly that of an open circuit (i.e., $\Gamma\approx1$).

If the voltage pulse is large enough to initiate avalanche breakdown, the diode will enter a reverse bias conducting state. During this transition, the impedance of the diode changes from high-impedance to low-impedance, thus causing the diode to exhibit a dynamic reflection coefficient. In this case, initially the pulser will be terminated by the high-impedance DUT ($\Gamma\approx1$), but as the system evolves in time and the diode impedance decreases, the pulser will be terminated by the impedance-matched oscilloscope (i.e., $\Gamma\approx0$). This results in a voltage collapse on the input to halve the applied voltage, while simultaneously a voltage rise on the DUT output indicates DUT conduction.

The response time of two types of power semiconductor diodes were evaluated: a vertical GaN device from Avogy Inc. and a vertical SiC device from On Semiconductor (OnSemi).

978-1-6654-8901-0/22 $31.00 © 2022 IEEE

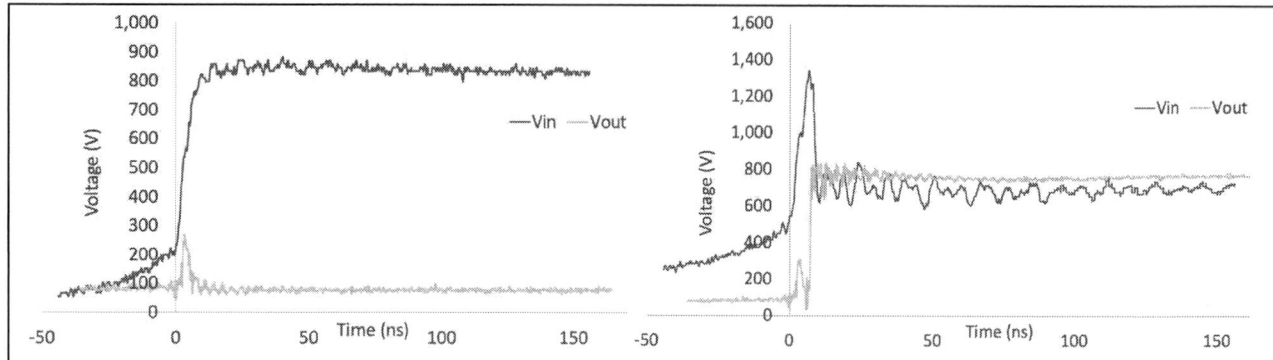

Figure 2: (Left) Input (blue) and output (yellow) voltages for SiC DUT in blocking mode ($V_{applied} < V_{br}$). **(Right).** Input (blue) and output (yellow) voltages for SiC DUT in reverse breakdown mode ($V_{applied} > V_{br}$).

The v-GaN devices from Avogy (AVD05120 series) are a true vertical structure, grown on conducting native GaN substrates. These PiN diodes consist of an n+ wafer, an n⁻ drift region, and a p+ anode as well as the required edge termination structures [20-22]. The diodes are rated for 1200 V and 100 A (pulsed).

The SiC devices from OnSemi (NGTD17R120F2) [23] are Schottky diodes nominally designed for flyback protection of insulated gate bipolar transistor (IGBT) switches. The diodes are rated for 1200 V and 35 A (DC).

III. RESULTS AND DISCUSSION

Figure 2 shows the results for a voltage pulse applied to a SiC diode for an applied voltage below breakdown (left) and just above breakdown (right). The voltage profile is reconstructed through integration of the signal of the V-dot probe located before the DUT (blue) and just after the DUT (yellow).

At voltages below breakdown (Figure 2, left), the PCSS is triggered via the triggering laser at t = 0. Prior to triggering at t = 0, there is a slight voltage rise at the input of the diode (blue) due to leakage through the Si PCSS. The actuation of the PCSS at t=0 applies a voltage step to the diode of 850 V. The applied voltage is not at a true steady-state during hold-off but is slowly decreasing (~ms) due to charge dissipation in the diode bridge and cable of the pulser. However, in the regime of interest (<< 1 μs), the voltage can be approximated by a step function.

The V-dot measurement on the output of the diode is shown in yellow. This trace shows a transient voltage pulse during the voltage step with a duration of ~4 ns. This transient voltage measurement indicates displacement current due to charging the junction capacitance of the diode [24] as voltage is applied to the diode.

At voltages above breakdown (Figure 2, right), the characteristic input (blue) and output (yellow) voltage traces look significantly different. When the PCSS is triggered (t = 0), the input voltage (blue trace) rises to a maximum value of 1.3kV and then collapses down to a voltage of ~750 V, approximately ½ of the full applied voltage. This voltage

collapse from full voltage to ½ of the full applied voltage indicates that there is no longer voltage doubling ($\Gamma = 1$) and that the device is now conducting with the pulser being terminated by the impedance-matched oscilloscope ($\Gamma = 0$).

Simultaneously, as the voltage at the input of the diode collapses, there is an increase in voltage measured at the output of the diode (yellow) to the applied voltage. This voltage increase on the output of the diode to approximately the same magnitude of the total voltage indicates the diode is in conduction with an intrinsic resistance that is much smaller than the pulser termination (i.e., 50 Ω).

To define the value the output voltage signal was approaching and the 90% threshold determining the end of the breakdown transients more accurately, the output voltage was normalized to the measured steady-state output voltage. For this normalized voltage measurement, a value of 0 indicates a fully blocking state while a value of 1 indicates fully conducting state. Figure 3 shows the normalized results for the GaN and SiC diodes at an applied voltage just above V_{br} at an

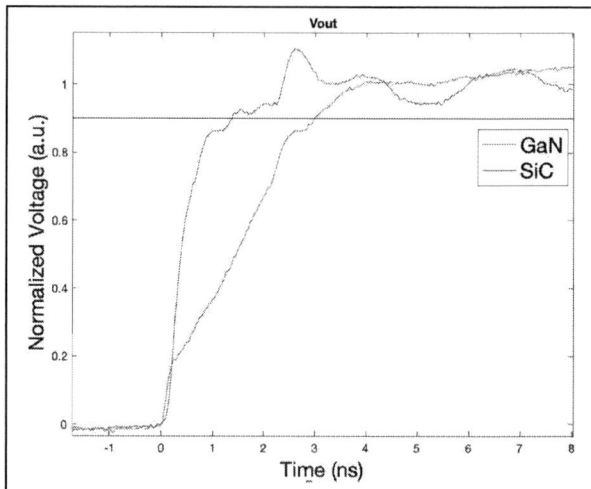

Figure 3: Output voltage of SiC (red) and GaN (blue) devices at breakdown.

applied voltage of 1.22 and 1.33 kV, respectively. Both devices cross the 90% threshold within 3 ns of the PCSS trigger pulse. The GaN device achieved breakdown significantly faster, with the 90% threshold being reached at 1.37 ns compared to 2.98 ns for SiC. Both devices respond within the nanosecond regime needed for protection elements. However, due to the difference in device materials stack, type, and footprint, it is difficult to make accurate apples-to-apples comparisons about the applicability of both materials systems.

Response times of both the GaN and SiC devices were measured at applied voltages ranging from just above V_{br} to $2*V_{br}$. The GaN device showed little dependence on the applied voltage. However, the SiC response time reduced significantly with increased voltage (TABLE I) from a value of 2.97 ns at 1.33 kV to 0.78 ns at 2.6 kV (a 3.8x reduction in response time). The applied voltage significantly increases the slope of the breakdown transient (Figure 4). The root cause of this and why this effect is not seen as dramatically in GaN is still under investigation to determine whether this effect is due to the intrinsic breakdown mechanisms within SiC, dynamics in the Schottky barrier, or the junction capacitance of the device.

TABLE I. SIC DIODE 90% RESPONSE TIME UNDER DIFFERENT APPLIED VOLTAGE

Applied Voltage (kV)	90% threshold time (ns)
1.33	2.98
1.7	2.42
1.8	2.17
2.35	0.82
2.5	0.90
2.6	0.78

IV. CONCLUSIONS

In this work, we have introduced a variation of a previously developed experimental pulser technique originally used to measure the fast reverse recovery time of WBG diodes with resolution of 130 ps. This variation evaluates the dynamic reflection coefficient (Γ) of the terminating DUT under applied reverse bias voltage pulses. When voltage pulses above the breakdown of the device are applied, the time evolution of device avalanche breakdown can be monitored from fully blocking through fully conducting.

This experimental technique was used to evaluate two 1200 V vertical WBG devices, a v-GaN pin diode from Avogy, Inc. and a SiC Schottky diode from OnSemi. At an applied voltage just above breakdown, both devices achieved the 90% conduction threshold within 3 ns. The v-GaN device demonstrated faster dynamics compared to the SiC device. The v-GaN device achieved 90% conduction within 1.37 ns compared to the SiC device response time of 2.98 ns. The v-GaN device did not demonstrate significant dependence of breakdown time with applied voltage. However, the SiC device breakdown time was strongly dependent on applied

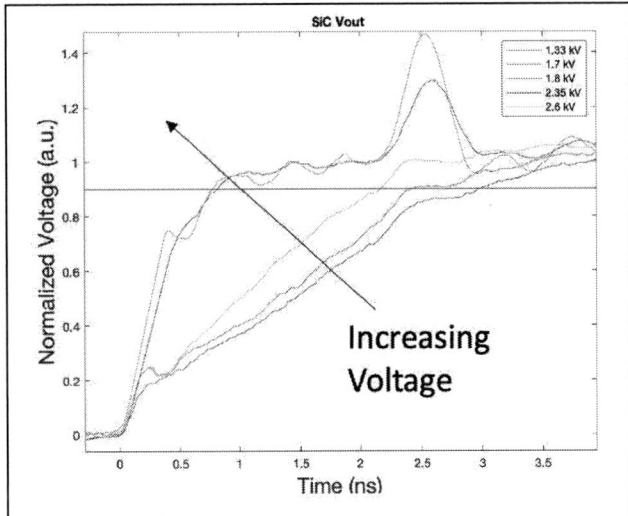

Figure 4: Avalanche response of SiC diode to applied voltages from 1.33 kV to 2.6 kV.

voltage, ranging from a value of 2.97 ns at 1.33 kV to 0.78 ns at 2.6 kV.

The high voltage hold-off of WBG diodes coupled with the repeatable, fast dynamics of avalanche breakdown make them appealing as high-speed protection devices in high voltage systems. An example of this is the EMP E1 transient mitigation on the electrical power grid. As the risetime of an E1 pulse is typically <10 ns, devices must exhibit response times faster than this to be considered as a protection element. The work presented here has demonstrated that both GaN and SiC diodes can meet these stringent response time requirements and may be appropriate for implementation as protection elements against E1 transients.

ACKNOWLEDGMENT

Sandia National Laboratories is a multimission laboratory managed and operated by National Technology & Engineering Solutions of Sandia, LLC, a wholly owned subsidiary of Honeywell International Inc., for the U.S. Department of Energy's National Nuclear Security Administration under contract DE-NA0003525.

REFERENCES

[1] J. Y. Tsao et al., "Ultrawide-Bandgap Semiconductors: Research Opportunities and Challenges," Advanced Electronic Materials, vol. 4, no. 1, pp. 1600501-n/a, 2018, Art no. 1600501, doi: 10.1002/aelm.201600501.

[2] K. Shenai, R. S. Scott, and B. J. Baliga, "Optimum semiconductors for high-power electronics," IEEE Trans. Electr. Dev., vol. 36, no. 9, pp. 1811-1823, 1989.

[3] A. Q. Huang, "New unipolar switching power device figures of merit," IEEE Electron Device Letters, vol. 25, no. 5, pp. 298-301, 2004, doi: 10.1109/LED.2004.826533.

[4] E. O. Johnson, "Physical limitations on frequency and power parameters of transistors," in Semiconductor Devices: Pioneering Papers: World Scientific, 1991, pp. 295-302.

[5] O. Slobodyan et al., "Analysis of the dependence of electric field on semiconductor bandgap," Journal of Materials Research, pp. 1-17, 2022.

[6] S. M. Sze and K. K. Ng, *Physics of semiconductor devices*. John wiley & sons, 2006.

[7] B. Ridley, "Lucky-drift mechanism for impact ionisation in semiconductors," *Journal of Physics C: Solid State Physics,* vol. 16, no. 17, p. 3373, 1983.

[8] P. Wolff, "Theory of electron multiplication in silicon and germanium," *Physical Review,* vol. 95, no. 6, p. 1415, 1954.

[9] W. Shockley, "Problems related to p-n junctions in silicon," *Solid-State Electronics,* vol. 2, no. 1, pp. 35-67, 1961/01/01/ 1961, doi: https://doi.org/10.1016/0038-1101(61)90054-5.

[10] A. G. Chynoweth, "Ionization rates for electrons and holes in silicon," *physical review,* vol. 109, no. 5, p. 1537, 1958.

[11] *IEEE Standard for Metal-Oxide Surge Arresters for AC Power Circuits, C62.11-2012,* IEEE Power and Energy Society, New York, NY, 2012.

[12] E. Savage, J. Gilbert, and W. Radasky, "The early-time (E1) high-altitude electromagnetic pulse (HEMP) and its impact on the US power grid," *Report Meta,* 2010.

[13] A. G. Tarditi *et al.*, "High Voltage Modeling and Testing of Transformer, Line Interface Devices, and Bulk System Components Under Electromagnetic Pulse, Geomagnetic Disturbance, and other Abnormal Transients," Oak Ridge National Lab.(ORNL), Oak Ridge, TN (United States), 2019.

[14] T. Bowman, M. Halligan, and R. Llanes, "High-Frequency Metal-Oxide Varistor Modeling Response to Early-time Electromagnetic Pulses," in *2020 IEEE International Symposium on Electromagnetic Compatibility & Signal/Power Integrity (EMCSI),* 2020: IEEE, pp. 466-471.

[15] A. Varpula, "Modeling of transient electrical characteristics for granular semiconductors," *Journal of Applied Physics,* vol. 108, no. 3, p. 034511, 2010.

[16] G. Xie, M. Wu, and W. Shi, "Boundary breakdown time of ZnO varistors under nanosecond pulse current," *AIP Advances,* vol. 9, no. 4, p. 045032, 2019.

[17] Daniel L. Mauch *et al.*, "Ultra-Fast Reverse Recovery Measurement for Wide-Bandgap Diodes," 2016 (*Submitted IEEE Transactions on Power Electronics).*

[18] R. H. Dean and C. J. Nuese, "A refined step-recovery technique for measuring minority carrier lifetimes and related parameters in asymmetric pn junction diodes," *IEEE Trans. Electr. Dev.,* vol. 18, no. 3, pp. 151-158, 1971.

[19] H. Bluhm, "Power Systems," 2006.

[20] C. Matthews *et al.*, "Switching characterization of vertical GaN PiN diodes," in *2016 IEEE 4th Workshop on Wide Bandgap Power Devices and Applications (WiPDA),* 2016: IEEE, pp. 135-138.

[21] I. C. Kizilyalli, A. P. Edwards, O. Aktas, T. Prunty, and D. Bour, "Vertical Power p-n Diodes Based on Bulk GaN," *IEEE Trans. Electr. Dev.,* vol. 62, no. 2, pp. 414-422, 2015, doi: 10.1109/TED.2014.2360861.

[22] I. C. Kizilyalli, A. P. Edwards, H. Nie, D. Disney, and D. Bour, "High Voltage Vertical GaN p-n Diodes With Avalanche Capability," *IEEE Trans. Electr. Dev.,* vol. 60, no. 10, pp. 3067-3070, 2013, doi: 10.1109/TED.2013.2266664.

[23] On Semiconductor, "NGTD17R120F2: Fast Switching Rectifier Die," March 2016. Available: https://www.onsemi.com/pdf/datasheet/ngtd17r120f2wp-d.pdf Accessed:Sept 8, 2022.

[24] T. Funaki, T. Kimoto, and T. Hikihara, "Evaluation of high frequency switching capability of SiC Schottky barrier diode, based on junction capacitance model," *IEEE transactions on power electronics,* vol. 23, no. 5, pp. 2602-2611, 2008.

Symmetrical V_{TH}/R_{ON} Drifts Due to Negative/Positive Gate Stress in p-GaN Power HEMTs

Nicolò Zagni[1,†], Marcello Cioni[1], Maria Eloisa Castagna[2], Maurizio Moschetti[2], Ferdinando Iucolano[2], Giovanni Verzellesi[3] and Alessandro Chini[1]

1. "Enzo Ferrari" Engineering Department, University of Modena and Reggio Emilia, Via P. Vivarelli 10, 41125 Modena, Italy
2. STMicroelectronics, Stradale Primosole, 50, 95121 Catania, Italy
3. Department of Sciences and Methods for Engineering and EN&TECH, University of Modena and Reggio Emilia, Via G. Amendola, 2, 42122, Reggio Emilia, Italy

†Corresponding Author E-mail: nicolo.zagni@unimore.it

Abstract—We investigate the drift of threshold voltage (V_{TH}) and on-resistance (R_{ON}) in p-GaN power HEMTs after being submitted to negative/positive gate stress. Negative (Positive) Gate Stress (NGS/PGS) was applied at a gate-to-source bias of $|V_{NGS}| = V_{PGS} = 6$ V up to a cumulative stress time of 8×10^3 s at room temperature. We found that during NGS both V_{TH} and R_{ON} increased over stress time, whereas during PGS both parameters decreased and stabilized to the values prior to stress application. This symmetric behavior was maintained after 5 full NGS/PGS stress cycles, indicating the absence of permanent degradation. To further characterize the V_{TH} and R_{ON} transients, the NGS/PGS stress cycles were repeated at different temperatures ($T=30\text{-}105$ °C). While V_{TH} exhibited a strong T-dependence ($E_A \approx 0.6$ eV) during NGS, a negligible variation of the V_{TH} transients with T was found during PGS ($E_A \approx 0$ eV). Instead, R_{ON} transients exhibited approximately the same T-dependence during both NGS and PGS ($E_A \approx 0.3\text{-}0.4$ eV).

Keywords—p-GaN HEMTs, NBTI, PBTI, Gate Stress, V_{TH} instability, dynamic R_{ON}

I. INTRODUCTION

p-GaN high electron mobility transistors (HEMTs) are widely adopted in the market as E-mode devices for power switching converters [1], [2]. However, stability issues of the gate stack lead to drift of key device parameters, i.e., threshold voltage, V_{TH}, and on-resistance, R_{ON}, which are still under intensive scientific investigation [3]–[6].The Schottky-type gate contact over the p-GaN region is employed to reduce gate leakage; however, this solution renders the p-GaN floating over a wide range of gate-to-source voltage (V_{GS}) which gives rise to unique instability features that are absent in more conventional GaN HEMTs or MIS-HEMTs.

Gate stress experiments are commonly used to investigate the instabilities related to the gate stack, either occurring in the p-GaN layer, AlGaN barrier, and/or GaN channel layer, mostly influencing V_{TH}. Different studies revealed that pGaN instabilities: *i*) can lead to fast dynamic V_{TH} shifts inducing hysteresis [5], [7]; *ii*) exhibit gate geometry dependence [4]; *iii*) are influenced by temperature [3]; and *iv*) have different features depending on the amount and origin of gate leakage [4], [8], [9]. While previous studies mainly focused on V_{TH} drifts, instabilities occurring under (or near) the gate region of the device can influence channel resistance (R_{CH}) as well – in turn affecting R_{ON}.

Here, we present the results of an experimental investigation of Negative/Positive Gate Stress (NGS/PGS)

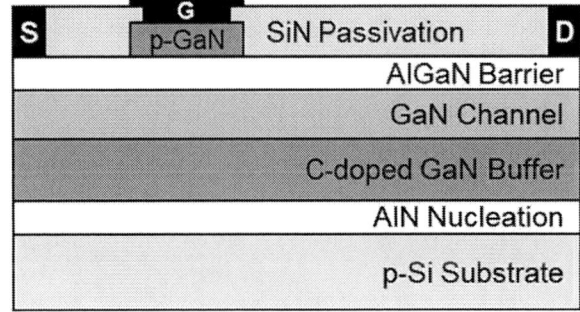

Fig. 1. Sketch of the cross-section of the p-GaN HEMT under test in this work.

drifts in both V_{TH} and R_{ON}. We found a previously undisclosed symmetric and cyclical increase/decrease of both V_{TH} and R_{ON} during NGS and PGS, respectively. Characterization of NGS/PGS at different temperature allow extracting different activation energies (E_A) for V_{TH} and R_{ON} drifts suggesting different underlying physical mechanisms.

II. DEVICES UNDER TEST

Devices tested in this work were packaged p-GaN Normally-OFF HEMTs grown on p-type Si substrate The p-type GaN gate layer was 100 nm thick with nominal Mg doping concentration $\approx 10^{19}$ cm^{-3}. The AlGaN barrier layer was 15 nm thick with 22% Al concentration. The channel was a 0.3-μm thick GaN layer . The GaN buffer was a C-doped 4.7-μm thick layer. Gate length was 1.5 μm. Tested devices presented a threshold voltage (V_{TH}) of approximately 1.4 V at room temperature. The device cross-section is sketched in Fig. 1.

III. STRESS CHARACTERIZATION SETUP

Device characteristics were acquired by means of Keithley 2400/2410 Source Measurement Units (SMUs). Negative/Positive gate stress was applied at a gate-source bias of $|V_{NGS}| = V_{PGS} = 6$ V up to a cumulative time of 8×10^3 s at room temperature. Stress was periodically removed to extract V_{TH} and R_{ON} from the full $I_D\text{-}V_{GS}$ characteristics. A sketch of the NGS/PGS sequences applied is shown in Fig. 2. In all cases, $V_{GS,meas}$ was swept between 0-6 V at $V_{DS,meas} = 50$ mV. V_{TH} was defined with the constant current method, i.e., $V_{TH} = V_{GS,meas}$ @ $I_D = 1$ mA. R_{ON} was evaluated as $V_{DS,meas}/I_D(V_{GS,meas} = 6$ V). PGS characterization was

978-1-6654-8901-0/22 $31.00 © 2022 IEEE

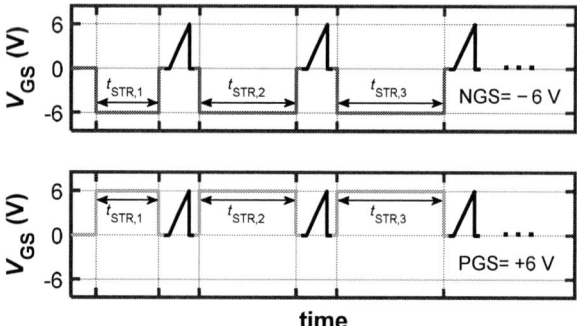

Fig. 2. Sketch of V_{GS} waveforms applied during NGS/PGS characterization. Stress time (t_{STR}) is increased after each I_D-V_{GS} characterization, during which stress is removed. PGS is performed sequentially after NGS is completed.

performed sequentially after NGS was completed. To exclude possible influence of fast dynamic effects leading to V_{TH} hysteresis [7] all I_D-V_{GS} characteristics were acquired with relatively long sweep times of few hundred milliseconds.

IV. EXPERIMENTAL RESULTS

Fig. 3 shows the I_D-V_{GS} curves obtained during NGS/PGS

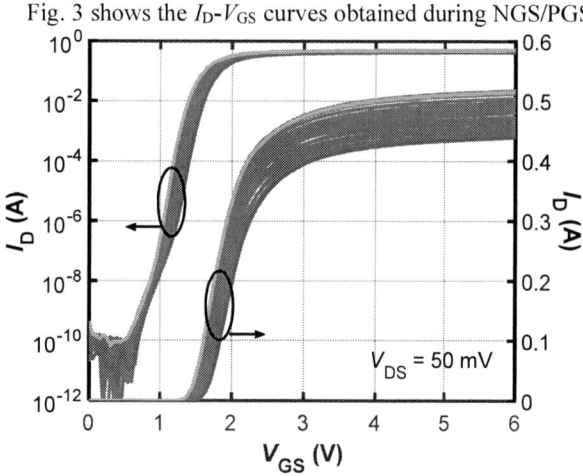

Fig. 3. I_D-V_{GS} curves in semi-log (left) and linear (right) scale extracted during NGS/PGS stress cycles. Blue (yellow) curve is I_D-V_{GS} at the end of 1st NGS (PGS) stress cycle.

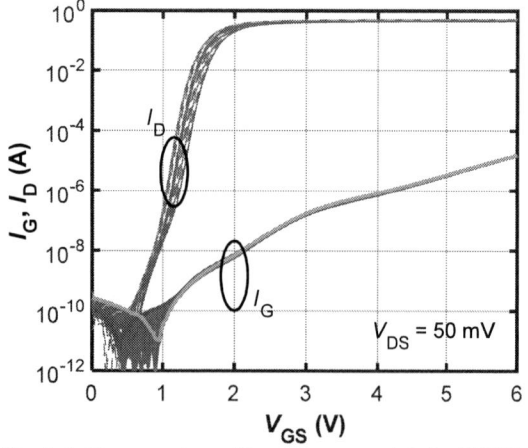

Fig. 4. I_G-V_{GS} curves in semi-log scale extracted during NGS/PGS stress cycles. Blue (yellow) curve is I_G-V_{GS} at the end of 1st NGS (PGS) stress cycle. I_D-V_{GS} curves (dashed) are also shown for comparison.

characterization, in both the semi-logarithmic and linear scale. Blue (yellow) lines are the I_D-V_{GS} curves at the end of the NGS (PGS). From Fig. 3 it can be observed how both V_{TH} and R_{ON} first increase during NGS and then decrease during PGS. This instability does not seem to be associated with a degradation of the gate stack, since gate current (I_G) is negligibly affected by the NGS/PGS characterization, see Fig. 4. Interestingly, we find that the total V_{TH} increase during NGS is the same as the V_{TH} decrease during PGS, as it can be appreciated in Fig. 5. Moreover, this positive/negative bouncing of V_{TH} is maintained over 5 consecutive NGS/PGS stress cycles, indicating no sign of permanent degradation of the gate stack. Similar observations apply to R_{ON} drifts as well, as depicted in Fig. 6.

To probe the possible mechanisms underlying V_{TH} and R_{ON} drifts, NGS and PGS characterization was carried out at different temperatures ($T = 30$–105 °C, $\Delta T = 15$ °C). To speed up characterization, fewer I_D-V_{GS} points are acquired with respect to measurements in Figs. 3-6. This way it is possible to acquire reliable data points even for t_{STR} as short as 1 s,

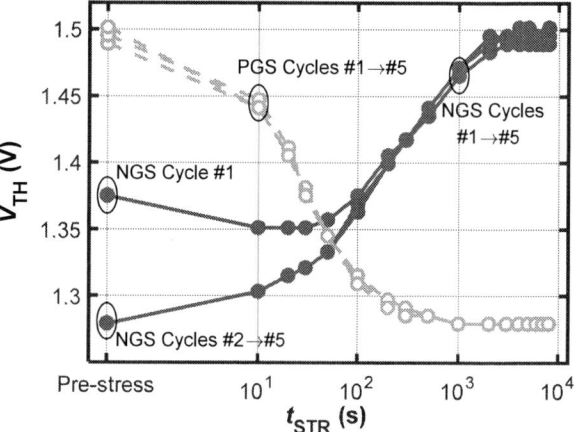

Fig. 5. V_{TH} drift vs cumulative stress time (t_{STR}) during NGS/PGS consecutive cycles. Blue solid (yellow dashed) curves correspond to V_{TH} during NGS (PGS). V_{TH} increases (decreases) during NGS (PGS) with symmetrical behavior over consecutive stress cycles (here shown up to 5).

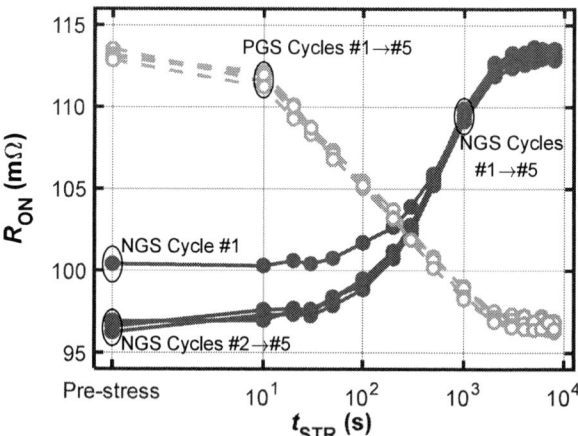

Fig. 6. R_{ON} drift vs cumulative stress time (t_{STR}) during NGS/PGS consecutive cycles. Blue solid (yellow dashed) curves correspond to R_{ON} during NGS (PGS). Similarly to V_{TH}, R_{ON} increases (decreases) during NGS (PGS) with symmetrical behavior over consecutive stress cycles (here shown up to 5).

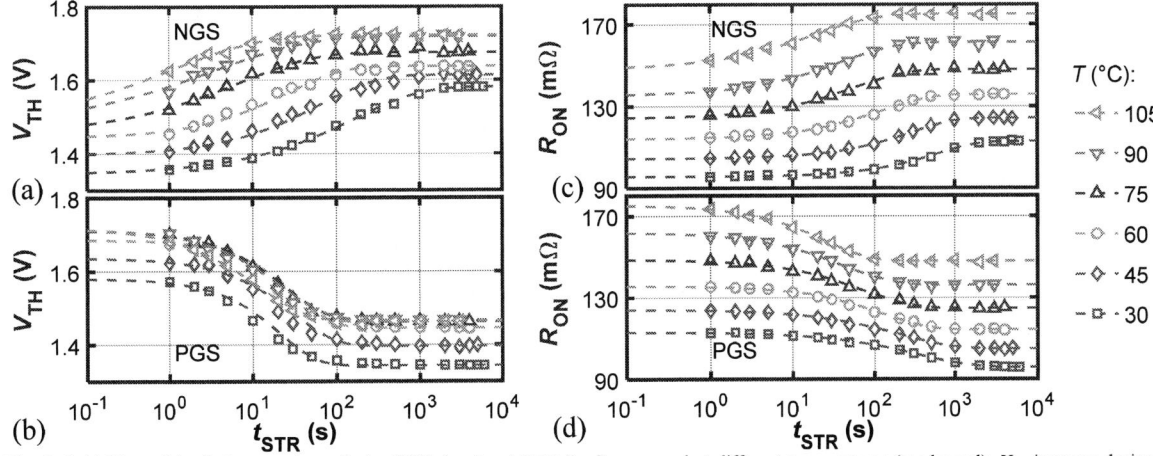

Fig. 7. (a-b) V_{TH} and (c-d) R_{ON} transients during NGS (a, c) and PGS (b, d) measured at different temperatures (see legend). V_{TH} increase during NGS is temperature accelerated, whereas V_{TH} decrease during PGS is not. Conversely, both R_{ON} increase and decrease during NGS and PGS, respectively, are found to be T-accelerated.

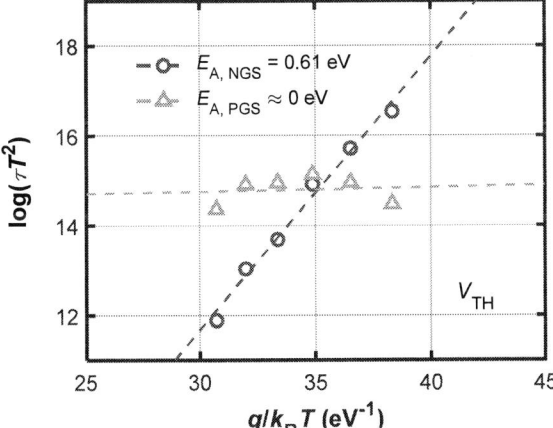

Fig. 8. Arrhenius Plots extracted from V_{TH} NGS/PGS transients, see Fig. 7(a-b). Activation energies, E_A's, obtained from linear fitting of data points are reported in the legend.

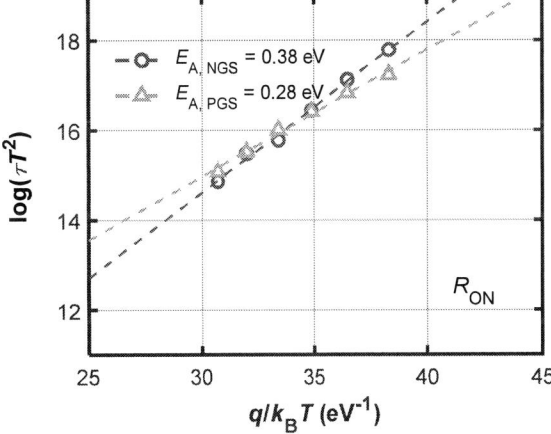

Fig. 9. Arrhenius Plots extracted from R_{ON} NGS/PGS transients, see Fig. 7(c-d). Activation energies, E_A's, obtained from linear fitting of data points are reported in the legend.

which is necessary at high temperatures to properly assess the V_{TH}/R_{ON} dynamics.

Fig. 7 shows the T-dependence of V_{TH}/R_{ON} vs stress time (t_{STR}) during NGS/PGS. It is found that V_{TH}/R_{ON} transients feature different *Arrhenius* signatures. To extract these, we fitted the data points with stretched exponential functions [10] from which then time constants (τ's) were obtained as the time instants at which $\partial V_{TH}(R_{ON})/\partial[\log 10(t_{STR})]$ reaches its maximum/minimum. The extracted τ's are then used to build the Arrhenius plots shown in Fig. 8 and 9. Fig. 8 shows that V_{TH} increase during NGS is temperature accelerated with an activation energy (E_A) of 0.61 eV, whereas V_{TH} decrease during PGS is negligibly accelerated by temperature ($E_A \approx 0$ eV). Fig. 9 shows that R_{ON} has clearly different T-dependence compared to V_{TH}. In this case in fact, similar E_A's were found for both NGS and PGS, i.e., 0.38 eV and 0.28 eV, respectively. The features of V_{TH}/R_{ON} transients suggest that the underlying physical mechanisms are inherently different.

V. CONCLUSIONS

In this work, we investigated the negative/positive gate bias stress effects on key parameters (V_{TH} and R_{ON}) of pGaN

HEMTs. We found a previously undisclosed symmetrical increase/decrease of both parameters during NGS/PGS. The behavior remains unaltered after 5 NGS/PGS cycles, suggesting that no permanent degradation of the gate stack occurs. Temperature dependence of V_{TH} and R_{ON} transients revealed different activation energies, suggesting inherently different underlying mechanisms.

REFERENCES

[1] J. A. del Alamo and E. S. Lee, "Stability and Reliability of Lateral GaN Power Field-Effect Transistors," *IEEE Trans Electron Devices*, vol. 66, no. 11, pp. 4578–4590, Nov. 2019, doi: 10.1109/TED.2019.2931718.

[2] K. J. Chen *et al.*, "GaN-on-Si power technology: Devices and applications," *IEEE Trans Electron Devices*, vol. 64, no. 3, pp. 779–795, Mar. 2017, doi: 10.1109/TED.2017.2657579.

[3] A. Stockman, E. Canato, M. Meneghini, G. Meneghesso, P. Moens, and B. Bakeroot, "Threshold Voltage Instability Mechanisms in p-GaN Gate AlGaN/GaN HEMTs," in *International Symposium on Power Semiconductor Devices and ICs (ISPSD)*, 2019, vol. 2019-May, pp. 287–290. doi: 10.1109/ISPSD.2019.8757667.

[4] E. S. Lee, J. Joh, D. S. Lee, and J. A. del Alamo, "Impact of Gate Offset on PBTI of p-GaN Gate HEMTs," in *IEEE International Reliability Physics Symposium (IRPS)*, Mar. 2022, pp. P21-1-P21-6. doi: 10.1109/IRPS48227.2022.9764442.

[5] X. Li *et al.*, "Observation of Dynamic VTH of p-GaN Gate HEMTs by Fast Sweeping Characterization," *IEEE Electron Device Letters*, vol. 41, no. 4, pp. 577–580, 2020, doi: 10.1109/LED.2020.2972971.

[6] L. Sayadi, G. Iannaccone, S. Sicre, O. Haberlen, and G. Curatola, "Threshold Voltage Instability in p-GaN Gate AlGaN/GaN HFETs," *IEEE Trans Electron Devices*, vol. 65, no. 6, pp. 2454–2460, Jun. 2018, doi: 10.1109/TED.2018.2828702.

[7] A. N. Tallarico *et al.*, "TCAD Modeling of the Dynamic VTH Hysteresis under Fast Sweeping Characterization in p-GaN Gate HEMTs," *IEEE Trans Electron Devices*, vol. 69, no. 2, pp. 507–513, 2022, doi: 10.1109/TED.2021.3134928.

[8] A. Stockman, E. Canato, M. Meneghini, G. Meneghesso, P. Moens, and B. Bakeroot, "Schottky Gate Induced Threshold Voltage Instabilities in p-GaN Gate AlGaN/GaN HEMTs," *IEEE Transactions on Device and Materials Reliability*, vol. 21, no. 2, pp. 169–175, Jun. 2021, doi: 10.1109/TDMR.2021.3080585.

[9] A. Stockman *et al.*, "On the origin of the leakage current in p-gate AlGaN/GaN HEMTs," in *IEEE International Reliability Physics Symposium Proceedings*, May 2018, vol. 2018-March, pp. 4B.51-4B.54. doi: 10.1109/IRPS.2018.8353582.

[10] M. Meneghini *et al.*, "GaN-based power devices: Physics, reliability and perspectives," *J Appl Phys*, vol. 130, no. 16, p. 227, 2021, doi: 10.1063/5.0061354.

[11] A. Stockman *et al.*, "On the origin of the leakage current in p-gate AlGaN/GaN HEMTs," in *IEEE International Reliability Physics Symposium Proceedings*, May 2018, vol. 2018-March, pp. 4B.51-4B.54. doi: 10.1109/IRPS.2018.8353582.

Peak channel temperature determination for an AlGaN/GaN HEMT with Raman Thermography and MTTF extraction for long term reliability

Cristina Miccoli
ADG R&D Power and Descretes
STMicroelectronics
Catania, Italy
cristina.miccoli@st.com

Leonardo Gervasi
ADG R&D Power and Descretes
STMicroelectronics
Catania, Italy
leonardo.gervasi@st.com

Viviana Cerantonio
ADG R&D Power and Descretes
STMicroelectronics
Catania, Italy
viviana.cerantonio@st.com

James Pomeroy
Centre for Device Thermography and
Reliability
University of Bristol
Bristol, UK
James.Pomeroy@bristol.ac.uk

Martin Kuball
Centre for Device Thermography and
Reliability
University of Bristol
Bristol, UK
Martin.Kuball@bristol.ac.uk

Ferdinando Iucolano
ADG R&D Power and Descretes
STMicroelectronics
Catania, Italy
ferdinando.iucolano@st.com

Abstract—**Reliability tests are a decisive step for GaN technology qualification. To evaluate the Mean Time to Failure (MTTF), it is necessary to know well the peak channel temperature (Tpeak) reached at the hottest point of the device. For an AlGaN/GaN High Electron Mobility Transistor (HEMT), this region lies at AlGaN/GaN interface where the two-dimensional electron gas (2DEG) is located near the gate contact. InfraRed (IR) Thermography was performed on our GaN on Si HEMT devices to measure Tpeak. Considering the correction in underestimation of Tpeak temperature, the MTTF was extracted. We also performed Raman measurements and Tpeak was evaluated with the aid of 3-Dimensional thermal simulations. The Tpeak difference between IR and Raman technique is more than 100°C and far from the evaluation of about 40-50°C often found in literature. The reliability measurements show a high robustness and from the Raman approach MTTF is approximately 10^9 hours instead of the 10^6 hours estimated with the IR method at Tpeak =200°C.**

Keywords—AlGaN/GaN, GaN, HEMT, MTTF, Raman Thermography, InfraRed Thermography, failure, reliability.

I. INTRODUCTION

GaN FETs are promising to be the workhorse of the future power-electronic and power-management industry, owing to the excellent material properties of the GaN [1]–[4], such as high values of low field mobility, saturation velocity, breakdown electric field, and so on, to name a few. As a result, the Baliga figure of merit [5], which is a benchmark metric of performance for power devices, is highly superior for GaN High Electron Mobility Transistor (HEMTs) [1]. GaN based devices have also a huge field of applications for power amplifiers, millimeter-wave integrated circuits (MMIC) [6], next generation information communication systems [7] and power switching. GaN technologies are attractive for base station equipment to power device industry and automotive. AlGaN/GaN devices, grown on Silicon substrates are the main candidates for RF and power applications in [8]-[9]. Their advantages compared with silicon (Si) or gallium arsenide (GaAs) devices are their ability to work at higher voltages and higher currents, which results in high power densities. One of the crucial effects that limit their performances, at high power densities, is the self-heating effect (SHE) associated with heat

generation due to the Joule heating. Thus, the elevated temperatures inside the GaN devices are a major concern and represent a serious reliability issue. As devices mature, thermal management will play an important role in further optimization of the GaN technology. In order to experimentally measure temperatures inside GaN-based devices, different methods have been established, as reported in [10]. Experimental methods are usually divided into three groups: physical contact, electrical, and optical methods. Long term reliability is a crucial point for the qualification of a new technology going towards the commercialization process. This work reports the comparison between the InfraRed (IR) and Raman thermography measurements for the peak channel temperature evaluation (Section III) and the Accelerated Life Test (ALT) conditions with the two different Mean-Time-To-Failure (MTTF) extrapolated values are reported in Section IV.

II. DEVICE UNDER INVESTIGATION

A. Device description and design

The device under examination is an AlGaN/GaN HEMT grown on Silicon substrate by Metal Organic Chemical Vapor Deposition with a gate length (Lg) of 0.4 µm, and gate-drain length (Lgd) under 5 µm and gate fingers spacing less than 60 µm. The gate/source offset length was 1 um and the unintentionally doped (UID) AlGaN layer thickness less than

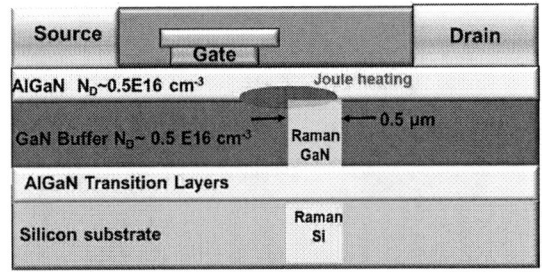

Figure 1 Two-dimensional scheme of AlGaN/GaN HEMT device under examination. Joule heating is localised in the high field region at the drain edge of the gate. Raman thermography measures the depth averaged GaN temperature in an 0.5 µm spot centred 0.5 µm from the gate edge.

20 nm, grown on UID GaN buffer layer (see Fig. 1). The vertical structure is formed by: a barrier of AlGaN thin layer, a channel layer of high-quality GaN with high mobility and high saturation velocity. At the interface between channel and barrier the two-dimensional electron gas (2DEG) is formed. Finally, there is a silicon substrate that provides mechanical support, heat spreading, and electromagnetic confinement. The substrate is silicon with high resistivity for RF applications.

B. Peak channel temperature extapolation method from Raman termography

Exact knowledge of channel temperature and its distribution on the die is a key factor for predicting operating temperatures and circuit performances. Channel temperature is fundamental for the correct evaluation of activation energies of the different failure mechanisms and for the extrapolation of high-temperature accelerated test results at different operating temperatures, for what concerns the mean time to failure (MTTF). Unfortunately, it is not easy to mapping surface and sub-surface device temperature with the adequate lateral and depth resolution and experimental evaluation has to be coupled with simulations in order to obtain useful thermal data. For example, Fig.1 illustrates that while Raman thermography measures close to the peak temperature location, with a spatial resolution of 0.5 µm, it nevertheless may underestimate the channel temperature due to the sub-micron Joule heating hot spot. Moreover, the metal source field plate over the gate prevents to measure the temperature closer to the gate foot where the peak temperature is located.

An experimentally calibrated 3D thermal simulation (Fig. 2) was used to extrapolate from the measured temperature, averaged in a well-defined region, the peak channel temperature which cannot be accessed directly by measurement. ANSYS 2022R1 was used for the 3-D finite element method (FEM) thermal simulation of the device under test. The model includes the die, die attach and package flange. Table I contains the known thermal conductivities of each layer in the device/package stage. The thermal conductivity of interfacial layers are assumed fixed. The thermal conductivities of the die attach and AlGaN transition layers are fitting parameters; model fitting is described in Section II D. Joule heating was applied in the model as a volumetric heat load at the AlGaN/GaN interface, representing the high electric field region in the 2DEG. TCAD simulation was used to determine the Joule heating spatial

TABLE I

Layer	Material	κ [W/m·K]
Contacts	Aluminum	235
Passivation	SiN	3[a]
Barrier	AlGaN	10[b]
Buffer	GaN	$160 \times (300/T[K])^{1.4}$ [c]
Substrate	Silicon	$148 \times (300/T[K])^{1.28}$ [d]
Flange	CuW	170

[a] [H. Ftouni, C. Blanc, D. Tainoff, A. D. Fefferman, M. Defoort, K. J. Lulla, J. Richard, E. Collin, O. Bourgeois "Thermal conductivity of silicon nitride membranes is not sensitive to stress", Phys. Rev. B 92, 125439, Sep. 2015] ,

[b] [A Filatova-Zalewska et al., "Anisotropic thermal conductivity of AlGaN/GaN superlattices", Nanotechnology 32, 2021],

[c] [A. Sarua, H. Ji, K. P. Hilton, D. J. Wallis, M. J. Uren, T. Martin, and M. Kuball, "Thermal boundary resistance between GaN and substrate in AlGaN/GaN electronic devices"m IEEE Trans. Elec. Dev. vol. 54, no. 12, dec. 2007],

[d] [C. J. Glassbrenner, G. A. Slack, "Thermal Conductivity of Silicon and Germanium from 3°K to the Melting Point", Phys. Rev. 134, A1058, may 1964].

distribution applied in the thermal model. An isothermal boundary condition was applied to the back side of the flange, corresponding to the experimental T_{ref}. Quadratic elements were used, and the mesh was refined until mesh independent solution was reached, i.e., the peak temperature changed less than 1°C when the element size was reduced further. The resulting mesh consisted of 0.14 million elements; Note that a symmetric ¼ of the total device structure was simulated to minimize computation time.

III. THERMAL MESUREMENT RESULTS: IR AND RAMAN

Two of the most important techniques for temperature evaluation are infrared (IR) thermography and Raman spectroscopy, we applied both to our GaN HEMT device samples with the same elec .

A. IR thermography and related results

IR thermography is based on the detection of thermal radiation emitted from the surface of the device in the 2-5 µm wavelength range for the InSb detector used in the QFI Infrascope, depending on device temperature. This technique requires calibration of the emissivity of the various materials present on the devices surface (pixel-by-pixel); optical resolution is limited due to the long wavelength of the emitted

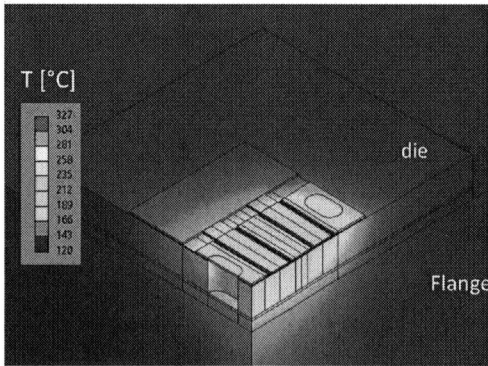

Figure 2 HEMT3-D FEM thermal simulation, showing a symmetric ¼ of the device.

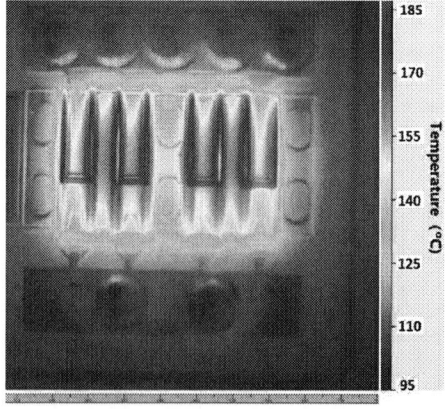

Figure 3 IR Thermal Microscopy at Tcase=95°C and at 4W/mm on packaged GaN HEMT.

978-1-6654-8901-0/22 $31.00 © 2022 IEEE

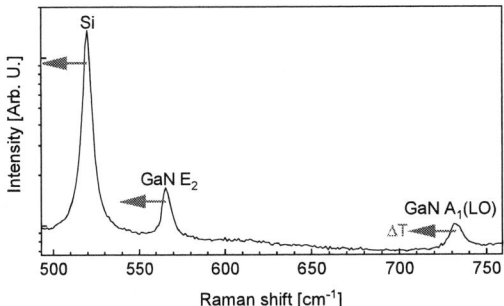

Figure 4. Typical Raman spectrum with the GaN and silicon phonon peaks indicated, with arrows illustrating the temperature induced peak shift, which is the principle of Raman thermography.

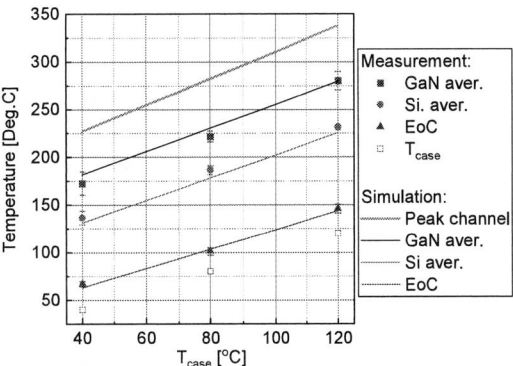

Figure 5. Raman thermography measured temperatures (GaN, Si, EoC) temperatures versus T_{case} measured by thermocouple. Simulated temperatures at the same locations (See Fig.1 for spatially averaged areas) are shown as solid lines, as well as the simulated peak channel temperature.

radiation. As most semiconductors are transparent in this wavelength range, temperature value is averaged over the entire thickness of the semiconductor, if the device is not coated. In general, this can produce a large underestimation of the device temperature [11]. On the other hand, use of the technique is relatively simple, and it is an imaging technique which allows direct measurements on large area devices. It is a very efficient method for mapping the temperature over large surface areas, making this technique especially suitable for comparative screening of identical devices where exact knowledge of peak temperatures may not be essential. In this work IR thermography was performed on 2mm packaged device, biased at 4W/mm and at T_{case} of 95°C (see Fig. 3). Considering the underestimation of temperature with this technique, that is evaluated in literature [12] e.g., around 40-50°C for GaN on SiC HEMT device, a shift of the peak temperatures of about 50°C was applied for the subsequential MTTF evaluation done in Section IV.

B. Raman thermography: condition and results

Raman thermography is a high spatial resolution temperature measurement method, but is a slower serial (point by point) measurement in contrast to the lower resolution quick imaging provided by IR thermography. Raman thermography measurements were made using a Renishaw inVia micro-Raman system using a 488 nm (λ) diode laser. Laser light is focused through a 50× 0.5 numerical aperture (NA) objective lens onto the region of interest; the same lens is used to collect Raman scattered light in the backscattering configuration. The lateral spatial resolution is given by $r_{lat}=0.51\cdot\lambda/NA=0.5$ µm, i.e. approximately a ~10× improvement over the typical spatial resolution of IR thermography. The depth of focus is greater than the epitaxial layer thickness and the epitaxial layers are transparent for the laser wavelength (negligible light absorption). Hence a depth average through the GaN buffer layer is measured, whereas light is absorbed withing the first 500 nm of the silicon substrate. Incident laser power was <4 mW to avoid heating the silicon substrate. The test fixture was fixed to a thermoelectric chuck with PID temperature control. The set ambient temperature, T_{ref}, is monitored and maintained to within 1°C using a thermocouple in contact with the back side of the flange.

The fixture and chuck were mounted on a computer controlled XYZ stage with 50 nm step precision, allowing the probe laser to be precisely positioned, for channel temperature measurements, at a distance of 0.5 µm from the drain edge of the gate, as illustrated in Fig. 1. Additionally, a point was measured at the edge of the chip (EoC) for comparison to the thermal simulation. Raman scattering is material specific, meaning that the GaN buffer and silicon substrate can be measured in the same spectrum; Fig. 4 shows a typical measured Raman spectrum for the devices under-test, indicating the characteristic GaN and silicon peaks.

The temperature of the GaN and silicon layers were determined by measuring the phonon peak shifts for the device powered with respect to pinched off (no self-heating at T_{ref}), decoupling the temperature induced shift from the small electric field induced phonon frequency shift [13]. The GaN E_2 and A_1(LO) are both measured, enabling the second order thermomechanical stress effect (temperature gradient around the hot spot) to be compensated following the method described in [14]; note that the temperature measured using the A_1(LO) alone, without stress compensation, is within 5°C of the actual temperature. More information about the Raman thermography method and the phonon shift temperature calibration process are given in [14]. Each measurement point is repeated 3 times. Furthermore, three devices were measured, and the results averaged. Plotted temperatures are the mean values and error bars represent the mean standard error; the measurement uncertainty is approximately ±5°C.

Measurement results, shown in Fig. 5, were used to adjust the unknown die attach and transition layer thermal conductivities in the following sequence: i) The die attach thermal conductivity was first adjusted to fit the simulated EoC temperature, since it is insensitive to the transition layer; ii) the transition layer thermal conductivity was adjusted to match the simulated GaN temperature, averaged over the area of the measurement (see Fig. 1). The simulated GaN, silicon and EoC temperatures after this fitting process are overlaid on the measured temperatures in Fig. 5, showing a good agreement. This gives us high confidence in the simulated peak temperatures, also shown in Fig. 5, which is used for the MTTF extraction.

978-1-6654-8901-0/22 $31.00 © 2022 IEEE

IV. ALT MEASUREMENTS AND MTTF EXTRACTION

One of the typical test required for GaN reliability is the Accelerated Life Test (ALT). It is a stress test performed on device in semi-on condition and at a predicted high junction temperature that is the maximum temperature present in the channel of the device. For this test we considered three different case temperatures: 160°C, 175 °C, 190 °C.

The test focuses on DC bias conditions at expected maximum operating levels of VDS = 50V and 4 W/mm power dissipation. In order to maintain the device temperature constant during the accelerated life tests, constant power dissipation is maintained during the stress. A power dissipation of 4 W/mm is considered, as this reflects the maximum average power dissipation expected for many RF applications with GaN device. Constant power dissipation is maintained by setting a common constant drain bias with VDS=50V for all stress conditions, and automatically adjusting the gate voltage VGS to maintain a constant drain current IDS = 80 mA/mm during these tests. Gate and drain voltages and currents are monitored throughout the stress periods. Stress is periodically interrupted for a more thorough set of swept voltage DC measurements. These measurements are made at ambient temperature and at fixed times. After DC measurements, the device temperature is returned to its stress condition and the reliability test starts again. In this work the test was performed on a set of samples at three different case temperatures for the extraction of activation energy linked to the physical mechanism that leads to the device failure. The tested device has 2 mm periphery and was assembled into air cavity package, as for IR and Raman thermography.

From a theoretical point of view for ALT measurements, a temperature range and bias condition that turn on the failure physical mechanism in the critical region of the device should be used; in our technology this region of degradation lies at gate metal and semiconductor interface. For this reason, the stress test and failure criteria are considered as following. Two DC parameters for the device failure criteria are recorded: saturated drain current (IDon) decreasing and off state leakage current (IDoff) increasing, otherwise a catastrophic failure occurs. In our study we choose a catastrophic fail that in our device occurs always between 15% and 20% Idon drop. The tests were performed in AARTS-36 Channels DC Reliability Test System, realized for reliability test on many packaged devices in parallel at different temperatures. In this equipment

TABLE II

Tcase [°C]	Number of samples	Tpeak Raman [°C]	Tpeak IR [°C]	MTTF [h]
190	10	423	336	116
175	8	402	321	327
160	4	382	306	933

each device can automatically run the stress and the DC parametric test for device monitoring. Due to the high periphery of the tested device a proper fixture board was used to avoid oscillations and keep stable the devices during the stress. During the life tests, stress is periodically interrupted, and the parts are returned to room temperature for parameter measurements. Change in IDon and IDoff are analyzed as a function of time to determine if devices have failed. After the monitoring the stress starts again for the next step. The step stress considered has an incremental time (1-5-10-20-50-100 hours and after every 100 hours) and during the stress only the IDon and IGon currents are monitored. The time when the devices stop working is captured and all the failure times are considered; these are used to analyze the failure distribution by the Weibull method (widely used in reliability life data analysis). Appling the best fit and considering all the observed failures at the three temperatures (Fig. 6), the median failure time at 50% for each temperature is then taken (Table 2). The temperature considered for IR analysis is corrected in order to compensate the underestimation of this technique [12]. The above outcome data is converted from median to mean failure time by using an Arrhenius model (Fig. 7) to predict the Mean-Time-To-Failure (MTTF) as a function of a specific device channel temperature:

$$MTTF = \exp(Ea/K_B T) \qquad (1)$$

where Ea is the energy activation, T is the absolute temperature and K_B is the Boltzmann's constant. From this analysis we extracted the mean failure time at Tpeak = 200°C (typical temperature value to be considered for RF applications [15]), and the activation energy for this failure mode. We conclude that at Tpeak=200°C the calculated MTTF final value depends on thermal analysis considered. The data are here reported based on Raman and IR technique, respectively, and show a difference of three orders of magnitude for the MTTF in terms of hours:

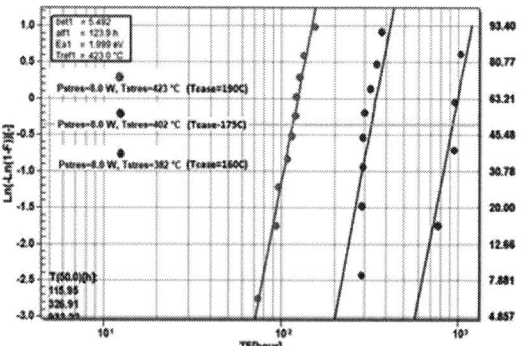

Figure 6. Weidbull fit based on observed failures at three temperature 160°C (brown), 175 °C (blu), 190 °C (red) . The light blu line represents the median failure time at 50% for each temperature.

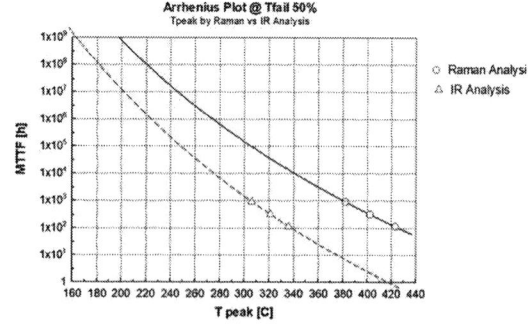

Figure 7. MTTF curves extration from IR (rel dotted line) and Raman (blu solid line) based on the thrree peak channel temperature values (triangle and cicle are referred to IR and Raman measurements, respctively) .

- 10^9 hours from Raman temperatures and with Ea = 2.0 eV
- 10^6 hours from IR temperatures and with Ea = 2.11 eV.

V. CONCLUSIONS

This paper reported the ALT Test measurements at 4W/mm at different temperatures (Tcase) carried out on AlGaN/GaN HEMT for RF application. The Weibuill distribution resulting from these data is shown in Fig.6. In order to correlate these failure tests with MTTF, it is necessary to know the peak device temperature that corresponds to the Tcase values considered. Fig. 3 showed IR thermography data; considering the underestimation of this technique, that is evaluated in the literature often to be around 40-50°C shift of the peak temperatures for GaN on SiC HEMT device, a 50°C shift was applied to subsequently evaluated the relative MTTF (see Fig. 7). Carrying out Raman measurements we found that the difference in the temperature peak between IR and Raman is 120-130°C at Tcase=190°C and the Tpeak reaches about 420°C. 50% of the population survives for 116 hours at this Tcase. Fig. 7 shows the MTTF corresponding to Raman Tpeak extrapolation. Based on Raman analysis we obtain an MTTF at Tpeak=200°C approximately 10^9 hours instead of 10^6 hours estimated with the IR Microscopy. These results underline the huge gap between the two different approach useful for long term reliability study applied to an AlGaN/GaN HEMT on Si substrate.

ACKNOWLEDGMENT

The authors would like to thank Giuseppe Privitera and Antonino Parisi from STMicroelectronics for useful discussions and technical support.

REFERENCES

[1] U. K. Mishra, P. Parikh, and Y.-F. Wu, "AlGaN/GaN HEMTs-an overview of device operation and applications," Proc. IEEE, vol. 90, no. 6, pp. 1022–1031, Jun. 2002.

[2] J. Millan, P. Godignon, X. Perpina, A. Perez-Tomas, and J. Rebollo, "A survey of wide bandgap power semiconductor devices," IEEE Trans. Power Electron., vol. 29, no. 5, pp. 2155–2163, May 2014.

[3] R. Mitova, R. Ghosh, U. Mhaskar, D. Klikic, M.-X. Wang, and A. Dentella, "Investigations of 600-V GaN HEMT and GaN diode for power converter applications," IEEE Trans. Power Electron., vol. 29, no. 5, pp. 2441–2452, May 2014.

[4] X. Huang, Z. Liu, Q. Li, and F. C. Lee, "Evaluation and application of 600 V GaN HEMT in cascode structure," IEEE Trans. Power Electron., vol. 29, no. 5, pp. 2453–2461, Apr. 2014.

[5] B. J. Baliga, "Power semiconductor device figure of merit for highfrequency applications," IEEE Electron Device Lett., vol. 10, no. 10, pp. 455–457, Oct. 1989, doi: 10.1109/55.43098.

[6] C.Potier, S. Piotrowicz, O. Patard, P. Gamarra, P. Altuntas, E. Chartier, C. Dua, , J.-C. Jacquet, C. Lacam, N. Michel, M. Oualli, S. Delage, C. Chang, J. Gruenenpuett, "First results on Ka band MMIC power amplifiers based on InAlGaN/GaN HEMT technology," in 2018 International Workshop on Integrated Nonlinear Microwave and Millimetre-wave Circuits (INMMIC), 2018, pp. 1–3.

[7] U. K. Mishra, P. Parikh and Yi-Feng Wu, "AlGaN/GaN HEMTs-an overview of device operation and applications," in Proceedings of the IEEE, vol. 90, no. 6, pp. 1022-1031, June 2002, doi: 10.1109/JPROC.2002.1021567.

[8] N. Ikeda et al., "GaN power transistors on Si substrates for switching applications," Proc. IEEE, vol. 98, no. 7, pp. 1151–1161, Jul. 2010.

[9] Y.-F. Wu et al., "30-W/mm GaN HEMTS by field plate optimization", IEEE Trans. Electron Devices, vol. 25, no. 3, pp. 117–119, Mar. 2004.

[10] D. L. Blackburn, "Temperature measurements of semiconductor devices—A review," in Proc. 20th Annu. IEEE Semiconductor Thermal Meas. Manage. Symp., Mar. 2004, pp. 70–80.

[11] I. Rossetto, M. Meneghini, T. Tomasi, D. Yufeng, G. Meneghesso, and E. Zanoni, "Indirect techniques for channel temperature estimation of HEMT microwave transistors: Comparison and limits," Microelectron. Reliab., vol. 52, no. 9–10, pp. 2093–2097, 2012.

[12] A. Sarua, Hangfeng Ji, M. Kuball, M. J. Uren, T. Martin, K. P. Hilton, and R. S. Balmer, "Combined Infrared and Raman temperature measurements on device structures," CS MANTECH Conference, April 24-27, 2006, Vancouver, British Columbia, Canada.

[13] K. R. Bagnall, C. E. Dreyer, D. Vanderbilt, E. N. Wang, "Electric field dependence of optical phonon frequencies in wurtzite GaN observed in GaN high electron mobility transistors", JAP vol. 120, 155104, 2016.

[14] M. Kuball, J. W. Pomeroy, "A review of Raman thermography for electronic and opto-electronic device measurement with submicron spatial and nanosecond temporal resolution" IEEE Trans. Dev. Mat. Reliab., vol. 16, no. 4, pp. 667-684, 2016.

[15] R. Menozzi, "Reliability of GaN-based HEMT devices", Conference on Optoelectronic and Microelectronic Materials and Devices, pp. 44-50, 2008.

High Temperature Robustness of Enhancement-Mode p-GaN-Gated AlGaN/GaN HEMT Technology

Mengyang Yuan[1*], Qingyun Xie[1*], John Niroula[1], Mohamed Fadil Isamotu[1,2],
Nitul S. Rajput[3], Nadim Chowdhury[4], and Tomás Palacios[1*]

[1] Microsystems Technology Laboratories, Massachusetts Institute of Technology, Cambridge, MA 02139, U.S.A.
[2] Department of Electrical and Computer Engineering, Morgan State University, Baltimore, MD 21251, U.S.A.
[3] Advanced Materials Research Center, Technology Innovation Institute, Abu Dhabi PO Box 9639, U.A.E.
[4] Department of Electrical and Electronic Engineering, Bangladesh University of Engineering and Technology,
Dhaka-1205, Bangladesh
[*]E-mail: {myyuan, qyxie, tpalacios}@mit.edu

Abstract—This paper reports on the high temperature (HT) robustness of enhancement-mode (E-mode) p-GaN-gated Al-GaN/GaN high electron mobility transistors (HEMTs), with an emphasis on the key transistor-level parameters for digital and analog mixed-signal applications. In-situ measurements from room temperature (RT) to 500 °C show that trends in V_{th}, R_{ON}, $I_{D,max}$ and $I_{G,max}$ are largely as expected based on first-order changes in the semiconductor properties. The fabricated transistors exhibited stable performance over 20 days at 500 °C. To the best of the authors' knowledge, this work is the first systematic study on the HT performance of E-mode p-GaN-gated AlGaN/GaN HEMTs, and sheds light on their use in mixed-signal and low-voltage power circuits.

Index Terms—GaN, p-GaN, transistor, high temperature, long-term survival

I. INTRODUCTION

Electronics that would operate reliably and robustly at high temperatures (HT, > 500 °C) are critical to push the frontiers of aeronautical engineering (e.g. hypersonic aircraft), resource extraction (e.g. deep well oil drilling), space exploration (e.g. Venus Rovers), and more. Electronic systems for these applications are all exposed to environments whose conditions well exceed the limits of silicon electronics, but offer a unique opportunity for wide band gap semiconductors, notably SiC and GaN. [1], [2] While early research on SiC high temperature power and high digital electronics has been conducted [3]–[6], GaN and other III-N materials offer potential for higher performance, in the domains of power [7]–[9], RF [10]–[12], MEMS [13], [14], digital circuits [15], [16] and sensors [17] over a wide temperature range. This wide range of potential applications allows GaN HT electronics to offer a promising path for the realization of a multi-functional, monolithically integrated HT electronics solution.

For HT digital circuits, the realization of enhancement-mode (E-mode) transistors is critical, otherwise it would be necessary to use a negative bias (V_{SS}) [18]. In the case of E-mode GaN transistors, the most commonly used technologies

include, recessed MIS and p-GaN gate stacks [19]–[24]. The performance of current recessed MIS-gate transistors, though competitive at RT [25], would unfortunately be limited at HT by gate dielectric degradation and the activation of the dielectric/semiconductor interface traps [26]. On the other hand, the p-GaN-gated AlGaN/GaN HEMT technology is an attractive option thanks to the easy realization of E-mode and lack of gate dielectric. Furthermore, for eventual realization of HT GaN complementary technology, the p-GaN-gated HEMT structure allows for monolithic integration with the E-mode p-FET [15], [27] and other devices/components like D-mode n-FET (Fig. 1) and 2DEG resistors [28], [29].

Early experiments have demonstrated p-GaN-gated HEMT operation up to 420 °C, as well as their integration in E/D-mode HT (up to 500 °C) digital circuits such as ring oscillators and memory cells [30], [31]. While these initial demonstrations are encouraging, it is equally important to understand the long-term robustness (beyond quick measurements in a laboratory setting) at HT (> 500 °C) of the transistor, in order to further evaluate their potential for HT electronics.

II. DEVICE TECHNOLOGY AND MEASUREMENT SETUP

E-mode HEMTs (devices under test, DUTs) were fabricated using a p-GaN/AlGaN/GaN-on-Si wafer as the starting material and the process flow reported elsewhere [31]. In particular, W was used as for the gate metallization due to its its properties as a refractory metal and its Schottky nature with p-GaN [32]. The bare die was packaged and placed in a furnace in N_2 ambient. The temperature was increased up to 500 °C (Fig. 2). In the measurement setup, special attention was paid to ensure all of the components are HT-rated.

III. RESULTS AND DISCUSSION

As shown in Fig. 3(a), at HT (in-situ measurement), the maximum drain current (I_{Dmax}) is reduced and ON-resistance (R_{ON}) is increased. This is primarily the result of the degradation of mobility due to the increase of phonon scattering. As shown in Fig. 3(b), the drain current in OFF-state is dominated by the gate leakage current. In the gate diode reverse bias (negative V_{GS}) and weak forward bias regime (small positive

---

This work was partially sponsored by the National Aeronautics and Space Administration (NASA) under grant no. 80NSSC17K0768 (Dr. Gary Hunter), Lockheed Martin Corporation under grant no. 025570-00036 (Dr. John Callahan), and the Air Force Office of Scientific Research (AFOSR) under grant no. FA9550-22-1-0367 (Dr. Kenneth Goretta).

978-1-6654-8901-0/22 $31.00 © 2022 IEEE

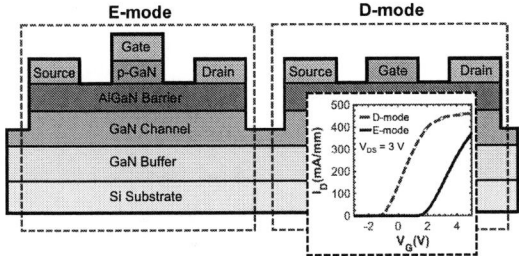

Fig. 1. Illustration of the E-mode p-GaN-gated AlGaN/GaN HEMT in this work. Its monolithic integration with a D-mode AlGaN/GaN HEMT is also illustrated. Typical transfer characteristics of both transistors are shown in the inset, with $V_{th\{E,D\}} = \{1.8, -1\}$ V.

Fig. 2. Experimental setup for HT (up to 500 °C) measurement, with (a)–(d) illustrating from the DUT to the overview, in the bottom-up hierarchy order. (a) DUT. The transistor electrodes are connected to Ti/Au bonding pads. (b) Printed circuit board (PCB) made of 98% alumina. The bare die is attached using a ceramic adhesive. (c) Custom-made packaging for HT measurement, showing the in-house machined connectors and wires which are wrapped in 3M™ Nextel™ sleeving. (d) Overview of the setup. The setup shown in (c) is placed in the furnace in N_2 ambient. The switching matrix is used if more than one DUT needs to be measured.

V_{GS}) regimes, the gate leakage current increases due to two-dimensional variable range hopping (2DVRH). In the strong forward bias (highly positive V_{GS}) regime, the gate current first decreases up to 300 °C due to reduced mobility of carriers, then increases at higher temperature due to leakage through the passivation or packaging (i.e., not intrinsic DUT). A detailed study of the high temperature behavior will be reported elsewhere.

The DUTs were subjected to HT (500 °C) robustness studies. Firstly, an in-situ measurement of the DUT was conducted over 24 hours. As shown in Fig. 4, the V_{th} and I_{Dmax} remained relatively stable ($< 5\%$ variation) for over 20 hours, before some slight degradation after 24 hours.

Having established the relative stability of the DUT using

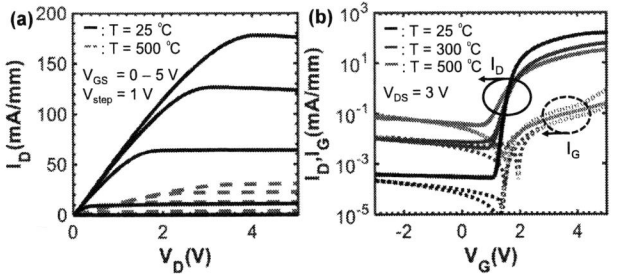

Fig. 3. (a) Output and (b) Transfer characteristics (double sweep) of the packaged DUT ($L_{SD} = 6$ μm, $L_G = 2$ μm) from RT up to 500 °C.

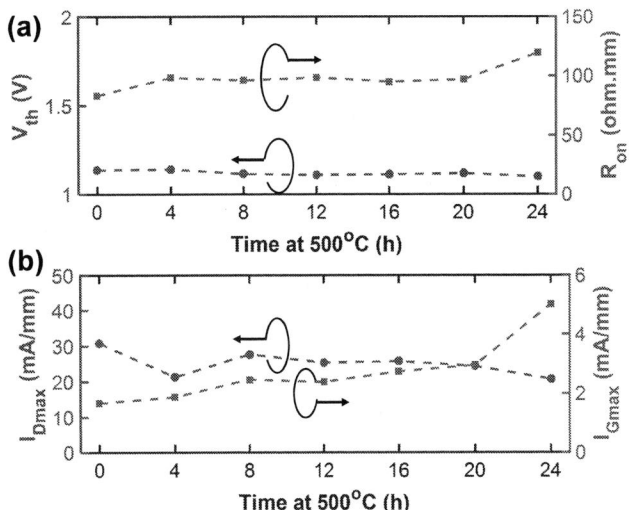

Fig. 4. In-situ measurement of packaged DUT at 500 °C in N_2 over 24 h: (a) V_{th} and R_{ON} ($V_{GS} = 5$ V), (b) I_{Dmax} ($V_{DS} = V_{GS} = 5$ V) and I_{Gmax} ($V_{DS} = 0$ V, $V_{GS} = 5$ V).

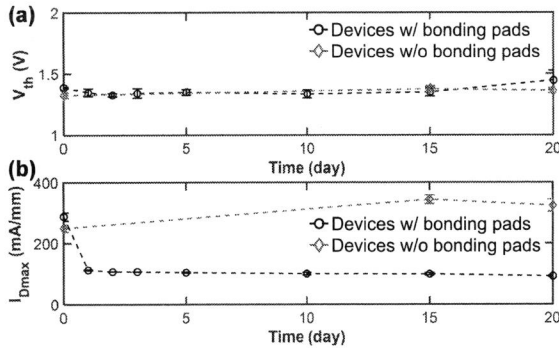

Fig. 5. Survival test of several DUTs with/without packaging (bonding pads) over 20 days in N_2 ambient. Ex-situ measurement was performed at RT: (a) V_{th}, (b) I_{Dmax} ($V_{DS} = V_{GS} = 5$ V). To the best of the authors' knowledge, this is the first report of long-term survivability of p-GaN-gated AlGaN/GaN HEMT at HT (500 °C). The results show great stability in both V_{th} and I_{Dmax}. However, the inclusion of bonding pads caused significant degradation to the measured I_{Dmax}.

in-situ measurement, a long-term survival test was conducted. For each measurement, the DUT was taken out of the furnace and cooled to RT for ex-situ characterization. It is recognized that, the packaging used in in-situ measurement might have contributed to some degradation. Therefore, in the survival test, DUTs with and without packaging were subjected to the same experiment. As presented in Fig. 5, the DUTs were found to exhibit stable DC performance over 20 days. An interesting observation is the difference between the DUTs with and without packaging, where the packaging caused a quick degradation of the measured current level on the first day.

At the end of the tests, the DUTs were inspected for structural degradation. Fig. 6(a)–(b) show that, after 1 day, minimal degradation was found to the gate structure. The DUT was no longer functional after 25 days of HT treatment. A cross-section (Fig. 6(c)) revealed that, a significant portion of the

978-1-6654-8901-0/22 $31.00 © 2022 IEEE

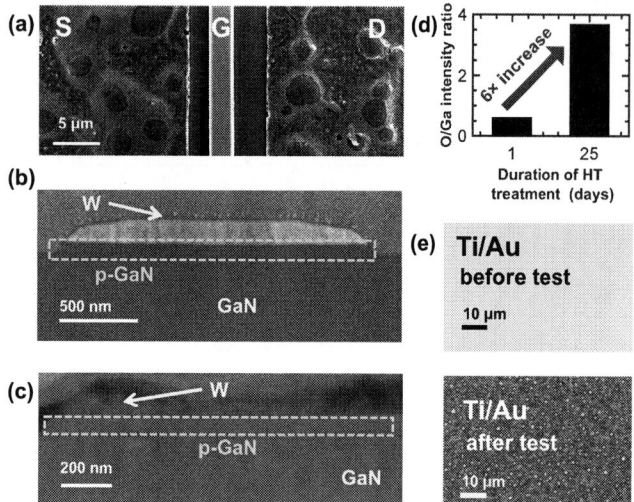

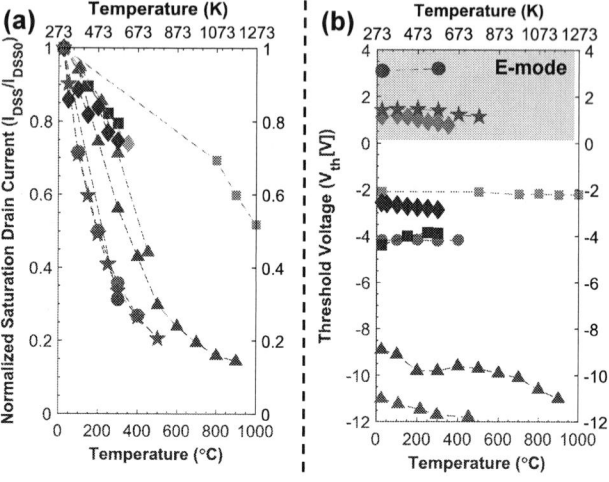

Fig. 6. Scanning electron microscopy (SEM) image of the gate region of DUT after survival test (500 °C in N_2 ambient) for 1 day, (a) top view (b) cross-sectional view. No noticeable degradation was observed in the alloyed ohmic contacts and the gate. (c) Cross-sectional view of the gate region of DUT after survival test for 25 days. A significant portion of the gate metal (W) was not present. (d) Results of the energy-dispersive X-ray spectroscopy (EDS) analysis of the 25 day HT treatment DUT (Fig. (c)). A comparison of the O/Ga intensity ratio shows an increase by a factor of 6 in the O composition relative to Ga from 1 day to 25 days of HT treatment. (e) Optical images of the Ti/Au bonding pad before and after HT survival test. The change in color indicates the change from a predominantly gold-based top layer to an inter-diffused Ti/Au alloy. The degradation of the Ti/Au bonding pad indicates the need for HT robust BEOL structures.

gate metal (W) was no longer present. In order to understand the degradation of the device from a materials perspective, energy-dispersive X-ray spectroscopy was conducted on the gate region of both DUTs. As shown in Fig. 6(d), the ratio of the intensity of the O peak to the intensity of the Ga peak increased from 0.61 (1 day HT treatment) to 3.68 (25 day HT treatment), which was an increase by a factor of 6. A likely cause for the significant introduction of oxygen is the leakage of atmospheric air into the furnace during the HT treatment. Nevertheless, no significant degradation to the epitaxial structure was found.

In terms of the back-end-of-line (BEOL) structures, significant inter-diffusion was observed in the Ti/Au layers (Fig. 6(d)), which serve as bonding pads (Fig. 2(d)). This is identified as the main cause of the current (I_{Dmax}) degradation in the packaged DUT after being exposed to HT. Metallization schemes which are less susceptible to inter-diffusion, such as Ni/Au, should be used in the future, considering that Ni is widely viewed as a diffusion barrier in Ti/Al/Ni/Au ohmic contacts [33].

A benchmarking of HT performance with published / commercially available wide band gap transistors is presented in Fig. 7. Two parameters, namely relative change in I_{Dmax} and V_{th}, are chosen due to their significance in circuit design. Fig. 7(a) indicates that, the reported p-GaN-gated AlGaN/GaN-on-Si HEMTs, exhibit similar HT current degradation as other transistors. It should be noted that, besides the epitaxial

Fig. 7. Benchmarking of performance variations at HT of the reported DUT with other published and commercially available wide band gap transistors, (a) saturation drain current as a function of temperature, normalized to value at RT. The DUT shows similar current degradation trend with typical AlGaN/GaN-on-Si HEMTs. (b) V_{th}, which is relatively stable and exhibits E-mode characteristics from RT to 500 °C. All of the listed devices are bare die devices [3], [5], [10], [36], [37]. The respective channel dimensions (L_G or L_{SD}), if published, are listed.

structure and the device type (conventional HEMT / MIS-HEMT / p-GaN-gated HEMT), an important factor in device degradation with temperature is the relative contribution of the contact resistance and channel resistance towards the ON-resistance (therefore drain current), because these two components of resistance vary at different rates with temperature. The transistor studied in this work is a long-channel transistor ($L_{SD} = 6$ µm), where $R_{ON} = 15$ Ω·mm, total contact resistance is 1 Ω·mm, therefore channel resistance is 14 Ω·mm (94 %) (estimated RT values). Channel length scaling would be a straightforward approach to reducing the HT degradation, if the DUT is intended to work over a wide temperature range [34], and the further scaling of the proposed transistor technology has been demonstrated [35]. Fig. 7(b) shows that, a relatively stable V_{th} (E-mode operation) is maintained from RT to 500 °C, which would greatly aid the design of mixed-signal circuits intended to work over a wide temperature range.

IV. CONCLUSION

The HT robustness of the p-GaN-gated AlGaN/GaN HEMT was systematically studied at 500 °C. The DUTs show stable performance over long-term survival tests, therefore attesting to the HT robustness of the proposed transistor technology for digital and analog mixed-signal applications. Nevertheless, several areas were identified to improve robustness, including choice of metallization scheme for bonding pads, and better passivation layer to protect the intrinsic DUT from moisture and reactive gases. Future research on the reliability of the DUT at high temperature, e.g. dynamic R_{ON}, would be valuable for a comprehensive understanding of the p-GaN-gated HEMT and their eventual large-scale application in HT electronics.

ACKNOWLEDGMENT

The authors gratefully acknowledge Enkris Semiconductor, Inc. (Dr. Kai Cheng) for providing the epitaxial wafers. Microfabrication was performed at MIT.nano.

REFERENCES

[1] P. Neudeck, R. Okojie, and L.-Y. Chen, "High-temperature electronics - a role for wide bandgap semiconductors?" *Proceedings of the IEEE*, vol. 90, no. 6, pp. 1065–1076, June 2002. doi: 10.1109/JPROC.2002.1021571

[2] K. H. Teo, Y. Zhang, N. Chowdhury, S. Rakheja, R. Ma, Q. Xie, E. Yagyu, K. Yamanaka, K. Li, and T. Palacios, "Emerging GaN technologies for power, RF, digital, and quantum computing applications: Recent advances and prospects," *Journal of Applied Physics*, vol. 130, no. 16, p. 160902, October 2021. doi: 10.1063/5.0061555

[3] P. G. Neudeck, S. L. Garverick, D. J. Spry, L.-Y. Chen, G. M. Beheim, M. J. Krasowski, and M. Mehregany, "Extreme temperature 6H-SiC JFET integrated circuit technology," *physica status solidi (a)*, vol. 206, no. 10, pp. 2329–2345, Oct 2009. doi: 10.1002/pssa.200925188

[4] J. Holmes, A. M. Francis, I. Getreu, M. Barlow, A. Abbasi, and H. A. Mantooth, "Extended High-Temperature Operation of Silicon Carbide CMOS Circuits for Venus Surface Application," *Journal of Microelectronics and Electronic Packaging*, vol. 13, no. 4, pp. 143–154, Oct 2016. doi: 10.4071/imaps.527

[5] P. G. Neudeck, D. J. Spry, L. Chen, N. F. Prokop, and M. J. Krasowski, "Demonstration of 4H-SiC digital integrated circuits above 800 °C," *IEEE Electron Device Letters*, vol. 38, no. 8, pp. 1082–1085, Aug 2017. doi: 10.1109/LED.2017.2719280

[6] M. Shakir, S. Hou, B. G. Malm, M. Östling, and C.-M. Zetterling, "A 600 °C TTL-based 11-stage ring oscillator in bipolar silicon carbide technology," *IEEE Electron Device Letters*, vol. 39, no. 10, pp. 1540–1543, October 2018. doi: 10.1109/LED.2018.2864338

[7] L. Nela, N. Perera, C. Erine, and E. Matioli, "Performance of GaN power devices for cryogenic applications down to 4.2 K," *IEEE Transactions on Power Electronics*, vol. 36, no. 7, pp. 7412–7416, July 2021. doi: 10.1109/TPEL.2020.3047466

[8] J. Liu, M. Xiao, R. Zhang, S. Pidaparthi, H. Cui, A. Edwards, M. Craven, L. Baubutr, C. Drowley, and Y. Zhang, "1.2-kV Vertical GaN FinJFETs: High-temperature characteristics and avalanche capability," *IEEE Transactions on Electron Devices*, vol. 68, no. 4, pp. 2025–2032, April 2021. doi: 10.1109/TED.2021.3059192

[9] M. Sadek, S.-W. Han, J. Song, J. C. Gallagher, T. J. Anderson, and R. Chu, "High-temperature static and dynamic characteristics of 4.2-kV GaN super-heterojunction p-n diodes," *IEEE Transactions on Electron Devices*, vol. 69, no. 4, pp. 1912–1917, April 2022. doi: 10.1109/TED.2022.3149453

[10] D. Maier, M. Alomari, N. Grandjean, J.-F. Carlin, M.-A. Diforte-Poisson, C. Dua, S. Delage, and E. Kohn, "InAlN/GaN HEMTs for operation in the 1000 °C regime: A first experiment," *IEEE Electron Device Letters*, vol. 33, no. 7, pp. 985–987, July 2012. doi: 10.1109/LED.2012.2196972

[11] Q. Xie, N. Chowdhury, A. Zubair, M. S. Lozano, J. Lemettinen, M. Colangelo, O. Medeiros, I. Charaev, K. K. Berggren, P. Gumann, D. Pfeiffer, and T. Palacios, "NbN-gated GaN transistor technology for applications in quantum computing systems," in *2021 Symposium on VLSI Technology*, June 2021, pp. T10–3.

[12] P. Palacios, T. Zweipfennig, A. Ottaviani, M. Saeed, C. Beckmann, M. Alomari, G. Lukens, H. Kalisch, J. Burghartz, A. Vescan, and R. Negra, "3D integrated 300°C tunable RF oscillator exploiting AlGaN/GaN HEMT for high temperature applications," in *2021 IEEE MTT-S International Microwave Symposium (IMS)*, June 2021. doi: 10.1109/IMS19712.2021.9574881 pp. 519–522.

[13] Q. Xie, N. Wang, C. Sun, A. B. Randles, P. Singh, X. Zhang, and Y. Gu, "Effectiveness of oxide trench array as a passive temperature compensation structure in AlN-on-silicon micromechanical resonators," *Applied Physics Letters*, vol. 110, no. 8, p. 083501, Feb 2017. doi: 10.1063/1.4976808

[14] W. Chen, W. Jia, Y. Xiao, Z. Feng, and G. Wu, "A temperature-stable and low impedance piezoelectric MEMS resonator for drop-in replacement of quartz crystals," *IEEE Electron Device Letters*, vol. 42, no. 9, pp. 1382–1385, Sep. 2021. doi: 10.1109/LED.2021.3094319

[15] N. Chowdhury, Q. Xie, M. Yuan, K. Cheng, H. W. Then, and T. Palacios, "Regrowth-free GaN-based complementary logic on a Si substrate," *IEEE Electron Device Letters*, vol. 41, no. 6, pp. 820–823, June 2020. doi: 10.1109/LED.2020.2987003

[16] L. Zhang, Z. Zheng, Y. Cheng, Y. H. Ng, S. Feng, W. Song, T. Chen, and K. J. Chen, "SiN/in-situ-GaON staggered gate stack on p-GaN for enhanced stability in buried-channel GaN p-FETs," in *2021 IEEE International Electron Devices Meeting (IEDM)*, Dec 2021. doi: 10.1109/IEDM19574.2021.9720653 pp. 5.3.1–5.3.4.

[17] T. Pu, X. Li, J. Wu, J. Yang, Y. Lu, X. Liu, and J.-P. Ao, "Recessed anode algan/gan schottky barrier diode for temperature sensor application," *IEEE Transactions on Electron Devices*, vol. 68, no. 10, pp. 5162–5166, Oct 2021. doi: 10.1109/TED.2021.3105498

[18] A. Hassan, J.-P. Noël, Y. Savaria, and M. Sawan, "Circuit techniques in GaN technology for high-temperature environments," *Electronics*, vol. 11, no. 1, Jan 2022. doi: 10.3390/electronics11010042

[19] Y. Cai, Z. Cheng, Z. Yang, C. W. Tang, K. M. Lau, and K. J. Chen, "High-temperature operation of AlGaN/GaN HEMTs direct-coupled FET logic (DCFL) integrated circuits," *IEEE Electron Device Letters*, vol. 28, no. 5, pp. 328–331, May 2007. doi: 10.1109/LED.2007.895391

[20] Y. Zhang, A. Zubair, Z. Liu, M. Xiao, J. Perozek, Y. Ma, and T. Palacios, "GaN FinFETs and trigate devices for power and RF applications: review and perspective," *Semiconductor Science and Technology*, vol. 36, no. 5, p. 054001, Mar 2021. doi: 10.1088/1361-6641/abde17

[21] J. He, Q. Wang, G. Zhou, W. Li, Y. Jiang, Z. Qiao, C. Tang, G. Li, and H. Yu, "Normally-OFF AlGaN/GaN MIS-HEMTs with low R_{ON} and V_{th} hysteresis by functioning in-situ SiN_x in regrowth process," *IEEE Electron Device Letters*, vol. 43, no. 4, pp. 529–532, April 2022. doi: 10.1109/LED.2022.3149943

[22] X. Liu, S. Zhang, K. Wei, Y. Zhang, H. Yin, X. Chen, S. Huang, G. Liu, Y. Zheng, T. Yuan, J. Niu, and X. Wang, "Improved stability of GaN MIS-HEMT with 5-nm plasma-enhanced atomic layer deposition SiN gate dielectric," *IEEE Electron Device Letters*, pp. 1–1, 2022. doi: 10.1109/LED.2022.3194136

[23] B. Zhang, J. Wang, X. Wang, C. Wang, C. Huang, J. He, M. Wang, J. Mo, Y. Hu, and W. Wu, "Improved performance of fully-recessed high-threshold-voltage GaN MIS-HEMT with in situ H_2/N_2 plasma pretreatment," *IEEE Electron Device Letters*, vol. 43, no. 7, pp. 1021–1024, July 2022. doi: 10.1109/LED.2022.3179136

[24] H. Liao, Z. Zheng, T. Chen, L. Zhang, Y. Cheng, S. Feng, Y. H. Ng, L. Chen, L. Yuan, and K. J. Chen, "Normally-OFF p-GaN gate double-channel HEMT with suppressed hot-electron-induced dynamic on-resistance degradation," *IEEE Electron Device Letters*, pp. 1–1, 2022. doi: 10.1109/LED.2022.3195489

[25] H. W. Then, M. Radosavljevic, P. Koirala, N. Thomas, N. Nair, I. Ban, T. Talukdar, P. Nordeen, S. Ghosh, S. Bader, T. Hoff, T. Michaelos, R. Nahm, M. Beumer, N. Desai, P. Wallace, V. Hadagali, H. Vora, A. Oni, X. Weng, K. Joshi, I. Meric, C. Nieva, S. Rami, and P. Fischer, "Advanced scaling of enhancement mode high-K gallium nitride-on-300mm-Si(111) transistor and 3D layer transfer GaN-silicon finfet CMOS integration," in *2021 IEEE International Electron Devices Meeting (IEDM)*, Dec 2021. doi: 10.1109/IEDM19574.2021.9720710 pp. 11.1.1–11.1.4.

[26] N. Sun, H. Huang, Z. Sun, R. Wang, S. Li, P. Tao, Y. Ren, S. Song, H. Wang, S. Li, W. Cheng, and H. Liang, "Improving gate reliability of 6-in E-Mode GaN-Based MIS-HEMTs by employing mixed oxygen and fluorine plasma treatment," *IEEE Transactions on Electron Devices*, vol. 69, no. 1, pp. 82–87, Jan 2022. doi: 10.1109/TED.2021.3131118

[27] J. Chen, Z. Liu, H. Wang, Y. He, X. Zhu, J. Ning, J. Zhang, and Y. Hao, "A GaN complementary FET inverter with excellent noise margins monolithically integrated with power gate-injection HEMTs," *IEEE Transactions on Electron Devices*, vol. 69, no. 1, pp. 51–56, Jan 2022. doi: 10.1109/TED.2021.3126267

[28] X. Li, N. Amirifar, K. Geens, M. Zhao, W. Guo, H. Liang, S. You, N. Posthuma, B. D. Jaeger, S. Stoffels, B. Bakeroot, D. Wellekens,

B. Vanhove, T. Cosnier, R. Langer, D. Marcon, G. Groeseneken, and S. Decoutere, "GaN-on-SOI: Monolithically integrated all-GaN ICs for power conversion," in *2019 IEEE International Electron Devices Meeting (IEDM)*, Dec 2019. doi: 10.1109/IEDM19573.2019.8993572 pp. 4.4.1–4.4.4.

[29] G. Lyu, J. Wei, W. Song, Z. Zheng, L. Zhang, J. Zhang, S. Feng, and K. J. Chen, "GaN on engineered bulk Si (GaN-on-EBUS) substrate for monolithic integration of high-/low-side switches in bridge circuits," *IEEE Transactions on Electron Devices*, vol. 69, no. 8, pp. 4162–4169, Aug 2022. doi: 10.1109/TED.2022.3178361

[30] C. Fleury, M. Capriotti, M. Rigato, O. Hilt, J. Würfl, J. Derluyn, S. Steinhauer, A. Köck, G. Strasser, and D. Pogany, "High temperature performances of normally-off p-GaN gate AlGaN/GaN HEMTs on SiC and Si substrates for power applications," *Microelectronics Reliability*, vol. 55, no. 9, pp. 1687–1691, 2015. doi: 10.1016/j.microrel.2015.06.010

[31] M. Yuan, Q. Xie, K. Fu, T. Hossain, J. Niroula, J. A. Greer, N. Chowdhury, Y. Zhao, and T. Palacios, "GaN ring oscillators operational at 500 °C based on a GaN-on-Si platform," *IEEE Electron Device Letters*, 2022. doi: 10.1109/LED.2022.3204566

[32] N. Chowdhury, Q. Xie, and T. Palacios, "Tungsten-gated GaN/AlGaN p-FET with $I_{max} > 120$ mA/mm on GaN-on-Si," *IEEE Electron Device Letters*, vol. 43, no. 4, pp. 545–548, April 2022. doi: 10.1109/LED.2022.3149659

[33] O. Odabasi, A. Ghobadi, T. G. U. Ghobadi, Y. Unal, G. Salkx00FD;m, G. Basar, B. Butun, and E. Ozbay, "Impact of the low temperature ohmic contact process on DC and forward gate bias stress operation of GaN HEMT devices," *IEEE Electron Device Letters*, 2022. doi: 10.1109/LED.2022.3199569

[34] A. Fontserè, A. Pérez-Tomás, M. Placidi, J. Llobet, N. Baron, S. Chenot, Y. Cordier, J. C. Moreno, P. M. Gammon, M. R. Jennings, M. Porti, A. Bayerl, M. Lanza, and M. Nafría, "Micro and nano analysis of 0.2 Ω mm Ti/Al/Ni/Au ohmic contact to AlGaN/GaN," *Applied Physics Letters*, vol. 99, no. 21, p. 213504, Nov 2011. doi: 10.1063/1.3661167

[35] Q. Xie, M. Yuan, J. Niroula, J. A. Greer, N. S. Rajput, N. Chowdhury, and T. Palacios, "Highly-scaled self-aligned GaN complementary technology based on a GaN-on-Si platform," in *2022 IEEE International Electron Devices Meeting (IEDM)*, 2022.

[36] S. Kargarrazi, A. S. Yalamarthy, P. F. Satterthwaite, S. W. Blankenberg, C. Chapin, and D. G. Senesky, "Stable operation of AlGaN/GaN HEMTs for 25 h at 400°C in air," *IEEE Journal of the Electron Devices Society*, vol. 7, pp. 931–935, 2019. doi: 10.1109/JEDS.2019.2937008

[37] Z. Xu, J. Wang, Y. Cai, J. Liu, Z. Yang, X. Li, M. Wang, M. Yu, B. Xie, W. Wu, X. Ma, J. Zhang, and Y. Hao, "High temperature characteristics of GaN-based inverter integrated with enhancement-mode (E-mode) MOSFET and depletion-mode (D-mode) HEMT," *IEEE Electron Device Letters*, vol. 35, no. 1, pp. 33–35, Jan 2014. doi: 10.1109/LED.2013.2291854

Effects of Oxide Electric Field Stress on the Gate Oxide Reliability of Commercial SiC Power MOSFETs

Limeng Shi, Tianshi Liu, Shengnan Zhu, Jiashu Qian, Michael Jin, Hema Lata Rao Maddi, Marvin H. White, and Anant K. Agarwal

Dept. of Electrical & Computer Eng., The Ohio State University, Columbus, OH, USA, shi.1564@osu.edu

Abstract— In this work, the influence of various oxide electric field (E_{ox}) stress conditions on the gate oxide lifetime, gate leakage current, and threshold voltage of 1.2 kV 4H-SiC power planar metal-oxide-semiconductor field-effect transistors (MOSFETs) is investigated. The results suggest that high E_{ox} stress (> 9.4 MV/cm) applied to the gate oxide of commercial SiC power MOSFETs for a certain period degrades the oxide lifetime due to high Fowler-Nordheim (F-N) electron tunneling current followed by hole trapping. Moreover, hole trapping enhances the gate leakage current and reduces the threshold voltage. Therefore, the generation of holes under high electric field conditions should be avoided to ensure the reliability of SiC power MOSFETs.

Keywords— Gate oxide reliability, electron and hole trapping, lifetime, gate oxide electric field stress, threshold voltage shift, leakage current

I. INTRODUCTION

SiC power devices are predominantly utilized in power supplies and electric vehicles, so the gate oxide reliability of SiC power MOSFETs influences the operational lifetime of these industrial applications. There are various issues that affect the reliability of gate oxide, such as higher defect density in the oxide limiting the lifetime [1], SiC/ SiO_2 interface traps resulting in charge trapping and a threshold voltage shift [2]. Compared with Si, SiC MOS structures have more early gate oxide failures. Therefore, an efficient screening method is required to screen out a significant number of extrinsic failures without degrading the intrinsic lifetime of the gate oxide. High gate voltage screening is a fast and effective method to screen out extrinsic defects [3], but impact ionization and/or anode hole injection (AHI) can be triggered under higher gate voltages which produce hole trapping in the gate oxide. The trapped holes increase the gate leakage current and decrease the threshold voltage of SiC power MOSFETs [4]. It has been shown that higher gate tunneling currents lead to reduced intrinsic gate oxide lifetimes [5]. Additionally, a negative shift of the threshold voltage can lead to increase in the OFF-state drain-leakage current. Therefore, it is valuable to find suitable oxide electric field stress and stress time that can satisfy application requirements without severely decreasing V_{th} and reducing gate oxide lifetime.

This work focuses on the effect of different oxide electric field stresses and stress times on the reliability of the gate oxide layer. The time-dependent dielectric breakdown (TDDB) results for commercial SiC power MOSFETs with and without stress are compared and discussed in terms of gate oxide lifetime prediction after being exposed to a variety of gate voltage stress conditions. The gate leakage currents and threshold voltage shifts of SiC power MOSFETs under high and low E_{ox} conditions are also monitored to investigate the gate oxide reliability.

II. DEVICE AND EXPERIMENTAL METHODS

A. Devices

The commercial 1.2 kV 4H-SiC planar power MOSFETs (packaged in TO-247) from vendor E are tested in this work. Table 1 contains the general characteristics of commercial 1.2 kV SiC planar MOSFETs. Fig. 1 depicts the cross-sectional view of a typical planar power MOSFET. A semiconductor parameter analyzer (B1506A, Keysight, Inc) is used to test the relevant parameters of SiC MOSFETs. The threshold voltages are extracted at V_{DS} = 0.1 V using the linear extrapolation approach [6]. The on-resistances are obtained with V_{DS} = 1 V and the maximum allowable gate voltage. When measuring gate oxide breakdown voltages (V_{BR}) at 150°C, the source and drain terminals of the devices under test (DUTs) are shorted to the ground while the gate voltage is ramped up and the gate leakage current is measured until dielectric breakdown. The oxide thicknesses are estimated based on the average oxide breakdown voltages at 150°C, with the assumption that the critical oxide breakdown electric field is 11 MV/cm (t_{ox} = V_{ox}/E_{ox}). The gate oxide breakdown voltage for vendor E is extracted to be ~ 49.46 V at 150°C. Thus, the oxide thickness is ~ 45 nm.

TABLE I.   TESTED COMMERCIAL 1.2KV SIC PLANAR MOSFETS

Properties	Vendor E
Mean V_{th} @RT (V)	5.94
Typical R_{on} @RT (mΩ)	280
Mean oxide V_{BR} @150°C (V)	49.46
Estimated t_{ox} (nm)	45

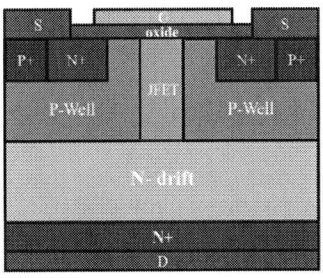

Fig. 1. A typical cross-sectional view of a commercial SiC planar power MOSFET.

B. EXPERIMENTAL METHODS

Constant-voltage TDDB measurements are performed on commercial 1.2 kV 4H-SiC power planar MOSFETs to obtain the gate oxide lifetime at 150°C. Ten devices as a group are carefully selected to make sure the threshold voltage variation among each group is less than 0.1 V. A constant gate voltage

978-1-6654-8901-0/22 $31.00 © 2022 IEEE

is applied to all 10 DUTs' gate electrodes at the same time, with their source and drain electrodes grounded. Failure times are recorded using a 10-channel digital multimeter (DMM 6500, Keithley, Inc), and the data is analyzed using Weibull statistics (method described in more details in [7, 8]). The Weibull distribution is obtained from the TDDB results to extract the 63% gate oxide lifetime ($t_{63\%}$) under different E_{ox} conditions. The oxide lifetime under typical operating conditions can be predicted based on the widely known thermo-chemical E-model [9].

To investigate the effects of different voltage stresses on the reliability of commercial SiC MOSFETs respectively, various gate oxide stress conditions are separately applied to commercial SiC power MOSFETs. After the stress, the following constant-voltage TDDB measurements are performed at 150°C with different gate voltages. The effect of gate stress conditions on the gate oxide reliability of SiC MOSFETs is investigated by comparing the changes in gate oxide lifetime before and after stress.

The threshold voltages and leakage currents have been recorded under different gate oxide voltages. The B1506A semiconductor parameter analyzer is utilized to provide the gate voltage and measure the curves of leakage current and threshold voltage over time to determine the influence of E_{ox} stress. Before the test and after every stress period, the threshold voltage at 150°C is extracted, as shown in Fig. 2. The gate leakage current is recorded during the stress period.

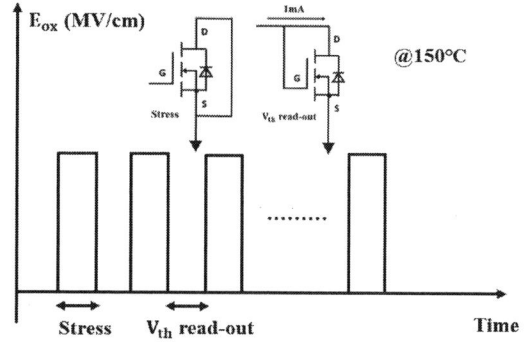

Fig. 2. Test procedure for gate leakage current and threshold voltage at 150°C

III. Experimental Results

A. TDDB measurements on un-stressed commercial SiC MOSFETs

The trend of gate leakage current vs. time varies depending on the electric fields of the gate oxide layer. Fig. 3 shows gate leakage current under different gate oxide electric fields of SiC MOSFETs from Vendor E. The applied E_{ox} varies from 8.5 MV/cm to 9.9 MV/cm. When the E_{ox} is less than 9 MV/cm, the gate leakage current of the DUTs decreases until the oxide layer breaks down. However, when the E_{ox} is higher than 9 MV/cm, the gate leakage current of the device rises initially then decreases until breakdown.

In our previous research, the variation of gate leakage current under different gate voltage conditions can be explained by hole trapping and electron trapping [10]. For the SiC MOSFETs from Vendor E, hole trapping originating from impact ionization and/or AHI dominates under high E_{ox} conditions, which reduces barrier width. As a result, electron injection and trapping of electrons in the gate oxide increase. However, when electron trapping dominates, the gate leakage current decrease due to the increase in barrier width. Therefore, under high E_{ox} (> 9 MV/cm) conditions, a large number of electrons and holes get trapped in the oxide. At a low E_{ox} (< 9 MV/cm), it is mainly electrons that are captured by traps both near the interface and in the oxide layer.

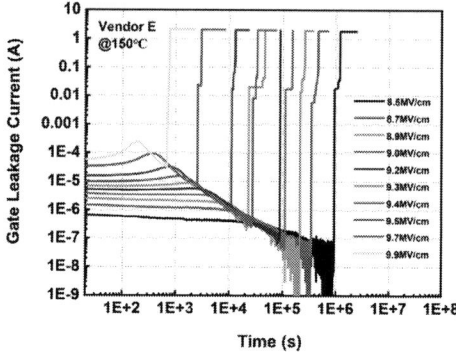

Fig. 3. Gate leakage current under different E_{ox} of SiC MOSFETs from Vendor E.

A group of gate oxide electric fields is chosen for the TDDB measurements: 8.5 MV/cm, 8.7 MV/cm, 8.9 MV/cm, 9.2 MV/cm, 9.5 MV/cm, 9.7 MV/cm, and 9.9 MV/cm. Ten devices with similar V_{th} are measured at each E_{ox}. Based on the Weibull plot, the failure time ($t_{63\%}$) for each E_{ox} is extracted and displayed in Fig. 4. Two different field acceleration factors (slopes of the fitted lines) are detected. An abrupt change in field acceleration factors is observed at E_{ox} of ~ 9.4 MV/cm. This is because when the E_{ox} exceeds 9.4 MV/cm, the hole trapping effect becomes more pronounced, causing higher F-N tunneling current thus degrading the gate oxide lifetime. Fig. 4 illustrates the extracted $t_{63\%}$ as a function of E_{ox} for commercial SiC MOSFETs from Vendor E. According to the thermo-chemical E-model, the predicted oxide lifetime at E_{ox} of 4 MV/cm and 150°C is above 3×10^{10} hours.

The transition from where electron trapping dominates to where hole trapping dominates happens at $E_{ox} = 9$ MV/cm. Gate oxide screening techniques are commonly used in the industry to remove extrinsic failures. If hole trapping dominates, the F-N tunneling current would increase and accelerate the degradation of the gate oxide. Therefore, in the subsequent tests, the effect of different E_{ox} stress on the gate oxide lifetime of commercial SiC MOSFETs is investigated.

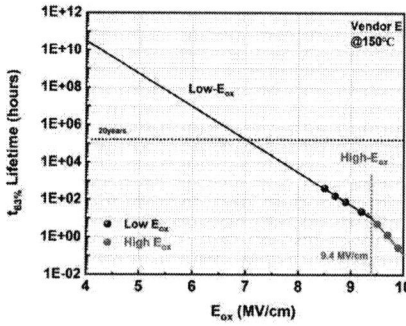

Fig. 4. The extracted $t_{63\%}$ as a function of E_{ox} for commercial SiC MOSFETs

B. TDDB measurements on commercial SiC MOSFETs after E_{ox} stress

In the previous study, it has been observed that high E_{ox} can induce a dominant effect of hole trapping and then reduce the lifetime of the gate oxide layer. Therefore, 1min @ 9.7 MV/cm, 1min @ 9.9 MV/cm, and 100 ms @ 9.9 MV/cm are chosen as the electric field stress conditions to investigate the impact of high voltage stress on the gate oxide lifetime.

In Fig. 5, the oxide lifetimes of SiC MOSFETs with and without E_{ox} stress at 150°C are compared. The lifetimes of some of the MOSFETs after stress are higher than the MOSFETs without any stress. This may be caused by the variation of the oxide lifetime due to fabrication process variation for different MOSFETs. Due to the limitation on the number of samples, no significant lifetime degradation is observed for devices exposed to 1 min at 9.7 MV/cm and 100 ms at 9.9 MV/cm stress. It is observed that the Weibull slop (β) reduces after 1 min 9.9 MV/cm stress, reflecting considerable variations in the oxide failure time. The reduction in β demonstrates the adverse effect of high E_{ox} stress on the gate oxide reliability.

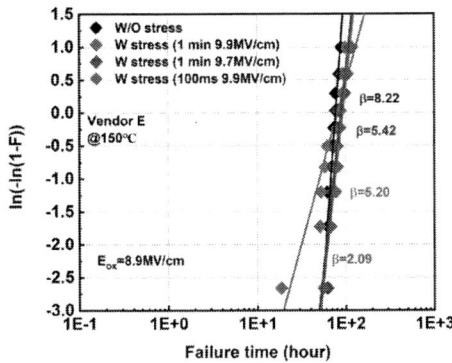

Fig. 5. Weibull distributions of measured lifetimes at 150°C for commercial SiC MOSFETs with stress and without stress.

To make the effect of stress more pronounced, initial electric field stress of 9.9 MV/cm is applied to the gate and maintained for three minutes at 150°C, and then the gate electric field is reduced to lower E_{ox} (8.5MV/cm, 8.7MV/cm, and 8.9MV/cm) for the TDDB test in order to extract the gate oxide lifetimes. The measured lifetimes for SiC MOSFETs with 3 min at 9.9 MV/cm stress and without stress are plotted in Fig. 6 (a). The intrinsic lifetime of the original device is 15 min at E_{ox} of 9.9 MV/cm and 150°C as shown in Fig. 4. When the applied stress time is 20% of the oxide lifetime at E_{ox} of 9.9 MV/cm, the oxide lifetimes of the device after the stress at low E_{ox} are reduced to ~10% of the original oxide lifetime. As shown in Fig. 6 (a), the predicted lifetime at 4 MV/cm for SiC MOSFETs without stress is ~ 10^{10} hours. However, SiC MOSFETs with 3 min 9.9 MV/cm stress have approximately two orders of magnitude reduction in lifetime prediction at 4 MV/cm.

A low E_{ox} stress condition of 8 MV/cm for 10 hours is used as a comparison. Fig. 6 (b) illustrates $t_{63\%}$ vs E_{ox} at 150°C for commercial SiC MOSFETs with 10 hours at 8 MV/cm stress and without stress. The measured oxide lifetimes indicate that a prolonged stress at a low electric field does not significantly degrade the gate oxide lifetime.

It is believed that high E_{ox} stress (> 9.4 MV/cm) has a harmful impact on the lifetime of oxide, while low electric field stress has less obvious influence on the lifetime of gate oxide since there is no enhanced F-N tunneling current caused by hole trapping.

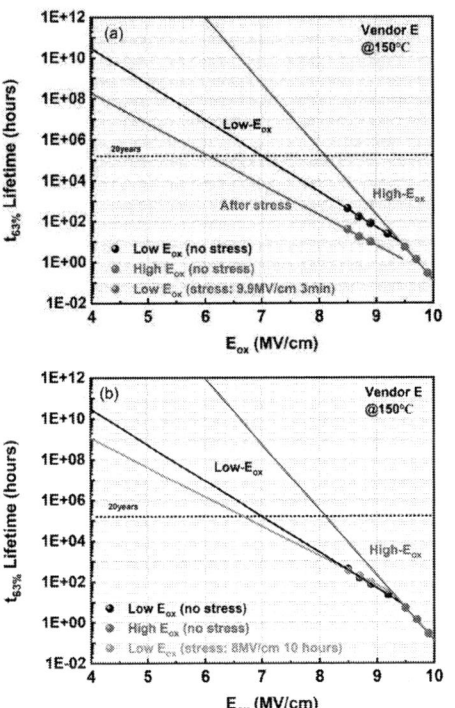

Fig. 6. $t_{63\%}$ vs E_{ox} at 150°C for commercial SiC MOSFETs with (a) 3 min 9.9 MV/cm stress and (b) 10 h 8 MV/cm stress.

C. Gate leakage currents and threshold voltage shifts

Under different E_{ox} conditions, the positive and negative charge trapping near the interface and inside the oxide layer affects the value of the threshold voltage, whereas the variation of the leakage current reveals the electron and hole distribution inside the oxide. The mechanisms of hole and electron trapping under varied oxide electric fields are discussed to explain the different gate leakage trends and related threshold voltage variations.

The threshold voltages and leakage currents under different E_{ox} conditions have been recorded at 150°C, as depicted in Fig. 7. When E_{ox} = 8 MV/cm, leakage current gradually decreases with time. So it can be assumed that electron trapping plays a dominant role under the electric field of 8 MV/cm. With the increase of stress time, a large number of electrons gradually get trapped in the oxide layer, and these electrons enhance the barrier width and shift threshold voltage in a positive direction. When E_{ox} = 8.9 MV/cm, similar trends can be observed for leakage current and threshold voltage. The threshold voltage shift values are 0.27 V and 0.13 V for 8 MV/cm and 8.9 MV/cm electric field stress conditions, respectively, when the stress time is 0.1 s. Hole generation diminishes part of the effect of electron trapping at E_{ox} of 8.9 MV/cm, so that a smaller threshold voltage shift value can be obtained when 8.9 MV/cm is applied for 0.1 s. After 1000 s of E_{ox} stress, more electrons are trapped in the gate oxide at 8.9 MV/cm, and the threshold voltage shift value is 1.84 V.

During the 9.9 MV/cm stress period, the barrier width decreases due to the prominent function of hole trapping, resulting in an increased leakage current over a period of 140

978-1-6654-8901-0/22 $31.00 © 2022 IEEE

s, while the threshold voltage decreases with time, as shown in Fig. 7. However, after 140 s of stress at 9.9 MV/cm, electron trapping is the primary effect, leading to a drop in leakage current and an increase in threshold voltage.

A higher screening voltage improves the screening efficiency and reduce the field-failure probability after screening. While the high gate voltage comes with the challenge of negative threshold voltage shift. Fig. 8 illustrates the threshold voltage shift after applying 100 ms E_{ox} stress to the gate oxide at 150°C. It can be observed that the threshold voltage shows a negative shift when E_{ox} is greater than 9 MV/cm. To avoid the negative threshold voltage shift, it is recommended to use a E_{ox} less than 9 MV/cm as the electric field for gate oxide screening.

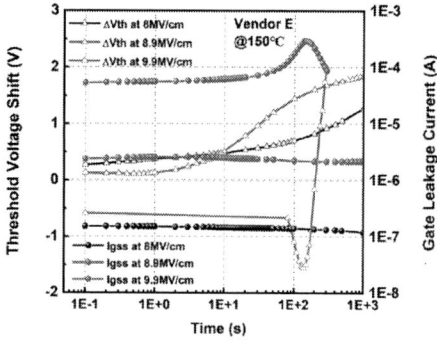

Fig. 7. Threshold voltage variations and gate leakage currents during gate voltage stresses at various E_{ox} for SiC MOSFETs from Vendor E.

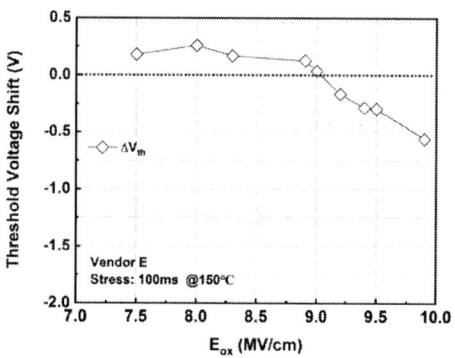

Fig. 8. Threshold voltage shift at various E_{ox} for SiC MOSFETs from Vendor E.

IV. CONCLUSION

The changes in gate oxide lifetime, leakage current, and threshold voltage after different E_{ox} stresses are shown in this paper. When E_{ox} is greater than 9.4 MV/cm, a large F-N tunneling current is generated due to hole trapping, thus degrading gate oxide lifetime. In addition, a negative threshold voltage shift is observed at E_{ox} value higher than 9 MV/cm. The transition from where electron trapping dominates to where hole trapping dominates happens at oxide electric field of 9 MV/cm. To avoid a reduction in threshold voltage and prevent degradation of gate oxide lifetime, the screening E_{ox} should be below 9 MV/cm.

ACKNOWLEDGMENT

This work is supported in part by the Ford Motor Company under the Ford Alliance 2019 Project to The Ohio State University and in part by the Block Gift Grant from II-VI Foundation.

REFERENCES

[1] J. Senzaki, K. Kojima, T. Kato, A. Shimozato, and K. Fukuda, "Correlation between reliability of thermal oxides and dislocations in n-type 4H-SiC epitaxial wafers," Appl.Phys.Letters, vol. 89, pp. 022909, 2006.

[2] C. U nger a nd M . P fost, " Determination of the Transient Threshold Voltage Hysteresis in SiC MOSFETs after Positive and Negative Gate Bias," in 2019 31st International Symposium on Power Semiconductor Devices and ICs (ISPSD), Shanghai, China, 2019, pp. 195–198.

[3] T. Aichinger and M. Schmidt, "Gate-oxide reliability and failure-rate reduction of industrial SiC MOSFETs," 2020 IEEE International Reliability Physics Symposium (IRPS), 2020, pp. 1-6

[4] S. Zhu, T. Liu, M. H. White, A. K. Agarwal, A. Salemi and D. Sheridan, "Investigation of Gate Leakage Current Behavior for Commercial 1.2 kV 4H-SiC Power MOSFETs," 2021 IEEE International Reliability Physics Symposium (IRPS), 2021, pp. 1-7.

[5] R. Singh, and A.R. Hefner, "Reliability of SiC MOS devices," Solid-State Electr. 48, pp.1717-1720, 2004.

[6] M. Tsuno, M. Suga, M. Tanaka, K. Shibahara, M. Miura-Mattausch, and M. Hirose, "Physically-based threshold voltage determination for MOSFET`s of all gate lengths," IEEE transactions on electron devices, vol. 46, no. 7, pp. 1429–1434, 1999.

[7] T. Liu et al., "Gate Leakage Current and Time-Dependent Dielectric Breakdown Measurements of Commercial 1.2 kV 4H-SiC Power MOSFETs," 2019 IEEE 7th Workshop on Wide Bandgap Power Devices and Applications (WiPDA), 2019, pp. 195-199.

[8] T. Liu, S. Zhu, M. H. White, A. Salemi, D. Sheridan and A. K. Agarwal, "Time-Dependent Dielectric Breakdown of Commercial 1.2 kV 4H-SiC Power MOSFETs," in IEEE Journal of the Electron Devices Society, vol. 9, pp. 633-639, 2021.

[9] J. W. McPherson and H. C. Mogul, "Underlying physics of the thermochemical E model in describing low-field time-dependent dielectric breakdown in SiO2 thin films," J. Appl. Phys., vol. 84, no. 3, pp. 1513–1523, 1998.

[10] P. Moens, J. Franchi, J. Lettens, L. D. Schepper, M. Domeij and F. Allerstam, "A Charge-to-Breakdown (QBD) Approach to SiC Gate Oxide Lifetime Extraction and Modeling," 2020 32nd International Symposium on Power Semiconductor Devices and ICs (ISPSD), 2020, pp. 78-81.

A Comparison of Ion Implantation at Room Temperature and Heated Ion Implantation on the Body Diode Degradation of Commercial 3.3 kV 4H-SiC Power MOSFETs

Jiashu Qian[1], Tianshi Liu[1], Jake Soto[2], Mowafak M. Al-Jassim[3], Robert Stahlbush[4], Nadeemullah Mahadik[4], Limeng Shi[1], Michael Jin[1], and Anant K. Agarwal[1]

[1]Dept. of Electrical and Computer Engineering, The Ohio State University, Columbus, OH 43210, USA
[2]Microchip Technology Inc., Bend, OR 97702, USA
[3]National Renewable Energy Laboratory (NREL), Golden, CO 80401, USA
[4]Naval Research Laboratory (NRL), Washington, DC 20375, USA

Abstract — It has been demonstrated that basal plane dislocations (BPDs)-induced stacking faults (SFs) cause body diode degradation in commercial 4H-SiC power MOSFETs, especially with higher voltage ratings. BPDs originate from 4H-SiC boule, epi growth, and ion implantation. Considering the lower cost of ion implantation at room temperature (RT), this work investigates the potential of RT ion implantation replacing heated (HT) ion implantation by comparing the influence of both ion implantations on the body diode degradation of commercial 3.3 kV 4H-SiC power MOSFETs. We demonstrate with long-term (up to 1000 hours) forward current stress that RT implantation can keep the body diode degradation of 3.3 kV 4H-SiC power MOSFETs within the specification limits compared with HT implantation.

Keywords—4H-SiC, MOSFET, body diode degradation, basal plane dislocation (BPD), ion implantation

I. INTRODUCTION

Basal plane dislocations (BPDs) can turn into stacking faults (SFs) in the drift layer of 4H-SiC power MOSFETs and keep expanding by absorbing the activation energy from the recombination of electron-hole pairs [1]. With the activated SFs, the carrier lifetime and mobility in the drift layer decrease, affecting the stable operation of the body diode. To reduce the density of BPDs, a low-dose high temperature Al implantation has been proved to be effective [2]. Besides, employing a recombination-enhanced buffer layer can suppress the expansion of BPDs-induced SFs [3]. Also, a molten KOH etching process before the conventional epitaxial growth is developed to enhance the conversion of BPDs to threading edge dislocations (TEDs) by creating BPD etch pits for low BPD density epilayers [4, 5]. However, K. Konishi., et al. have demonstrated that SFs originating from BPDs converted into TEDs within the buffer layer still expand under high current stress [6]. With these improvements in the material, it is not expected to find significant body diode degradation for low-voltage 4H-SiC power MOSFETs (600-1200 V). However, it is still a challenge in high-voltage 4H-SiC power MOSFETs ($\geq$ 6.5 kV) because the longer growth time of the thicker drift layer ($\geq$ 60 μm) significantly enhances BPDs formation and the spatial size of BPDs-induced SFs [7, 8]. Current process can already grow an epitaxial thickness of up to 30 μm from a 4H-SiC substrate with a negligible density of BPDs (0.5 BPDs on average per wafer in the substrate) for the fabrication of 3.3 kV 4H-SiC power MOSFETs [8]. To further minimize the density of BPDs in the drift layer, heated ion implantation (HT implantation) is employed to reduce generation of new BPDs in industry. Considering the costly heated ion-implantation systems and limited throughput, the main objective of this study is to investigate the possibility of ion implantation at room temperature (RT implantation) replacing HT implantation in the fabrication of 3.3 kV 4H-SiC power MOSFETs.

	RT Implanted Devices		HT Implanted Devices	
	Avg.	Std.	Avg.	Std.
V_F (V)	-3.802	0.040	-3.816	0.044
V_{th} (V)	5.954	0.103	5.179	0.041
R_{ON} (Ω)	0.071	0.001	0.066	0.001

Table. 1 Pre-test results of 3.3 kV 4H-SiC power MOSFETs under test. V_F is extracted at I_{DS} = -20 A under the 3rd quadrant operation. R_{ON} is extracted at V_{GS} = 20 V, V_{DS} = 0.5 V. V_{th} is extracted with the linear extrapolation method at V_{DS} = 0.1 V.

II. EXPERIMENTS

A. *Device Under Test (DUT)*

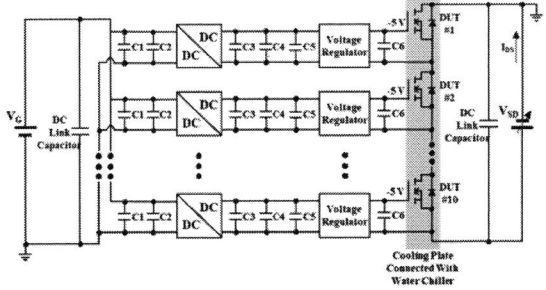

Fig. 1 Schematic of the body diode test setup for ten 3.3 kV 4H-SiC MOSFETs in series.

The 127 DUTs are provided by Microchip Technology Incorporated (MCHP), which are

978-1-6654-8901-0/22 $31.00 © 2022 IEEE

engineering prototypes of their commercial 3.3 kV 4H-SiC MOSFETs (MSC080SMA330B).

Among these devices, 61 are RT implanted devices and 66 are HT implanted devices. The HT implantation is conducted at 500 °C. Both types of devices are from wafers of the same fab lot. All devices are either all implanted at RT, or all implanted at HT, using equivalent doses and energy. The specific doses and energy are MCHP proprietary information. The active area of the die for both types of devices is 14.3 mm^2.

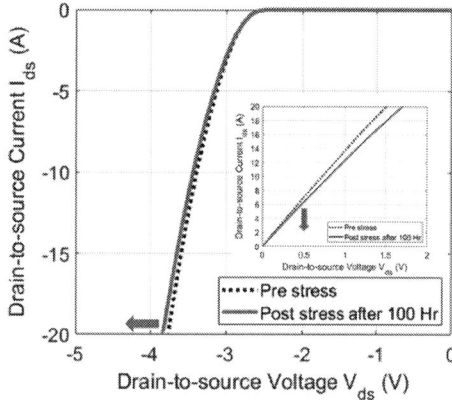

Fig. 2 The criterion ($\geq 2\%$ shift for V_F and $\geq 10\%$ shift for R_{ON}) for degradation is reflected in the 1st quadrant I_{DS}-V_{DS} output characteristics (V_{GS} = -5 V) and 3rd quadrant I_D-V_D characteristics (V_{GS} = 20 V).

B. Body Diode Stress Test

To analyze the influence of the body diode degradation on the performance of devices under test, the 3rd quadrant I_D-V_D characteristics, the 1st quadrant I_D-V_D output characteristics, the 1st quadrant I_D-V_G transfer characteristics, and the 1st quadrant forward blocking I_D-V_D characteristics are measured by using a Keysight B1506A power device analyzer before and after the stress. Table I shows the pre-stress measurement results of the 61 RT implanted devices and 66 HT implanted devices. Both types of devices have similar average on-state resistance (R_{ON}) and average forward voltage drop (V_F) of the body diode with equivalent doses and energy for all the implants. Only the average threshold voltage (V_{th}) of the RT implanted devices is slightly higher than that of the HT implanted devices. This can be explained by the higher P-type doping concentration within the depth of the p-well in RT implanted devices than that in HT implanted devices [9]. 10 DUTs are connected in series as shown in Fig. 1 to conduct a constant 10 A DC (the stress current density is ~70 A/cm^2) through the forward-biased body diodes. Isolated DC-DC converters and voltage regulators help maintain stable gate-source voltages for all devices in series. The gate-source voltages in this study are set to -5 V to make

sure the MOS channels of all the devices are pinched-off. The body diodes of all the devices in series are stressed for 100 hours, followed by a 10-min cool-down to room temperature. Then, the electrical measurements mentioned above are performed on each device for comparison with the pre-stress data. This process is repeated every 100 hours until the total stress time of 1000 h is achieved. During the stress, the metal sides of the DUTs' TO-247 packages are fixed on a heat sink with an industrial-grade water chiller connected to dissipate the generated heat and maintain the case temperature at a typical operating level.

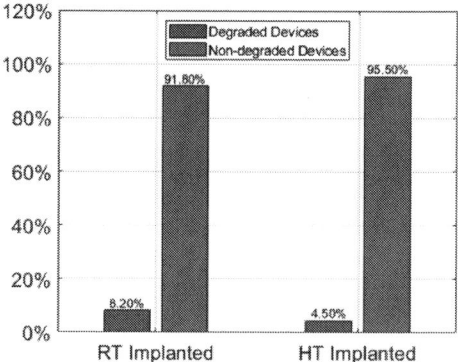

Fig. 3 Degraded vs. non-degraded device ratio of RT and HT implanted devices after 100-hr body diode stress.

III. RESULTS AND DISCUSSION

A. 100-hr Body Diode Stress Test

After 100-hr stress, the body-diode-degraded parts are counted for both types of devices. The criterion of body diode degradation in this study is defined as $\geq 2\%$ shift in V_F and $\geq 10\%$ shift in R_{ON} as shown in Fig. 2. The shift percentage is calculated by the ratio of the shifted value and the pre-stress value. Based on this criterion, all 127 DUTs are binned into two parts (degraded devices vs. non-degraded devices). Fig. 3 illustrates that after 100-hr stress there is no significant difference between RT implanted devices and HT implanted devices on the body diode degradation with a similar sample size. The cumulative frequency plots in Fig. 4 for both RT implanted and HT implanted devices on the V_F shift and the R_{ON} shift give the information that most of both RT implanted and HT implanted devices have the V_F shift and the R_{ON} shift concentrated in 0-0.9% and 0-9% after 100-hr stress, which do not meet the criterion of a 'body-diode-degraded device'. However, RT implanted devices can reach a maximum V_F shift and R_{ON} shift in the range of 1.8%-2.1% and 18%-21%, a little larger than 1.2%-1.5% and 12%-15% of HT implanted devices. There is reason to believe that the RT implantation has the potential to replace the HT implantation in the fabrication of 3.3 kV 4H-SiC MOSFETs based on 100-hr test.

978-1-6654-8901-0/22 $31.00 © 2022 IEEE

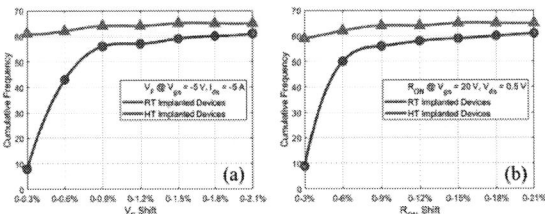

Fig. 4 The cumulative frequency plots for (a) the V_F shift; (b) the R_{ON} shift; of 61 RT implanted devices and 66 HT implanted devices after 100-hr stress. The Y-axis is the cumulative number of devices.

B. 1000-hr Body Diode Stress Test

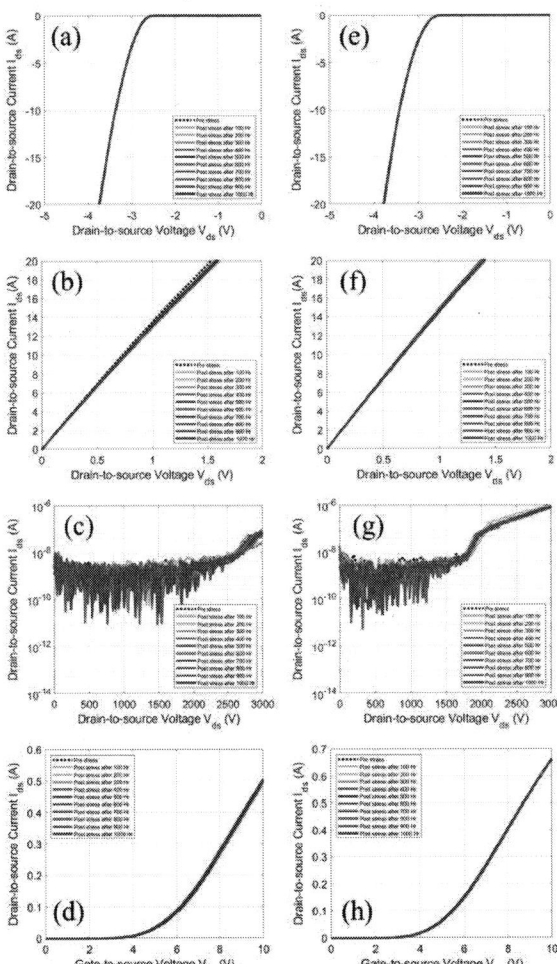

Fig. 5. The 3rd quadrant I_{DS}-V_{DS} characteristics (V_{GS} = -5 V) of the least degraded (a) RT; (e) HT; implanted device. The 1st quadrant I_{DS}-V_{DS} output characteristics (V_{GS} = 20 V) of the least degraded (b) RT; (f) HT; implanted device. The 1st quadrant forward blocking I_{DS}-V_{DS} characteristics (V_{GS} = 0 V) of the least degraded (c) RT; (g) HT; implanted device. The 1st quadrant I_{DS}-V_{GS} characteristics (V_{DS} = 0.1 V) of the least degraded (d) RT; (h) HT; implanted device. The stress hours are 1000h.

To further investigate, 20 DUTs each from both types of devices after 100-hr stress are selected for extra 900-hr stress in order to approach the practical lifetime requirement for the body diode of the commercial 4H-SiC power MOSFETs.

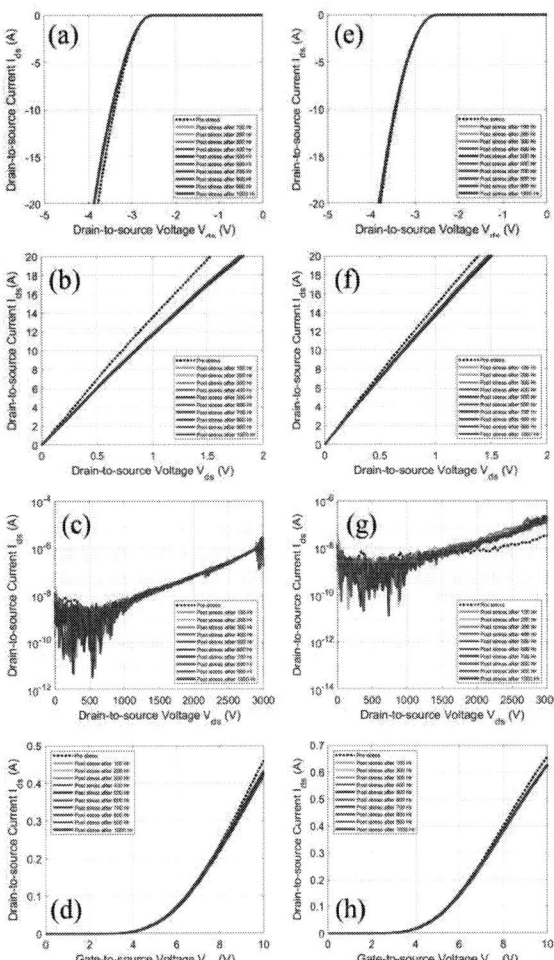

Fig. 6. The 3rd quadrant I_{DS}-V_{DS} characteristics (V_{GS} = -5 V) of the most degraded (a) RT; (e) HT; implanted device. The 1st quadrant I_{DS}-V_{DS} output characteristics (V_{GS} = 20 V) of the most degraded (b) RT; (f) HT; implanted device. The 1st quadrant forward blocking I_{DS}-V_{DS} characteristics (V_{GS} = 0 V) of the most degraded (c) RT; (g) HT; implanted device. The 1st quadrant I_{DS}-V_{GS} characteristics (V_{DS} = 0.1 V) of the most degraded (d) RT; (h) HT; implanted device. The stress hours are 1000h.

Fig. 5 shows the degradation evolution with 1000 hours of the least degraded parts for both RT implanted and HT implanted devices. The SMUs of the power device analyzer used in this study have the measurement limitations for the current and voltage of 20 A and 3 kV, respectively. Thus, the 3rd quadrant I_{DS}-V_{DS} characteristics, the 1st quadrant I_{DS}-V_{DS} output characteristics at V_{GS} = 20 V, and the 1st quadrant forward blocking I_{DS}-V_{DS} characteristics can be only observed within the above limits. Almost no visible degradation can be observed in the HT implanted device. Only slight and acceptable degradations can be observed in the output characteristics and the forward

978-1-6654-8901-0/22 $31.00 © 2022 IEEE

blocking leakage current of the RT implanted device after 1000-hr stress. Most of the DUTs show a similar degradation level to the least degraded parts, regardless of implantations.

Fig. 6 shows the degradation evolution with 1000 hours of the most degraded parts for both RT implanted and HT implanted devices. Although the V_{th} of both RT implanted and HT implanted devices shows a slight variation after 1000-hr stress, the influence on the degradation is negligible. In this case, the degradation reflected on the V_F and R_{ON} of the HT implanted device is still acceptable, however, it can be considered as a 'degraded' part for the RT implanted device. Nevertheless, few RT implanted devices can reach the 'degraded' level, which can be considered an allowable industrial spread. It is worth noting that the HT implanted device has significant blocking leakage current degradation, but the RT implanted device does not, even the RT implanted device is a 'degraded' part. It could be due to the fact that breakdown in the HT implanted device happens in the active area caused by the activated SFs, and breakdown in the RT implanted device occurs in the edge termination region where no SFs form [10].

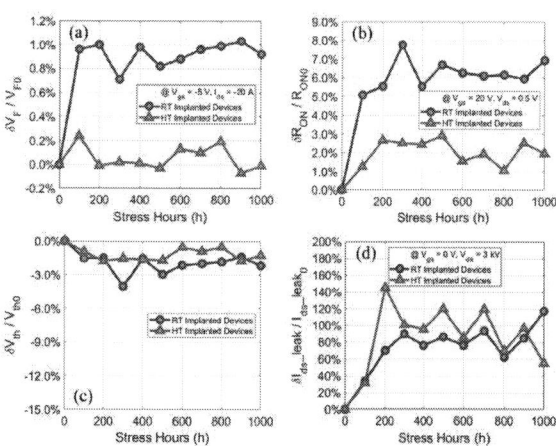

Fig. 7. Comparison between RT and HT implanted devices in terms of the influence of the 1000-hr body diode stress on (a) V_F degradation. (b) R_{ON} degradation. (c) V_{th} degradation. (d) Degradation of the forward blocking leakage current.

The average degradation level of both the 20 RT implanted devices and the 20 HT implanted devices is reflected by the shift percentages in Fig. 7. Although the V_F and R_{ON} degradations of the RT implanted devices in Fig. 7 (a) and Fig. 7 (b) are slightly higher than the HT implanted devices, the degradations are still within an acceptable range based on the criterion above. Fig. 7 (c) shows that the V_{th} shift induced by the gate oxide defects under 1000-hr negative bias (-5 V) is similar for both types of devices Fig. 7 (d) illustrates that the forward blocking leakage current degradations of both types of devices have similar

trends and fluctuations at a similar level within the 1000-hr stress. It is worth mentioning that both the V_F and R_{ON} degradations of the RT implanted devices and the HT implanted devices increase significantly at the early 100-hr stage and eventually tend to saturate. The saturated degradation level is very close to the degradation level after 100-hr stress. Therefore, the conclusion that the RT implantation has the potential to replace the HT implantation in the fabrication of 3.3 kV 4H-SiC MOSFETs is also believed to be valid for 1000-hr stress conditions.

IV. CONCLUSION

This paper has compared the body diode degradation of commercial 3.3 kV 4H-SiC power MOSFETs with the heated ion implantation and the ion implantation at room temperature, with equivalent doses and energy of the same implants for both types of devices. The results indicate that the ion implantation at room temperature has the potential to replace the heated ion implantation in the fabrication of commercial 3.3 kV 4H-SiC power MOSFETs, with a highly qualified 4H-SiC substrate and the latest generation of process. This replacement can help the industry eliminate the purchase and maintenance costs of heating equipment, and the step of device cooling in the fabrication to shorten the production cycle. Further investigation is needed to check if the conclusion applies to higher-voltage (>3.3 kV) devices since the thicker drift layer of the higher-voltage devices enhances the BPDs formation, whether this conclusion applies to higher-voltage (>3.3 kV) devices still needs to be investigated.

ACKNOWLEDGEMENT

This research is supported in part by the US Department of Energy (DOE), the Office of Energy Efficiency and Renewable Energy (EERE), the Advanced Manufacturing Office (AMO), and technical support from Microchip Technology Inc. The authors would also like to thank National Renewable Energy Laboratory (NREL) for the helpful discussions.

REFERENCES

[1] Galeckas., et al., "Recombination-enhanced Extension of Stacking Faults in 4H-SiC p-i-n Diodes under Forward Bias", *Applied Physics Letters*, Vol. 81, issue 5, pp. 883-885, 2002.

[2] K. Konishi., et al., "Investigation of Forward Voltage Degradation due to Process-Induced Defects in 4H-SiC MOSFET", *Materials Science Forum*, Vol. 924, pp. 365-368, 2018.

[3] T. Tawara., et al., "Suppression of the Forward Degradation in 4H-SiC PiN Diodes by Employing a Recombination-

978-1-6654-8901-0/22 $31.00 © 2022 IEEE

Enhanced Buffer Layer", *Materials Science Forum*, Vol. 897, pp. 419-422, 2016.

[4] Z. Zhang., et al., "Basal Plane Dislocation-free Epitaxy of Silicon Carbide", *Applied Physics Letters*, Vol. 87, issue 15, 2005.

[5] Z. Zhang., et al., "Mechanism of Eliminating Basal Plane Dislocations in SiC Thin Films by Epitaxy on an Etched Substrate", *Applied Physics Letters*, Vol. 89, issue 8, 2008.

[6] K. Konishi., et al., "Stacking Fault Expansion from Basal Plane Dislocations Converted into Threading Edge Dislocations in 4H-SiC Epilayers under High Current Stress", *Journal of Applied Physics*, Vol. 114, issue 1, 2013.

[7] A. Agarwal., et al., "A New Degradation Mechanism in High-Voltage SiC Power MOSFETs," *IEEE Electron Device Letters*, Vol. 28, no. 7, pp. 587-589, 2007.

[8] R. Stahlbush., et al., "Defects in 4H-SiC Epilayers Affecting Device Yield and Reliability," *IEEE International Reliability Physics Symposium (IRPS)*, Mar. 2022.

[9] J.F. Michaud., et al., "Aluminum Implantation in 4H-SiC: Physical and Electrical Properties," *Materials Science Forum*, Vol. 740-742, pp. 581-584, 2013.

[10] M. Kang et al., "Body Diode Reliability of Commercial SiC Power MOSFETs", *IEEE 7th Workshop Wide Bandgap Power Devices Appl. (WiPDA)*, Oct. 2019.

A Comparison of Short-Circuit Failure Mechanisms of 1.2 kV 4H-SiC MOSFETs and JBSFETs

Dongyoung Kim, Skylar DeBoer, Seung Yup Jang, Adam J. Morgan, and Woongje Sung
College of Nanoscale Science and Engineering, State University of New York Polytechnic Institute,
Albany, NY, 12203, USA, kimd1@sunypoly.edu

Abstract— This paper presents a comparison of 1.2 kV 4H-SiC MOSFETs and Ti JBSFETs with deep P-well structures. For a fair comparison of the short-circuit characteristics between the MOSFETs and JBSFETs, an innovative design approach for the JBSFETs was implemented to obtain the same specific on-resistance to the MOSFETs. To improve the short-circuit characteristics of the MOSFETs and JBSFETs, channeling implantation was conducted to form a deep P-well structure, that helps reduce the maximum saturation current during the short-circuit event. Using this approach superior short-circuit characteristics are achieved in the MOSFETs and JBSFETs. However, the JBSFETs provide a shorter short-circuit withstand time than the MOSFETs due to the high leakage current from Schottky contact. Sentaurus 2D TCAD was used to understand and clarify the short-circuit mechanisms of the MOSFETs and JBSFETs. It was discovered that the MOSFETs failed due to the high current in the channel region, but the failure of JBSFETs happens in the Schottky contact. Moreover, solutions to improve the short-circuit characteristics of the JBSFETs are proposed; a narrow Schottky width and high work function metal.

Keywords— 4H-SiC, MOSFETs, JBSFETs, Short-circuit capability, Channeling implantation, Schottky metal, Schottky width, Deep P-well, Failure mechanism

I. INTRODUCTION

Junction Barrier Schottky diode integrated MOSFETs (JBSFETs) have been demonstrated to overcome issues associated with the inherent PN body diode of 4H-SiC MOSFETs as well as MOSFET issues pertaining to poor reverse recovery characteristics [1–3]. In previous literatures, it was difficult to fairly compare current-voltage characteristics of planar 1.2 kV 4H-SiC MOSFETs and JBSFETs due to the larger cell pitch of JBSFETs, resulting in a higher specific on-resistance ($R_{on,sp}$). Thanks to an innovative layout approach, the identical cell pitch was achieved, and thus the identical $R_{on,sp}$ was obtained in the recent generation of the planar JBSFETs [3]. Moreover, novel channeling implants were implemented to improve the short-circuit characteristics [3], [4].

This paper presents a comparison of short-circuit failure mechanisms of 1.2 kV 4H-SiC MOSFETs and Ti JBSFETs,

The information, data, or work presented herein was funded in part by the Office of Energy Efficiency and Renewable Energy (EERE), U.S. Department of Energy, the Vehicle Technologies Program Office Award Number DE-EE0008710.

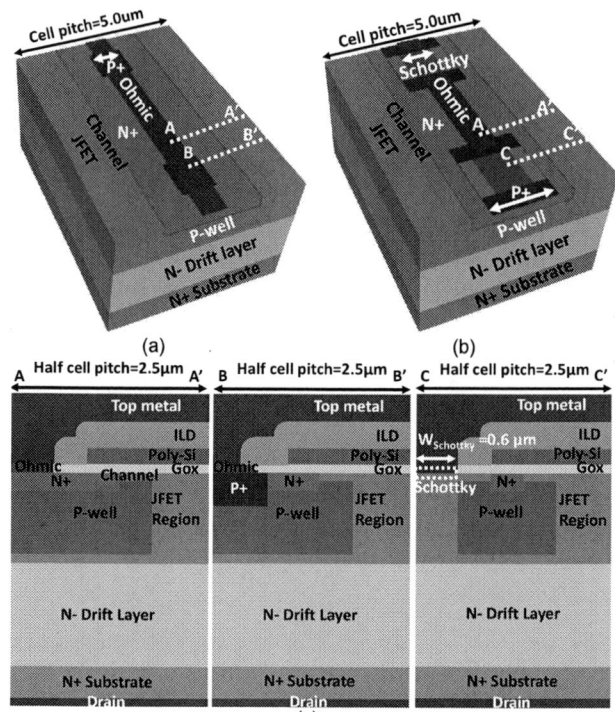

Fig. 1. (a) The layout approach of the MOSFET and (b) JBSFET, and (c) The cross-sectional views of A-A', B-B' from MOSFET (see (a)), and A-A', C-C' from JBSFET (see (b)).

which have the same cell pitch, and thus same $R_{on,sp}$. By adopting a deep P-well structure, superior short-circuit characteristics were achieved in both structures. However, at the same channel density, different behaviors for short-circuit characteristics were obtained due to the leakage current from the Schottky contact in the JBSFETs. In order to elucidate the short-circuit failure mechanisms, non-isothermal mixed-mode 2D TCAD device simulations were utilized. Moreover, based on the experimental results and analyses, solutions to further improve the short-circuit capabilities of JBSFETs are proposed.

II. DEVICE DESIGN

Fig. 1 shows the layout approach and cross-sectional views of the MOSFETs and Ti JBSFETs with deep P-well (~1.8 μm deep) implemented by channeling implantations [3]. The removal of the P+ source implant in the horizontal direction in JBSFETs allows the same cell pitch as the MOSFETs [3]. The tight cell pitch of 5.0 μm for both MOSFET and JBSFET was

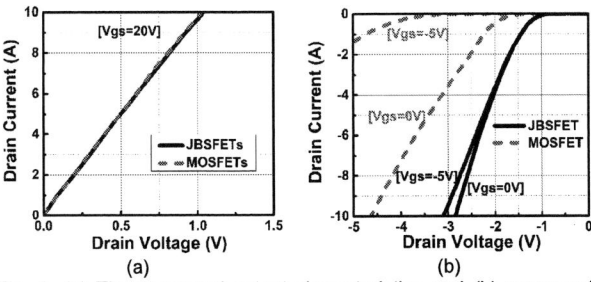

Fig. 2. (a) The measured output characteristics and (b) measured third quadrant characteristics of the fabricated 1.2 kV 4H-SiC MOSFETs and JBSFETs.

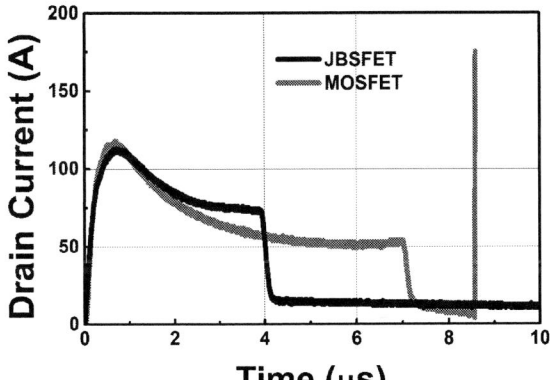

Fig. 3. The measured drain current of the fabricated MOSFETs and JBSFETs when the device is failure under short-circuit conditions. The short-circuit characteristics were evaluated at R_g of 20 Ω, V_{gs} of 20 V, and V_{ds} of 800 V.

accomplished through enhanced doping concentrations in the JFET region.

III. FABRICATION TECHNOLOGY

The MOSFETs were fabricated by SiCamore Semi, OR, U.S.A, using the same base process line [3], [4]. A 10 μm thick drift layer with N- type doping concentration of 8×10^{15} cm^{-3} on a N+ 4H-SiC substrate was used for the fabrication of the 1.2 kV MOSFETs and JBSFETs. The MOSFETs and JBSFETs were fabricated on the same, 4-inch wafer using the same mask set. Random implantation for Aluminum and Nitrogen were used to form the P-well/P+ source/JTE, and JFET/N+ source, respectively. Additional channeling implantation for Aluminum and Phosphorus during the P-well and JFET implantation were implemented for the formation of the deep P-well and JFET regions, respectively. At the conclusion of all the implantation steps, a 1650 °C, 10-min activation anneal with a carbon cap was conducted. A 50 nm thick gate oxide was formed, followed by a post oxidation anneal (POA) in NO ambient. A N-type polysilicon was deposited and patterned for the formation of the gate. After interlayer dielectric (ILD) was deposited, the ILD was then patterned and etched to make ohmic contact regions. Nickel (Ni) was deposited on the frontside, followed by an RTA for the silicidation process. Next, unsilicided Ni metals were removed and annealed by RTA. Ni was then deposited on the backside, followed by the same RTA process. After the

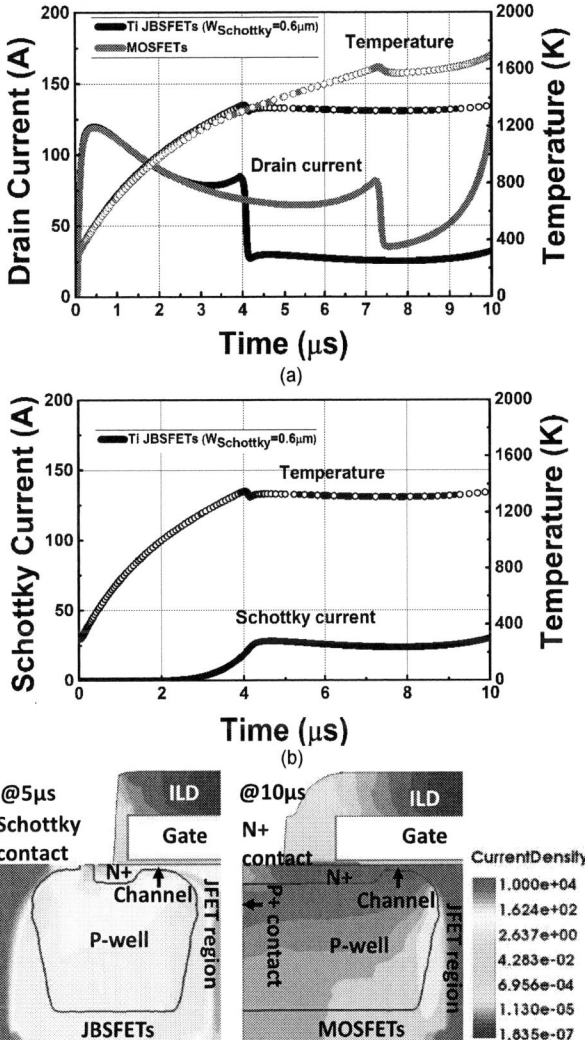

Fig. 4. (a) Simulated drain current and maximum junction temperature of MOSFETs and JBSFETs under a short-circuit condition. (b) Simulated Schottky current and maximum junction temperature of JBSFETs under a short-circuit condition. (c) Current density distribution of the JBSFET (left) and MOSFET (right) under short-circuit conditions. The short-circuit characteristics were evaluated at V_{gs} of 20 V and V_{ds} of 800 V.

formation of ohmic contacts to both N+ and P+ sources, the ILD (interlayer dielectric) on the Schottky area was etched while etching the ILD on Polysilicon, making a contact to the top metal (Ti/TiN/AlCu; Ti forms Schottky on SiC). Thus, there is no additional process required to make the JBSFET along with pure MOSFETs. The Schottky, source, and gate metal (Ti/TiN/AlCu) were deposited, patterned, and etched. For the passivation, Silicon nitride was deposited, patterned, and etched. Finally, a solderable metal stack was deposited on the backside.

IV. RESULTS AND DISCUSSIONS

Fig. 2 (a) shows the typical output characteristics of the fabricated 1.2 kV 4H-SiC MOSFETs and JBSFETs. Due

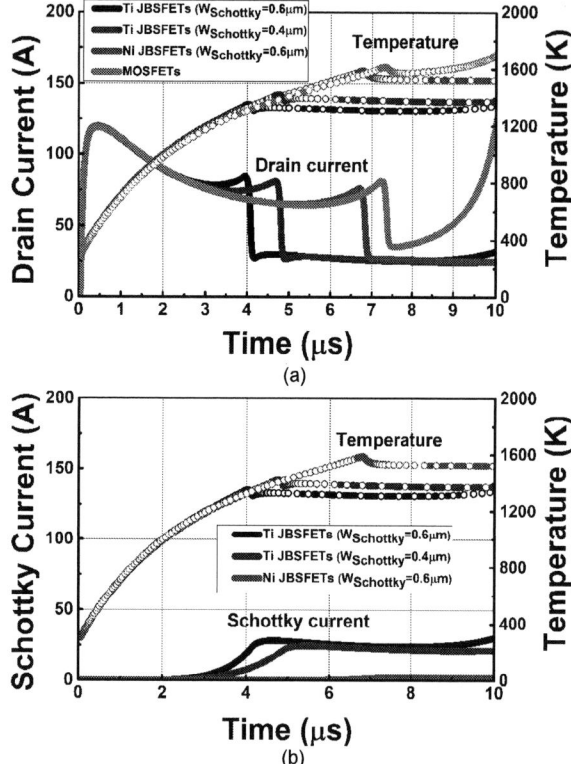

(a)

(b)

Fig. 5. (a) Simulated drain current and maximum junction temperature of MOSFETs and JBSFETs with different Schottky widths and metals under short-circuit conditions. (b) Simulated Schottky current and maximum junction temperature of JBSFETs with different Schottky widths and metals under short-circuit conditions. The short-circuit characteristics were evaluated at V_{gs} of 20 V and V_{ds} of 800 V.

to the same cell pitch, identical conduction behaviors were achieved; the specific on-resistance ($R_{on,sp}$) of the MOSFETs and JBSFETs are 4.23 and 4.24 mΩ-cm², respectively (On-wafer measurement; substrate resistance is included). $R_{on,sp}$ was extracted at V_{gs} of 20 V and V_{ds} of 0.1 V.

The measured third quadrant characteristics of the fabricated MOSFETs and JBSFETs are shown in Fig. 2 (b). Low knee voltage was achieved in the JBSFETs thanks to the operation of Ti Schottky contact before the body diode turned on. As a result, low forward voltage drop under third quadrant condition was achieved in the JBSFETs.

Fig. 3 shows the measured drain current of the fabricated MOSFETs and JBSFETs under short-circuit conditions. The short-circuit characteristics were evaluated at R_g of 20 Ω, V_{gs} of 20 V, and V_{ds} of 800 V. In order to obtain short-circuit withstand time (SCWT), the gate pulse width was increased until the devices fail. The MOSFETs achieved superior short-circuit characteristics due to the formation of the deep P-well structure [4]. Although the JBSFETs still provide a short circuit withstand time of ~4 µs, which is a reasonably high value when compared to other 4H-SiC planar MOSFETs, a shorter SCWT was obtained when compared to the MOSFETs fabricated on the same wafer. After the maximum drain current occurs, the drain current starts decreasing due to the reduction in the electron mobility at high temperatures.

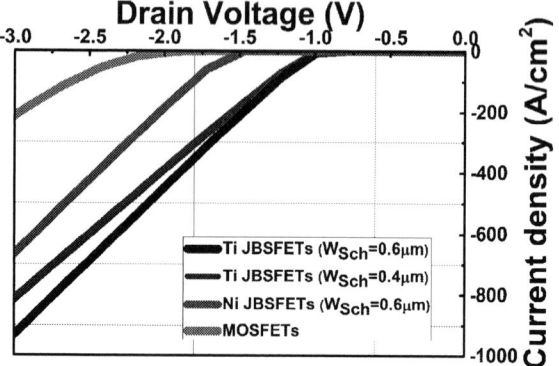

Fig. 6. Simulated third quadrant characteristics of the simulated MOSFETs and JBSFETs with different Schottky widths and metals.

However, the reduction in the drain current of the JBSFETs is smaller than that of the MOSFETs. This is due to additional leakage current from the Schottky contact. The short-circuit current in the MOSFET flows only through the n+ source. However, for the JBSFETs, the current flows not only through the n+ source, but also through the Schottky contact (leakage current) under the short-circuit condition.

In order to further elucidate the short-circuit failure mechanisms, non-isothermal device simulations were conducted. Fig. 4 (a) shows the simulated SC characteristics of the MOSFETs and Ti JBSFETs. Thanks to well optimized thermal-related simulation models [5], [6], the simulated results match well with the experimental results (i.e. Fig. 3). As shown in Fig. 4, different short-circuit failure mechanisms happen in the MOSFETs and JBSFETs. The junction temperature in the JBSFETs at the failure is lower than that observed in the MOSFETs. The Schottky leakage current is strongly governed by Schottky barrier height, electric field at the Schottky surface, and the junction temperature [7–9]. Due to the low Schottky barrier height obtained from Ti contact, high leakage current from Schottky contact occurs at a relatively low temperature.

Fig. 4 (b) shows the simulated Schottky current and the maximum junction temperature of the JBSFETs under short-circuit conditions. The Schottky current of the JBSFETs increases due to the increase of the junction temperature during the short-circuit condition. Even after the gate bias is removed, the Schottky current of JBSFETs does not diminish, as shown in Fig. 4 (b) and (c). Fig. 4 (c) shows the current density distribution of the JBSFETs and MOSFETs under short-circuit conditions. In the JBSFETs, the high leakage current flows through the Schottky regions as mentioned above. High temperatures cause the increase in the thermionic field emission (TFE), resulting in the increase of the leakage current from the Schottky contact during the short-circuit condition. In contrast to the JBSFETs, it should be noted that the MOSFETs failed due to the high current through the channel region, as shown in Fig. 4 (c). Due to the low potential barrier provided by the Schottky contact, and due to the high thermionic field emission through the Schottky contact at high temperatures, the JBSFETs failed at the lower junction temperature than the MOSFETs.

In order to further improve the short-circuit characteristics of JBSFETs, the leakage current from the Schottky contact should be suppressed. Higher work function metals such as Ni, Au, or Pt can be utilized as a Schottky metal [10], [11]. Moreover, the narrower Schottky width can be designed to reduce the electric field on the Schottky surface.

Fig. 5 (a) shows the simulated drain current and maximum junction temperature of the MOSFETs and JBSFETs with different Schottky widths and metals under short-circuit conditions. The short-circuit withstand time increases in the JBSFETs with narrow Schottky width or high work function metal due to the reduction in the electric field on the Schottky surface or the increase of SBH. Especially, the Ni JBSFETs achieved similar SCWT to the MOSFETs by suppressing the leakage current from Schottky region. The simulated Schottky current of the JBSFETs with different Schottky widths and metals under short-circuit conditions is shown in Fig. 5 (b). Thanks to the reduction in the TFE, the failure junction temperature increases in the Ti JBSFETs with the Schottky width of 0.4 μm and Ni JBSFETs, resulting in longer SCWTs.

It should be noted that there is a trade-off relationship between the forward voltage drop during the third quadrant operation and the SC leakage current when either using a different Schottky metal or design in the Schottky region. Fig. 7 shows simulated third quadrant characteristics of the MOSFETs and JBSFETs with different Schottky widths and metals. Ni for the Schottky results in a high knee voltage, and thus a high forward voltage drop. Ti JBSFETs with narrow Schottky width shows the identical knee voltage when compared to Ti JBSFETs with Schottky width of 0.6 μm, but the forward voltage drop increases due to the increase of JFET resistance.

Due to the trade-off relationship between the third quadrant and SC characteristics, optimization of the Schottky region is required; Schottky metals with too high work function and narrow Schottky width should be avoided to achieve the lowest forward voltage drop.

V. CONCLUSION

The short-circuit failure mechanisms of 1.2 kV 4H-SiC MOSFETs and Ti JBSFETs were compared. To fairly compare the short-circuit characteristics of the planar 1.2 kV 4H-SiC MOSFETs and JBSFETs, an innovative layout approach was applied in the JBSFETs to achieve the same cell pitch to the MOSFETs. Moreover, channeling implantation was implemented to form a deep P-well structure to improve the short-circuit characteristics. Thanks to the deep P-well structure, long SCWT was achieved in both MOSFETs and JBSFETs. However, with the same channel density, the JBSFETs provide shorter SCWT compared to equivalent MOSFETs. The difference in the short-circuit characteristics between the MOSFETs and JBSFETs is due to the different short-circuit failure mechanisms; the JBSFETs failed in the Schottky region under short-circuit characteristics while the failure of the MOSFETs happens in the channel region. In order to improve the short-circuit characteristics of the JBSFETs, a narrow Schottky width and high work function metal are proposed.

ACKNOWLEDGMENT

The authors would like to thank SiCamore Semi, Bend, OR for the fabrication of the devices. The authors acknowledge that the channeling implantations for the proposed devices were conducted by NISSIN ION EQUIPMENT CO.,LTD., Kyoto, Japan. The authors would like to thank Mr. Takashi Kuroi and Mr. Nobuhiro Tokoro, and the team for the valuable discussion on the principle of channeling implantation.

REFERENCES

[1] W. Sung, and B. J. Baliga, "On Developing One-Chip Integration of 1.2 kV SiC MOSFET and JBS diode (JBSFET)," *IEEE Transactions on Industrial Electronics*, vol.64, no.10, pp. 8206-8212, October, 2017. DOI: 10.1109/TIE.2017.2696515

[2] K. Han, A. Kanale, B. J. Baliga, B. Ballard, A. Morgan, and D. C. Hopkins, "New Short Circuit Failure Mechanism for 1.2kV 4H-SiC MOSFETs and JBSFETs," *in 2018 IEEE 6th Workshop on Wide Bandgap Power Devices and Applications (WiPDA)*, Oct. 31 – Nov. 2, 2018. DOI: 10.1109/WiPDA.2018.8569178.

[3] D. Kim, S. Y. Jang, S. DeBoer, A. J. Morgan, and W. Sung, "An Optimal Design for 1.2kV 4H-SiC JBSFET (Junction Barrier Schottky Diode Integrated MOSFET) With Deep P-Well," *IEEE Electron Device Letters*, vol. 43, no. 5, pp. 785–788, May 2022, doi: 10.1109/LED.2022.3162156.

[4] D. Kim, and W. Sung, "Improved Short-circuit Ruggedness for 1.2kV 4H-SiC MOSFET using a Deep P-well implemented by Channeling Implantation," *IEEE Electron Device Letters*, vol. 42, no. 12, Oct., 2021.
DOI: 10.1109/LED.2021.3123289

[5] D. Kim, A. J. Morgan, N. Yun, W. Sung, A. Agarwal, and R. Kaplar, "Non-Isothermal Simulations to Optimize SiC MOSFETs for Enhanced Short-circuit Ruggedness," *in 2020 IEEE International Reliability Physics Symposium (IRPS),* April 28 – May 30, 2020.
DOI: 10.1109/IRPS45951.2020.9128324

[6] Synopsys Inc., SentaurusTM Device User Guide, ver. K-2015.06, June 2015.

[7] B. J. Baliga, Fundamentals of Power Semiconductor Devices, 2nd ed. Cham, Switzerland: Springer, 2019, Chap. 4, pp. 171–206.

[8] T. Hatakeyama and T. Shinohe, "Reverse Characteristics of a 4H-SiC Schottky Barrier Diode," *Materials Science Forum*, vol. 389–393, pp. 1169–1172, 2002, doi: 10.4028/www.scientific.net/MSF.389-393.1169.

[9] R. Aiba, K. Matsui, M. Baba, S. Harada, H. Yano, and N. Iwamuro, "Demonstration of Superior Electrical Characteristics for 1.2 kV SiC Schottky Barrier Diode-Wall Integrated Trench MOSFET With Higher Schottky Barrier Height Metal," *IEEE Electron Device Letters*, vol. 41, no. 12, pp. 1810–1813, Dec. 2020, doi: 10.1109/LED.2020.3031598.

[10] A. B. Renz, V. A. Shah, O. J. Vanvasour, G. W. C. Baker, Y. Bonyadi, Y. Sharma, V. Pathirana, T. Trajkovic, P. Mawby, M. Antoniou, and P. M. Gammon, "The Optimization of 3.3 kV 4H-SiC JBS Diodes," *IEEE Transactions on Electron Devices*, vol. 69, no. 1, pp. 248–251, Jan. 2022. DOI: 10.1109/TED.2021.3129705.

[11] D. Perrone, M. Naretto, S. Ferrero, L. Scaltrito, and C. F. Pirri, "4H-SiC Schottky Barrier Diodes Using Mo-, Ti- and Ni-Based Contacts", *Materials Science Forum*, vol. 615-617, pp. 647–650, March, 2009.
https://doi.org/10.4028/www.scientific.net/MSF.615-617.647

978-1-6654-8901-0/22 $31.00 © 2022 IEEE

Failure Rate Calculation Due to Neutron Flux with SiC MOSFETs and Schottky Diodes

Dennis Meyer, Xuning Zhang, Reenu Garg, Bruce Odekirk, Steve Chenetz, Ehab Tarmoom, Kevin Speer

Microchip Inc.

Chandler, AZ, 85224 USA
Dennis.Meyer@microchip.com

Abstract—Neutron-induced failures in power electronics are a source of concern in many high-reliability applications. This paper describes how to estimate the device's decreased lifetime expectancy based on its usage. The physical mechanism, Failure in Time (FIT) statistics and FIT rate calculation methodology are presented. Several Microchip SiC MOSFETs and Schottky barrier diodes at different voltage ratings were tested, and the results for the excess FIT rate calculation based on various factors such as operation voltage, altitude, device area, latitude, and duty factor are presented.

Keywords—*Silicon Carbide, SiC MOSFET, SiC Diode, Neutron Flux, and Failure Rate Calculation.*

I. INTRODUCTION

Silicon Carbide (SiC) power devices have emerged as ideal candidates to replace Si in the high-power arena to help meet increasing demands for efficiency, power density, reliability, and lower system cost. SiC MOSFETs and diodes outperform Si devices in terms of electrical performance and offer advantages for potential high-density high-power converter design The benefits of SiC increase with device voltage rating, making SiC more suitable for high-current, high-voltage applications. However, the failure rate characterization, modeling, and calculation due to neutron flux with SiC MOSFETs and diodes are limited [1–7].

In practical applications, the terrestrial neutron radiation hardness of these SiC power devices must be examined to prevent unexpected system failures [8]. The neutron-induced failures start with high-energy particles entering the Earth's atmosphere. The very high-altitude terrestrial neutrons are byproducts of cosmic rays such as high-energy protons, alphas and heavy ions interacting with Earth's atmosphere. These interactions result in many different particles, such as neutrons, protons, pions, muons, electrons, and electromagnetic waves. Among these, the relative abundance and charge neutrality of the newly created neutrons give rise to a high flux traveling vast distances in the atmosphere, even reaching low altitudes [9]. The terrestrial neutron flux reaches peaks at around 60,000 feet. The integral neutron flux drops to about one-tenth of its peak value at 30,000 feet. It drops by another two orders of magnitude at sea level. However, the neutron flux at sea level, which is approximately <25 n/cm²/hr for E > 1 MeV, can still cause upsets and failures for electronics and power switches.

Furthermore, at sea level, approximately 95% of the cosmic shower constituents are neutrons [10].

Neutron flux-related failure mechanisms are well-studied in Si devices, and ample data available for numerous applications such as datacenter and railway traction applications [11–14]. The problem is predominantly associated with devices functioning at high altitudes due to the exponentially rising neutron flux levels as elevation increases. In certain applications where high-voltage high-power semiconductor devices are used at high altitudes, it is important to evaluate the terrestrial neutron radiation hardness of the devices to prevent unexpected system failures.

This paper presents the characterization results of several Microchip SiC MOSFETs and diodes at different voltage ratings for the excess FIT rate modeling and calculation in neutron failure sensitive applications. Section II presents the failure mechanism and test methodology, while Section III presents the modeling method of neutron induced FIT rate and the use of the FIT rate chart. Section IV presents the FIT rate test results of six different Microchip SiC MOSFETs and diodes. Section V compares the results of different devices, and Section VI concludes the whole paper.

II. PHYSICAL MECHANISM AND TEST METHODOLOGY

Particles entering the atmosphere are composed of numerous nuclear fragments such as protons, helium, heavier nuclear ions, high-energy gamma rays, and so on. Various nuclear interactions take place early in the process of entering the atmosphere. Energy is spread to other gas atoms. This reduces the energy of individual interactions and increases the number of particles involved, subject to the conservation of energy and momentum for the overall system. Deeper in the atmosphere, the most significant reaction product is neutrons.

Most of these particles are charged and are deflected by the Earth's magnetic field. For this reason, neutron-induced failures are latitude dependent. Neutrons are not charged. This allows them to penetrate through any solid without interaction with other atoms. Charged particles are subject to what is referred to as the Coulomb barrier. Neutrons, not being charged, can easily penetrate the barrier.

When a neutron penetrates through the semiconductor crystal, it sometimes collides with a lattice nucleus, kicking the

978-1-6654-8901-0/22 $31.00 © 2022 IEEE

atom out of the lattice and ionizing the atom. This collision generates free particles. These free particles ionize other atoms that are still fixed to the atom. As a result, a stream of holes/electron pairs along the path of the atom. If the free particles are in a voltage blocking region of a semiconductor, they will accelerate toward their terminal atoms. Electrons to the positive side, holes to the negative side. As they accelerate, they can generate more hole/electron pairs, multiplying their effect. By this multiplication, an avalanche is formed. Avalanching generates heat and can destroy the device.

The avalanche process is voltage dependent. The probability of a device capturing a neutron is dependent on the volume of the device. Five factors influence the probability of neutron-induced failure:

- Operating voltage in relation to breakdown voltage
- Altitude
- Device volume, with a slight dependence upon device structure
- Latitude
- Duty factor, the device must be blocking voltage for a failure to occur

Testing was performed by Dr. Akin Akturk, CoolCAD LLC, at Los Alamos Neutron Science Center (LANSCE). Devices are irradiated with a high-density neutron beam while biased. Their time to failure is measured. This time is then scaled to a nominal 13 n/cm^2/hr in the following sections.

Neutron flux is generated by irradiating a tungsten target with 800 MeV protons. Neutrons are generated through neutron spallation, resulting in various reaction products within the target. The neutrons are released in a non-directional manner, that is, isotropic, which means the flux density falls as the square of the distance from the tungsten target. Charged reaction products are filtered out, then the neutron flux is then presented through a 3-inch window to the devices under test.

Figure 1 shows the energy spectra of the neutron flux in relation to atmospheric flux at 40K feet multiplied by 3*10^5 as measured at the test point.

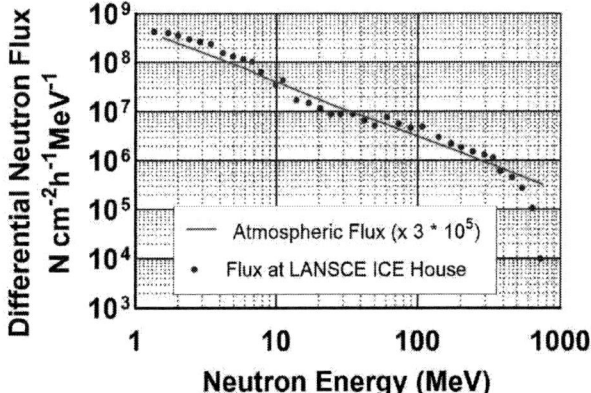

Figure 1: Neutron flux energy density

As can be seen, there is good agreement between the test energy levels and natural energy levels.

Figure 2 shows a placement sketch using TO247 devices, with a test setup in the rear view.

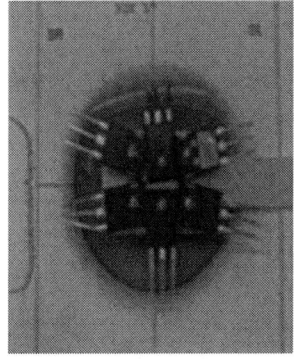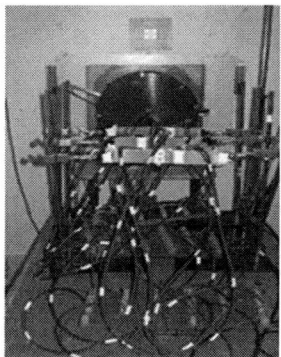

Figure 2: Placement sketch using TO247 devices and test setup rear view

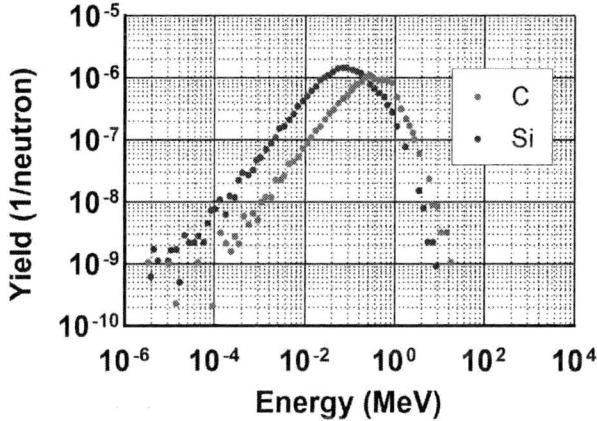

Figure 3: Relative energy required to displace atoms from a SiC crystal, IEEE TNS 59(4), 880, 2012.

The neutron flux described in the preceding section induces collisions with atoms within the SiC crystal. This interaction is dependent upon the energy and flux of the neutrons. Figure 3 shows the relative energy required to displace atoms from a SiC crystal. Silicon and carbon are the elements with the highest probability of a collision. As can be seen, collisions occur over a wide energy range.

III. FIT RATE MODELING AND USING FIT RATE CHARTS

A. FIT Rate Statistics

The semiconductor industry normalizes the FIT (Failure in Time) rate at over 10^9 hours. For example, a FIT rate of one means that at an average of 10^9 device-hours of functioning, one device is expected to fail. It can also mean that, out of 10^9 devices, it is expected that an average of one device will fail in one hour. Neutron-induced failures are a random occurrence that fits with this type of model. There are other sources of failure, which are not considered in this document:

- Infant mortality aspect of reliability curves
- Extended usage outside of reliability curves
- Failures during normal usage related to stress caused by humidity or heat. These stresses may be expressed as a FIT rate or a wear out mechanism. These depend on the way stress results in failure

Some properties of FIT rates:
- FIT rates can be added. In a system with multiple devices, the FIT rate of the individual components is added
- FIT rates can be summed over time. An average FIT rate can be assigned by weighing the FIT rate under various operating conditions multiplied by the fraction of time the device is at that operating condition. Neutron induced failures are an example of this. If an aircraft spends 30% of its time at an altitude of 10 Km and 70% at 1 Km, the profile adjusted FIT rate equals 0.3 * FIT(10 Km) + 0.7 * FIT(1 Km)
- The system MTBF (Mean Time Between Failures) = 1 / the total FIT rate of the system

B. FIT Rate Calculation

There are many ways to calculate neutron flux density. This is due to the many sources of information, the variation of measured results, and the overall complexity of the problem. The equations for latitude and altitude dependence are described here.

LATITUDE DEPENDENCE

IEC TS 62396-1 2006, section 5.3.4 provides neutron flux as a function of latitude. At a latitude of 45 degrees, this was curve fit to a tanh function and normalized to a value of one. 45 degrees is used to match the latitude of NASA (National Aeronautics and Space Administration) and Boeing altitude dependence measurements. The latitude correction formula is:

$$Flat\_norm = 0.857 + 0.559 * \tanh\left(\frac{latitude - 42}{11.5}\right) \quad (1)$$

The range of this function is approximately 0.30 to 1.40, domain 0 to 90. The result is unitless. The function plots are as follows.

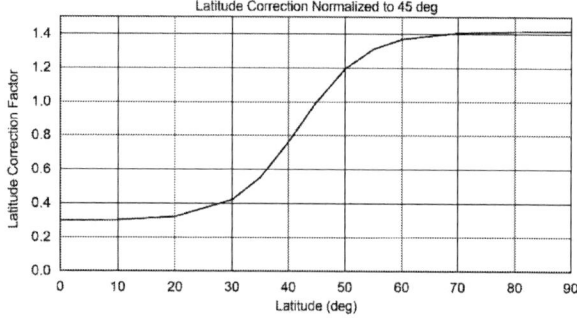

Figure 4: Latitude correction normalized to 45 degrees.

ALTITUDE DEPENDENCE

There are two methods to calculate altitude dependence. Due to high altitude effects, both are only useful up to approximately 15,000 meters (50,000 feet). The first is by ABB referenced in ABB application notes 5SYA 2046-03 and 5SYA 2042-09.

$$Falt\_norm = \exp\left[\frac{1-\left(1-\frac{h}{44300}\right)^{5.26}}{0.143}\right] \quad (2)$$

The altitude "h" is measured in meters with a range from 0 to 15 km. This equation is unitless and is normalized to a value of one at sea level.

Equation 1 and Equation 2 can be used together to represent absolute flux density. For this, Equation 2 is modified with a multiplier of 13.3 to express flux density in measurement units of n/cm²/hr. This matches roughly with Boeing data presented in IEC TS 62396-1 2006. The resulting equation is presented below.

$$Falt = 13.3 * \exp\left[\frac{1-\left(1-\frac{h}{44300}\right)^{5.26}}{0.143}\right] \quad (3)$$

The altitude "h" is measured in meters with a range from 0 to 15 km. As was explained above, the measurement units of this resulting function are n/cm²/hr.

A second equation provided by CoolCAD follows. CoolCAD literature provides the equation in measurement units of n/cm²/s with the altitude in units of feet. The final form provided here is adjusted to measurement units of n/cm²/hr and altitude in meters:

$$Falt = 3600 \times$$
$$10^{0.2093 - 2.41511*log10(H) + 0.50386*log10(H)*log10(H)} \quad (4)$$

The altitude "H" is measured in feet. This equation is discontinuous at zero. When calculating FIT below 1000 feet it is recommended to use 1000 feet.

The graphic below compares the Boeing simplified model presented in IEC TS 62396-1 with the two equations above by ABB and CoolCAD. As can be seen in the NASA data, the flux density is very nonlinear above about 15,000 meters, which is the reason these equations should not be used in that case. The raw NASA data is available in IEC TS 62396-1.

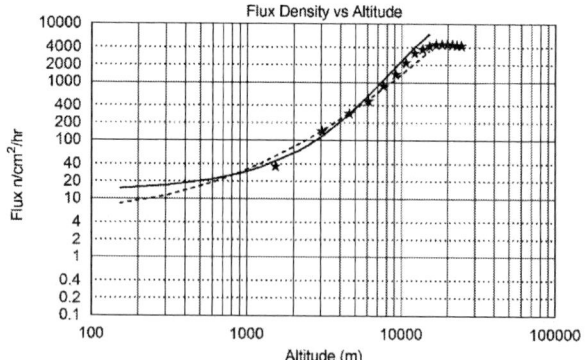

Figure 5: Altitude correction comparison. Star: NASA data, solid line: ABB equation, dotted line: CoolCAD equation

IV. TEST RESULTS

The following six graphs present experimental data for one device from each of the three voltage MOSFET and diode categories. These devices are part of Microchip's current generation of devices. Experimental data is presented together with an average FIT equation. The FIT rate graphics and equations presented are all normalized to a measurement unit of 13 n/cm²/hr. Based on the previous section, the average flux density is calculated by association with the application using Equation 1 together with Equation 3 or Equation 4.

The overall FIT rate = the measured FIT rate of the device family x a die size scaling factor x × the application flux density divided by 13 n/cm²/hr. For details regarding the use of the normalized FIT rate charts for actual FIT rate calculations for specific part numbers, please refer to [15].

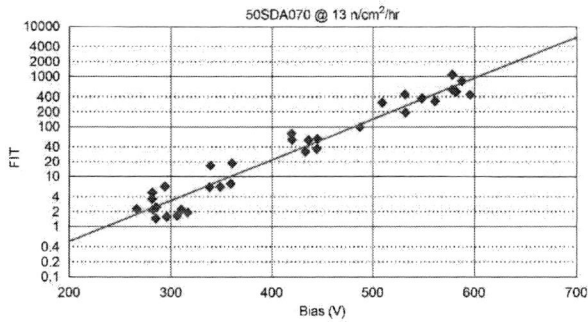

Figure 6.1 FIT measurements as a function of operation voltage for 700V 50A diode, part number MSC050SDA070

The curve fit shown above is
$$FIT = \exp(0.0188 \times V - 4.44)$$

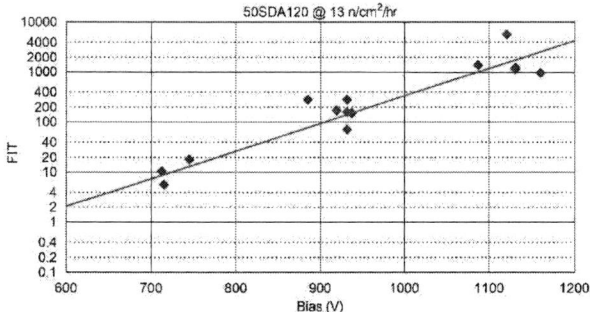

Figure 6.2 FIT measurements as a function of operation voltage for 1200V 50A diode, part number MSC050SDA120

The curve fit shown above is
$$FIT = \exp(0.0128 \times V - 6.88)$$

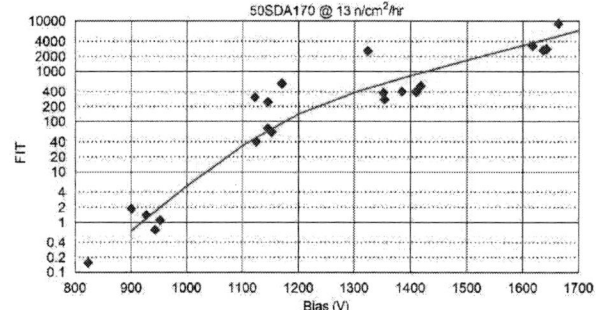

Figure 6.3 FIT measurements as a function of operation voltage for 1700V 50A diode, part number MSC050SDA170

The curve fit shown above is
$$\frac{1}{FIT} = \frac{1}{FIT1} + \frac{1}{FIT2}$$
$$FIT1 = \exp(0.0067 \times V - 2.72)$$
$$FIT2 = \exp(0.0211 \times V - 19.4)$$

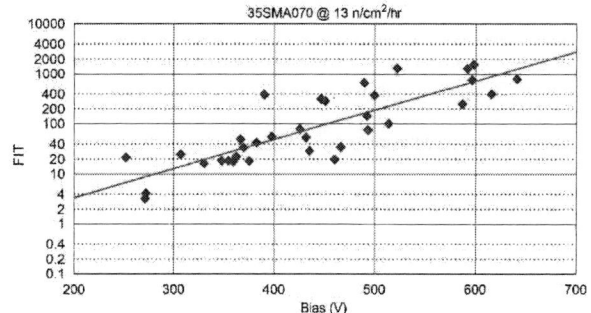

Figure 6.4 FIT measurements as a function of operation voltage for 700V 35mohm MOSFET, part number MSC035SMA070

The curve fit shown above is
$$FIT = \exp(0.0134 \times V - 1.473)$$

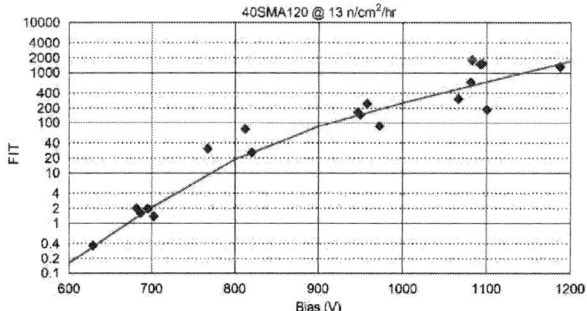

Figure 6.5 FIT measurements as a function of operation voltage for 1200V 40 mΩ MOSFET, part number MSC040SMA120

The curve fit shown above is
$$\frac{1}{FIT} = \frac{1}{FIT1} + \frac{1}{FIT2}$$

978-1-6654-8901-0/22 $31.00 © 2022 IEEE

$$FIT1 = \exp(0.0093 \times V - 3.73)$$
$$FIT2 = \exp(0.0266 \times V - 17.7)$$

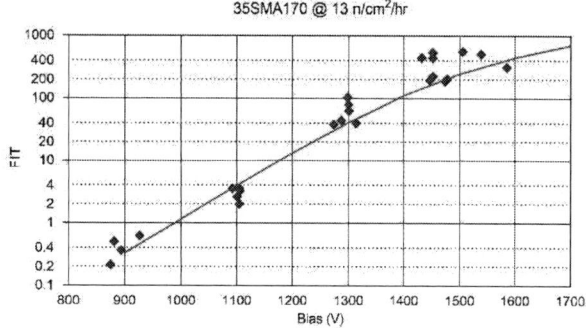

Figure 6.6 FIT measurements as a function of operation voltage for 1700V 35mohm MOSFET, part number MSC035SMA170

The curve fit shown above is

$$\frac{1}{FIT} = \frac{1}{FIT1} + \frac{1}{FIT2}$$
$$FIT1 = \exp(0.0126 \times V - 12.4)$$
$$FIT2 = \exp(0.00326 \times V + 1.1)$$

V. Result Comparison among Different Devices

Due to different material selection and different design details, the FIT rate shows different voltage dependence across the product line. The three product voltage lines, 700V, 1200V and 1700V MOSFETs and diodes, were designed at different times with differing design rules. The following graph plots the FIT rates of the various devices normalized to a space charge region size of 1 cm². Dotted lines are diodes; solid lines are MOSFETs. The vertical lines represent the average measured breakdown voltage of the devices tested. Breakdown voltage is commonly thought to be directly associated with the electric field stress of a device.

The change in slope of the curves occurs at about the same voltage as the punch through. This is probably not a coincidence.

The evolution of product designs involves trade-offs of various parameters. As SiC devices have evolved, the average breakdown voltage headroom overrated voltage between product generations has typically been reduced to provide better on resistance ratings. The radiation dependence of a family is determined by an average of the FITs. Reducing the average breakdown voltage will result in a higher average FIT. The current generation of 1200V MOSFETs has a nominal breakdown voltage of around 1400V. They are margined over 1200V to optimize yield at the expense of somewhat higher resistance. Improvements make it possible to reduce the design margin, but the impact on increased FIT due to neutron flux needs to be considered.

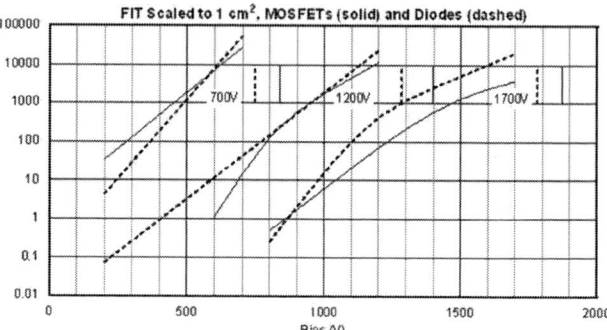

Figure 7. FIT rates of the various devices normalized to a space charge region size of 1 cm²

VI. Conclusions

Neutron-induced failures in power electronics are of concern in applications where high reliability is a must-have requirement. This paper describes the method by which the decreased lifetime expectancy can be estimated based upon the usage of the device. The physical mechanism and Failure in Time (FIT) statistics and FIT rate calculation methodology are presented. Several Microchip SiC MOSFETs and Schottky barrier diodes of different voltage classes are tested, and the FIT rate is calculated based on factors including operational voltage, altitude, device area, latitude, and duty factor.

References

[1] K. Rashed, R. Wilkins, A. Akturk, R. C. Dwivedi, and B. B. Gersey, "Terrestrial neutron induced failure in silicon carbide power MOSFETs," in Proc. IEEE Radiat. Effects Data Workshop, Jul. 2014, pp. 1–4.

[2] A. Akturk, R. Wilkins, and J. McGarrity, "Terrestrial neutron induced failures in commercial SiC power MOSFETs at 27C and 150C," in Proc. IEEE Radiat. Effects Data Workshop, Jul. 2015, pp. 115–119.

[3] H. Asai et al., "Tolerance against terrestrial neutron-induced single-event burnout in SiC MOSFETs," IEEE Trans. Nucl. Sci., vol. 61, no. 6, pp. 3109–3114, Dec. 2014.

[4] H. Asai et al., "Terrestrial neutron-induced single-event burnout in SiC power diodes," IEEE Trans. Nucl. Sci., vol. 59, no. 4, pp. 880–885, Aug. 2012.

[5] A. Griffoni et al., "Neutron-induced failure in silicon IGBTs, silicon super-junction and SiC MOSFETs," IEEE Trans. Nucl. Sci., vol. 59, no. 4, pp. 866–871, Aug. 2012.

[6] C. Abbate, G. Busatto, F. Iannuzzo, A. Sanseverino, and F. Velardi, "Irradiation tests on SiC power devices," in Proc. Apollo Workshop, Rome, Italy, Dec. 2013.

[7] Galloway, Kenneth F., Arthur F. Witulski, Ronald D. Schrimpf, Andrew L. Sternberg, Dennis R. Ball, Arto Javanainen, Robert A. Reed, Brian D. Sierawski, and Jean-Marie Lauenstein. 2018. "Failure Estimates for SiC Power MOSFETs in Space Electronics" Aerospace 5, no. 3: 67.

[8] A. Akturk, R. Wilkins, J. McGarrity and B. Gersey, "Single Event Effects in Si and SiC Power MOSFETs Due to Terrestrial Neutrons," in IEEE Transactions on Nuclear Science, vol. 64, no. 1, pp. 529-535, Jan. 2017.

[9] J. L. Barth, C. S. Dyer, and E. G. Stassinopoulos, "Space, atmospheric, and terrestrial radiation environments," IEEE Trans. Nucl. Sci., vol. 50, no. 3, pp. 466–481, Jun. 2003.

[10] J. F. Ziegler, "Terrestrial cosmic rays," IBM J. Res. Develop., vol. 40, no. 1, pp. 19–39, Jan. 1996.

[11] D. L. Oberg, J. L. Wert, E. Normand, P. P. Majewski, and S. A. Wender, "First observations of power MOSFET burnout with high energy neutrons," IEEE Trans. Nucl. Sci., vol. 43, no. 6, pp. 2913–2920, Dec. 1996.

[12] R. Sheehy, J. Dekter, and N. Machin, "Sea level failures of power MOSFETs displaying characteristics of cosmic radiation effects," in Proc. IEEE 33rd Annu. Power Electron. Specialists Conf., vol. 4. Jun. 2002, pp. 1741–1746.

[13] J. L. Titus, "An updated perspective of single event gate rupture and single event burnout in power MOSFETs," IEEE Trans. Nucl. Sci., vol. 60, no. 3, pp. 1912–1928, Jun. 2013.

[14] E. Normand et al., "Single event upset and charge collection measurements using high energy protons and neutrons," IEEE Trans. Nucl. Sci., vol. 41, no. 6, pp. 2203–2209, Dec. 1994.

[15] Dennis Meyer, "Excess Failure Rate Calculation Due to Neutron Flux with SiC MOSFETs and Schottky Diodes", application note, AN4589/ DS00004589A, Microchip Inc, 2002.

[16] K. Matocha, I. -H. Ji, X. Zhang and S. Chowdhury, "SiC Power MOSFETs: Designing for Reliability in Wide-Bandgap Semiconductors," 2019 IEEE International Reliability Physics Symposium (IRPS), 2019, pp. 1-8.

Scaling of EPC's 100 V Enhancement-Mode Power Transistors

Gordon Stecklein, Jordan Green, Christopher Wong, Joe Cao and Bob Beach
Efficient Power Conversion Corp.
El Segundo, CA, USA

Email: {gordon.stecklein, jordan.green, christopher.wong, joe.cao, bob.beach}@epc-co.com

Abstract— **The same core device model is shown to accurately reproduce the current-voltage and capacitance-voltage characteristics of enhancement-mode GaN 100 V power transistors of various sizes. Using linear scaling, excellent agreement with measurements is achieved over an order of magnitude variation in total gate width. Fractional variation in on-resistance is shown to decrease with increasing transistor size, with implications for integrated circuit-sized transistors where gate width decreases by up to 5 orders of magnitude.**

Keywords—HEMT, E-mode, GaN, Scaling, Variation, Device Modeling

I. Introduction

The product portfolio offered by Efficient Power Conversion, Corp. (EPC) includes 100 V power transistors tailored for different applications. Various application requirements are at times mutually exclusive, such as needing either fast switching ($Ciss$ < 0.3 nF) or high maximum currents ($Idmax$ > 400 A). To achieve these results, each device was designed by scaling a fixed unit cell according to the required specification. Scaling is achieved by increasing the transistor's total gate width, which is the dimension perpendicular to current flow, either by direct extension or by placing multiple transistor fingers in parallel. Electrical device properties are expected to follow linear scaling, e.g., on-resistance will vary inversely with total gate width while capacitances will increase linearly with total gate width.

The effect of device size on the variation of electrical characteristics informs expected device performance and may be useful in identifying process improvements. While relationships between device size and matching are known for MOSFETs [1], the variation of enhancement-mode GaN FET device parameters with size has not received as much attention. Such studies are crucial for the continued development of integrated circuit (IC)-sized GaN transistors.

In this study, the electrical characteristics of transistors fabricated with a fixed process and unit cell are shown to exhibit the expected linear scaling over a wide range of total gate width. The scaling of median properties shown here enables future investigations into how variation depends on device size.

II. Measurement Methodology

A. Power FET measurements

Power transistors in both chip-scale and packaged form factors were soldered to mini-PCBs in preparation for measurement. Parametric measurements used a Keithley 2612A Source Meter. Typical sample sizes for parametric measurements were ~10,000 transistors per gate width. Current-Voltage (IV) measurements used a Tektronix 371A High Power Curve Tracer at a low duty cycle with 300 us pulse width. Capacitance-Voltage (CV) measurements used a Keithley HP 4192A LF Impedance Analyzer at 1 MHz, with DC bias provided by a Keithley 237 High Voltage Source-Measure Unit. Data were collected in uncontrolled ambient (~25 C). Typical sample sizes for detailed IV and CV measurements were 15 transistors per gate width.

B. IC FET measurements

IC FETs were measured in wafer form using a Cascade Summit 12000B semi-automated prober and a Keysight B1505A Power Device Analyzer at 25 C. Typical sample sizes were 100 transistors per gate width.

III. Power FETs

A. Results

Recent 100 V enhancement-mode power transistors released by EPC include EPC2070, EPC2071, EPC2088, EPC2204, EPC2302, and EPC2306. Example IV and CV measurements are shown in Figs. 1-3, where output capacitance is plotted as a function of drain-source voltage Vds with $Vg = Vs = 0$ and gate capacitance is plotted as a function of gate-source voltage Vgs with $Vd = Vs = 0$, where d, g, and s denote the drain, gate, and source ports, respectively. Output capacitance $Coss$ is the sum of two-port capacitances $Cds + Cdg$. Similarly, gate capacitance $Cg = Cgs + Cgd$.

978-1-6654-8901-0/22 $31.00 © 2022 IEEE

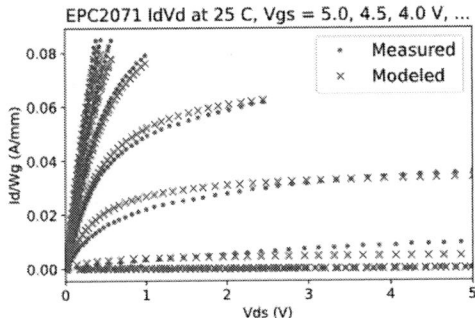

Fig. 1 Example output curve data and fit.

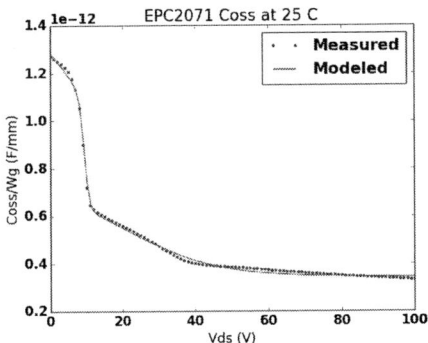

Fig. 2 Example Coss data and fit.

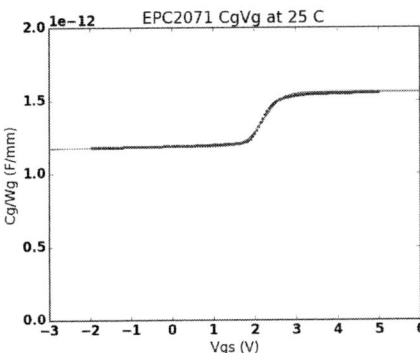

Fig. 3 Example Cg data and fit.

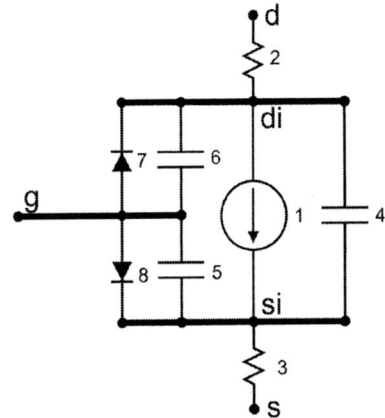

Fig. 4 Sub-circuit device model.

In this empirical model, total gate width Wg enters as

$$I_1 = Wg * a(Vg, Vdi, Vsi, T), \qquad (1)$$

$$R_2 + R_3 = b / Wg, \qquad (2)$$

$$C_4 = Wg * f_4(Vdi, Vsi), \qquad (3)$$

$$C_5 = Wg * f_5(Vg, Vsi), \qquad (4)$$

$$C_6 = Wg * f_6(Vg, Vdi), \qquad (5)$$

where Vx is the voltage at node x, T is temperature, and b is a constant.

The gate width-independent IV relation of element 1 is given by

$$a(Vg, Vdi, Vsi) = \frac{A1*Vdisi*\ln\left[1+\exp\left(\frac{Vgsi-k2}{k3}\right)\right]}{1+[x0_0+x0_1*Vgsi]*Vdisi}, \qquad (6)$$

where nominal temperature and $Vdi > Vsi$ are assumed for clarity.

Each voltage-dependent capacitance follows the general form

$$f(Vx, Vy) = c_0 + \sum_i \frac{c_i*\exp[(Vxy-u_i)/w_i]}{1+\exp[(Vxy-u_i)/w_i]}, \qquad (7)$$

B. Fitting

While physics-based compact models are available for GaN high electron mobility transistors (HEMTs) [2][3], EPC's power FET devices were fit to an empirical subcircuit model to ensure compatibility with open-source circuit simulators.

A median device was identified for each part. Its IV and CV data were fit to the subcircuit model shown in Fig. 4, where element 1 is a nonlinear voltage-dependent current source representing the transistor channel, elements 2 and 3 are parasitic resistances representing the contact and access region resistances, elements 4-6 are voltage-dependent capacitors, and elements 7 and 8 are gate leakage diodes.

978-1-6654-8901-0/22 $31.00 © 2022 IEEE

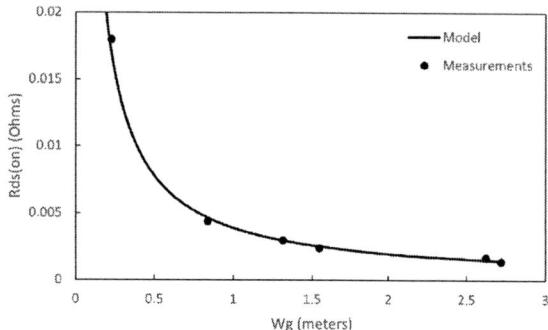

Fig. 5 Scaling of power FET on-resistance at Vgs = 5 V.

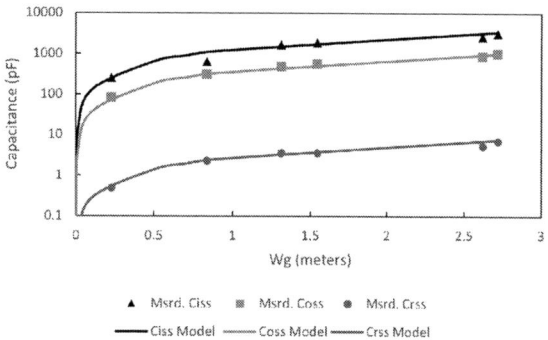

Fig. 6 Scaling of power FET capacitances at Vds = 50 V.

where c_0 is a constant and c_i is the magnitude of a feature in the CV data that occurs at voltage $Vxy = u_i$, e.g., corresponding to the voltage at which the channel is enhanced or a field-plate pinches off. The width of each feature is described by fitting parameter w_i.

Additional details, including device models, are provided on EPC's website.

C. Scaling

A single set of parameters accurately models the median device's IV behavior over the full range of Wg. Fig. 5 compares on-resistance $Rds(on)$ and device size, where on-resistance is measured at Vgs = 5 V and Id/Wg = 20 A/m. Similarly, the median device's CV behavior is accurately simulated over the full range of Wg. Fig. 6 compares $Ciss$, $Coss$, and $Crss$ with device size, where these capacitances are measured at Vds = 50 V. At all values of Wg, agreement is observed between the model and measurements for IV and CV characteristics.

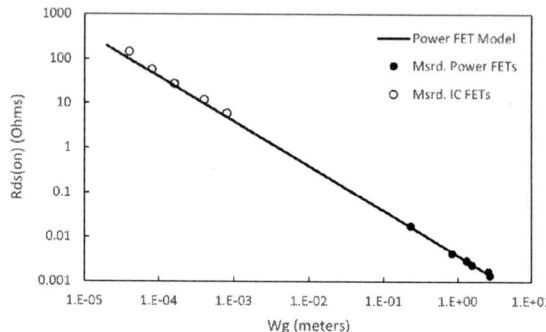

Fig. 7 Comparison of IC FET on-resistance to the scaled power FET model.

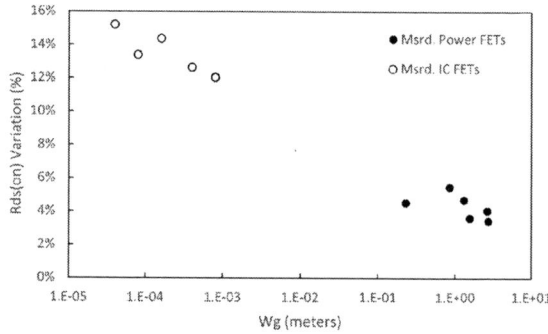

Fig. 8 Scaling of fractional variation in on-resistance.

IV. COMPARISON TO IC FETS

A. Scaling

100 V FETs with relatively small Wg are used in EPC ICs. The cell design is similar to but not exactly the same as for the power FETs. As described elsewhere, the ASM-HEMT compact model can be used to accurately model EPC's IC FETs [4]. One rough gauge of the quality of the 100 V power FET scaling model is to compare it to measurements of the IC FETs, as shown in Fig. 7. Reasonable agreement is observed.

B. Variation

Given reasonable scaling of median device behavior, future studies may focus on how variation in electrical properties scales with device size. For example, Fig. 8 compares the conventional Gaussian standard deviation of the measured on-resistance across all FET sizes, where fractional variation is quantified as a percentage of median. A modest size dependence is observed, where the standard deviation decreases by a factor of ~3x as gate width increases 5 orders of magnitude.

V. CONCLUSIONS

Linear scaling is shown to apply to enhancement-mode GaN 100 V power FETs. IV and CV data are well-described by a single device model, which can be extrapolated to IC-sized FETs with reasonable agreement. These results enable future studies of size-dependent variation.

REFERENCES

[1] M. J. M. Pelgrom, A. C. J. Duinmaijer, and A. P. G. Welbers, "Matching properties of MOS transistors," IEEE J. Solid-State Circuits, vol. 24, no. 5, pp. 1433–1440, Oct. 1989.

[2] S. Khandelwal et al., "Robust surface-potential-based compact model for GaN HEMT IC design," IEEE Trans. Electron Devices, vol. 60, no. 10, pp. 3216–3222, Oct. 2013.

[3] U. Radhakrishna, T. Imada, T. Palacios, and D. Antoniadis, "MIT virtual source GaNFET-high voltage (MVSG-HV) model: A physics based compact model for HV-GaN HEMTs," Phys. Status Solidi C, vol. 11, nos. 3–4, pp. 848–852, Mar. 2014.

[4] S. Khandelwal, G. Stecklein and T. Herman, "Modeling Substrate Voltage Effects on GaN I-V Characteristics with ASM-HEMT model," 2022 IEEE Applied Power Electronics Conference and Exposition (APEC), 2022, pp. 1731-1734.

Advancement in Integration for GaN Power ICs

Tom Ribarich
Strategic Marketing
Navitas Semiconductor
El Segundo, CA 90245, USA
tom.ribarich@navitassemi.com

Stephen Oliver
Corporate Marketing
Navitas Semiconductor
El Segundo, CA 90245, USA
tom.ribarich@navitassemi.com

Marco Giandalia
IC Design
Navitas Semiconductor
El Segundo, CA 90245, USA
tom.ribarich@navitassemi.com

Llew Vaughan-Edmunds
Marketing
Navitas Semiconductor
El Segundo, CA 90245, USA
tom.ribarich@navitassemi.com

Abstract—**In this paper, we discuss how advancements in the integration of gallium nitride (GaN) power ICs leads to faster switching, higher efficiencies, simplifier designs and improved reliability. Navitas' integrated GaNSense™ technology features provide real-time sensing and protection to achieve these new levels.**

INTRODUCTION

Gallium Nitride (GaN) is a next-generation 'wide-bandgap' power semiconductor which have superior characteristics compared to older silicon equivalents, including the ability to switch up to 20x faster and increase power density by over 3x times. GaN power devices are capable of extremely high switching speeds, therefore monolithically integrating the driver stage and power stage is key to achieve highest switching frequencies and performance, whilst keeping the gate drive signal clean, and eliminating any unwanted noise to effect the control and reliability of the device.

GaNFast™ power ICs have achieved unprecedented application power densities (2 W/cc) due to their integrated and regulated gate drive technology. Discrete GaN FETs continue to have slow adoption due to ruggedness & reliability concerns, plus difficulties with designing robust gate drive circuits. In addition to very low ESD tolerance, the exposed gate of the discrete GaN FET has a high susceptibility to device, package and PCB parasitics. These 'hidden' parasitics are the well-known enemy of fast switching and can cause transient ringing, glitching and destruction. Unregulated on- and off-state gate voltage levels are also concerning due to wide voltage variations that can cause poor reliability and drastically reduce lifetime. As a result, discrete GaN FET solutions must add large filter circuits that prevent high frequency operation and prohibit the benefits of GaN. With GaNFast power ICs, the GaN power FET, gate drive circuit and gate voltage regulator are monolithically-integrated on the same IC chip. These circuit blocks provide excellent immunity against switching noise and parasitics, eliminates gate voltage variations, and delivers a rugged and robust solution that is easy-to-use and unlocks high frequency operation for next generation power-supply designs. This article exposes these 'hidden' parasitics, explains why they are concerning, and compares the robustness and reliability of exposed gate discrete GaN approaches versus GaNFast integrated gate drive solutions during different switching conditions.

I. NECESSITY OF MONOLITHIC GATE DRIVE

A. Hidden Resonant Parasitic Network

For enhancement-mode (eMode, normally off) discrete GaN FETs, a high level of sensitivity exists at the gate due to a low threshold voltage and a narrow gate voltage operating range. To make things worse, there is a 'hidden' resonant network of parasitic resistors, inductors and capacitors around the FET formed by the PCB, GaN FET package, and GaN device itself. An example of this parasitic resonant network consists of a discrete GaN FET mounted on a standard FR4 PCB (Figure 1a). This 'deceptively simple' circuit seems like it will easily work for generating the necessary square-wave gate voltage. A more detailed 'extracted' circuit including parasitics reveals a much more complex network surrounding the discrete GaN FET. PCB material, PCB copper traces, GaN FET package leadframe and wire bonds, and external component parasitics all form parasitic resistors, inductors and capacitors at the gate, drain and source. The intrinsic capacitors of the GaN FET device itself (CGS, CGD, CDS) also form part of the parasitic network.

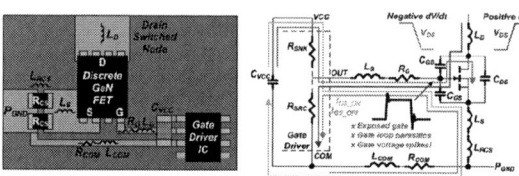

(a) Discrete GaN FET on PCB and 'hidden' parasitic resonant network with switching current paths

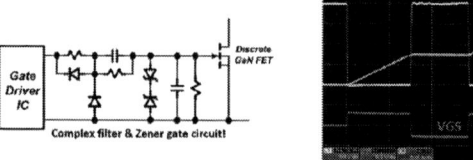

(b) Complex filter and Zener gate circuit, and, gate voltage ringing and glitching waveforms

Fig. 1. Discrete GaN FET on PCB, parasitic resonant network, and ringing & glitching waveforms

This parasitic resonant network becomes excited during each dV/dt switching transition of the drain voltage causing currents to flow around various parasitic resonant circuit loops. These resonant currents (Figure 1a) will cause excessive voltage spikes and ringing to occur at different parasitic circuit nodes, including the gate of the GaN FET. During turn-on, a negative dV/dt occurs at the drain node causing current to flow into the gate node and through C_{GD} to the drain. During turn-off, a positive dV/dt occurs at the drain node causing current to flow through C_{GD} and out of the gate node. The external gate drive signal itself will also excite the gate and source loop parasitics. All of these switching conditions will cause voltage spikes and ringing (Figure 1b) at the gate that will turn the discrete GaN FET on or off unexpectedly (glitching) and can cause poor reliability, reduced lifetime or destruction of the GaN FET. These constraints make these devices very susceptible to high-frequency and high dv/dt noise from the surrounding switched-mode converter circuit. To mitigate these problems, external components are placed at the gate that include a gate capacitor to filter the voltage spikes, a series capacitor and Zener diodes to generate positive and negative gate voltage levels, and additional resistors to help dampen the gate ringing induced by the gate loop inductance (Figure 1b). Adding external components will not completely eliminate these problems due to component tolerances, temperature variation, and internal package parasitics, so the internal gate of the device can still have spikes and ringing even if the external gate pin waveform looks clean. Also, the gate is still exposed on the PCB making it always susceptible to external voltage surges and transients, or dV/dt occurring elsewhere in the circuit, creating high risk for field failures. Dealing with all these problems, while trying to increase switching frequency and increase power density, can be a major roadblock to bringing a product out of the lab, into production, and into the field. A simpler and more reliable solution is needed without adding additional design difficulties or robustness concerns.

B. Integration of Gate Driver

It is well known that GaN offers the ability to integrate additional circuits monolithically on the same IC as the power device, so it makes good sense to include the gate drive circuitry in order to eliminate these external gate drive challenges. GaNFast power ICs integrate the GaN power FET together with the gate driver, the gate voltage regulator, and voltage reference (Figure 2). During turn-on, current flowing through C_{GD} due to the negative dV/dt at the drain no longer flows through the gate loop as it does with external gate drive circuits and exposed gates. Instead, the current flows from gate drive supply voltage (V_{DD}), through the pull-up transistor of the gate driver, and through C_{GD} to the drain. During turn-off, current flowing through C_{GD} due to the positive dV/dt at the drain flows through the pull-down transistor of the gate driver and then to the source. The sink and source currents are

supplied on-chip directly from the gate drive supply voltage (V_{DD}), not from the external gate driver as with discrete GaN FETs. The integrate gate drive circuit bypasses the external gate drive loop completely and eliminates gate voltage spikes and ringing. The gate voltage is now decoupled from the parasitic resonant network and has well-controlled on and off voltage levels that are unaffected by PCB and package parasitics, component tolerances, temperature variations, and external surge or burst transient conditions. GaNFast power ICs also include programmable dV/dt control and a wide-range supply voltage input (V_{CC}) that makes them highly flexible and compatible with many popular power supply controllers on the market.

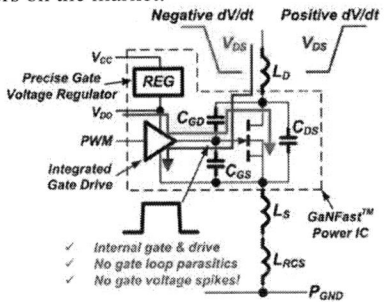

(a) GaNFast technology highlights

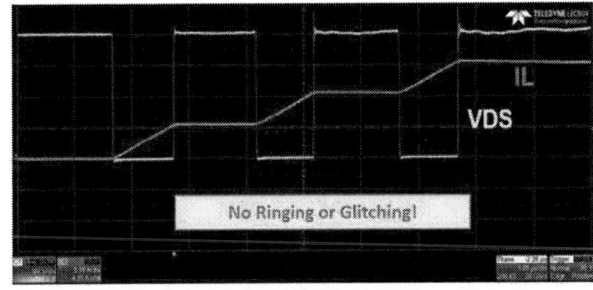

(b) Switching waveforms (Boost CCM, Fsw = 400kHz)

Fig. 2. GaNFast power IC with integrated gate, no gate loop parasitics, and no ringing or glitching

II. GANSENSE™ TECHNOLOGY

The latest Navitas GaNFast power IC product family now includes GaNSense technology. GaNSense technology strengthens immunity against parasitics, voltage spikes and ringing even further by integrating loss-less current sensing. By integrating the current sensing function into the IC (Figure 3), the external current-sensing resistors, typically placed between the discrete GaN FET Source and PGND, are now eliminated. This eliminates the parasitic inductance typically formed by the PCB trace connecting the Source to the external current sensing resistor, and the current sensing resistor itself, that can

978-1-6654-8901-0/22 $31.00 © 2022 IEEE

also cause gate voltage spikes and ringing with discrete GaN FETs. This gives further decoupling of parasitics from the GaNFast power IC resulting in even more robust and reliable switching performance without the need for additional filter components or bead inductors to suppress ringing and slow down the switching frequency. Removing the external current sensing resistors also increases system efficiency by +0.5% and removes a PCB hotspot (typically > 85C). GaNSense technology also includes additional protection features such as over-current protection and over-temperature protection, making it the most robust, reliable and efficient solution available today.

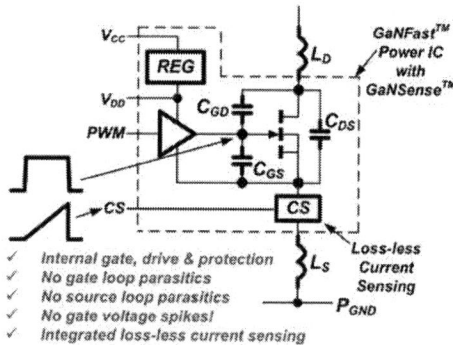

(a) GaNSense technology highlights

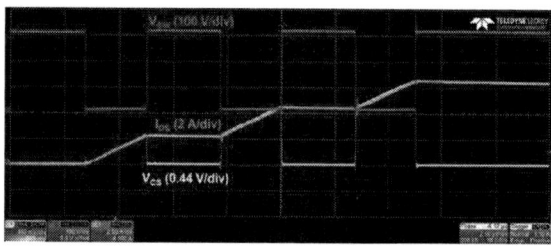

(b) No ringing or glitching and clean current sensing waveforms (Boost CCM, Fsw = 200 kHz)

Fig. 3. GaNSense technology integrates loss-less current sensing and eliminates source loop parasitics

A. Avoiding the Danger Zones

The GaN FET gate voltage levels (transient and continuous) during each on and off period are also critical for proper robustness and reliability. During the on-time period, the 'on' gate voltage level must stay above a minimum value (6V) to prevent RDS_{ON} from increasing, and, must stay below a maximum value (7V) to prevent accelerated lifetime degradation or destruction of the gate. During the off-time period, the 'off' gate voltage level (GND) must stay a safe margin away from the gate threshold level (1V) to prevent false turn-on and must not go below a maximum negative voltage rating to prevent destruction (-20V). To avoid these danger zones (Figure 4), the gate voltage levels must be well controlled at all times and should not have excessive variation over tolerances and temperature. External gate drive components used to control the gate voltage of discrete GaN FETs will still exhibit voltage spikes and will vary with temperature. Both plateau or transient gate voltage levels in the upper danger zone area will have a dramatic effect on reliability and will cause the lifetime of the device to degrade rapidly (as can be seen by the logarithmic scale of the y-axis for time-to-fail, Figure 4). The integrated voltage regulator of the GaNFast™ power IC ensures that the voltage levels always stay within a safe range (6 to 7V), and the integrated gate driver prevents gate voltage spikes from entering these danger zones resulting in excellent reliability and lifetime.

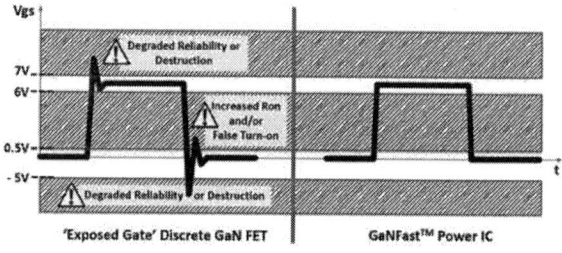

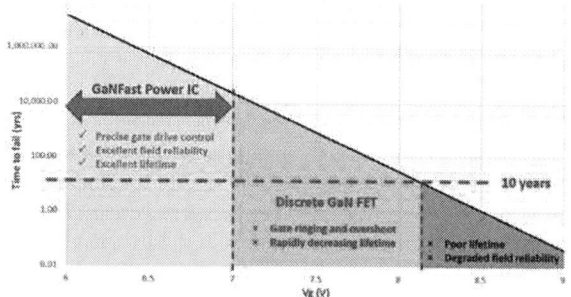

Fig. 4. Gate voltage danger zones and gate voltage vs lifetime graph

978-1-6654-8901-0/22 $31.00 © 2022 IEEE

III. GaNSense Half-Bridge

GaNSense half-bridge power ICs integrate two GaN FETs with drive, control, sensing, autonomous protection, and level shift isolation, to create a fundamental power-stage building block for power electronics. This revolutionary single component solution reduces component counts and footprints by over 60% compared to existing discretes, which cuts system cost, size, weight and complexity. The integrated GaNSense technology enables autonomous protection for increased reliability & robustness, combined with loss-less current sensing for higher levels of efficiency and energy savings. The high integration levels also serve to eliminate circuit parasitics and delays, making MHz frequency operation a reality for a broad range of ac/dc-power applications including LLC resonant, asymmetric half-bridge (AHB), and active-clamp flyback (ACF). The GaNSense half-bridge ICs are also a perfect fit for totem-pole PFC, as well as motor drive applications.

Simplified Schematic

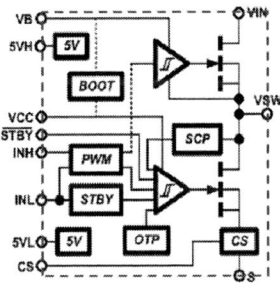

Fig. 5. GaNSense half-bridge ICs offer complete integration of a half-bridge stage with integrated sensing and autonomous protection.

A. Soft-switching optimizes power-conversion performance

High-switching-frequency operation shrinks the size, weight and cost of 'passive' elements (transformers, capacitors, EMI filters, etc.) within a power system. However, simply running a standard topology at high speed means extreme losses and reliability risks due to silicon's highly-capacitive material properties. 'Soft-switching' is a control technique in which excess voltage and/or current across a power device are eliminated before the device is switched on or off, avoiding capacitive- or switching-speed-related losses. Figure 6 details the efficiency benefits of soft- vs. hard-switching, and also highlights the material advantage of GaN power ICs vs. legacy Si discrete FETs.

Primary Switch Power Loss using Silicon FETs:

$$P_{FET} = P_{COND} * k + P_{DIODE} + P_{T\text{-}ON} + P_{T\text{-}OFF} + P_{DR} + P_{QRR} + P_{QOSS}$$

Primary Switch Power Loss using GaN Power ICs:

$$P_{FET} = P_{COND} * k + P_{DIODE} + P_{T\text{-}ON} + P_{T\text{-}OFF} + P_{DR} + P_{QRR} + P_{QOSS}$$

Fig. 6. Soft-switching and GaN ICs eliminate turn-on & reverse recovery losses & minimize drive, deadtime, and device charging losses

B. Loss-less Current Sensiung

For many applications it is necessary to sense the cycle-by-cycle current flowing through the low-side GaN power FET. Existing current-sensing solutions include placing an external current-sensing resistor in between the source connection of the low-side power FET and PGND. Using external current-sensing resistors increases system conduction power losses, creates a hot-spot on the PCB, and lowers overall system efficiency. To eliminate external resistors and hot-spot, and increase system efficiency, the IC integrates accurate and programmable loss-less current-sensing. The I_{DS} current flowing through the lowside GaN power FET is sensed internally (Fig. 6) and then amplified, trimmed and converted to a current at the current-sensing output pin (CS). An external resistor (R_{SET}) is connected from the CS pin to the P_{GND} pin and is used to set the amplitude of the CS pin voltage signal. This allows for the CS pin signal to be programmed to work with different controllers with different current-sensing input thresholds.

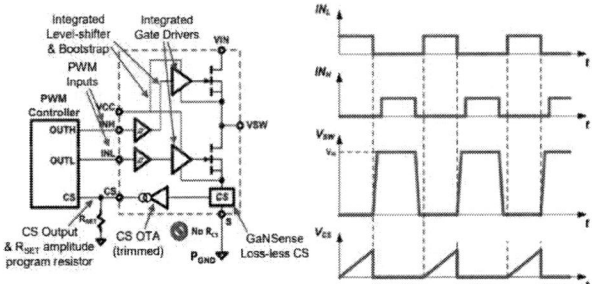

Fig. 7. IC simplified internal block diagram & timing diagram

978-1-6654-8901-0/22 $31.00 © 2022 IEEE

When comparing GaNSense technology versus existing external current sensing resistor method, the total ON-resistance, $R_{ON(TOT)}$, can be substantially reduced. For a 65 W, high-frequency ACF circuit, for example, $R_{ON(TOT)}$ is reduced from 320 mW to 160 mW. The power loss savings by eliminating the external resistor results in a +0.5% efficiency benefit for the overall system.

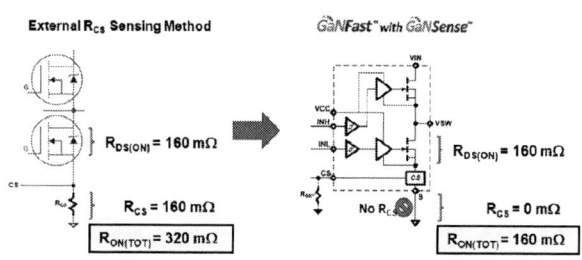

Fig 8. External resistor-sensing vs. GaNSense loss-less current sensing

Fig 9. GaNSense Half-bridge ICs can detect and protect 6x faster than silicon or GaN discretes - improving system-level reliability

C. Autonomous Over-Current Protection (OCP)

GaNSense half-bridge ICs integrated over-current protection (OCP) provides cycle-by-cycle over-current detection and protection circuitry to protect the low-side GaN power FET against high current levels. Compared to discrete silicon or discrete GaN approaches, GaNSense technology can 'detect and protect' in only 30 ns - 6x faster than silicon or GaN discretes - improving system-level reliability. During the on-time of each low-side switching cycle, should the peak current exceed the internal OCP threshold, then the internal low-side gate drive will turn the low-side GaN power FET off quickly and truncate the low-side on time period to prevent damage from occurring to the IC. The IC will continue to function normally, and the high-side GaN power FET will then turn on at the next rising edge of the input pulse for the duration of the high-side on-time.

IV. CONCLUSION

GaNSense half-bridge power ICs, with integrated drive, control and protection, enable the design of next generation high-density power converters without the difficulties associated with high-frequency switching environments, 'hidden' parasitic networks, and gate drive ringing and glitching. The parasitic gate and source loops have been completely eliminated to prevent gate voltage spikes, and the gate voltage levels are well-controlled in order to achieve excellent reliability and lifetime standards. Integration of loss-less current sensing increases system efficiency and eliminates hot spots, and integrated protection features provide fast and autonomous immunity against over-current and over-temperature fault conditions.

REFERENCES

"Breaking Speed Limits with GaN Power ICs", Kinzer, keynote at APEC, March2 01

Threshold Voltage Behavior and Short-Circuit Capability of p-Gate GaN HEMTs Depending on Drain- and Gate-Voltage Stress

Thorsten Oeder, Martin Pfost

Chair of Energy Conversion, TU Dortmund

Martin-Schmeißer-Weg 4, Dortmund, Germany - email: thorsten.oeder@tu-dortmund.de

Abstract—**In this study, the impact and correlation of drain- and gate-voltage stress on the threshold voltage of commercially available p-gate GaN HEMTs are investigated. This is based on single-pulse measurements acquired with a custom pulse setup, thus the transient behavior can be determined and subsequently translated into dc characteristics. As a novelty, measurements of the threshold voltage for short-circuit stress of up to 600 V are shown. As a result, a constant threshold voltage shift is observed for on-state drain-voltage stress, while off-state drain-voltage and gate-voltage stress lead to a threshold voltage instability. Finally, drain stress and gate stress appears to superimpose, consequently already application-relevant gate-driving conditions show a drastic impact of the short-circuit capability.**

Index Terms—**GaN, HEMT, p-GaN, threshold voltage instability, threshold voltage shift, short circuit**

I. INTRODUCTION

Gallium nitride (GaN) based high electron mobility transistors (HEMTs) enable the development of high-power and high-frequency applications [1]. To obtain a device suitable for fail-save applications, the two-dimensional electron gas (2DEG) is capped on the gate stack with a p-doped GaN layer (p-gate) resulting in a normally-off operation [2]. Commercially available p-gate GaN HEMTs are affected by the drain voltage [3]–[5] and the gate voltage [6]–[8] that was previously applied to the device, which can be interpreted as an instability (a transient change) of the threshold voltage V_{th}.

Recent studies on the reliability of p-gate GaN HEMTs indicate that important degradation mechanisms are accompanied by a shift of V_{th} [9], [10]. Many studies investigate the impact of off-state V_{DS} stress [3]–[5], as a result a positive V_{th} shift is observed, as illustrated in Fig. 1. These investigations typically apply a high drain voltage V_{DS} during the off-state ($V_{GS} \leq 0\,\mathrm{V}$), while measuring V_{th} during the on-state ($V_{GS} > 0\,\mathrm{V}$) utilizing a low V_{DS}. However, already a short-duration of V_{DS} stress applied during the on-state can yield a negative V_{th} shift, which is subject of this work.

In this study, the impact of on-state V_{DS} stress affecting the V_{th} behavior is investigated for voltages up to 600 V. Besides V_{DS} stress, V_{th} is also affected by V_{GS} stress, which has been shown to eventually affect the short-circuit capability [11], [12]. Thus, we show a comprehensive investigation to correlate the impact of V_{DS} and V_{GS} stress with the V_{th} behavior as well as their influence on the short-circuit capability.

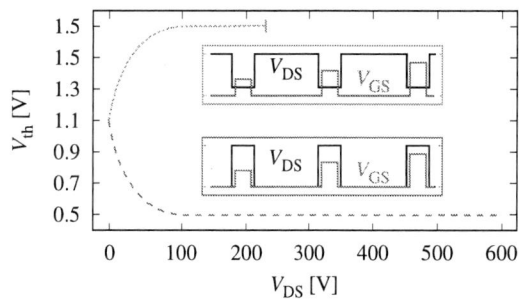

Fig. 1. Exemplary threshold voltage (V_{th}) behavior for applied drain-voltage (V_{DS}) stress that depends on the driving conditions of V_{DS} and V_{GS}. Applying V_{DS} stress during off-state (green) yields a positive V_{th} shift, while on-state V_{DS} stress (blue) leads to a negative V_{th} shift.

II. EXPERIMENTAL SETUP

A. Devices Under Test

The devices under test (DUTs) are commercially available 600-V class p-gate GaN HEMTs from different manufacturers. In both DUTs, the gate metal/p-GaN interface is realized as a Schottky contact, but differs in leakage current [6]. Here, the high gate-leakage DUT is labeled *ohmic gate* and the low gate-leakage DUT as *Schottky gate*. Besides this, both DUTs have comparable ratings.

B. Measurement Setup

In this study, a custom pulse setup is used offering accurate time domain measurements for pulse lengths from $t \geq 100\,\mathrm{ns}$ up to a duration of $t = 100\,\mathrm{s}$, as depicted in Fig. 2. The setup contains an in-house developed multi-stage gate driver that can apply three voltages (V_{ntr}, V_{bias} and V_{meas}) to the gate of the DUT. Here, all voltages can be adjusted independently. The DUT is operated in a half bridge as the low-side (LS) switch with four paralleled Si-MOSFETs as the high-side (HS) switch. The HS switch is used to apply V_{DS} on the DUT independently of V_{GS}. The V_{DS} is provided by a dc power supply V_{DC} buffered with a dc link capacitance C_{DC} and is low-inductively connected to the HS switch and DUT. Finally, the drain current I_D is acquired using a hall-sensor based current probe offering a current sensitivity of 1 mA with a bandwith of 100 MHz.

978-1-6654-8901-0/22 $31.00 © 2022 IEEE

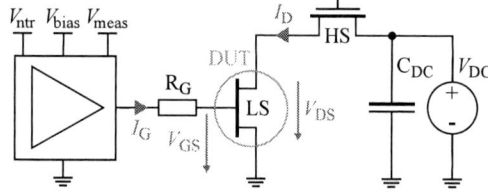

Fig. 2. Setup used for single pulse measurements. Driving the DUT in a half bridge allows for independent turn-on of V_{DS} and V_{GS}. The custom gate driver can switch between three different voltages (V_{ntr}, V_{bias}, V_{meas}), see Fig. 3(b).

C. Driving conditions

To investigate the impact of V_{DS} stress during on-state two types of turn-on are considered, both resulting in a short circuit (SC) of the DUT, as shown in Fig. 3(a). If $V_{DS} > 0\,V$ is already applied during the off-state ($V_{GS} \leq 0\,V$), a turn-on of the DUT results in a short circuit type I (SC I) resembling a hard-switching fault. On the other hand, the second type (SC II) represents the occurrence of a short circuit during on-state ($V_{GS} > 0\,V$) or rather a fault under load. Note that for SC I the bias $V_{GS} = V_{bias}$ needs to be $0\,V$ or negative and for SC II the on-state $V_{GS} = V_{meas}$ is applied $100\,ns$ prior to V_{DS}.

Besides this, for investigating V_{GS} stress each gate pulse is divided into three phases, see Fig. 3(b). During *neutralization*, $V_{ntr} = 0\,V$ is applied for $t_{ntr} = 100\,s$ to allow the DUT to revert to a steady-state value of V_{th} [5], [13]. In the *biasing* phase, the gate bias V_{bias} is applied for a duration of t_{bias} prior to the *measurement* with $V_{GS} = V_{meas}$ for a duration of t_{meas}.

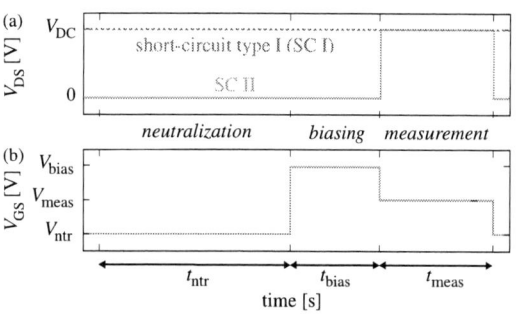

Fig. 3. Illustration of the DUT driving waveforms showing (a) the drain voltage, which is either constantly applied $V_{DS} = V_{DC}$ (short-circuit type I, SC I) or simultaneously applied with the gate voltage $V_{GS} = V_{meas}$ (SC II). In (b) the gate voltage is shown, which switches between the levels for neutralization (V_{ntr}), biasing (V_{bias}), and measurement (V_{meas}) of the DUT.

For the investigation of the short-circuit capability, various V_{GS} stress conditions are defined that are relevant for operation in power electronic applications, see Tab. I. These conditions are applied during the biasing phase with V_{bias} and t_{bias}. As a *reference* no bias is applied $V_{bias} = 0\,V$. Depending on the gate driver design and operation, a turn-on *overshoot* can occur, which is why the maximum gate voltages of both DUTs are applied for a short duration. The nominal *ON-state* condition is used to demonstrate the impact of the on-state voltage recommended by the manufacturer, applied for an increased duration. Finally, the excessive *OFF-state* condition

demonstrates the turn-on from the maximum negative V_{GS} after a long off-state duration.

TABLE I
APPLICATION-RELEVANT GATE-DRIVING CONDITIONS.

driver condition	ohmic gate	Schottky gate	
	V_{bias}	V_{bias}	t_{bias}
reference	$0\,V$	$0\,V$	$0\,s$
turn-on *overshoot*	$6\,V$	$10\,V$	$100\,ns$
nominal *ON-state*	$3.5\,V$	$6\,V$	$10\,\mu s$
excessive *OFF-state*	$-10\,V$	$-10\,V$	$10\,s$

D. V_{th} extraction procedure

For the investigation of the V_{th} behavior, the values of V_{th} are extracted from dc transfer characteristics that are acquired from many single-pulse measurements. In Fig. 4(a) the transient I_D waveforms of two exemplary pulse measurements are shown. Both are measured with an on-state duration of $t_{meas} = 10\,\mu s$, while one serves as reference and the other as result of V_{DS} stress. To acquire the transfer characteristics, the dc values of I_D are determined by averaging from the point in time where V_{DS} settles until the end of the pulse at $10\,\mu s$. Therefore, sweeping through many V_{GS}, the transfer characteristics are acquired, see Fig. 4(b). Finally, assuming V_{th} to be defined as V_{GS} for $I_D = 10\,mA$ the observed changes of I_D, cf. Fig. 4(a), can be translated into a V_{DS} stress dependent V_{th} shift, as illustrated in Fig. 4(c). Note that this is similarly done for V_{GS} stress, too.

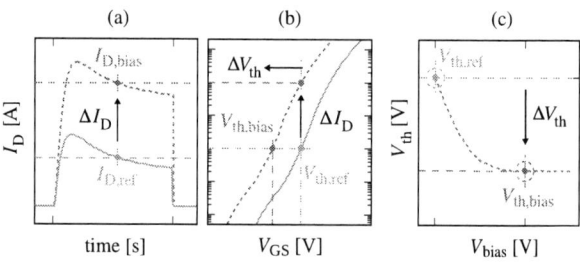

Fig. 4. Extraction procedure of the V_{th} behavior for exemplary bias conditions. (a) Drain current over time. (b) Semilogarithmic transfer characteristics. (c) Bias-depended V_{th} shift.

III. THRESHOLD VOLTAGE

A. Drain-Voltage Stress

Comparing on-state V_{DS} stress that is applied with SC I and SC II, cf. Fig. 3(a), a discrepancy in the I_D waveforms is observed, as shown for the Schottky-gate DUT in Fig. 5(a). Here, SC I and SC II use the same V_{GS} and V_{DS}. Apart from the initial discrepancy, the two I_D waveforms converge over time, which has similarly been observed for V_{GS} stress induced V_{th} instabilities [13]. Since SC I and SC II have the same on-state duration, the difference in I_D is certainly caused by V_{DS} stress already applied during the off-state of SC I for about $100\,s$. Off-state V_{DS} stress can lead to hole deficiency

978-1-6654-8901-0/22 \$31.00 © 2022 IEEE

in the gate stack due to ionization of acceptor traps with subsequent hole depletion [3], [4]. Another possibility is the field-assisted trapping of electrons at the drain-sided gate edge or in the buffer region [4]. However, both mechanisms result in a temporal reduction of charge in the 2DEG, thus affecting I_D at the turn-on. Therefore, applying off-state V_{DS} stress (SC I) results in a temporal reduction (instability) of I_D.

Apart from the SC type of V_{DS} stress, the applied V_{DS} significantly affects the I_D waveform, as shown in Fig. 5(b). To exclude interactions with mechanisms induced by off-state V_{DS} stress, here only on-state stress (SC II) is used. For higher voltages a persistent increase (shift) of I_D is observed, which has shown to be reversible. Since this is not a temporal phenomenon, mechanisms based on depletion or trapping of carriers can most likely be neglected. As stated in [14], the behavior can be explained by a drain-voltage induced lowering of the gate-metal/p-GaN barrier. High values of V_{DS} can yield an potential elevation of the p-GaN and subsequently of the 2DEG resulting in an increase of the carrier density below the gate stack. Therefore, less V_{GS} is needed to turn on the DUT which corresponds to a reduction of V_{th}.

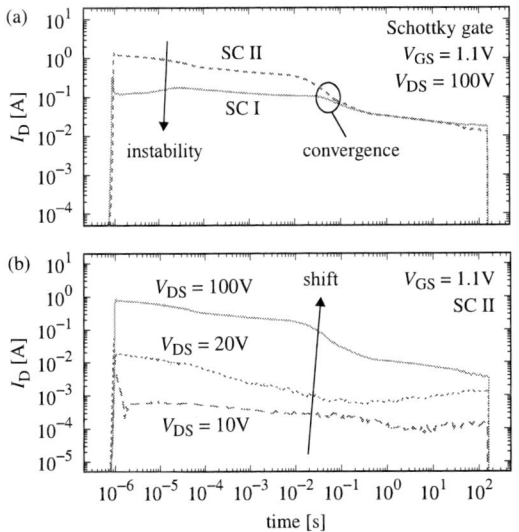

Fig. 5. Transient drain current waveforms of the Schottky-gate DUT at $V_{GS} = V_{th}$, comparing different V_{DS} stress conditions. (a) Comparison of SC I and SC II with equal V_{DS}, (b) different V_{DS} levels at SC II. Note that, a similar behavior is observed for the ohmic-gate device, although less pronounced.

The behavior of V_{th} for V_{DS} stress voltages up to 600 V is shown in Fig. 6, comparing the different short-circuit types in case of both DUTs. Overall, the impact on the ohmic-gate DUT is much less pronounced compared to the Schottky-gate DUT. Due to a more ohmic-like behavior of the gate-metal/p-GaN contact the potential elevation of the p-GaN layer is reduced, alleviating the impact of V_{DS} [14]. However, in case of both DUTs increasing V_{DS} stress leads to a negative V_{th} shift. In case of the SC II operation, the on-state stress results in a saturation of the V_{th} shift at $V_{DS} \approx 50 V - 100 V$ due to clamping or weakening of the gate-metal/p-GaN contact [14].

Additionally, the off-state stress of SC I yields a superimposed positive V_{th}, almost neglecting the impact of the on-state stress for the ohmic-gate DUT.

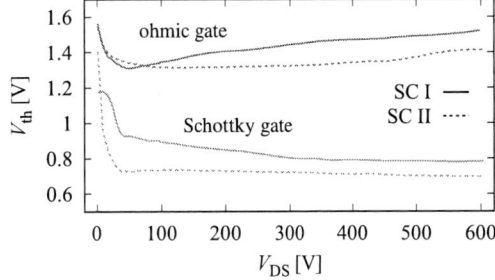

Fig. 6. Threshold voltage V_{th} of both DUTs, comparing the impact of on-state V_{DS} stress with (SC I) and without (SC II) additional off-state V_{DS} stress.

B. Gate-Voltage Stress

In [13] the impact and mechanisms of gate-voltage V_{GS} stress induced V_{th} instabilities are summarized. These are mostly based on the depletion and trapping of electrons as well as on the accumulation and trapping of holes in the gate stack that temporarily affect the charge in the 2DEG. To clarify the correlation of V_{DS} and V_{GS} stress, application-relevant gate-driving conditions, cf. Tab. I, are used that lead to an instability of I_D, as shown in Fig. 7. Here, the I_D waveforms are acquired using a constant on-state V_{DS} and $V_{GS} = V_{meas}$, but different off-state gate-driver bias conditions $V_{GS} = V_{bias}$, cf. Fig. 3(b). Similar to the findings comparing SC I and SC II, cf. Fig. 5(a), carrier deficiency and trapping effects appear to superimpose the gate-potential elevation of on-state V_{DS} stress.

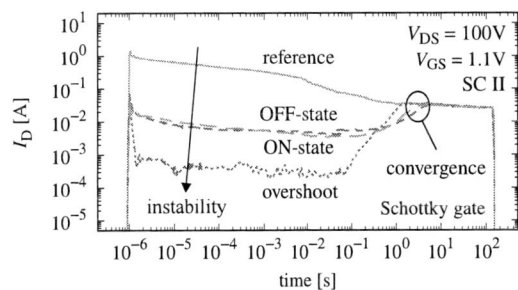

Fig. 7. Transient drain current waveforms of the Schottky-gate device at $V_{GS} = V_{th}$, comparing application-relevant gate conditions ,cf. Tab. I).

The correlation of V_{DS} and V_{GS} stress on the V_{th} behavior of both DUTs is shown in Fig. 8. As indicated by the transient I_D measurements, cf. Fig. 7, the carrier-based mechanisms of V_{GS} stress result in an superimposed V_{th} shift. Here, an additional negative V_{th} shift is present for the ohmic-gate device, while for the Schottky-gate DUT it appears to be positive. However, the additional impact of V_{GS} stress occurs to be independent of the applied V_{DS} stress for voltages above $50 V - 100 V$ (V_{th} saturation), in case of both DUTs. On the other hand, for low V_{DS} stress the impact of V_{GS} stress drastically changes the course of the V_{th} shift.

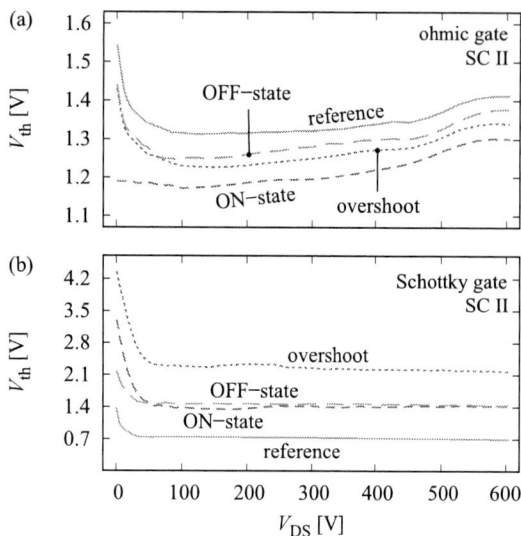

Fig. 8. Threshold voltage V_{th} of the (a) ohmic-gate and (b) Schottky-gate device for SC II, comparing different gate driving conditions, see Tab. I.

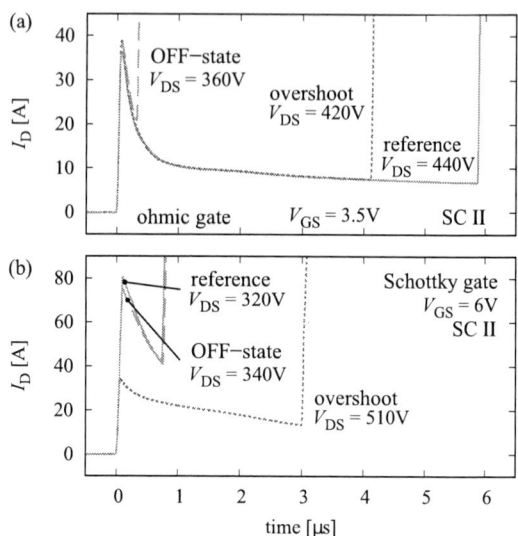

Fig. 9. Short circuit current waveforms of the (a) ohmic-gate and (b) Schottky-gate device. Here, the SC II breakdown is shown for different gate driving conditions, cf. Tab. I. Furthermore, both DUTs are driven with their nominal V_{GS}, recommended by the manufacturer, during the on-state.

C. Short-Circuit Capability

A correlation of V_{DS} and V_{GS} stress is present that induces a combined V_{th} shift up to very high V_{DS} voltages. Therefore, the maximum short-circuit (SC II) capabilities at the nominal on-state of both DUTs are shown in Fig. 9. Here, a difference of about 80 V for the ohmic-gate and up to 190 V for Schottky-gate DUT is observed, depending on the gate driving condition. For the ohmic-gate DUT, the turn on from a long *OFF-state* with high negative V_{GS} is most problematic. On the other hand, driving the Schottky-gate DUT with a short V_{GS} *overshoot* can even increase the short-circuit capability.

IV. CONCLUSION

In this study, the correlation of the drain- and gate-voltage stress and its impact on the threshold voltage V_{th} of commercially available p-gate GaN HEMTS is investigated. Here, V_{th} measurements are shown for voltages up to $V_{DS} = 600$ V. On-state V_{DS} stress leads to a gate-barrier lowering that results in a constant negative V_{th} shift. On the other hand, off-state V_{DS} and V_{GS} stress results in a superimposed temporal V_{th} shift (instability) that is based on carrier deficiency and trapping. Finally, the short-circuit capability is drastically affected by a combined V_{DS} and V_{GS} stress-induced V_{th} shift, while already application-relevant short duration V_{GS} stress can significantly impact the short-circuit capability.

REFERENCES

[1] K. J. Chen *et al.*, "GaN-on-Si power technology: Devices and applications," *IEEE Trans. Electron Devices*, vol. 64, pp. 779–795, March 2017, doi: 10.1109/ted.2017.2657579

[2] X. Hu *et al.*, "Enhancement mode AlGaN/GaN HFET with selectively grown pn junction gate," *Electronics Letters*, vol. 36, pp. 753–754(1), April 2000, doi: 10.1049/el:20000557

[3] L. Efthymiou *et al.*, "Understanding the threshold voltage instability during off-state stress in p-GaN HEMTs," *IEEE Electron Device Letters*, vol. 40, pp. 1253–1256, Aug 2019, doi: 10.1109/led.2019.2925776

[4] J. Chen *et al.*, "Off-state drain-voltage-stress-induced VTH instability in schottky-type p-GaN gate HEMTs," *IEEE Journal of Emerging and Selected Topics in Power Electronics*, pp. 1–1, 2020, doi: 10.1109/jestpe.2020.3010408

[5] K. Zhong *et al.*, "A bootstrap voltage clamping circuit for dynamic vth characterization in schottky-type p-GaN gate power HEMT," in *2021 33rd International Symposium on Power Semiconductor Devices and ICs (ISPSD)*, May 2021, pp. 39–42, doi: 10.23919/ispsd50666.2021.9452232

[6] L. Sayadi *et al.*, "Threshold voltage instability in p-gan gate AlGaN/GaN HFETs," *IEEE Trans. Electron Devices*, vol. 65, pp. 2454–2460, June 2018, doi: 10.1109/ted.2018.2828702

[7] A. Stockman *et al.*, "Threshold voltage instability mechanisms in p-GaN gate AlGaN/GaN HEMTs," in *2019 31st International Symposium on Power Semiconductor Devices and ICs (ISPSD)*, May 2019, pp. 287–290, doi: 10.1109/ispsd.2019.8757667

[8] K. Murukesan *et al.*, "Gate stress induced threshold voltage instability and its significance for reliable threshold voltage measurement in p-GaN HEMT," in *2019 IEEE 7th Workshop on Wide Bandgap Power Devices and Applications (WiPDA)*, Oct 2019, pp. 177–180, doi: 10.1109/wipda46397.2019.8998859

[9] S. Li *et al.*, "Reliability concern of 650-V normally-off GaN devices under reverse freewheeling stress," *IEEE Trans. Electron Devices*, vol. 67, pp. 3492–3495, Aug 2020, doi: 10.1109/ted.2020.2999026

[10] Y. Q. Chen *et al.*, "Degradation behavior and mechanisms of e-mode GaN HEMTs with p-GaN gate under reverse electrostatic discharge stress," *IEEE Trans. Electron Devices*, vol. 67, pp. 566–570, Feb 2020, doi: 10.1109/ted.2019.2959299

[11] T. Oeder *et al.*, "Experimental study of the short-circuit performance for a 600V normally-off p-gate GaN HEMT," in *2017 29th International Symposium on Power Semiconductor Devices and IC's (ISPSD)*, May 2017, pp. 211–214, doi: 10.23919/ispsd.2017.7988925

[12] J. Sun *et al.*, "Short circuit capability characterization and analysis of p-GaN gate high-electron-mobility transistors under single and repetitive tests," *IEEE Transactions on Industrial Electronics*, vol. 68, pp. 8798–8807, Sep. 2021, doi: 10.1109/tie.2020.3009603

[13] T. Oeder *et al.*, "Gate-induced threshold voltage instabilities in p-gate GaN HEMTs," *IEEE Trans. Electron Devices*, vol. 68, pp. 4322–4328, Sep. 2021, doi: 10.1109/ted.2021.3098254

[14] M. Nuo *et al.*, "Gate/drain coupled barrier lowering effect and negative threshold voltage shift in schottky-type p-GaN gate HEMT," *IEEE Transactions on Electron Devices*, vol. 69, pp. 3630–3635, July 2022, doi: 10.1109/ted.2022.3175792

Series Compensation for DC Microgrid Stabilization Utilizing T-Type Modular Dc Circuit Breaker

Faisal Alsaif, Yue Zhang, Nihanth Adina, Khalid Alkhalid, and Jin Wang

Center for High Performance Power Electronics (CHPPE)
The Ohio State University, Columbus, OH, USA

Abstract—Dc microgrids during voltage transients or load power changes are prone to instability due to the negative impedance characteristics of the constant power loads (CPLs). The newly proposed T-Type Modular Dc Circuit Breaker (T-Breaker) integrates current breaking, current limiting and grid transient compensation functions into one device thanks to its integrated energy storage and modular multilevel converter structures. Much like the current and voltage compensations in ac systems, the T-Breaker can help the ride-through of grid transients by implementing shunt (current) and series (voltage) compensations. This paper focuses on the series compensation function of the T-Breaker which can insert or absorb voltage. Related system modeling, small signal analysis, large signal analysis, simulation are presented here. Experimental validation result on a scaled down system is also presented to show how the compensation improves the microgrid's stability.

Index Terms—Solid state circuit breaker, T-Breaker, Voltage compensation, constant power load CPL, dc microgrid, Silicon Carbide (SiC)

I. INTRODUCTION

Power quality of dc microgrids can be challenged as a result of the bus voltage transient. Sudden load changes or faults in the microgrids are common causes of dc bus's voltage transients [1]. Constant Power Loads (CPL) could destabilize the system under large bus voltage transient due to their negative impedance behavior. Abundant research have been conducted to examine the stability issues in dc microgrids when CPLs are present [2].

Solutions to overcome the CPLs instability are introduced in literature. First solution to ensure stability of the dc microgrid is adding more passive elements [3]. However, bulky capacitors and inductors would decrease the power density in weight-size-sensitive applications such as aviation. Second approach is controlling the source side or load side converters to overcome these transient [4], [5]. Since this solution can be designed for a specific converter, it can not be generalized. Third solution suggests adding auxiliary circuit at the load side and utilize energy storage to change CPL's equivalent impedance to improve stability [6], [7]. In this work, voltage compensation is proposed by inserting/absorbing voltage to improve the system's stability by maintaining the CPL's input voltage under bus voltage transients. To achieve that, the newly proposed T-Type Modular Dc Circuit Breaker (T-Breaker) [8], [9] will be used.

This work is based upon research supported by The Advanced Research Projects Agency-Energy (ARPA-E) under award number DE-AR0001110.

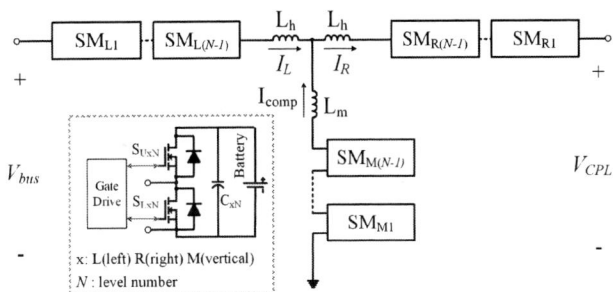

Fig. 1. A generic T-Breaker system diagram with half bridge submodules.

T-Breaker installation in a microgrid helps with achieving protection functions and the compensation functions. Generic T-breaker system diagram is shown in Fig. 1. T-Breaker's modular structure qualifies it to be implemented in low voltage and medium voltage microgrid applications. Compensation functions can be achieved when its integrated energy storage units are utilized. the T-Breaker can be controlled to implement shunt current injection/absorption using the vertical arm [10], and series voltage insertion with the two horizontal arms. The T-Breaker Achieving these two compensation functions qualifies it imitate the shunt/series compensation devices in the flexible alternating current transmission system (FACTs) in ac systems [11]. The principle, operation and analysis of the series compensation function using the T-Breaker will be discussed in this paper. Small signal analysis, large signal analysis, simulation and the experimental results are presented.

II. T-BREAKER'S SERIES COMPENSATION IMPLEMENTATION IN DC MICROGRID SUBSECTION

T-Breaker is introduced in a single-source single-load dc microgrid subsection as shown in Fig. 2. When the T-Breaker's vertical arm's switches are off ($I_{comp} = 0$) and the right arm is used for the voltage compensation, the simplified circuit is shown in Fig. 3. Where R_L represents the combination

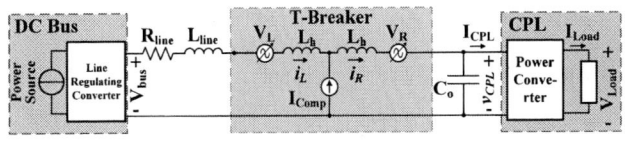

Fig. 2. T-Breaker implemented in a dc microgrid subsection.

978-1-6654-8901-0/22 $31.00 © 2022 IEEE

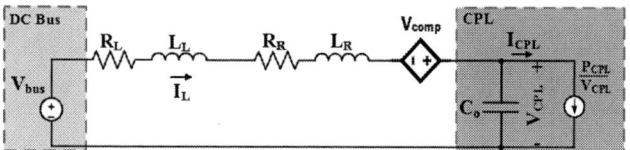

Fig. 3. Simplified model of the SISO dc microgrid with the T-Breaker as a series voltage compensator.

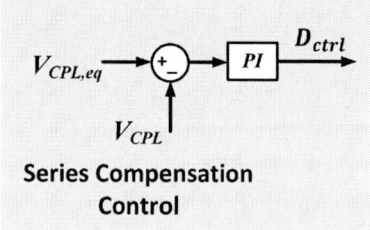

Fig. 4. Series voltage compensation control.

of left arm resistance R_h and the line resistance R_{line}. L_L represents the combination of the left arm inductance L_h and the line inductance L_{line}. R_R is the right arm resistance. L_R is the right arm inductance.

Compensation voltage (V_{comp}) is generated using the proportional gain (K_p) of a PI control as shown in Fig. 4, hence V_{comp} can be defined as:

$$V_{comp} = K_p * [V_{CPL(ref)} - v_{CPL}]. \tag{1}$$

The system description can be written as:

$$
\begin{aligned}
\dot{x}_1 &= -\frac{R_L + R_R}{L_L + L_R} x_1 - \frac{(1 + K_p)}{L_L + L_R} x_2 + \frac{V_{bus}}{L_L + L_R} \\
\dot{x}_2 &= \frac{x_1}{C_o} - \frac{P_{CPL}}{C_o x_2}
\end{aligned} \tag{2}
$$

State variable x_1 is the line current i_L and x_2 is the v_{CPL}. V_{bus} is the input and P_{CPL} is assumed constant. The non-linear term $\frac{P_{CPL}}{C_o x_2}$ is caused by the CPL.

The small signal analysis of the system can be carried out by linearizing the system in (2) around its equilibrium point X_0, (X_{10}, X_{20}).. The linearized system around the equilibrium point is given in (3) and (4). Locations of the poles of the system can be found using (4).

$$
\begin{bmatrix} \triangle \dot{x}_1 \\ \triangle \dot{x}_2 \end{bmatrix} = A * \begin{bmatrix} \triangle x_1 \\ \triangle x_2 \end{bmatrix} + \begin{bmatrix} \frac{1}{L_L + L_R} \\ 0 \end{bmatrix} * \begin{bmatrix} \triangle U \end{bmatrix} \tag{3}
$$

$$
A = \begin{bmatrix} -\frac{R_L + R_R}{L_L + L_R} & -\frac{(1 + K_p)}{L_L + L_R} \\ \frac{1}{C_o} & \frac{P_{CPL}}{C_o * X_{20}^2} \end{bmatrix} \tag{4}
$$

Large signal stability of the system is investigated using the region of attraction (ROA) method. Where ROA is an estimation of the stability region and can be calculated using Takagi–Sugeno (T–S) fuzzy model method [12]. The (T–S) fuzzy model of the system can be formulated around the equilibrium point before the disturbance occurs. The system description becomes:

$$
\begin{aligned}
\dot{\tilde{x}}_1 &= -\frac{(R_L + R_R)\tilde{x}_1}{(L_L + L_R)} - \frac{(1 + K_p)\tilde{x}_2}{L_L + L_R} \\
\dot{\tilde{x}}_2 &= \frac{1}{C_o}\tilde{x}_1 - \frac{1}{C_o}\tilde{f}\tilde{x}_2
\end{aligned} \tag{5}
$$

where $x = (x_1, x_2)$ and $\tilde{x} = x - X_o$

$$\tilde{f} \triangleq \frac{P_{CPL}}{X_{20}} \frac{1}{\tilde{x}_2 + X_{20}} . \tag{6}$$

Since this system has one CPL, two linear equations can be defined as $\dot{\tilde{x}} = A_{LS,i}\ \tilde{x}$ ($i = 1, 2$) based on (5). The $A_{LS,i}$ matrices represent the minimum and the maximum of $\tilde{f}$:

$$
A_{LS,1} = \begin{bmatrix} -\frac{(R_L + R_R)}{(L_L + L_R)} & -\frac{1 + K_p}{(L_L + L_R)} \\ \frac{1}{C_o} & \frac{1}{C_o}\tilde{f}_{min} \end{bmatrix} \tag{7}
$$

$$
A_{LS,2} = \begin{bmatrix} -\frac{(R_L + R_R)}{(L_L + L_R)} & -\frac{1 + K_p}{(L_L + L_R)} \\ \frac{1}{C_o} & \frac{1}{C_o}\tilde{f}_{max} \end{bmatrix} . \tag{8}
$$

With the $A_{LS,i}$ matrices, system's stability can be tested by examining the feasibility of the following Linear Matrix Inequality (LMI):

$$
\begin{aligned}
P &> 0 \\
A_{LS,i}^T P + P A_{LS,i} &< 0 \ for \ i = 1, 2
\end{aligned} \tag{9}
$$

where P is a variable ($P = P^T$) that is needed to satisfy (9).

III. CASE STUDY

A case study is conducted to analyze both small signal stability and large signal stability of a SISO dc microgrid with a CPL load. Stability of both uncompensated ($V_{comp} = 0$) and compensated cases are analyzed and compared. Case study's parameters are shown in Table I. The case study can represent a dc distribution system with a typical motor drive load, which often works in the constant power mode [13].

The small signal analysis is performed around the system's equilibrium point X_0. Fig. 5 shows the small signal stability by examining the location of the system's poles without compensation. When the input capacitor at the load side $C_o = 0.25\ mF$ the poles are on the right hand side of the plane, hence the system is unstable. Though by increasing the C_o, the system stability can be improved, the size and weight of the system would be increased. Fig. 6 illustrates the stability improvement with compensation. The T-Breaker's

TABLE I
SYSTEM SPECIFICATIONS

Parameter	Variable	Value
Dc bus voltage	V_{bus}	400 V
CPL power	P_{CPL}	50 kW
Line resistance	R_{line}	100 mΩ
Line inductance	L_{line}	100 μH
T-Breaker horizontal arm resistance	R_h	1 mΩ each
T-Breaker horizontal arm inductance	L_h	2 μH each
CPL bus capacitance	C_o	0.25 mF

978-1-6654-8901-0/22 $31.00 © 2022 IEEE

voltage compensation made the system stable while keeping the CPL capacitance C_o at $0.25\ mF$ ($K_p = 2$).

Large signal analysis is also conducted to examine the system's response with and without compensation. Matrices (7), (8) are used to estimate the ROA of the system. Fig. 7 shows that the uncompensated system has no ROA (unstable) at $C_o = 0.25\ mF$. Increasing the CPL capacitance shows an increase in the ROA. On the other hand, the compensated system exhibits a ROA when the original 0.25-mF C_o as shown in Fig. 8. Large signal simulation using the switching model in Simulink is shown in Fig. 9 to show how the compensation can stabilize the system when a 10 % source voltage sag occurs.

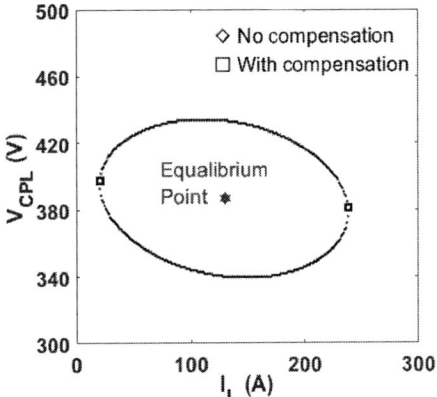

Fig. 8. ROA of the compensated system.

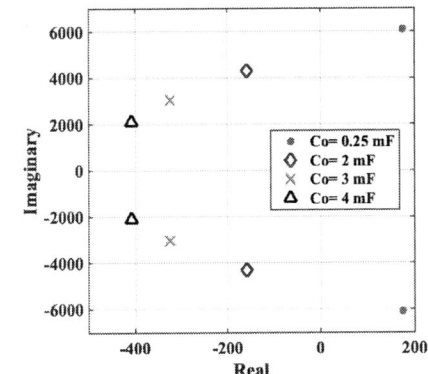

Fig. 5. Small signal stability: C_o effect.

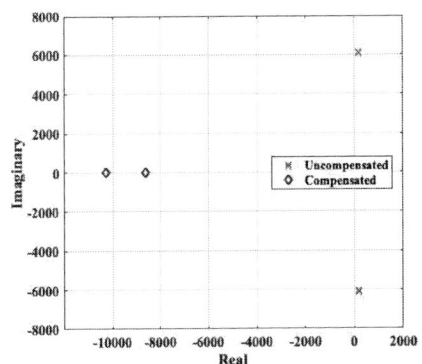

Fig. 6. Small signal stability: effect of compensation.

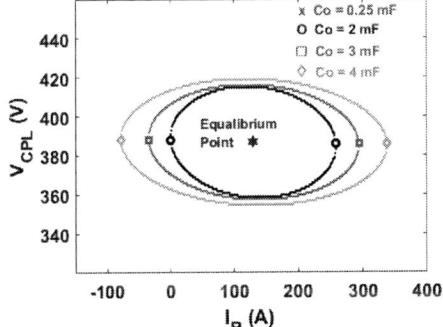

Fig. 7. ROA of uncompensated system.

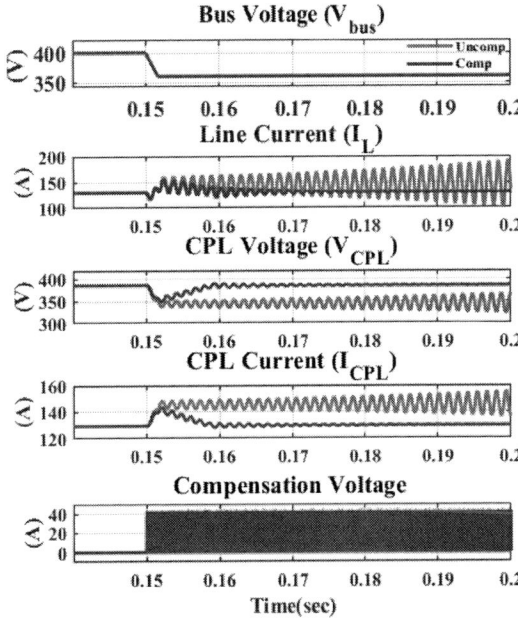

Fig. 9. Simulated system response to a 10 % source voltage sag.

IV. EXPERIMENTAL VERIFICATION

A scaled down system is built to demonstrate the T-Breaker's shunt compensation. The scaled down circuit model is shown in Fig. 10. The parameters of the test setup are shown in Table II. V_{bus} is maintained with a dc power supply simulating an active rectifier with a constant output voltage. An electronic load is configured as a CPL.

Fig. 11 shows the test setup. The system first operates at 4 kW and then is exposed to a (20%) bus voltage drop. The tests were performed with and without the T-Breaker based series compensation. Fig. 12 contains waveforms of $V_{bus}, I_L, V_{CPL}, I_{CPL}$ and the compensation voltage $V_{ds(S_{LR})}$ of the system's response without compensation. Fig. 13 shows the waveforms when the the compensation was enabled to realize the SR control. It can be seen that the compensation improved the stability of the system by maintaining the load bus's voltage. Also, the line current oscillations reduced by

978-1-6654-8901-0/22 $31.00 © 2022 IEEE

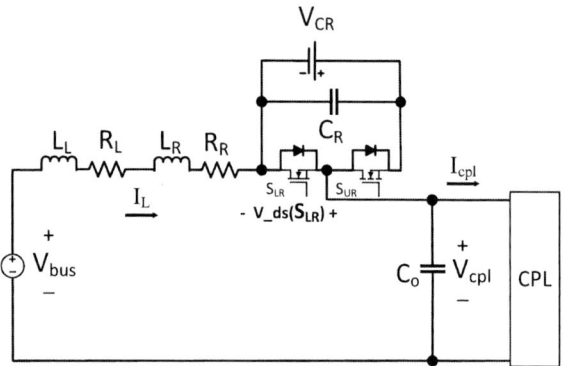

Fig. 10. Scaled-down hardware test circuit.

50%.

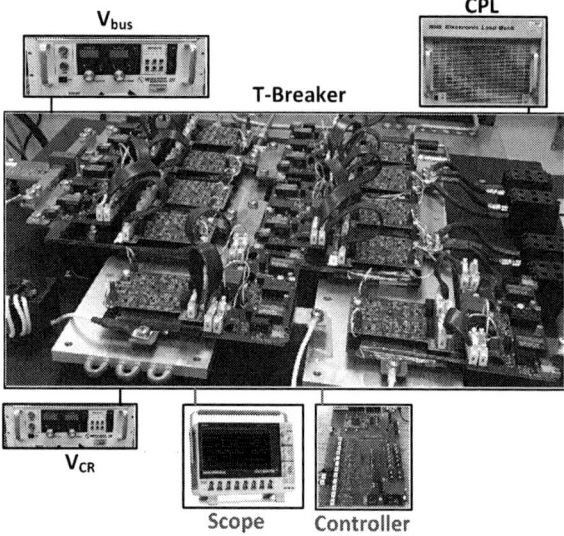

Fig. 11. Hardware test setup.

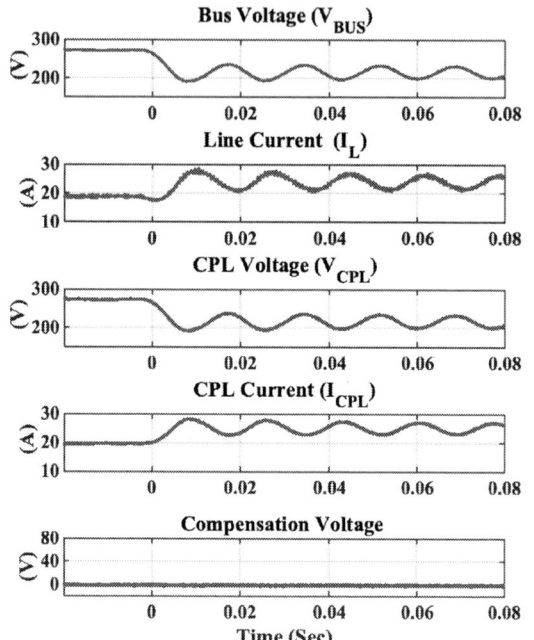

Fig. 12. Measured uncompensated system response to a 50 % power step.

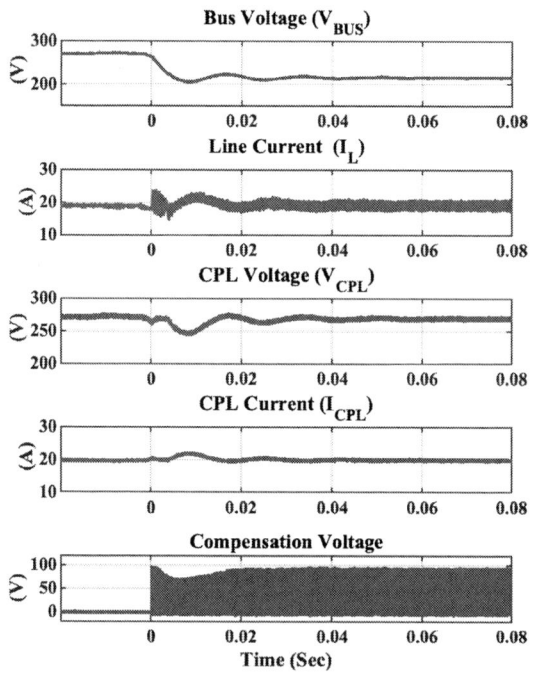

Fig. 13. Measured compensated system response to a 50 % power step.

V. CONCLUSION AND FUTURE WORK

The series compensation function of the T-Breaker system is presented in this work. To realize the series compensation, integrated energy storage in the horizontal arms of the T-Breaker can insert/absorb voltage to/from the dc microgrid. Both small signal and large signal analysis are investigated to show how a T-Breaker's compensation can improve the dc microgrid stability. A down-scaled experimental validation is presented. CPL voltage is maintained, and line current oscillitions are reduced by 50%. T-Breakers can function as conventional circuit breakers and behave as compensators in dc microgrids to improve their stability when energy storage is integrated.

TABLE II
SCALED-DOWN SYSTEM PARAMETERS

Parameter	Variable	Value
Dc bus voltage	V_{bus}	270 V
CPL power	P_{CPL}	4 kW
Line resistance	R_{line}	10 mΩ
Line inductance	L_{line}	65 μH
T-Breaker horizontal arm resistance	R_h	1 mΩ
T-Breaker horizontal arm inductance	L_h	2 μH
CPL capacitance	C_o	220 μF

REFERENCES

[1] D. Kumar, F. Zare, and A. Ghosh, "Dc microgrid technology: system architectures, ac grid interfaces, grounding schemes, power quality,

communication networks, applications, and standardizations aspects," *Ieee Access*, vol. 5, pp. 12 230–12 256, 2017.

[2] A. Emadi, A. Khaligh, C. H. Rivetta, and G. A. Williamson, "Constant power loads and negative impedance instability in automotive systems: definition, modeling, stability, and control of power electronic converters and motor drives," *IEEE Transactions on Vehicular Technology*, vol. 55, no. 4, pp. 1112–1125, 2006.

[3] M. Cespedes, L. Xing, and J. Sun, "Constant-power load system stabilization by passive damping," *IEEE Transactions on Power Electronics*, vol. 26, no. 7, pp. 1832–1836, 2011.

[4] A. Kwasinski and C. N. Onwuchekwa, "Dynamic behavior and stabilization of dc microgrids with instantaneous constant-power loads," *IEEE Transactions on Power Electronics*, vol. 26, no. 3, pp. 822–834, 2010.

[5] J. Siegers, S. Arrua, and E. Santi, "Stabilizing controller design for multibus mvdc distribution systems using a passivity-based stability criterion and positive feedforward control," *IEEE Journal of Emerging and Selected Topics in Power Electronics*, vol. 5, no. 1, pp. 14–27, 2016.

[6] K. A. Potty, E. Bauer, H. Li, and J. Wang, "Smart resistor: Stabilization of dc microgrids containing constant power loads using high-bandwidth power converters and energy storage," *IEEE Transactions on Power Electronics*, vol. 35, no. 1, pp. 957–967, 2020.

[7] K.-T. Mok, M.-H. Wang, S.-C. Tan, and S. R. Hui, "Dc electric springs—a technology for stabilizing dc power distribution systems," *IEEE Transactions on Power Electronics*, vol. 32, no. 2, pp. 1088–1105, 2016.

[8] Y. Zhang, F. Alsaif, X. Li, R. Na, and J. Wang, "T-type modular dc circuit breaker (t-breaker) for future dc networks," in *2021 IEEE Applied Power Electronics Conference and Exposition (APEC)*, 2021, pp. 1146–1152.

[9] Y. Zhang, X. Li, D. Ma, Y. Cong, F. Alsaif, Z. Zhang, R. Borjas, B. Hu, J. Wang, B. Riar *et al.*, "Development of a 1 kv, 500 a, sic-based t-type modular dc circuit breaker (t-breaker)," in *2021 IEEE 8th Workshop on Wide Bandgap Power Devices and Applications (WiPDA)*. IEEE, 2021, pp. 199–204.

[10] F. Alsaif, Y. Zhang, X. Li, B. Hu, N. Adina, D. Ma, K. Alkhalid, and J. Wang, "Shunt compensation for dc microgrid stabilization utilizing t-type modular dc circuit breaker," in *2022 IEEE Applied Power Electronics Conference and Exposition (APEC)*. IEEE, 2022, pp. 1538–1542.

[11] F. Z. Peng and J. Wang, "Flexible transmission and resilient distribution systems enabled by power electronics," *Power Electronics in Renewable Energy Systems and Smart Grid: Technology and Applications*, pp. 271–314, 2019.

[12] L. Herrera, W. Zhang, and J. Wang, "Stability analysis and controller design of dc microgrids with constant power loads," *IEEE Transactions on Smart Grid*, vol. 8, no. 2, pp. 881–888, 2017.

[13] P. Magne, D. Marx, B. Nahid-Mobarakeh, and S. Pierfederici, "Large signal stabilization of a dc-link supplying a constant power load using a virtual capacitor: Impact on the domain of attraction," *IEEE Transactions on Industry Applications*, vol. 48, no. 3, pp. 878–887, 2012.

3kV 6.7mΩ·cm² 4H-SiC BJT with An Effective Junction Termination Extension (JTE)

Xixi Luo
Semiconductor Power Electronics Center (SPEC)
the University of Texas at Austin
Austin, Texas, USA
xixiluo@utexas.edu

Alex Q. Huang
Semiconductor Power Electronics Center (SPEC)
the University of Texas at Austin
Austin, Texas, USA
aqhuang@utexas.edu

Abstract— **In this paper, an implantation-free 3 kV 4H-SiC Bipolar Junction Transistor (BJT) is designed, fabricated, and characterized. With a 40μm-wide Four-step Junction Termination Extension (JTE), an open base breakdown voltage (BV$_{CEO}$) and an open emitter breakdown voltage (BV$_{CBO}$) of more than 3000V are measured. The total width of the JTE is less than two times of the drift thickness (23μm), which can be considered as highly area efficient. The designed BJT has a 1.2μm narrow base width with 1×10^{17}cm^{-3} doping, where implantation-free Ohmic contact was achieved. The BJT exhibits an excellent on-resistance of 6.7mΩ·cm² for small size device and an on-resistance of 39.7mΩ·cm² for large size device. The measured current gain for devices with additional anneal process is 21.**

Keywords—BJT, 4H-SiC, JTE, Breakdown, carrier lifetime, TCAD.

I. Introduction

While 4H-SiC MOSFET is the preferred structure for high voltage power devices, 4H-SiC BJT has several advantages such as better short circuit capability and reliability due to the absence of the gate dielectric layer [1] [2]. The device ruggedness with a very large RBSOA has also been demonstrated [1] which is very different from silicon BJT. SiC BJT on-resistance is also lower than MOSFET due to better current conduction path and modest conductivity modulation.

Over the past few years, higher voltage 4H-SiC BJTs with implantation-free multi-step JTE structure have been proposed. However, the total JTE widths reported are typically three to five times the drift layer thickness [3] [4]. Various current gain improvement process were also proposed [5] [6]. Many 4H-SiC BJTs demonstrate on-resistance smaller than unipolar limit of the drift layer, meaning conductivity modulation is achievable in 4H-SiC BJTs. For 3kV BJT, specific on-resistances around 4mΩ·cm² are reported for devices with drift layers three to four times higher doping than the one used in this study [7].

II. Design and Simulations

A. Blocking capability

The 4H-SiC BJTs were fabricated on an 8° off-axis n-type 4H-SiC substrate with a 1μm thick 1×10^{19} cm^{-3} nitrogen-doped epitaxial N-emitter layer, a 1.2μm thick 1×10^{17} cm^{-3} aluminum-doped epitaxial P-base layer and a 23μm thick 2×10^{15} cm^{-3} nitrogen-doped epitaxial N-drift layer (**Fig.1**).

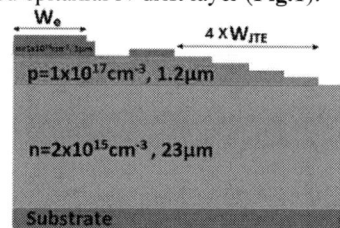

Fig. 1. Schematic cross-section of the fabricated 4H-SiC BJT.

A relatively light drift doping of 2×10^{15} cm^{-3} was selected, so that it could achieve a desirable breakdown voltage with thinner thickness and reduce the epi-growth cost. Therefore, the thickness of the drift layer can be calculated by using punch-through electric field distribution:

$$E_c W_d - \frac{q N_d W_d^2}{2\varepsilon} = BV_{pt} \qquad (1)[8]$$

In which E$_c$, the 4H-SiC critical electric field, was assumed to be 2MV/cm. The drift thickness W$_d$ was selected to be 23μm to guarantee a theoretical breakdown voltage of at least 3.6kV.

For higher current gain, a narrower and lightly doped base layer was preferable. On the other hand, it needs to have enough charges so that it is not fully depleted when the BV is applied. Therefore, base thickness W$_b$ has:

$$W_b \geq \sqrt{\frac{2\varepsilon BV}{q} \frac{N_d}{(N_d+N_a)N_d}} \qquad (2)[8]$$

The base doping was selected to be 1×10^{17} cm^{-3}, which required W$_b$ to be at least 0.87μm according to (2). Adding a bit of the process margin, the base thickness was determined to be 1.2μm, giving a total base charge density of 1.2×10^{13}cm^{-2} which is lighter than previous published BJTs (1.75×10^{13}cm^{-2}) [3] [4] [5].

978-1-6654-8901-0/22 $31.00 © 2022 IEEE

According to TCAD simulations, these selected epitaxial n-p-n layers would have a BV$_{CBO}$ of 4320V and a BV$_{CEO}$ of 3710V. The base region minority carrier lifetime was set to be 1µs for BV$_{CEO}$ simulation.

A four-step Junction Termination Extension (JTE) structure was utilized to terminate the BJT (**Fig.1**). The steps will be formed by multiple Reactive Ion Etch (RIE). TCAD simulations predicted that this structure could block 98% ideal BV$_{CBO}$ with W$_{JTE}$=20µm step width and 99% of ideal BV$_{CEO}$ with 10µm step width (**Fig.2**). To estimate the sensitivity to step heights, it was simulated under four different designs (**Table I**).

TABLE I.   SIMULATED JTE DESIGNS

	Step height (µm)			
	Step1	*Step2*	*Step3*	*Step4*
ideal	*1.2*	*0.9*	*0.6*	*0.3*
overcharge	*1.2*	*1.0*	*0.7*	*0.4*
ETCH1 overcharge	*1,2*	*1.0*	*0.6*	*0.3*
undercharge	*1.2*	*0.8*	*0.5*	*0.2*

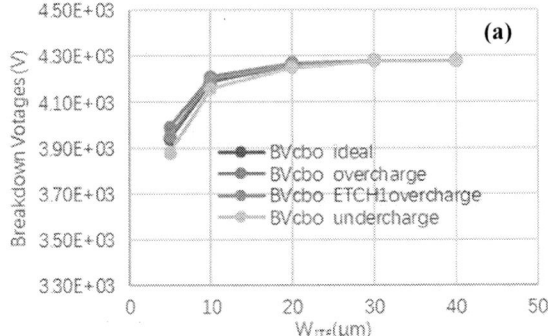

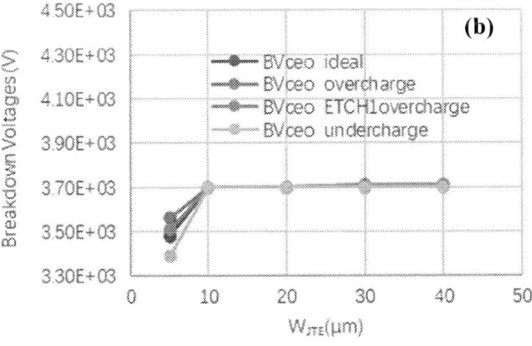

Fig. 2.   Simulated (a) BV$_{CBO}$ and (b) BV$_{CEO}$ under different JTE step height and W$_{JTE}$ variations.

Although JTE designs demonstrate insensitivity to step height variations (**Fig.2**), it was expected to have a better performance with over-charged JTE, especially over-charged first two steps. This over-charged JTE performance will be later verified by experimental results in part IV (**Fig.6**).

B. Conduction performance

Since a narrow and lightly doped base was selected, the BJT was expected to have a high current gain. TCAD simulations indicated a maximum DC current gain of about 189 when base minority carrier lifetime was set to be 500ns. When the maximum minority carrier lifetime was below 200ns, maximum current gain would quickly decline and drop to almost unity (**Fig.3**).

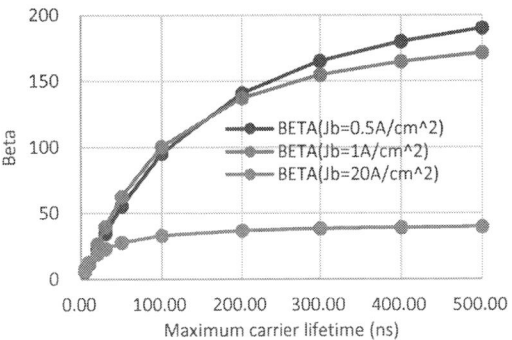

Fig. 3.   Simulated DC current gain (beta) of BJTs under different base current density and maximum minority carrier lifetime (W$_e$ = 50µm, beta extracted from V$_{ce}$=20V, T=300K).

These current gain TCAD simulations did not include any surface trapping model and were reflecting a pure bulk material determined current gain. Therefore, it is certain that, actual current gain will be lower than Fig.3 due to additional surface recombination [9].

Epilayer carrier lifetime in 4H-SiC is highly process related. Depending on epi-growth and enhancement technique, the reported minority carrier lifetime measured in p-type SiC can vary from as long as 2.6µs [10] to as short as 20ns [11]. The minority carrier lifetime in the epilayers of our device is not provided by the vendor, therefore it is not clear if this will be the major factor in determining the measured current gain below.

For the on-resistance of BJT, the bulk material resistance of 8.06 mΩ·cm² was calculated using (3). Any measured on-resistance smaller than this value indicates a conductivity modulation.

$$R_{on-sp} = \frac{W_D}{q\mu_{nD}N_D} + \frac{W_B}{q\mu_{pB}N_{AB}} + \frac{W_E}{q\mu_{nE}N_{DE}} \quad (3)$$

III. FABRICATION

The fabrication of the BJT follows a top-down process. Emitter island was first defined with ICP (Inductively-coupled-plasma) dry etch and served as alignment markers for the next few steps. Multiple ICP etchings with thick photoresist masks were performed to form the JTE. The ICP etching process had an 800W power level and an etching speed of about 50nm/min, giving a JTE structure shown in **Fig.4**. To reveal the breakdown voltage sensitivity to JTE step height, two different processes were conducted for two JTE steps combinations (**Table II**).

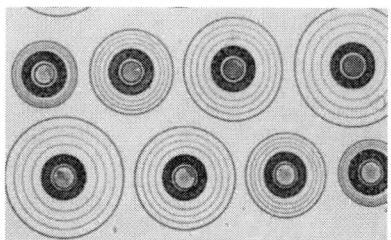

Fig. 4. Microscope image of fabricated 4H-SiC BJTs (small size)

After the etching, a surface passivation protective layer of 120nm SiO_2 was deposited by PECVD (plasma enhanced chemical vapor deposition). To explore the impact of post-deposition recovery process, some samples were annealed in a 1100°C furnace for 3 hours in O_2 ambient, while other samples are going directly to metallization. For emitter and collector, a 30/70nm Ti/Ni metal stack was deposited. For the P base, a 15/10/125nm Ni/Ti/Al metal stack was utilized to achieve low contact resistance. All emitter, collector and base metal were sent for a 1.5 minutes 900°C RTA (Rapid Thermal Annealing) in N_2 ambient. After RTA, another 30/120nm Ti/Ni was re-deposited on collector, emitter and base to enhance the metal layer conductivity which was reduced by the previous RTA process.

IV. RESULTS AND DISCUSSION

A. Breakdown characterization

The fabricated BJTs were characterized with a probe station and the Keithley 2651a/2657 high power source meter system. For breakdown voltage measurement, devices were immerged in dielectric fluid FC-40 to prevent air breakdown(**Fig.5**).

Fig. 5. Dielectric fluid FC-40 high-voltage measure environment.

The width of each JTE steps vary from 10/20/30/40μm (total width 40/80/120/160μm), and JTE step heights have two different combinations (**Table II**). For Process1, the two steps closer to the base metal were under-charged than original design (**Table I**). For Process2, it was the furthest two steps under-charged. Both process1/2 devices present BV_{CBO} higher than 3000V (the measurement limit of Keithley 2657 high voltage SMU) when W_{JTE} is larger than 20μm, which indicates the four-step JTE structure has a good process error tolerance. However, for 10μm W_{JTE}, only Process2 devices have high BV_{CBO} while Process1 device breaks down at 2902V (**Fig.6**). Therefore, TCAD simulations were verified that over-charged first two steps of the four-step JTE improved the breakdown performance.

TABLE II.   FABRICATED JTE COMBINATIONS

	Step height (μm)			
	Step1	*Step2*	*Step3*	*Step4*
Process 1	*1.07*	*0.775*	*0.674*	*0.330*
Process 2	*1.14*	*0.942*	*0.592*	*0.173*

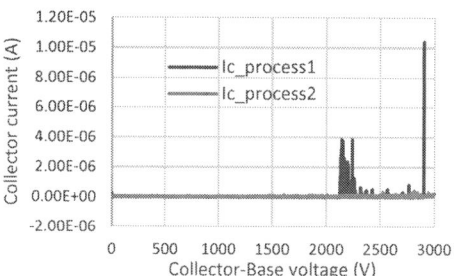

Fig. 6. Meadured BV_{CBO} of BJTs with different JTE step heights. (W_e=50μm W_{JTE}=10μm, JTE total width 40μm)

No obvious evidence shows that BJTs have BV_{CEO} smaller than BV_{CBO}. The BJTs with process2 JTE gave BV_{CEO} larger than 3000V for all 10/20/30/40μm W_{JTE}. A maximum collector-emitter leakage current of about 0.4μA (5mA/cm^2) at 2.8kV was measured from the small size BJT with 10μmW_{JTE} JTE (**Fig.7**).

B. On-resistance characterization

Normalized by the size of the emitter A_E, a minimal differential on-resistance of 6.7mΩ·cm^2 can be extracted from the conduction measurement for the small size devices (**Fig.7**). The forward drop at 200A/cm^2 is 1.75V. The differential on-resistance includes TLMs estimated 0.89 mΩ·cm^2 emitter contact resistance, 0.5 mΩ·cm^2 collector contact resistance and at least 0.5 mΩ·cm^2 resistance calculated from the 350μm thick N+ substrate. Although, due to the current spreading of the small size device, the normalized on-resistance may be underestimated, the small extracted on-resistance indicates the existence of a weak conductivity modulation in the saturation region.

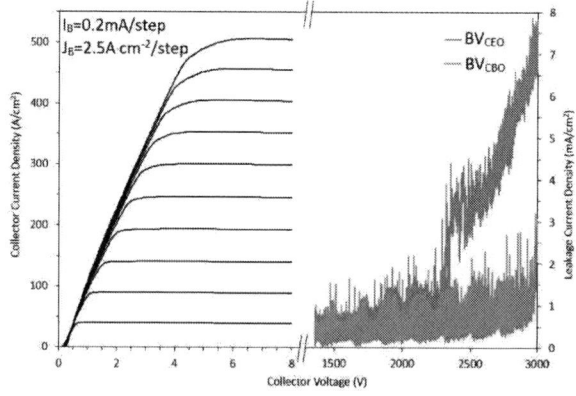

Fig. 7. Conduction and Blocking characteristics of fabricated BJT with 100μm emitter diameter (W_e = 50μm) and total JTE width 40μm under room-temperature. All normalized by the size of emitter.

The re-deposited 150nm conductive metal is important to enhance the metal conductivity. Small size devices without post-

978-1-6654-8901-0/22 $31.00 © 2022 IEEE

anneal metal re-deposition have an on-resistance of 64.2 $m\Omega \cdot cm^2$, which means re-deposited 150nm conductive metal can improve on-resistance by about 10 times.

There are larger size ($0.19mm^2$) finger devices with same emitter size ($W_e=50\mu m$, total width $100\mu m$) for comparison, which gave minimal differential on-resistance of 39.7 $m\Omega \cdot cm^2$. It is believed that the finger metal resistance contributed to the increased on-resistance. Much thicker metal will be needed for larger size devices.

C. Current gain

According to the Gummel plots, the BJTs after 3hr high temperature anneal process have maximum current gain of about 21, while devices without high temperature enhancement process only gave a current gain of about 10(**Fig.8**).

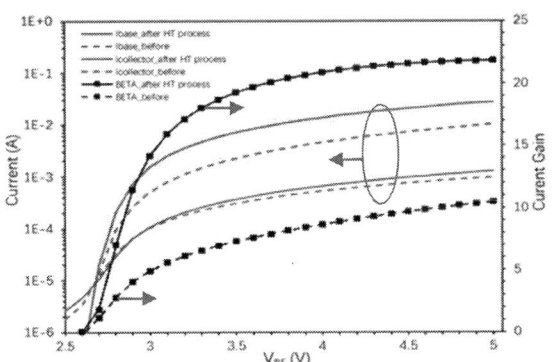

Fig. 8. Gummel plots of fabricated small size BJTs before/after High Temperature process under V_{CB}=5V.

This measured current gain is obviously much smaller than the TCAD simulated maximum current gain. The reason causing this small current gain could be a poor base epi bulk carrier lifetime, or surface post-etching damages that failed to be recovered from the 3hr 1100°C O_2 anneal. According to published SiC BJT papers [5] [6], a post-oxide nitridation anneal in N_2O or NO is very necessary to minimize the interface defects and improve the current gain. However, due to the facility limitation, this kind of nitridation anneal process was not able to be conducted presently.

V. CONCLUSIONS

In this paper, a small size 3kV 4H-SiC BJT with 6.7m$\Omega \cdot cm^2$ on-resistance was designed, fabricated and characterized. A four-step implantation-free JTE structure was utilized and analyzed, which proved to have a good process error tolerance. It is also highly area efficient since the minimal JTE width needed for 3kV breakdown was only 40μm, less than two times as drift thickness (23μm). Extracted differential on-resistance indicated conductivity modulation exist in this device and a maximum current gain of 21 was achieved with a simple 3hr high temperature enhancement process.

ACKNOWLEDGMENT

The work was partly done at the Texas Nanofabrication Facility supported by NSF grant NNCI-2025227.

REFERENCES

[1] Gao, Y., Huang, A. Q., Chen, B., Agarwal, A. K., Krishnaswami, S., & Scozzie, C. (2006, June). Analysis of SiC BJT RBSOA. In 2006 IEEE International Symposium on Power Semiconductor Devices and IC's (pp. 1-4). IEEE.

[2] Kimoto, T., & Watanabe, H. (2020). Defect engineering in SiC technology for high-voltage power devices. Applied Physics Express, 13(12), 120101.

[3] Elahipanah, H., Salemi, A., Zetterling, C. M., & Östling, M. (2014). 5.8-kV implantation-free 4H-SiC BJT with multiple-shallow-trench junction termination extension. IEEE Electron Device Letters, 36(2), 168-170.

[4] Salemi, A., Elahipanah, H., Jacobs, K., Zetterling, C. M., & Östling, M. (2017). 15 kV-class implantation-free 4H-SiC BJTs with record high current gain. IEEE Electron Device Letters, 39(1), 63-66.

[5] Lanni, L., Malm, B. G., Östling, M., & Zetterling, C. M. (2014). Influence of passivation oxide thickness and device layout on the current gain of SiC BJTs. IEEE Electron Device Letters, 36(1), 11-13.

[6] Miyake, H., Kimoto, T., & Suda, J. (2011). Improvement of Current Gain in 4H-SiC BJTs by Surface Passivation With Deposited Oxides Nitrided in N_2O or NO. IEEE electron device letters, 32(3), 285-287.

[7] Ghandi, R., Buono, B., Domeij, M., Zetterling, C. M., & Ostling, M. (2011). High-voltage (2.8 kV) implantation-free 4H-SiC BJTs with long-term stability of the current gain. IEEE transactions on electron devices, 58(8), 2665-2669.

[8] Baliga, B. J. (2010). Fundamentals of power semiconductor devices. Springer Science & Business Media.

[9] Gao, Y., Huang, A. Q., Krishnaswami, S., Agarwal, A. K., & Scozzie, C. (2006, August). Emitter size effect in 4H-SiC BJT. In 2006 CES/IEEE 5th International Power Electronics and Motion Control Conference (Vol. 1, pp. 1-4). IEEE.

[10] Hayashi, T., Asano, K., Suda, J., & Kimoto, T. (2012). Enhancement and control of carrier lifetimes in p-type 4H-SiC epilayers. Journal of Applied Physics, 112(6), 064503.

[11] Murata, K., Tawara, T., Yang, A., Takanashi, R., Miyazawa, T., & Tsuchida, H. (2021). Carrier lifetime control by intentional boron doping in aluminum doped p-type 4H-SiC epilayers. Journal of Applied Physics, 129(2), 025702.

978-1-6654-8901-0/22 $31.00 © 2022 IEEE

Thermal Design and Experimental Evaluation of a 1kV, 500A T-Type Modular DC Circuit Breaker

Baljit Riar
Raytheon Technologies Research Center (RTRC)
East Hartford, CT, USA
baljit.riar@rtx.com

Jeffrey Ewanchuk
Raytheon Technologies Research Center (RTRC)
East Hartford, CT, USA
jeffrey.ewanchuk@rtx.com

Hailing Wu
Raytheon Technologies Research Center (RTRC)
East Hartford, CT, USA
hailing.wu@rtx.com

Yue Zhang
Center for High Performance Power Electronics (CHPPE)
The Ohio State University
Columbus, OH, USA

Xiao Li
Center for High Performance Power Electronics (CHPPE)
The Ohio State University
Columbus, OH, USA

Dihao Ma
Center for High Performance Power Electronics (CHPPE)
The Ohio State University
Columbus, OH, USA

Rob Borjas
Center for High Performance Power Electronics (CHPPE)
The Ohio State University
Columbus, OH, USA

Jin Wang
Center for High Performance Power Electronics (CHPPE)
The Ohio State University
Columbus, OH, USA
wang.1248@osu.edu

Abstract—**This paper presents thermal design of a 1kV, 500A T-type modular DC circuit breaker. Along with the thermal design, which determines the trip/protection curve of the breaker, experimental results of the breaker are also presented in this paper. The breaker is over 99.5% efficient that is evaluated using both thermal measurements and millivolt meter for improving the accuracy of measurements. In addition, experimental results shows that T-breaker can break a fault current of up to 4.24 kA, which is approximately equal to 8.5 times its rated current.**

Keywords—*DC circuit Breaker, T-Breaker, SiC*

I. INTRODUCTION

With the increasing demand for high power in ships, aircrafts and commercial buildings, electrical power distribution with high DC voltage in the range of few kVs is beneficial for weight reduction and high efficiency [1], [2]. There are, however, challenges associated with breaking DC current during faults. For breaking a fault current, solid state circuit breakers are favored over hybrid circuit breakers because of their short interruption time, high power density, high reliability, and increased safety [3]. With aviation in perspective, most of the circuit breakers have been presented for a low bus voltage (270 V) and are not suitable for high voltage DC buses of the modern aircrafts as also highlighted by FOA of DOE [4].

Recently, a T-Breaker that features a modular multilevel "T" structure with integrated energy storage devices was proposed for protecting a DC bus rated at 1 kV and 500 A [5], [6], and the layout of the breaker is shown in Fig. 1. The circuit comprises of two half-bridge sub-modules in each of the three arms. The sub-module capacitors not only absorb the fault energy while disrupting the fault current but also alleviates device synchronization related overvoltage issues, which can be expected from a series connection of power modules. The stored energy in each sub-module could be used to realize series/shunt compensation for improving power transferability, quality, and stability of the dc bus.

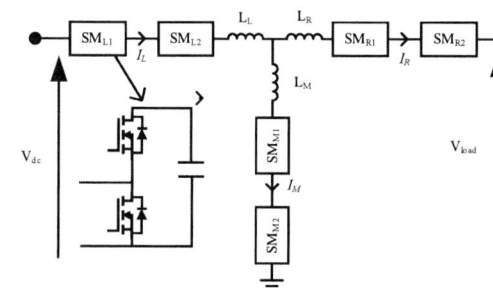

Fig. 1: T-Breaker with two half-bridge sub-modules per arm.

II. DESIGN OF A T-BREAKER

A typical design approach of a breaker is based on the requirements of an underlining system, such as maximum fault current, I_{brk}, fault disruption time, T_{brk}, system DC link voltage, V_{dc}. As shown in Fig. 2, the line current increase linearly as $di/dt = V_{dc}/L_{line}$ when the line inductance, $L_{line} = L_L + L_R$, is dominant over other impedances in the path of a fault current. This inductance results in a fault development time, T_{dev}, i.e. time taken by the current to increase from its nominal value, I_{nom}, to I_{brk}.

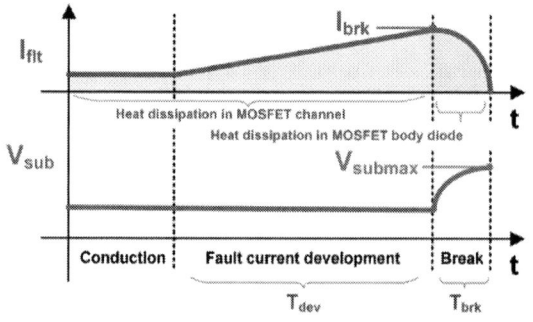

Fig. 2: A typical waveform of a fault current development, and breaking process.

The work presented in this paper was financially supported by Advanced Research Projects Agency-Energy (ARPA-E) under the award DEAR0001110.

978-1-6654-8901-0/22 $31.00 © 2022 IEEE

A. Selection of devices

Among Si, SiC and GaN semiconductor technologies, which have their own advantages and disadvantages, SiC MOSFET is chosen as a fault breaking device and its performance during continuous conduction and active fault scenarios is analyzed. Particularly, the analysis is based on HT3234-R-VB, a pre-released prototype of the HM3 module from Wolfspeed. Two SiC power modules with $R_{DS(on)}$ of 2.5mΩ each are connected in parallel for meeting the efficiency target. With two sub-modules, SM, in each arm, the steady state voltage across each module would be 500 V with the given DC link voltage of 1 kV. A capacitor was selected to absorb fault energy and to compensate for any variation in the DC bus. The electrical design and selection procedure of components is already discussed in [5], [6] and is not further discussed in this paper.

III. THERMAL ANALYSIS

At the basic level, the fault handling capability of the electronic breaker, shown in Fig. 1, is determined by electrical loading due to the fault itself and the thermal handling capability of the specific power module used within the breaker. Hence, the analysis of the fault handing capability is reduced to the analysis of a given power module. Owing to the arm inductance of the breaker, the fault has the electrical characteristic profile as shown in Fig. 2. In this case, the electrical loading can be defined by a peak breaking current, I_{brk}, that evolves over the fault current development time, T_{dev}, also termed as fault duration time, from its steady state current, I_{ss}.

In order to understand the fault handling capability of a power module, the work focused on the development of a thermal model that is applicable for time domain simulations. In order to achieve this objective, a high fidelity 3D thermal FEA model was developed using 3D CAD model and mechanical properties of the power module, and the transient thermal response characteristics were derived. Given the FEA data, a 1-D time-based compact thermal model was developed. In order to improve the confidence of this derived thermal model, the impact of the internal die to die stray electrical inductance and thermal distribution was derived and compared against the 1-D compact model. Finally, with a validated 1-D model, the thermal performance under increasingly long fault durations is established to determine the breaking current limit of the selected power module.

A. Thermal Analysis of Power Module

A high fidelity 3D FEA based thermal model was developed for estimating the steady state thermal resistance of the power module and its transient response. For the thermal model, temperature dependent properties were used for achieving high fidelity in the predictions [7]. At first, a steady state thermal FEA model was developed to predict the thermal resistance of the power module. The converged model has ~500,000 mesh elements which was determined after a thorough mesh convergence study and captures the temperature distribution in the layers with the minimal thicknesses in the stack-up. A heat loss of 500W, which was deduced from the total power loss when the breaker is 99.5% efficient, was specified for all the dies in one switch position in the thermal model. As a first step, this

loss was distributed evenly across all the dies. A boundary condition imitating very high convective heat dissipation rate (h = 10,0000 W/m2-K @ 22 °C) was applied on the base plate. The resulting temperature distribution prediction in the power module and the associated temperature profile in the power module stack-up is shown in Fig. 3. A typical stack up can be found in [7]. The calculated steady state thermal resistance of the power module is ~10% lower than the value in the datasheet,

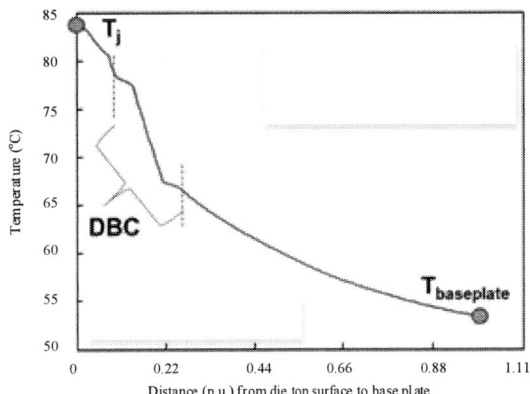

Fig. 3: Temperature profile in a power module stack-up under hottest die.

but this difference in results is acceptable from thermal analysis perspective.

The converged thermal model was used to develop a transient thermal model for estimating the power module response to an abrupt thermal load. The steady state thermal condition was used as the initial condition and the thermal load was abruptly set to a value of 0 W and the cool down behavior of the power module dies were simulated. The time step size chosen for the simulation varied between 10^{-5} to 10^{-3} seconds. The μs level time step resolution was employed to capture the die specific time constants as they would be excited by the breaking event and the ms level time step resolution was used closer towards the expected asymptote. The converged temperature evolution of the hottest die at the steady state condition as it undergoes a step change in thermal loading is shown in Fig. 4. It is found that 10% of the temperature change

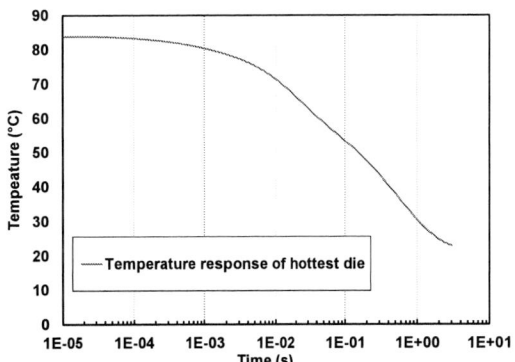

Fig. 4: Thermal response of power module from 3D FEA thermal analysis.

occurs below 1ms, 72% at 21ms, and the remaining temperature change occurs beyond that point.

B. 1-D Compact Thermal Model

As a first approach, the power module is assumed to have uniform heating as a function of the breaking current duration and magnitude, and hence, can be described by a single input, single output model. Given the heating and cooling curves obtained from the FEA program during the transient thermal impedance derivation, a 1D thermal model is fit based on 4th order foster thermal model of the heat propagation, i.e. $Z_{jc} = \sum_{i=1}^{4} r_i \cdot (1 - e^{-t/\tau_i})$. A simple least squares fitting algorithm is used to minimize the error between the FEA derived transient thermal impedance and the fit thermal model. The normalized thermal impedance, Z_{th}, response of the power module based on the temperature response is shown in Fig. 5. It is observed that there is an infinitesimal change in thermal impedance of the power module immediately after the step and the impedance approaches its steady state value at ~ 2 secs. This transient thermal impedance of the power module forms the basis for analyzing thermal performance during transients.

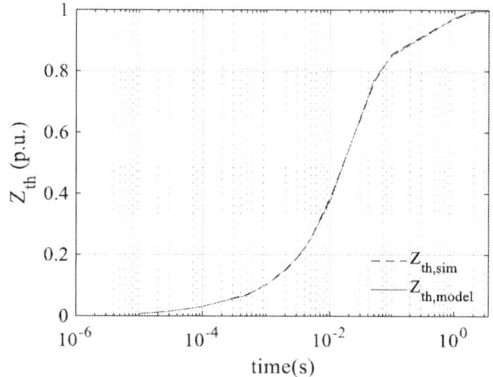

Fig. 5: Transient thermal impedance.

The error between this fit model and the FEA is also shown in Fig. 6, where the normalized error is shown to be less than 1% throughout the transient time response. As the derived 1D compact thermal model is shown to be a good fit with the FEA data obtained from the power module, the power module can be examined in the context of the time-loss relationship for the breaker application. However, before presenting this investigation, several factors are related to improving the confidence of the derived 1D compact module: current sharing from die to die and the thermal dispersion within the dies of the power module, and the impact of sub-100 µs fault currents on the peak internal temperature of the die.

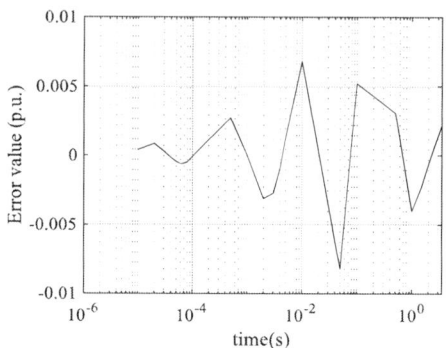

Fig. 6: Error between fitting of compact 1D thermal model and FEA derived data.

C. Die to Die Loading During Fault Current Evolution

In the previous section, the fault is assumed to uniformly heat the power module due to a uniformly distributed current across the composite dies. However, in reality, the current within the power module distributes according to the frequency content of the current, and the parasitic impedance of the power module at that frequency. Furthermore, the thermal boundary conditions on each die is not equal within the power module, as dies on the edge of the power module will experience less cross coupling of the heating from the neighboring dies as compared to those in the center. Hence, in order to study the impact of the internal die layout of the selected power module on the fault handling capability of the breaker, the individual die thermal impedance and impedance network needs to be modelled.

As previously presented, the thermal impedance of the power module can be characterized by applying a heat load and evaluating the step change in temperature for a given loss. Similarly, at the die level, the self-heating thermal impedance and die-die thermal impedance is characterized by applying a single heat load to a die and measuring the resulting temperature rise in the applied die as well as the neighboring dies. In this fashion, each of the n dies can be characterized with the diagonal impedance matrix with $Z_{n,n}$ of die 'n', being the self-heating thermal impedance, and the off-diagonal component being the thermal coupling, $Z_{m,n}$ of die 'n' and die 'm', as per (1).

$$Z_{jc} = \begin{bmatrix} Z_{1,1} & \cdots & Z_{1,n} \\ \vdots & \ddots & \vdots \\ Z_{n,1} & \cdots & Z_{n,n} \end{bmatrix} \quad (1)$$

The thermal FEA system is solved at a 100 µs time step as a compromise between accuracy and convergence speed, and the resultant $n{\times}n$ sets of data is fit to a 1D compact time based model, as per previous section. As an example, Fig. 7 (a) & (b) illustrates thermal response of die 1 (the furthest from the center) to die 2 (closest to die 1) when the loss is applied to die 1. As can be seen from Fig. 7 (a) and (b), the thermal coupling between die 1 and die 2 occurs at a slower time scale than the self-heating response of the die 1. The dies that are further away from the heating die would show minimal thermal interaction with die 1. Least square fitting algorithm was used to obtain an analytical equation and results are presented as $Z_{th,model}$ in Fig. 7.

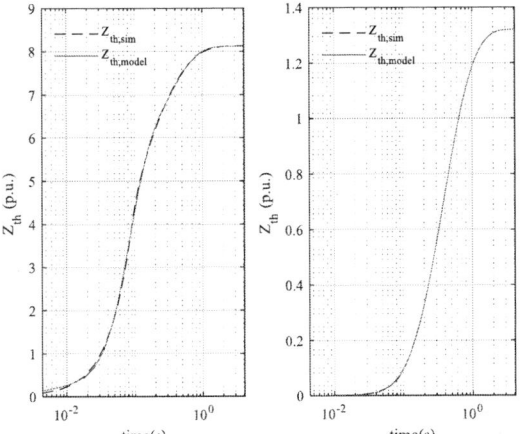

Fig. 7: Thermal response characteristic of the internal dies within the power module, (a), with die 1, and (b) die 2 when heat is applied to die 1.

As the current sharing within the power module is dictated by the parasitic impedances from the DC terminal to the output terminal, a Q3D parasitic extraction method is applied to the 3D CAD model. The individual parasitic paths are determined by applying the input current at each die surface. Due to the minimum fault current duration of 10 μs, the simulation is solved with a skin depth of 100 kHz. The resulting analysis shows that while the inductance from dies on the edge to the center differs by up to 50%, the magnitude of this inductance is in the order of few nH, and with a frequency of 100 kHz the layout provides a relatively small impedance compared to the dc resistance of the power module and on-state resistance of the dies, concluding that the current is well balanced withing the module. The above derived $Z_{th,model}$ is used to determine the power loss handling capability. Here maximum time duration is evaluated for each power level such that the junction temperature can be kept below 150°C and the results are shown in Fig. 8. The power loss ratio is normalized to the module's dc loss rating. This result can be further transformed into maximum current as a function of time by using VI characteristics of the module.

IV. EXPERIMENTAL EVALUATION

A. Conduction Test

The efficiency of the circuit is calculated using two methods: electrical conduction test and thermal coolant temperature analysis, where the electrical test offers high accuracy over the thermal test. The biggest challenge was to determine the efficiency with high accuracy when the breaker is operated at its rated current of 500A and rated voltage of 1kV. Most of the measurement devices, which are rated for high voltage and current measurements, do not offer high enough accuracy. We therefore used a millivolt meter for capturing the small voltage drop across the breaker and used this loss for evaluating efficiency. The output terminals of the breaker are shorted with which the lab power supply is operated in the current limit mode, allowing to measure conduction losses in breaker with a high accuracy. Here the vertical or middle arm of the breaker, which has very low losses in the range of 10W per sub module, is disabled for evaluating the conduction losses. This approach does not compromise efficiency calculations because of the very low losses in the vertical arm. The accuracy of used current probe is + 1% and accuracy of the millivolt meter is + (0.0035% of measurement + 0.0005% of range i.e. 10V). The accuracies of the temperature probes and micro motion flow meter are + 0.015 °C and + 0.1%, respectively. The efficiency is then calculated as $\eta = 1 - \frac{\Delta V_{meter} \times I_{dc}}{V_{dc,rated} I_{dc}} = 1 - \frac{\Delta V_{meter}}{V_{dc,rated}}$.

With the water coolant temperature set at its nominal value, the value of the breaker current is varied in steps in the range of $\{100, 200, 300, 400, 500\}$ A and voltage drop across the breaker input terminals is recorded. For each breaker current, measurements namely voltage, current, coolant inlet and outlet temperatures, and flow rate are recorded after five minutes giving enough time for the temperature dependent variables to reach their equilibrium. All the possible uncertainties related to % accuracy of current, voltage, temperature and flow rate measurements are used for evaluating efficiency of the circuit and results are presented in Fig. 9 and Fig. 10. All the possible measurement uncertainties, which are plotted using "o" asterisk, show that the efficiency of the breaker is higher than 99.5 %.

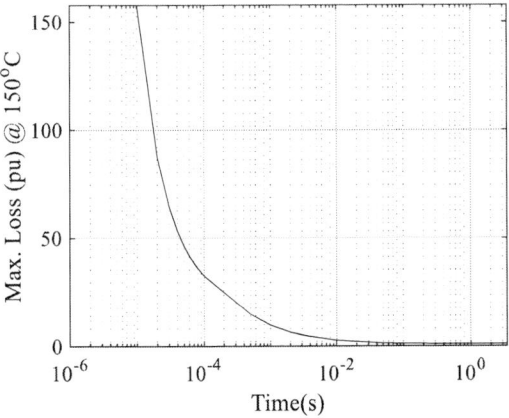

Fig. 8: The FEA derived power loss ratio versus fault duration from a fixed junction case temperature of 150°C.

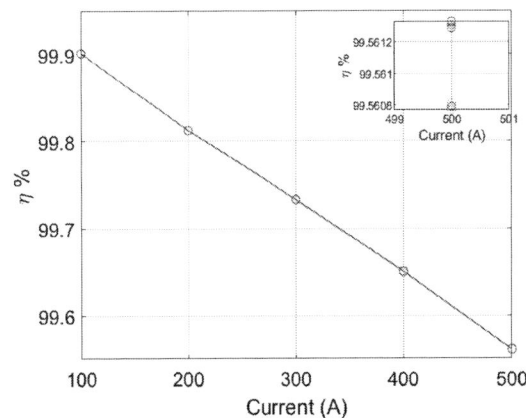

Fig. 9: Efficiency evaluation based on electrical measurements. The inset shows efficiency variations at 500A resulting from uncertainties in voltage and current measurements.

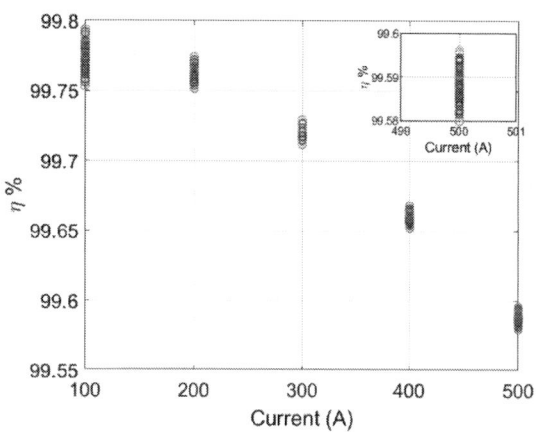

Fig. 10: Efficiency evaluation based on thermal measurements. The inset shows efficiency variations at 500A resulting from uncertainties in temperature and flow rate measurements.

B. Transient Performance

The setup for the short circuit test is shown in Fig. 11. Here relays S_1 and S_2 i.e. GX110BE Giavac are used to connect the capacitors C_1 and C_2 to the breaker during normal operation. Couple of IGBTs (S_3) FZ2400R12HP4 where each is rated for a repetitive pulse current of 4800A is used to create fault current upto 5000A. The gate drivers of the IGBTs are tuned to open the IGBTs once the fault current reaches around 5500A and thereby only interrupt the fault if T-breaker fails to interrupt the current. For initiating the fault, switches S_1 and S_2 are opened and then S_3 is closed. The gate driver protection of the IGBT was tested before connecting the setup to the breaker

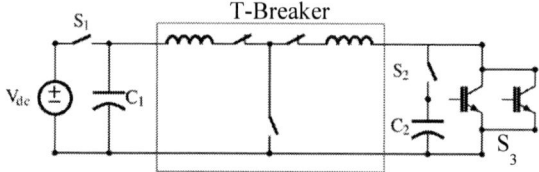

Fig. 11: Electrical setup for a short circuit test.

The T-breaker is connected to a test circuit for breaking test and here DC link voltage is increased incrementally while performing the test at each step. The T-breaker successfully opened the circuit while limiting the fault current below 4.24 kA Fig. 12. It takes about 45 us for the current to reach 4.24 kA. This current is less than the planned value of 5000A because the desaturation detection of the T-breaker gate drivers prevented the fault current from reaching its maximum value. Voltage of a submodule in the left arm and gate source voltage of two submodule are also shown in the same figure. As expected, the voltage across the submodules builds up as soon as T-breakers starts interrupting the fault current. As predicted by thermal models, the temperature rise in the coolant is measured ~ 3 °C when the breaker carries the full load current of 500 A, and there is no noticeable change in the coolant temperature during the fault scenario.

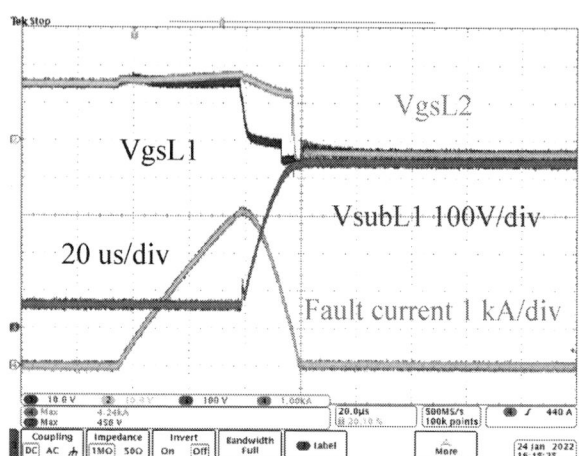

Fig. 12: System breaking test where fault current is interrupted by desaturation-based protection.

V. Conclusions

The paper presented a thermal design framework for a 1-kV, 500- A T-Breaker that use SiC power modules. Thermal models were used to predict that the T-breaker can handle ten times its rated current i.e. 5 kA of fault current within a 150 µs fault clearing time. The experimental results have been presented to show that the breaker can break current up to 4.24 kA, where a de-sat protection of the gate drivers prevented the fault currents to reach 5kA. In addition, efficiency of the breaker has been presented using measurements of a millivolt meter and thermal measurements for showing that the breaker is 99.5% efficient at its rated current of 500A.

References

[1] A. Barzkar and M. Ghassemi, "Electric Power Systems in More and All Electric Aircraft: A Review," in IEEE Access, vol. 8, pp. 169314-169332, 2020, doi: 10.1109/ACCESS.2020.3024168.

[2] Z. Dong, R. Ren and F. Wang, "Development of High-power Bidirectional DC Solid-state Power Controller for Aircraft Applications," IEEE Journal of Emerging and Selected Topics in Power Electronics., doi: 10.1109/JESTPE.2021.3139903.

[3] R. Rodrigues, Y. Du, A. Antoniazzi and P. Cairoli, "A Review of Solid-State Circuit Breakers," in IEEE Transactions on Power Electronics, vol. 36, no. 1, pp. 364-377, Jan. 2021, doi: 10.1109/TPEL.2020.3003358.

[4] DE-FOA-0001953, ARPA-E funding opportunity announcements, 11 Jan 2022, [Online]. Available: https://arpa-e-foa.energy.gov/

[5] Y. Zhang, F. Alsaif, X. Li, R. Na and J. Wang, "T-Type Modular Dc Circuit Breaker (T-Breaker) for Future Dc Networks," in 2021 IEEE Applied Power Electronics Conference and Exposition (APEC), 2021, pp. 1146-1152.

[6] Y. Zhang et al., "Development of a 1 kV, 500 A, SiC-Based T-Type Modular DC Circuit Breaker (T-Breaker)," 2021 IEEE 8th Workshop on Wide Bandgap Power Devices and Applications (WiPDA), 2021, pp. 199-204.

[7] M. Chen, H. Wang, D. Pan, X. Wang and F. Blaabjerg, "Thermal Characterization of Silicon Carbide MOSFET Module Suitable for High-Temperature Computationally Efficient Thermal-Profile Prediction," IEEE Journal of Emerging and Selected Topics in Power Electronics, Vol.

Source Turn-off (STO) MOSFET: A New Driving Architecture for Smart SiC Module

Zhicheng Guo
Semiconductor Power
Electronics Center
The University of Texas at Austin
Austin, USA
zcguo@utexas.edu

Alex Q. Huang
Semiconductor Power
Electronics Center
The University of Texas at Austin
Austin, USA
aqhuang@utexas.edu

Abstract—High-speed and intelligent gate driver is the critical interface between power semiconductor devices and control signals to achieve low switching loss and intelligent protection. The proposed novel Source Turn-Off (STO) MOSFET is a driver integrated MOSFET architecture that can achieve ultra-fast turn-on and turn-off operation beyond traditional voltage source gate driver approach. 1200V SiC STO half bridge was developed to demonstrate the concept. Faster switching speed and lower switching losses are demonstrated. The STO can be readily realized by using commercially available Si power MOS IC which can further enable intelligent functionalities such as build-in current monitoring, temperature monitoring and over current protection which are important for high power SiC modules.

Keywords— SiC MOSFET, gate driver

I. INTRODUCTION

Wide-bandgap (WBG) semiconductor devices such as SiC MOSFET is a unipolar power switch which inherently can switch much faster than Si bipolar devices such as IGBT, resulting in lower switching loss in power converter applications. However, switching loss is not negligible particularly if the switching frequency is high. The situation is getting worse for a high current power module due to the increased gate capacitance and the need to slow down the gate driver to achieve proper current sharing among parallel dies. So, there is a need to develop new techniques to achieve low noise and ultra-fast gate driving. In [1], a driver integrated SiC MOSFET was demonstrated to achieve close to zero turn-off loss by utilizing the so called hard-driven concept. The gate driving loop is faster than the drain loop. However, it is hard to scale this to higher current devices.

To achieve higher switching speed, current source [2] and resonant gate driver [3] have been developed. Another attractive hard-driven turn-off concept is the cascode architecture used in emitter turn-off (ETO) [4] and SiC JFETs [5]. In the case of ETO, hard-driven condition is achieved even at 4000A.

This paper proposes the Source Turn-off (STO) MOSFET concept to achieve ultra-fast driving for SiC MOSFET with the driving speed proportional to the load current. To realize the STO, a commercial power IC chips with compact package is utilized to minimize the loop parasitic inductances. The detailed switching performance of the proposed STO MOSFET is

demonstrated and compared with datasheet with conventional driving scheme. On top of the improvements, the build-in current monitoring, temperature monitoring and over current protection are also considered in the design.

II. STO MOSFET CONCEPT

Fig. 1 shows STO concept and simplified switching process. The driver is formed by a low voltage half bridge S1 and S2 and its output is connected to the source of the SiC MOSFET. Con and Coff provide the turn-on voltage and turn-off voltage respectively. For the proof-of-concept demonstration, a 80A buck IC from Vishay (SiC830) is used which has an extremely small package (5.0 x 6.0 x 0.75 mm^3).

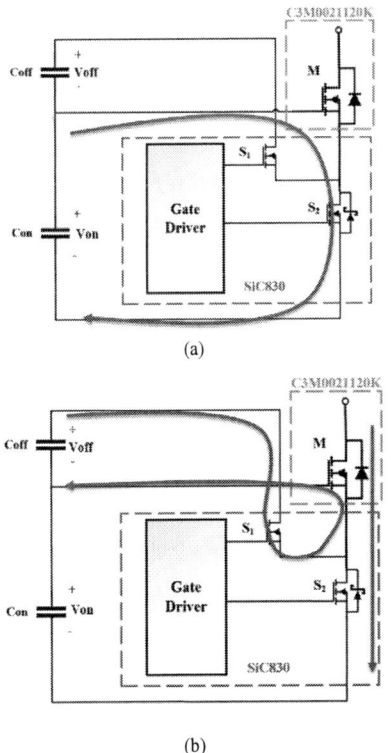

(a)

(b)

Fig. 1. Switching process of the source switched gate driver (a) turn-on operation; (b)turn-off operation.

The gate of the SiC MOSFET is connected between Coff and Con. The voltages applied on Con and Coff are Von and Voff, which control the positive and negative driver voltages. The integrated power stage IC and bypass capacitors are placed close to the gate and source terminals of the high voltage SiC MOSFET to minimize the parasitic inductance.

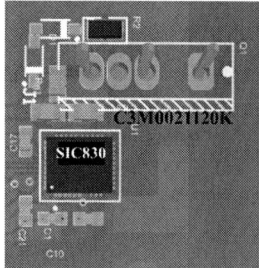

Fig. 2. Conceptual layout of the STO MOSFET.

The turn-on process starts with the turn-on of the source MOSFET S2, then Von charges the gate of the SiC MOSFET. Compare with conventional gate driver IC, the STO has a much higher peak charge current of up to 80A (SiC830 capability). During normal conduction, the load current of the SiC MOSFET also flows through S2. During the turn-off process, the S2 is turned off and S1 turns on first. The inductive load current charges the Coss of S2 so the dynamic voltage Vds(S2) becomes higher than Von voltage, effectively creating a negative voltage for SiC MOSFET turn-off. If inductance is ignored, the SiC MOSFET gate to source voltage

$$V_{gs}(M) = V_{on} - V_{ds}(S2) \qquad (1)$$

which will become negative quickly. At higher load currents, Vds(S2) rises faster hence the STO MOSFET turns off faster.

The negative voltage is clamped by Coff to

$$V_{gs}(M) = -V_{off} \qquad (2)$$

The Von and Voff can be easily adjusted in the design to match the requirements of different devices.

III. PROTPTYPE AND EXPERIMENTAL RESULTS

A. STO Prototype

A half-bridge SiC STO module with four paralleled STO MOSFET branches was developed, and the hardware is shown in Fig. 3. The power module has three high power connections (VDC+, VDC- and SW node) and four optical fiber connections for two PWM inputs and two current monitor outputs. Each module is formed by eight 1200V SiC MOSFETs (C3M0021120K), utilizes PCB to form low inductance power loop and internal decoupling capacitance [6]. To verify the performance of the STO MOSFET and analysis above, one half-bridge branch was used in the double pulse test(DPT).

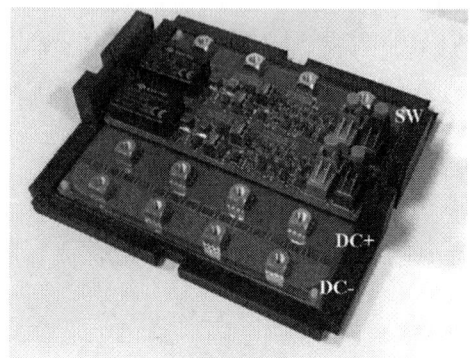

Fig. 3. 1200V/5mohm half-bridge SiC STO MOSFET power module.

B. STO Double Pulse Test

Fig. 4 shows the double-pulse test circuit which includes the STO half-bridge module, a 60 μH air-core inductor, a 60 uF DC capacitor and a high-voltage DC power supply.

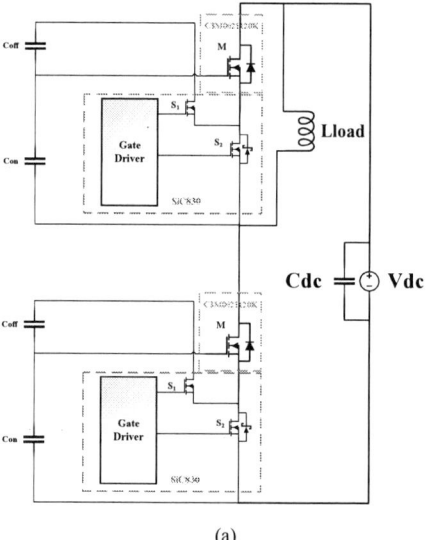

(a)

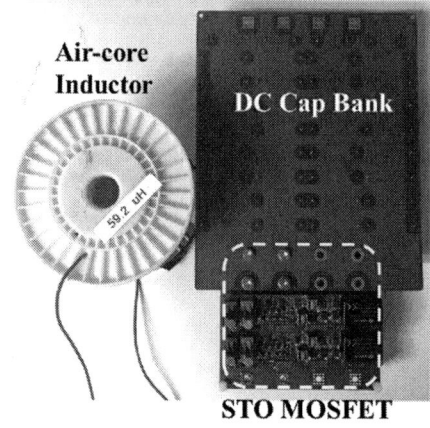

(b)

Fig.4. (a) STO MOSFET double-pulse test circuit diagram. (b) DPT hardware.

A group of DPT results were captured with different load current (from 10 A to 80 A) at the same voltage (800 V). Drain-source voltage (Vds) of the SiC MOSFET M and the drain to source current (Id) waveforms are captured. Fig. 5 shows the eight groups DPT test waveforms marked with different colors. The switching times with different Id are plotted in Fig. 6. The fall time decreases from 37.5 ns to 20.3 ns with the Id increase from 10 A to 80 A. The higher the load current the faster the turn-off. The switching energy which include turn-on energy (E_{on}), turn-off energy (E_{off}) and total energy (E_{total}) under different drain to source current are plotted in Fig.7.

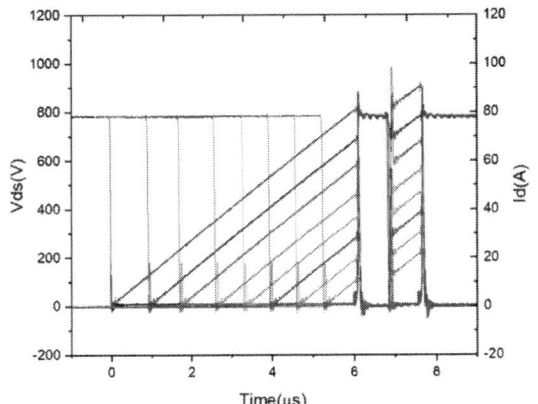

Fig. 5. Double pulse switching waveforms with 800 V drain-source voltage and different load current (10 A-80 A).

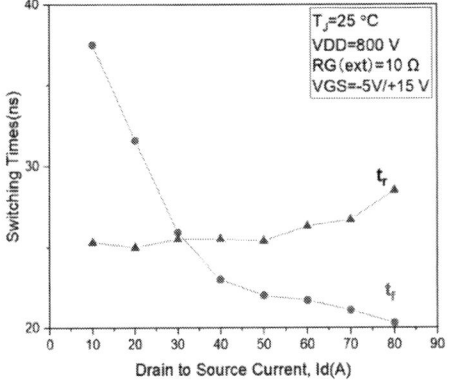

Fig. 6: Switching times at different drain current (Junction temperature 25 °C, drain-source voltage 800 V, external gate resistance 10 Ω, gate-source voltage -5/+15 V).

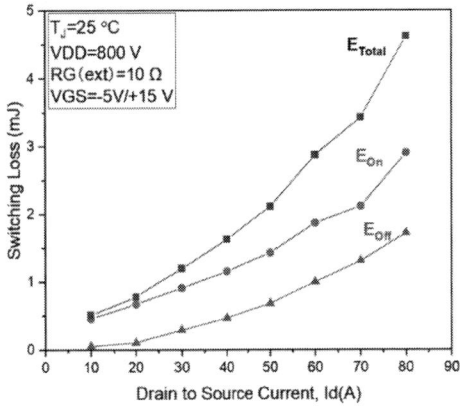

Fig. 7: Switching losses at different drain current (Junction temperature 25 °C, drain-source voltage 800 V, external gate resistance 10 Ω, gate-source voltage -5/+15 V).

C. Switching Performance Comparision with Datasheet

The datasheet of C3M0021120K SiC MOSFET presents detailed device parameters under different test conditions. Table I compares the switching speed and switching energy value between C3M0021120K datasheet and the proposed STO under same drain-source voltage, drain to source current and external gate resistance. The gate-source voltage has minor differences with -4/+15 V for datasheet and -5/+15 V for STO. The proposed source turn off MOSFET archives faster switching speed and lower switching loss with 13.9 % reduction.

TABLE 1: SWITCHING PERFORMACE COMPARISION

Parameters	Datasheet*	Proposed STO**
Rise Time tr (ns)	37.5	25.4
Fall Time tf (ns)	29	22
Turn-on Energy Eon (mJ)	1.65	1.43
Turn-off Energy Eoff (mJ)	0.8	0.68
Total Energy Etotal (mJ)	2.45	2.11

* TJ = 25 °C , V_{DD}= 800 V, I_{DS}= 50 A, $R_{G(ext)}$= 10 Ω, V_{GS}= -4/+15 V.

** TJ= 25 °C , V_{DD}= 800 V, I_{DS}= 50 A, $R_{G(ext)}$= 10 Ω, V_{GS}= -5/+15 V.

IV. CONCLUSIONS

This paper has demonstrated a novel STO MOSFET concept. A high power half-bridge SiC STO MOSFET power module was developed to demonstrate the concept. Double-pulse test (DPT) has been conducted which confirms the fast switching speed of the proposed STO MOSFTE. Faster switching speed and lower switching losses have been proved by compare with datasheet value under similar test conditions. For the proposed STO MOSFET, the integrated Si power MOS IC which can further enable intelligent functionalities such as build-in current monitoring, temperature monitoring and over current protection which are important for high power SiC module.

978-1-6654-8901-0/22 $31.00 © 2022 IEEE

REFERENCES

[1] S. Guo et al., "3.38 Mhz operation of 1.2kV SiC MOSFET with integrated ultra-fast gate drive," 2015 IEEE 3rd Workshop on Wide Bandgap Power Devices and Applications (WiPDA), 2015, pp. 390-395, doi: 10.1109/WiPDA.2015.7369298.

[2] J. Fu, Z. Zhang, Y. Liu, P. C. Sen and L. Ge, "A New High Efficiency Current Source Driver With Bipolar Gate Voltage," in IEEE Transactions on Power Electronics, vol. 27, no. 2, pp. 985-997, Feb. 2012, doi: 10.1109/TPEL.2010.2077741.

[3] Z. Yang, S. Ye and Y. -F. Liu, "A New Dual-Channel Resonant Gate Drive Circuit for Low Gate Drive Loss and Low Switching Loss," in IEEE

Transactions on Power Electronics, vol. 23, no. 3, pp. 1574-1583, May 2008, doi: 10.1109/TPEL.2008.920877.

[4] Bin Zhang, A. Q. Huang, Xigen Zhou, Yunfeng Liu and S. Atcitty, "The built-in current sensor and over-current protection of the emitter turn-off (ETO) thyristor," 38th IAS Annual Meeting on Conference Record of the Industry Applications Conference, 2003., 2003, pp. 1264-1269 vol.2, doi: 10.1109/IAS.2003.1257712.

[5] T. McNutt et al., "Silicon carbide JFET cascode switch for power conditioning applications," 2005 IEEE Vehicle Power and Propulsion Conference, 2005, pp. 499-506, doi: 10.1109/VPPC.2005.1554617.

[6] Z. Guo, L. Zhang, S. Sen and A. Q. Huang, "A Novel 3.6kV/400A SiC Intelligent Power Module (IPM)," 2021 IEEE Applied Power Electronics Conference and Exposition (APEC), 2021, pp. 39-43, doi: 10.1109/APEC42165.2021.9487034.

A 5 to 50 V, −25 to 225 °C, 0.065%/°C GaN MIS-HEMT Monolithic Compact 2T Voltage Reference

Ziqian Li[1], Yi Shen[1], Ang Li[1,2], Wen Liu[*1,2]

[1]*School of Advanced Technology, Xi'an Jiaotong-Liverpool University, Suzhou, China*
[2]*Department of Electrical Engineering and Electronics, University of Liverpool, Liverpool, UK*
*Email: Wen.Liu@xjtlu.edu.cn

Abstract—Wide-bandgap GaN devices are emerging as promising candidates for high switching frequency, high-voltage applications, and high-temperature operation in smart power conversion systems. This work introduces a monolithic GaN MIS-HEMT-based voltage reference consisting of two transistors (2T), a D-mode and an E-mode device. Experimental results have been obtained that a 2.5 V reference voltage (V_{REF}) is produced for a supply voltage ranging from 5 to 50 V, achieving a maximum V_{REF} line sensitivity of 0.065%/V and a temperature coefficient of 24.6~28.6 ppm/°C from −25 to 225 °C. This structure also shows competitive levels of uniformity, power consumption and flexibility. With the superior characteristics of the GaN device, the presented results will provide stable reference voltage in various sensing and biasing circuits, especially for GaN power ICs.

Keywords—*GaN, MIS-HEMT, monolithic integrated circuit, voltage reference.*

I. INTRODUCTION

Wide band gap GaN HEMTs are promising candidates for high switching frequencies, high voltage applications and high-temperature operation in smart power conversion systems [1]. In these power conversion systems, GaN is widely used, exhibiting high-performance conversion and showing higher efficiency than conventional Si-based designs [2], [3]. In electric vehicle applications, IC designs place higher demands on the supply voltage and temperature range, as shown in Figs. 1 (a) and (b). In GaN IC function blocks, larger/greater operating margins for supply voltage and temperature are required to achieve high-performance designs [4].

For GaN monolithic ICs, reference voltages are important and can be used as bias or reference for other functional blocks and feedback systems, such as sensors and power conversion systems [5]–[7]. In GaN technology, there have been attempts for such a voltage reference function blocks realized by using GaN E-mode HEMT with F-ion injection gate and Schottky barrier diodes (SBDs) [8], but its output voltage is negative, which circumscribes its practical application. Recently reported voltage references are supply voltage and temperature insensitive with feedback mechanisms, however the design can be further simplified [9].

In this work, a simple connected GaN voltage reference structure is presented. With the connection of D-mode and E-mode MIS-HEMTs, the proposed voltage reference has a wide temperature and supply voltage range without the requirement for an additional pre-modulator. The structure also exhibits good stability, uniformity and low energy consumption. This structure enables a simple and stable design that provides input voltages for bias and sensing ICs and has applications, such as for advanced 48-to-1 V conversions [10].

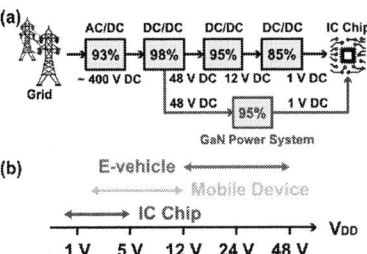

Fig. 1. (a) Schematic diagram [2] and (b) typical source voltages of power conversion systems.

II. WORKING PRINCIPLES OF 2T VOLTAGE REFERENCE

Based on commercial Si based GaN wafer, the proposed GaN MIS-HEMT structure and voltage reference block are shown in Figs. 2(a) and (b). The voltage reference is formed by a D-mode MIS-HEMT M1 whose gate terminal is grounded to ensure a constant gate voltage. M1 is in series with a diode-connected E-mode transistor M2, leading to a turn-on voltage of this reference block to be approximately equivalent to the threshold voltage of the E-mode HEMT. This structure empowered the proposed voltage reference to have a constant output voltage independent of supply voltage and temperature. The used D/E-mode transistor has a gate width of 100 μm and a gate length of 3 μm, and the block ends up with a size of 340×220 μm^2 with the testing pad.

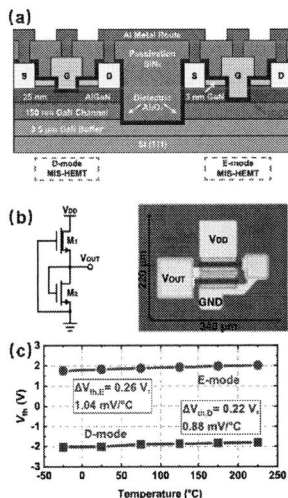

Fig. 2. (a) Cross-section view of GaN D/E-mode MIS-HEMTs. (b) Circuit schematic and micrograph of the proposed voltage reference. (c) Threshold voltage-temperature relationship of D- and E-mode HEMTs.

978-1-6654-8901-0/22 $31.00 © 2022 IEEE

The voltage reference performs a tolerance to changes in the supply voltage. As the supply voltage V_{DD} increments from 0 V, the output voltage firstly increases with it. This state is because the gate-to-source voltage of M2 is insufficient to turn it on, and M1 works in the linear mode to pass V_{DD} as the V_{out}. The increasing V_{DD} will eventually turn on M2, which constantly works in saturation mode, during which M1 will also become saturated with a decreasing gate voltage V_{GS1} ($V_{GS1} = -V_{out}$) and an increasing drain current I_D. This process could be fast with the proposed well-matched threshold voltages of D- and E-mode transistors as $|V_{th,D}| \approx |V_{th,E}|$, and the large static power consumption can be avoided since the ($V_{GS1} - V_{th,D}$) of M1 is at the level of millivolt s, so the total current increment is limited. Once M1 enters the saturation region with a certain V_{GD1} ($-V_{out}$), it operates as a constant current source. The M2 will convert this constant current into a voltage, thus achieving a constant voltage that can be output as a voltage reference which is independent of V_{DD}.

The voltage reference is expected to be temperature independent through the conjunction with temperature-positively and negatively related characteristics of threshold voltages as shown in Fig. 2 (c). The approximately equal threshold voltage variations ΔV_{th} can lead to suppression of V_{out} variation. In the working region of the proposed voltage reference, both D- and E-mode are in their saturation region and have the same current flow from drain to source, $I_{D,D} = I_{D,E}$, then the reference voltage can be derived as [11]

$$V_{REF} = (\sqrt{\frac{k_{n,E}}{k_{n,D}}} V_{th,E} - V_{th,D})/(1 + \sqrt{\frac{k_{n,E}}{k_{n,D}}}) \qquad (1)$$

where $k_{n,E}/k_{n,D} = \mu_{eff,E} C_{eff,E} / \mu_{eff,D} C_{eff,D}$, $V_{th,E}$ is proportional to absolute temperature (PTAT) and $-V_{th,D}$ is complementary to absolute temperature (CTAT). A slight temperature-related variation of the output can be expected according to the report of similar thermal degradation of both D-mode and E-mode MIS-HEMTs [12]. Furthermore, this structure is not limited to only one specific fabrication technology; even with other process flows which perform negative temperature drifts of V_{th} for both D/E-mode devices [13], as long as they show a similar trend of V_{th} temperature-related drift, the compensation can still be achieved.

III. Experimental Results

Referring to the experiment results shown in Fig. 3, this circuit can output a stable reference voltage regardless of the supply voltage (5~50 V) and temperature (−25~225 °C) variations. At room temperature (25 °C), its output is 2.5 V with a line sensitivity of 0.044 %/V, which proves the negligible supply voltage dependence. By selecting a V_{DD} of 10/20/30/40/50 V, the output gives a small temperature coefficient (TC) of 24.6/26.8/27.1/28.3/28.6 ppm/°C, respectively, which validates the temperature drift suppression of the proposed voltage reference.

The statistical results of V_{REF} through 30 samples of 10 dies are shown in Fig. 4, which exhibit an average V_{REF} of 2.55 V and a standard deviation of 0.0292 V. The power consumption of the voltage reference varies with temperature from −25 to 225 °C with V_{DD} = 5 V, as shown in Fig. 5. The rated power consumption of the voltage reference is 75 μW at room temperature and decreases to 42 μW at 225 °C because the current of the GaN MIS-HEMT decays as the temperature increases. Compared to our previous work, by optimizing

layout structure and fabrication processes, this voltage reference operates around threshold voltage [11], then the power consumption is reduced to a competitive level.

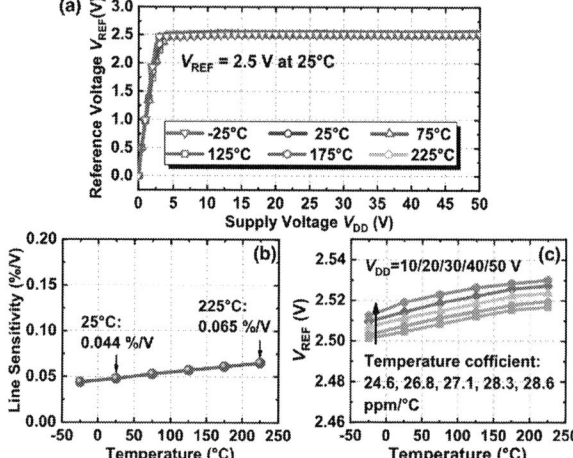

Fig. 3. (a) Measured output voltage V_{REF} versus supply voltage V_{DD} at temperatures from −25 to 225 °C. (b) Measured line sensitivity versus temperature at V_{DD} from 5 to 50 V. (c) V_{REF} versus temperature at V_{DD} = 10/20/30/40/50 V.

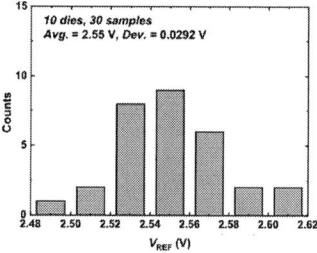

Fig. 4. Statistic results of V_{REF} through 30 samples of 10 dies. The average generated reference voltage is 2.55 V, and the deviation is 0.0292 V.

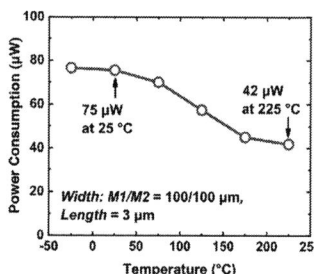

Fig. 5. Rated power consumption of the voltage reference at temperatures from −25 to 225 °C under V_{DD} = 5 V. At room temperature, the rated power consumption is 75 μW and 42 μW at 225 °C.

In order to generate different reference voltages, variants based on this 2T voltage reference are proposed. The first variant is a higher voltage reference that produces a higher output by stacking two 2T voltage references, as shown in Fig. 6(a). The measured results (Fig. 6(b)) show that this 4T structure doubles the output voltage of the 2T structure. Moreover, the lower output voltage can be generated by replacing the bottom E-mode device (M2 in Fig. 2) in the

978-1-6654-8901-0/22 $31.00 © 2022 IEEE

original 2T voltage reference with multiple E-mode devices; the circuit structure and test results are shown in Figs. 6(c) and (d). The different variants extend the range of the output voltage, and so the applications of the 2T structure and demonstrate the flexibility and applicability of it.

In Table I, several voltage reference implementations in both GaN and Si CMOS technologies are compared. The 2T structure shows a wider operating range in terms of the supply voltage (5 to 50 V) and temperature (−25 to 225 °C) with a smaller line sensitivity (0.065 %/V) among these works. This 2T structure uses only 0.078 mm² of chip area, achieving a better area efficiency.

TABLE I.   PERFORMANCE SUMMARY AND COMPARISON

	This work	[8] (2009)	[9] (2020)	[14] (2016)	[15] (2013)
Process	GaN HEMT	GaN HEMT	p-GaN	0.18 µm Si CMOS	0.18 µm Si CMOS
V_{DD} Range (V)	5~50	−4.3~10	3.9~24	1.2~2.2	1.2~1.8
V_{REF} (V)	2.5	−2.1	3.19	0.9862	1.9
Line Sensitivity (%/V)	0.065	N/A	0.32	0.38	0.737
Temp. Range (°C)	−25~225	25~250	−50~200	−40~85	−40~120
Temp. Coefficient	26.8 ppm/°C	70 ppm/°C	23.6 ppm/°C	N/A	147 ppm/°C
Area (mm²)	0.078	0.315	0.52	0.0488	0.0294

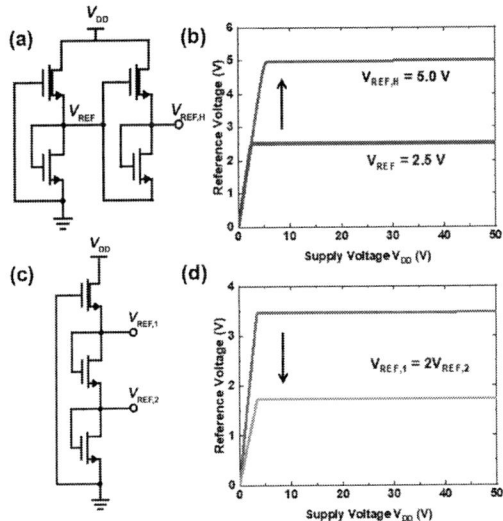

Fig. 6. (a) Schematic of the structure of the higher voltage reference, (b) measured results at supply voltages from 5 to 50 V as $V_{REF,H} = 2V_{REF}$. (c) and (d) schematic and measured results for the lower voltage reference as $V_{REF,2} = 1/2 V_{REF,1}$.

IV. CONCLUSION

A monolithic voltage reference has been implemented by AlGaN/GaN MIS-HEMTs. The design consists of only two transistors and exhibits superior stability over a wide range of operating voltages and temperatures Extensions of the circuit were attempted and fabricated, voltage references with a higher or lower output voltage were obtained. With its simple structure and high performances, the voltage reference can be designed as a high-density functional module in an all-GaN smart power IC.

ACKNOWLEDGEMENT

This work was supported in part by the Suzhou Science and Technology Program under Grant SYG201923 and Grant SYG202131, and in part by the Key Program Special Fund in Xi'an Jiaotong–Liverpool University (XJTLU) under Grant KSF-T-07.

REFERENCES

[1] K. J. Chen et al., "GaN-on-Si Power Technology: Devices and Applications," IEEE Transactions on Electron Devices, vol. 64, no. 3, pp. 779–795, 2017.

[2] D. Reusch, J. Strydom, and J. Glaser, "Improving high frequency DC-DC converter performance with monolithic half bridge GaN ICs," in 2015 IEEE Energy Conversion Congress and Exposition (ECCE), 2015, pp. 381–387.

[3] D. Kinzer and S. Oliver, "Monolithic HV GaN Power ICs: Performance and application," IEEE Power Electronics Magazine, vol. 3, no. 3, pp. 14–21, 2016.

[4] A. Lidow, "The Path Forward for GaN Power Devices," in 2020 IEEE Workshop on Wide Bandgap Power Devices and Applications in Asia (WiPDA Asia), 2020, pp. 1–3.

[5] R. Sun, Y. C. Liang, Y.-C. Yeo, C. Zhao, W. Chen, and B. Zhang, "All-GaN Power Integration: Devices to Functional Subcircuits and Converter ICs," IEEE Journal of Emerging and Selected Topics in Power Electronics, vol. 8, no. 1, pp. 31–41, Mar. 2020.

[6] O. Trescases, S. K. Murray, W. L. Jiang, and M. S. Zaman, "GaN Power ICs: Reviewing Strengths, Gaps, and Future Directions," Dec. 2020, p. 27.4.1-27.4.4.

[7] Y. Yamashita, S. Stoffels, N. Posthuma, S. Decoutere, and K. Kobayashi, "Monolithically Integrated E-mode GaN-on-SOI Gate Driver with Power GaN-HEMT for MHz-Switching," in 2018 IEEE 6th Workshop on Wide Bandgap Power Devices and Applications (WiPDA), 2018, pp. 231–236.

[8] K.-Y. Wong, W. Chen, and K. J. Chen, "Integrated voltage reference and comparator circuits for GaN smart power chip technology," in 2009 21st International Symposium on Power Semiconductor Devices & IC's, 2009, pp. 57–60.

[9] C.-H. Liao et al., "3.8 A 23.6ppm/°C Monolithically Integrated GaN Reference Voltage Design with Temperature Range from −50°C to 200°C and Supply Voltage Range from 3.9 to 24V," in 2020 IEEE International Solid-State Circuits Conference - (ISSCC), 2020, pp. 72–74.

[10] E. A. Jones, M. de Rooij, and S. Biswas, "GaN Based DC-DC Converter for 48 V Automotive Applications," in 2019 IEEE Workshop on Wide Bandgap Power Devices and Applications in Asia (WiPDA Asia), 2019, pp. 1–6.

[11] A. Li et al., "A Monolithically Integrated 2-Transistor Voltage Reference With a Wide Temperature Range Based on AlGaN/GaN Technology," IEEE Electron Device Letters, vol. 43, no. 3, pp. 362–365, 2022.

[12] J. Zhu et al., "Impact of Recess Etching on the Temperature-Dependent Characteristics of GaN-Based MIS-HEMTs With Al2O3/AlN Gate-Stack," IEEE Transactions on Electron Devices, vol. 64, no. 3, pp. 840–847, 2017.

[13] S. Yang, Y. Lu, H. Wang, S. Liu, C. Liu, and K. J. Chen, "Dynamic Gate Stress-Induced VTH Shift and Its Impact on Dynamic RON in GaN MIS-HEMTs," IEEE Electron Device Letters, vol. 37, no. 2, pp. 157–160, 2016.

978-1-6654-8901-0/22 $31.00 © 2022 IEEE

[14] Q. Dong, K. Yang, D. Blaauw, and D. Sylvester, "A 114-pW PMOS-only, trim-free voltage reference with 0.26% within-wafer inaccuracy for nW systems," in *2016 IEEE Symposium on VLSI Circuits (VLSI-Circuits)*, 2016, pp. 1–2.

[15] Y. Osaki, T. Hirose, N. Kuroki, and M. Numa, "1.2-V Supply, 100-nW, 1.09-V Bandgap and 0.7-V Supply, 52.5-nW, 0.55-V Subbandgap Reference Circuits for Nanowatt CMOS LSIs," *IEEE Journal of Solid-State Circuits*, vol. 48, no. 6, pp. 1530–1538, 2013.

Embedding Solutions for vertical SiC and GaN Power Devices

Hoang Linh Bach[a], Anqi Huang[a], Yue Teng[a], Hubert Rauh[a], Andreas Schletz[a], Michael P. M. Jank[a], Martin März[b]

Email: Linh.Bach@iisb.fraunhofer.de, Anqi.Huang@iisb.fraunhofer.de, Yue.Teng@iisb.fraunhofer.de, Hubert.Rauh@iisb.fraunhofer.de, Andreas.Schletz@iisb.fraunhofer.de, Michael.Jank@iisb.fraunhofer.de, Martin.Maerz@fau.de

[a] Hybrid Integration, Fraunhofer Institute for Integrated Systems and Device Technology IISB, Erlangen, Germany
[b] Institute for Power Electronics, Friedrich-Alexander-Universität Erlangen-Nürnberg, Erlangen, Germany

Abstract—This paper presents further development of a module concept for Wide Bandgap (WBG) power devices. By embedding WBG power devices in ceramic substrates, high performance of the complete package can be achieved to fully make use of their potentials. In this work, optimizations for design concept and manufacturing processes have been performed to eliminate risk of failures, especially when embedding thin and sensitive devices with fine pad structures. Warpage of package and deformation of devices have been successfully minimized by adjusting volume and layer thickness of die attach and embedded components. Furthermore, processes such as die attach applying, pre-drying and die bonding have been evaluated and improved. Thus, a significant increase of solder and Ag sinter joint quality, especially in fine chip pad areas, have been achieved. Embedded WBG packages with single chip and half bridge topology have been successfully produced and tested in terms of their functionality.

Keywords—High performance packages, chip embedding, WBG packaging, high temperature, multilayer DBC, plug-and-play adaption, cost-effective rapid prototyping

I. INTRODUCTION

WBG power semiconductors based on silicon carbide (SiC) or gallium nitride (GaN) surpass silicon-based (Si) devices in terms of thermal and electrical properties. Nevertheless, they place increased demands for surrounding components. When making use of WBG advantages such as high current density at significant smaller chip size (compared to Si), power density and power losses will be strongly increased, resulting in a higher chip operating temperature that greatly exceeds 200 °C. Therefore, standard power module components such as plastic housings, organic potting compounds and Sn-based solder joints will be a performance bottleneck due to their limited temperature stability. Furthermore, aluminium bond wires as chip top contact are vulnerable to failure at thermal cycles. Advanced packaging techniques have been investigated and introduced in recent years to improve the performance of conventional power modules. Double-sided cooling concepts eliminate bond wires and decrease the R_{th} of the power module [1-3]. The devices are soldered or Ag sintered on both side between two ceramic substrates (sandwich concept) and filled or coated with encapsulation materials. However, the manufacture of such packages is very demanding. The metallization of ceramic substrate has to be etched properly in high resolution to form joints with fine chip electrode

pads. When applying spacers, the number of joints rises and thus the risk of packaging failure increases. Furthermore, warpage, reliability and lifetime need to be addressed further with regards to WBG power devices. For applications in very harsh environment, several approaches for hermetically sealed ceramic packages exist [4, 5]. The packages are mainly designed for Si devices and poorly for WBG performance. Acquirable hermetically sealed packages on the market are mostly used for niche applications. A more established technology is known as Printed-Circuit-Boards (PCB) Embedding, where the devices are mounted onto Cu plates and laminated in organic multilayers [6]. This concept has several key advantages in terms of electrical performance and design flexibility. However, when operating at high temperature ranges over 200 °C, the organic PCB material represents the bottleneck. To overcome the main drawbacks and limitations of existing advanced power modules, a novel packaging concept have been developed and evaluated.

In this approach, WBG devices are embedded in ceramic substrates with thick copper metallization such as Direct Copper Bonded (DBC) or Active Metal Brazed (AMB) to achieve high thermal and electrical performance of the complete package. The key features of this approach are summarized in Table I.

TABLE I. KEY FEATURES OF CERAMIC EMBEDDING TECHNOLOGY

Features	Description (realized by)
Miniaturization[a]	No housing, 3D-integration, reduction of connection points or joints
High switching performance[a]	Short current paths, 3D-integration, vias instead of bond wires
Efficient cooling[a]	Double-sided cooling, thermal vias
High temperature capability of complete package[b]	Al_2O_3, AlN or Si_3N_4 for insulation, no plastic housing, no bond wires, high temperature capable filling or inorganic encapsulation
High electrical and thermal performance[b]	Al_2O_3, AlN or Si_3N_4 for insulation, thick copper metallization ≥ 300 μm
High corrosion resistance[b]	Al_2O_3, AlN or Si_3N_4 for insulation, no plastic housing

[a]. Key features of state-of-the-art PCB-Embedding

[b]. Added values of ceramic embedding technology

As mentioned before, organic PCB materials limit the module performance due to its low corrosion resistance, high CTE, low thermal conductivity and stability. Therefore, by embedding chips in ceramic substrates, key features of PCB technology can be transferred and additional values are added for the end user. The concept and feasibility of this technology have been investigated and presented in previous

This project has received funding from the ECSEL Joint Undertaking (JU) under grant agreement No 101007229. The JU receives support from the European Union's Horizon 2020 research and innovation programme and Germany, France, Belgium, Austria, Sweden, Spain, Italy.

978-1-6654-8901-0/22 $31.00 © 2022 IEEE

work [7-9]. The ceramic substrates were produced out of master cards by laser technology. For the embedding process, cavities and fine structures in ceramic and Cu layers are required.

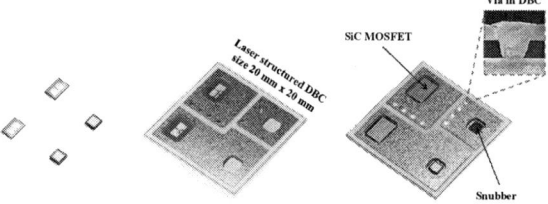

Figure 1. Ceramic embedding approach, embedded SiC half bridge in DBC substrate

The manufacturing of DBCs and interconnections by using laser technology has been evaluated in [7, 8]. DBCs fulfilled requirements for embedding have been successfully produced. Especially for rapid prototyping and small-batch, DBCs and AMBs with high design flexibility can be manufactured in very short time and thus, enabling cost savings in common procurement processes such as tooling, shipping time and minimum order quantity. The embedding of SiC devices have been successfully performed by soldering and Ag sintering in [9]. The pre-packages consist of sealed SiC Schottky diodes showed high voltage capability up to 3.3 kV and high temperature stability up to 250 °C. Furthermore, the pre-packages were successfully stacked together and characterized with high blocking characteristics up to 12.5 kV. Thus, multichip topologies can be highly 3D-integrated in such embedded packages and when constructing the circuit paths appropriately, high switching speed can be achieved.

Considering the performance of the tested pre-packages, this approach can be promising for future power electronics applications. However, to be fully implementable into application systems, aspects such as manufacturability, reliability and lifetime still need to be addressed. In the current state, the produced packages exhibit strong warpage. This can impact reliability negatively and complicates system integration. A further challenge is the embedding of sensitive WBG devices with fine electrode contacts such as transistors. The small gate pads highly increase the risk of packaging failure. This also relates to chip deformation, when embedding thin die structures. Therefore, relevant packaging processes will be evaluated in this work to improve the embedding quality of compact and sensitive WBG devices. Optimized solder and sinter processes will be performed on various SiC and GaN chip samples. The produced packages will be characterized regarding joint quality, warpage and functionality. Based on the results, strategies will be discussed to manufacture embedded packages without any failures.

II. EMBEDDING WBG DEVICES IN DBC SUBSTRATE

A. Topology of Embedded Pre-Packgage

When embedding vertical WBG transistors, a customized layout of the ceramic substrate is required to match the electrodes of the devices such as gate, drain and source. Therefore, isolation trenches and fine ceramic structures are realized by lasering. The topology of an embedded package including a SiC MOSFET is depicted in Figure 2.

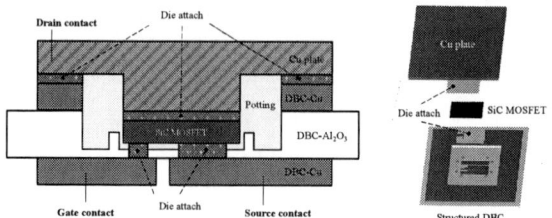

Figure 2. Topology of an embedded SiC MOSFET in a DBC, cross sectional view (left) and exploded view (right)

The SiC MOSFET is flipped and attached into the DBC substrate, whereby its gate and source electrodes are contacted to the bottom DBC-Cu. Here, an isolation trench separates the potentials. The backside of the chip is connected to a Cu plate, which operates as drain contact. To ensure high voltage capability, the chip surroundings are filled with an encapsulation material. All electrical joints are realized by soldering or Ag sintering.

B. Preliminary Studies - Main Packaging Challenges of Embedding Thin Chips and Fine Structures

Produced pre-packages in preliminary studies exhibited strong warpage. The measured height differences along the package surface were up to several 100 µm (Figure 3). Since ceramic and Cu layers are processed and bonded at very high temperatures over 1000 °C, the DBCs possess high residual stress. Warpage of the master cards themselves are common. In addition, the package consists of various bonded materials with different CTE-ranges and unequal distributed Cu layers (Fig. 2). Considering these aspects, chip bonding processes such as soldering and Ag sintering intensifies warpage.

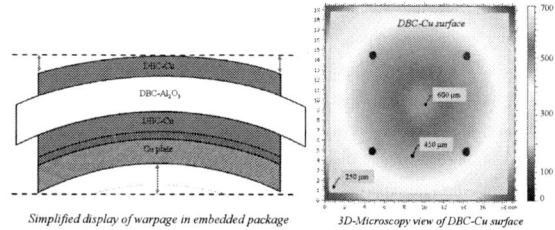

Figure 3. Typical warpage of embedded packages

A further packaging challenge is the processing of thin devices. After the embedding process, deformation or warpage of the chips can be observed. Depending on the bond quality, die attach volume and contact area of joint partners, thermomechanical stress appears to different extents in chip region.

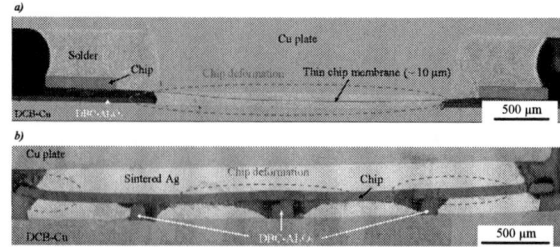

Figure 4. Deformation of thin chips in embedded packages, a) 100 µm thin chip with 10 µm thin membrane after solder process, b) 70 µm thin chip after pressure-assisted Ag sinter process

978-1-6654-8901-0/22 $31.00 © 2022 IEEE

When applying pressure during Ag sintering, the deformation even becomes stronger, resulting in chip cracks. Figure 4 depicts typical deformation or warpage of chip samples after embedding. Soldered chips with very thin membrane of approx. 10 µm showed strong warpage caused by solder joints and embedded components such as Cu plate (Figure 4, a). Other embedded packages with 70 µm chips showed similar appearance after Ag sintering. Here, cracks were also observed caused by the high sinter pressure.

In addition to these issues, bonding transistors with multi-pad structures is demanding due to gate pad dimensions of approx. 1 mm² or smaller. The fine joints are more vulnerable to a lift-off of the die attach caused by thermomechanical stresses. Typical joint failures are depicted in Figure 5. When embedding SiC MOSFETs, most of the packages showed insufficient bonding strength of the die attach on the DBC-Cu surface.

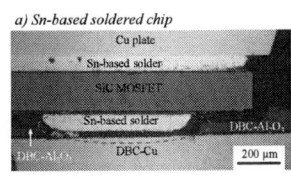

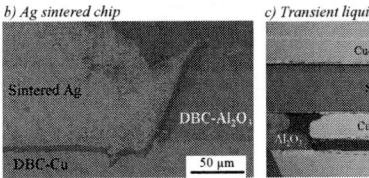

Figure 5. Examples of insufficient gate joints in embedded packages, die attach lifted off from DBC-Cu surface

In summary, preliminary studies showed critical packaging aspects that lead to a malfunction of embedded packages. Based on the findings, optimization approaches have been developed and evaluated to prevent or decrease the mentioned risks to a minimum. The performed tests and results are discussed in the main experiments.

III. MAIN EXPERIMENTS

The devices used for the main experiments consisted of SiC MOSFETs such as QPM3-1200-0013D and CPM3-1200-0016A. IGBTs (SIGC100T65R3E) and GaN dummy chips were added to investigate the embedding of thin devices.

TABLE II. DEVICES USED FOR CERAMIC EMBEDDING EXPERIMENTS

Devices	Die size (L × W × H) in mm³	Metallization
SiC MOSFET QPM3-1200-0013D	4.36 × 7.26 × 0.18	Gate & Source: Au Drain: Au
SiC MOSFET CPM3-1200-0016A	4.04 × 6.44 × 0.18	Gate & Source: Ag Drain: Au
IGBT SIGC100T65R3E	9.73 × 10.23 × 0.07	Gate & Emitter: Ag Collector: Ag
GaN dummy chip with membrane	7 × 7 × 0.1 5 × 5 × 0.01	Top side: Cu Bottom side: Cu

Especially the GaN dummy chips are structured with a very thin membrane of approx. 10 µm in the centre. Properties relevant for packaging are summarized in Table II.

Considering the chip structures and dimensions, DBCs with a size of 20 mm x 20 mm have been laser manufactured. The functions of a produced DBC are explained in Figure 6 for a better understanding of the embedding process and characterizations in the later stage of work.

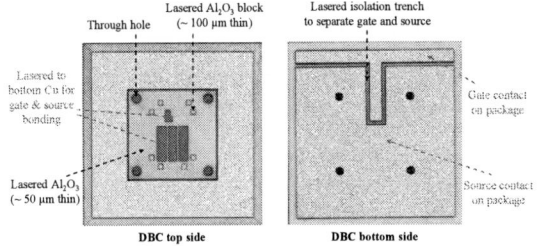

Figure 6. Lasered DBC for the embedding of SiC MOSFETs, standard thickness of 0.3 mm / 0.38 mm / 0.3 mm for Cu / Al₂O₃ / Cu

The centre of the DBC is ablated and the remained Al_2O_3 layer is approx. 50 µm. The Al_2O_3 blocks near chip region support the die placing and prevent uncontrolled shifting of the chip during solder or sinter process. Short circuit between gate and source are prevented by an isolation trench. After sealing by the Cu plate, the package can be filled with encapsulation materials through the drilled holes.

A. Minimizing Warpage of Ceramic Embedded Packages

In the first series of experiments, embedded packages with different thicknesses have been produced. The goal was to observe the influence of thick Cu components on warpage. Therefore, the thickness of Cu plates used for sealing was varied from 0.1 mm up to 2 mm. Dimensions of other components such as DBC, chip and die attach were identical. Mechanical dies (QPM3-1200-0013D) were reflow-soldered with a Sn-based alloy (SnAgCu305) at 220 °C in nitrogen atmosphere. The samples were analysed by scanning acoustic microscopy (SAM) to detect critical voids. Then, only samples with similar high solder quality were measured by using an optical profilometer to define warpage. Based on the height profile, the average height difference between the centre and the corners of the package has been calculated. Both sides of the package, upper Cu plate surface and bottom DBC surface, were measured. The summarized results are depicted in Figure 7. A high value represents strong warpage, whereas a lower height difference indicates significant reduction of it.

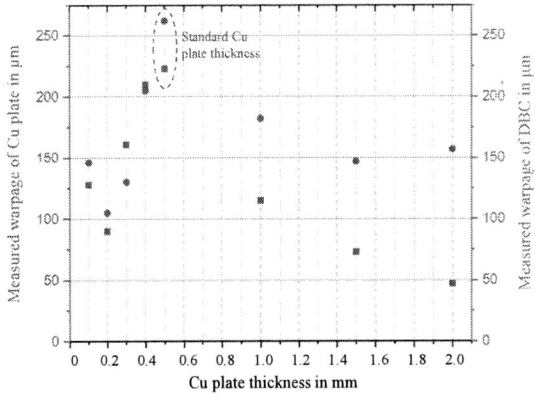

Figure 7. Influence of Cu plate thickness on warpage of embedded package

The 0.5 mm thick Cu plates have been used by default in previous studies. As it can be seen, the extent of warpage changed significantly when varying Cu plate thickness. By decreasing the Cu plate thickness to 0.3 mm or lower, warpage was significantly reduced. Packages with very thin Cu plates even showed up to 50 % lower warpage. This can be explained by the tensile stress of the package components. The package is stressed by the upper and lower Cu layers during cooling. Here, tensile stresses of upper Cu layers are dominating due to the thick Cu plate, causing a bending of the bonded DBC edges in direction towards the Cu plate. Thus, reducing Cu plate thickness results in lesser warpage. A similar trend was observed when increasing the Cu plate thickness greatly to 1 mm or higher. The reason for this occurrence can be explained by the geometrical dependent stiffness of the component. Cu plates thicker than 1 mm are stiffer and thus when bonded to the DBC, the complete package is more resistance to deflection. Therefore, design-oriented strategies can be applied to minimize warpage. Either by reducing layer thickness of dominant metallization layers to balanced thermomechanical stresses in the complete package or increasing it greatly for higher mechanical stability. For the subsequent investigations, packages with thinner Cu plates ($\leq$ 0.2 mm) were produced.

B. Preventing Deformation of Thin Dies during Ceramic Embedding Process

As presented in the pretests, sensitive devices pose a higher risk of chip deformation during embedding. Especially when applying pressure on flipped chips, cracks occur. To address this issue, experiments have been performed to optimize die attach, solder and sinter process. Sinter tests of thin IGBTs (~ 70 μm) showed that chip deformation already appeared at low pressure ranges.

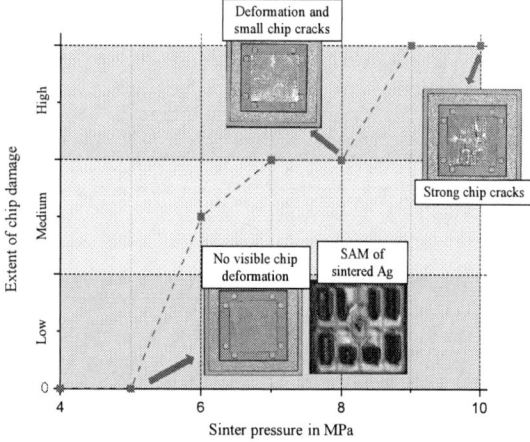

Figure 8. Influence of sinter pressure on IGBTs during embedding

All samples were sintered at 250 °C for 180 s in air atmosphere. Only sinter pressure was varied from 4 to 10 MPa. A micro-scaled Ag paste compatible with Ag, Au and Cu surface has been used and applied by jet-dispensing. In the first step, chips were sintered in DBCs without Cu plates. As depicted in Figure 8, the chips cracked strongly at 10 MPa. Significant deformation and cracks were also observed in pressure ranges below 8 MPa. Nevertheless, non-damaged samples were successfully produced when applying a sinter pressure of 5 MPa or lower. The samples were sealed by a Cu plate in a second sinter steps with the same sinter pressure. Recorded SAM images showed a sufficient distribution of sintered Ag layers in chip electrode areas. No critical voids were observed. Overall, sintered Ag bonds at electrode pads were sufficient according to cross section and scanning electron microscopy analyzation (SEM).

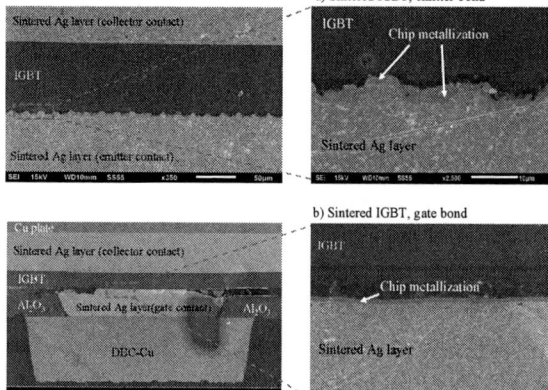

Figure 9. SEM analyzation of Ag sintered IGBT in embedded package, processed at 250 °C, 5 MPa, 180 s

This was also proved by electrical characterization which has been performed on produced packages. Measured current and blocking characteristics were in the range of the IGBT data sheet and gate leakage currents were below 1 nA. Thus, it can be said that pressure-assisted Ag sintering is applicable for the embedding of such thin dies. Although the sinter pressure is limited to a lower range, sufficient bonding quality of the embedded components can be achieved. If the dies are thinner, the maximal applicable pressure range is expected to be significantly lower. This case was tested in the following experiments.

GaN chips have been used for the embedding in DBCs. The 100 μm thin dummy chips consisted of a very thin membrane structure of appr. 10 μm. Due to this fine topology, the chips are very vulnerable to mechanical impacts during the die placing and die bonding process. Solder tests have been performed with Sn-based alloy (SnAgCu). 80 μm thick solder preforms were tailored and placed in the fine chip membrane area. The packages were reflow-soldered at 220 °C in nitrogen atmosphere. In the first soldered batch, pressure in the range of below 1 MPa was applied on the packages. SAM and cross section analyzation showed large-scaled solder joints on chip front and back side. Especially on the back side, the solder spread over the whole chip area and formed thick bond layers to the Cu plate up to 1 mm (Figure 10, a). The molten solder was pressed out of the membrane area by the applied force, which can be observed by the mechanical contact between Cu plate and chip. Warpage was observed in chip membrane area, but it did not lead to obvious chip cracks. The second batch were soldered without any mechanical pressure. Furthermore, the die attach volume in the chip membrane area was reduced by approximately 40 %. Compared to the first batch, samples of the second batch showed significant smaller solder joints in chip region (Figure 10, b).

978-1-6654-8901-0/22 $31.00 © 2022 IEEE

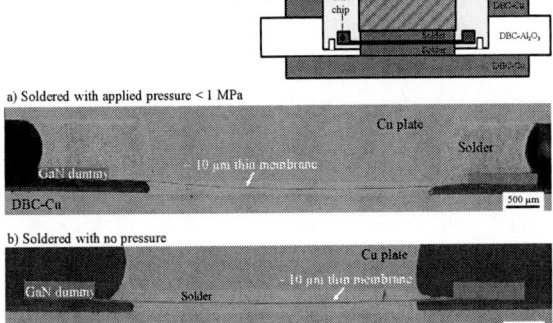

Figure 10. Soldered GaN dummy chips in embedded packages with SnAgCu alloy at 220 °C in nitrogen atmosphere

The solder did not spread extensively over chip and Cu plate surfaces. Only contact areas in membrane region were wetted by molten solder. Furthermore, warpage of the chip membrane was significant lower. Based on the results, very thin chip structures can be utilized for ceramic embedding. Nevertheless, pressure should be avoided or reduced to a minimum if applicable, and the appropriate amount of die attach should be adjusted.

C. Improving Joint Quality of Small Contact Areas in Ceramic Embedded Packages

In considering the warpage of the complete package, the risk of failure increases when bonding smaller chip area. A high failure rate was observed in preliminary studies, especially in gate contact areas of embedded transistors (Figure 5). The size of the gate pads was below 1 mm². Thus, the gate joints were more vulnerable to mechanical or thermomechanical impacts. By decreasing warpage of the complete package, gate joint failures were significantly reduced. However, when performing sinter process, pre-processing of Ag paste is an additional relevant aspect for high bond quality. An appropriate escape of organic solvents and a uniform shrinkage of the dried Ag paste are required. The main challenges for the embedding of vertical transistors are depicted in Figure 11.

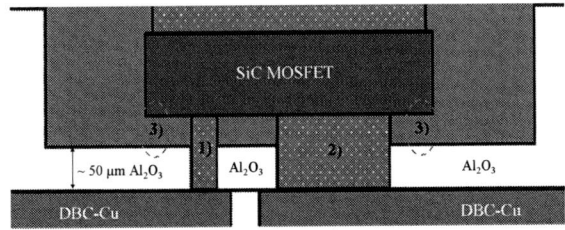

Figure 11. Challenges when embedding a vertical transistor by Ag sintering, 1) Sintered Ag joint at gate pad, 2) Sintered Ag joint at Source pad, 3) Critical location with regards to electrical breakdown

In this work, the thickness of ablated Al_2O_3 layer is approx. 50 µm. Therefore, applied die attach materials must be at least thicker to guarantee a bonding with the chip. Furthermore, it is mandatory to fill the gap between chip and ceramic with an appropriate encapsulation material to prevent electrical breakdowns during high voltage operation (Fig. 11, 3). When perform a solder process, required solder amount can be simply adjusted by the preform thickness.

Warpage optimized packages with soldered SiC MOSFETs showed increased joint quality when using preforms of 80 µm or thicker. By applying pressure of several MPa, the wetting of molten solder was improved. However, Ag sinter process requires more effort to adjust an appropriate die attach volume. Since the chips are double-side sintered and embedded in sealed packages, the escape of organic solvents and the supply of process atmosphere can be impeded. Furthermore, depending on the chip layout, high deviation of dried paste heights between gate and source contacts can occur, resulting in an uneven distribution of the sinter pressure. This was observed during the embedding of WBG transistors. A micro-scaled Ag paste was jet dispensed in DBC contact areas and dried at 120 °C for 20 min. Although all parameters for jet dispensing and drying were set identically, the measured height of dried Ag paste in the gate contact area (1.18 mm × 0.5 mm) was 22 % lower than that in the source contact area (4.89 mm × 1.08 mm). The sintered samples showed a high failure rate of gate joint. Therefore, jet dispensing and drying process have been adjusted to minimize height deviation of dried Ag paste before sintering. An overlap function of 30 % has been applied for dispensing Ag paste dots in the gate area. The measured heights of applied Ag paste in all contact areas were approx. 165 µm after drying, and approx. 85 µm after sintering at 260 °C, 16 MPa for 180 s. The sintered packages showed a significant improvement in terms of gate joint quality (Figure 12).

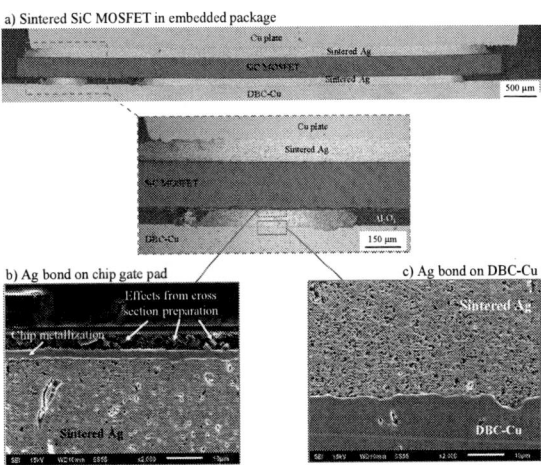

Figure 12. Improved Ag joint quality in gate contact area by adjusting appropriate paste height before sintering

Large-scaled Ag joints have been realized in all electrodes contact areas. Especially in the gate area, no critical voids were observed between the interfaces of chip metallization, sintered Ag layer and DBC-Cu. The measured porosity in the centre of the Ag layers were below 10 %.

D. Characterization of Ceramic Embedded Packages

Packages produced under optimized conditions have been electrically characterized. The samples consisted of Ag sintered SiC MOSFETs (CPM3-1200-0016A). After sealing, a high temperature capable silicone gel (Semicosil 915HT) was dispensed to ensure high blocking characteristics. In parallel, a thin film coating was performed to cover the enclosed areas in the package with a vacuum deposited dielectric material. The laser drilled through holes in the

978-1-6654-8901-0/22 $31.00 © 2022 IEEE

packages served as small openings. The utilized material was Parylene-C.

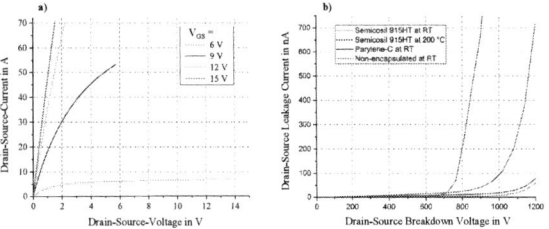

Figure 13. Measured electrical characteristics of embedded SiC MOSFETs, a) Output characteristics, b) Blocking characteristics

All produced samples showed appropriate output characteristics according to the device data sheet. The measured $R_{DS(on)}$ was below 20 mΩ (V_{GS} = 15 V, I_D = 70 A). Measured gate-source leakage currents were in the range of several 100 pA (V_{GS} = 15 V). The encapsulated packages were capable of blocking voltages up to 1200 V. No indications of electrical breakdown were observed. Thus, both applied encapsulation materials, Semicosil 915HT and Parylene-C, were successfully filled in the required areas near chip edge structures. In addition, packages filled with silicone gel were heated up to 200 °C. The drain-source leakage currents increased by a factor of 12 (~ 720 nA). However, the values were still in the permissible range of the data sheet ($\leq$ 100 μA). Several non-encapsulated packages have been heated up to 300 °C to observe failure of joints and warpage at high temperature operation. The packages were successfully tested in terms of output characteristics. Measured values for $R_{DS(on)}$ increased by over 100 % to 45 mΩ and gate-source leakage currents rose from several 100 pA to approx. 5 nA or higher. This can be explained by the chip behaviour itself and does not indicate a failure of the die attach. The same packages were stored at 300 °C for 10 hours and characterized again. Also here, electrical tests showed no failure indications. A half-bridge topology was embedded in the next step to demonstrate the packaging feasibility at circuit level. The circuit consisted of two embedded SiC MOSFETs (QPM3-1200-0013D) in a DBC with integrated electrical vias.

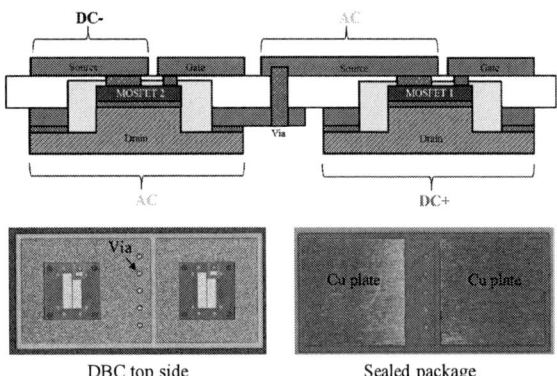

Figure 14. Topology of embedded SiC half bridge and laser manufactured DBC substrate

The vias were filled with Ag paste and the complete package was soldered with Sn-based alloy at 220 °C in nitrogen atmosphere. After that, silicone gel Semicosil 915HT was applied as encapsulation material. The SAM analyzation showed a sufficient distribution of the molten solder without critical voids (Figure 15, a). Both embedded devices were electrically characterized.

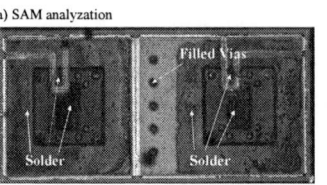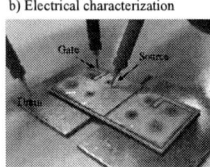

Figure 15. Characterization of embedded SiC half bridge

The measured $R_{DS(on)}$ was 13.6 mΩ for high-side and 14.2 mΩ for low-side (V_{GS} = 15 V, I_D = 70 A). Measured gate-source leakage currents were below 200 pA (V_{GS} = 15 V). Both high- and low-side showed sufficient blocking characteristics up to 1200 V (I_{DSS} < 50 nA). In addition, the measured electrical resistance of the Ag filled vias was below 1 mΩ. According to the measurements, the embedding process of the SiC half bridge has been successfully performed.

IV. CONCLUSIONS

Solder and Ag sinter processes have been evaluated to manufacture ceramic embedded packages. The performed optimizations showed a significant improvement in terms of warpage, chip deformation and joint failure. Functional packages with single embedded SiC MOSFETs and a SiC half bridge with integrated electrical vias were successfully produced. The results demonstrated the packaging feasibility of the ceramic embedding concept at circuit level. Further tests and demonstrators for system applications will be presented in future work.

REFERENCES

[1] C. Buttay, R. Riva, B. Allard, M. Locatelli, V. Bley, "Packaging with double-side cooling capability for SiC devices, based on silver sintering," In Annual Conference of IEEE Industrial Electronics Society (IECON), pp. 5753-5759, 2018.

[2] Z. Liang, "Planar-Bond-All: A Technology for Three-dimensional Integration of Multiple Packaging Functions into Advanced Power Modules," In IEEE International Workshop on Integrated Power Packaging (IWIPP), pp. 115-118, 2015.

[3] H. Zhang, S. S. Ang, H. A. Mantooth, S. Krishnamurthy, "A high temperature, double-sided cooling SiC power electronics module," In IEEE Energy Conversion Congress and Exposition, pp. 2877-2883, 2013.

[4] H. H. Yuan et al., "Extreme High Pressure and High Temperature Package Development," In IEEE 15th Electronics Packaging Technology Conference (EPTC), pp. 379-383, 2013.

[5] K. Y. Au, D. M. Zhi, V. Chidambaram, B. Lin, K. Piotr, C. KaiLiang, "High temperature endurable hermetic sealing material selection and reliability comparison for IR gas sensor module packaging," In IEEE Electronics Packaging Technology Conference (EPTC), 2016.

[6] T. Gottwald, C. Rößle, "p² Pack - the Paradigm Shift in Interconnect Technology," In PCIM Europe, pp. 1054-1062, 2014.

[7] H. L. Bach et al., "Vias in DBC Substrates for Embedded Power Modules," In International Conference on Integrated Power Electronics Systems (CIPS), pp. 144-148, 2018.

[8] H. L. Bach et al., "Ceramic Embedding as Packaging Solution for Future Power Electronic Applications," In International Power Electronics Conference (IPEC), pp. 2410-2415, 2018.

[9] H. L. Bach et al., "Stackable SiC-Embedded Ceramic Packages for High-Voltage and High-Temperature Power Electronic Applications," In Journal of Microelectronics and Electronic Packaging, Volume 16, Issue 4, pp. 176-181, 2019.

Three-level ANPC Inverter Common-mode Voltage Analytical Characterization

Yang Huang
Department of Electrical Engineering and Computer Science
University of Tennessee
Knoxville, TN, USA, 37996
yhuang65@vols.utk.edu

Xin Xia
Department of Electrical Engineering and Computer Science
University of Tennessee
Knoxville, TN, USA, 37996
xxia6@vols.utk.edu

Hua (Kevin) Bai
Department of Electrical Engineering and Computer Science
University of Tennessee
Knoxville, TN, USA, 37996
hbai2@utk.edu

Fanning Jin
Mercedes Benz Research and Development Center
Mercedes Benz North America
Redford, MI, USA, 48239
fanning.jin@daimler.com

Xiaodong Shi
Mercedes Benz Research and Development Center
Mercedes Benz North America
Redford, MI, USA, 48239
xiaodong.shi@daimler.com

Bing Cheng
Mercedes Benz Research and Development Center
Mercedes Benz North America
Redford, MI, USA, 48239
bing.cheng@daimler.com

Abstract—Evident in previous literature, in the motor drive system, the research on the analytical models for common-mode (CM) performance evaluation is inadequate but needed. Especially for the 3-level inverter system. The majority of work focuses on simulations and experiments. In this paper, an analytical model of CM voltage (CMV) in a 3-level inverter is presented based on Double Fourier Integral (DFI). The model could be extended to different 3-level modulation schemes such as conventional space vector PWM (CSVM), nearest three space vectors modulation (NTSVM), and reduced common-mode voltage modulation (RCMVM). The impact of these three modulation schemes on the CMV is comprehensively compared across varying modulation indices. An 800V/50kW 3L inverter will be built using off-the-shelf automotive-qualified 650V/60A GaN HEMTs from GaN Systems.

Keywords—common-mode voltage, three-level inverter, space vector modulation

I. INTRODUCTION

Recently, high voltage batteries became popular in EVs because of the benefit of fast charging, longer mileage, and lower loss. In [1], a comparison has been made between multiple EVs and traditional petrol-driven vehicles in terms of inter-city travel. To have comparable travel time as petrol-driven vehicles, it is required to have a charger power of higher than 400 kW. With extreme fast charging techniques, the output voltage should be at least 800 V [2]. The high DC-link voltage not only enables the extremely fast charging but also leads to a smaller motor current for higher efficiency and smaller cables [3]. This requires either the device to have high voltage blocking capability, or the inverter to be multiple levels, such as a 3-level (3L) neutral point clamped inverter (NPCI) and active neutral point clamped inverter (ANPCI). Considering the cost, using lower voltage blocking devices in the multilevel inverter is more preferred, which not only lowers the device voltage stress but also enhances the output power quality because of the increase

in voltage levels. GaN device has the merits of small package and high current capability, which is a perfect candidate for 3-level (3L) application. However, as the battery voltage increases, the motor drive system also suffers the increased common-mode voltage (CMV). In addition, the GaN device enables a higher switching frequency, thereby potentially higher CM current (CMC).

The CMV is generated through modulation and switching actions, hence, there is a possibility that by properly assigning the space vector combinations, the CMV can be reduced [4-6]. Reference [7] has validated this concept to achieve zero CMV generation with only 7 vectors, however, the maximum modulation index (MI) is reduced to 86.6% of the conventional SVM for the 2-level inverter, and the neutral point balancing issue will appear as these vectors are middle-length vectors. Therefore, CMV elimination is not the practical goal in the 3L system. As there are multiple 3L modulation schemes, each causes different CM performance, a proper model for the CMV evaluation is important. In [8-11], the double Fourier integral (DFI) method is used to model the CMV in a 2-level (2L) drive system, which shows excellent accuracy in CM profile prediction. However, few have been validated in the 3L system.

In this paper, a three-level ANPCI CMV modeling work for different 3L modulation schemes will be presented. The analytical model offers an approach to mathematically calculate the CM performance for a specific modulation method. Such an analytical model provides us with a powerful tool, allowing a comprehensive vision for further evaluations. Compared to the traditional simulation approach, the proposed analytical model could help designers to summarize the principle of CMV effectively. The analytical result of CMV could be coupled with the CM impedance, thereby conducting the CMC. The impact of all other variables, such as motor and controller parameters, could be actively evaluated through analytical equations instead of repeating the simulation.

978-1-6654-8901-0/22 $31.00 © 2022 IEEE

II. THREE-LEVEL INVERTER CMV MODELING

An ANPCI topology is shown in Fig. 1, compared with a two-level system, a three-level system alleviates the common-mode noise by providing more vector redundancy, and some vector combinations even null the CMV. The phase output voltage of a three-level inverter has 3 voltage levels ($V_{dc}/2$, 0, -$V_{dc}/2$). One conventional modulation method is, to use space vector modulation waveform to compare with the level-shifted carriers to generate the high-frequency PWM. The modulation mechanism is shown in Fig. 2.

$$V_{CM} = \frac{\left(V_{ao} + V_{bo} + V_{co}\right)}{3} \quad (1)$$

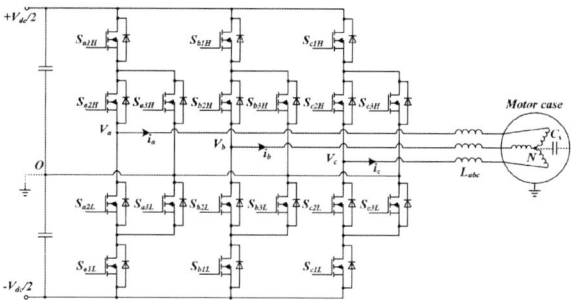

Fig. 1. Three-level ANPC inverter motor drive system

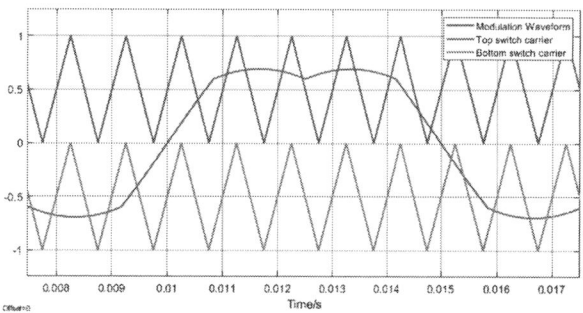

Fig. 2. 3L-ANPC CSVM modulation waveform

It is essential to use a low-frequency modulation waveform to compare with a high-frequency carrier waveform in any PWM method. The phase leg outputs a series of pulses switching between upper and lower DC bus, which not only have the fundamental component but also incorporate a series of unwanted harmonics due to switching processes. This indicates that based on the modulation waveform and carrier frequency, the output voltage spectrum could be predicted analytically. In some literature, the CMV analytical model for two-level inverters for various modulation schemes is proposed, however, for three-level inverters, the related modeling work is inadequate, the modeling complexity does exist, but a high integrated modeling approach could save the effort and extended to various three-level modulation methods.

DFI analysis is a mathematical tool that can provide analytical solutions to precisely identify harmonic components of a PWM signal. In inverter applications, assume the target function $f(t)$ is the voltage between the middle point of the phase leg and the middle point of the DC bus. $f(t)$ is a cyclic signal related to both $x(t)$ and $y(t)$ with period=2π. Here, $x(t)=\omega_s t+\theta_s$ and $y(t)=\omega_0 t+\theta_0$, representing the time variation of the high-frequency modulating wave and low-frequency modulated wave, respectively. By doing the DFI analysis on the PWM waveform, $f(t)$ can be expanded as (2). Where m is the carrier index variable, and n is the baseband index variable.

$$\begin{aligned} f(x,y) = \frac{A_{00}}{2} &+ \sum_{n=1}^{\infty}\left\{A_{0n}\cos\left(n\left(\omega_0 t+\theta_0\right)\right)+B_{0n}\sin\left(n\left(\omega_0 t+\theta_0\right)\right)\right\} \\ &+ \sum_{m=1}^{\infty}\left\{A_{m0}\cos\left(m\left(\omega_s t+\theta_s\right)\right)+B_{m0}\sin\left(m\left(\omega_s t+\theta_s\right)\right)\right\} \\ &+ \sum_{m=1}^{\infty}\sum_{\substack{n=-\infty\\n\neq0}}^{\infty}\left\{A_{mn}\cos\left(m\left(\omega_s t+\theta_s\right)+n\left(\omega_0 t+\theta_0\right)\right)+B_{mn}\sin\left(m\left(\omega_s t+\theta_s\right)+n\left(\omega_0 t+\theta_0\right)\right)\right\} \end{aligned} \quad (2)$$

A. Conventional Space Vector Modulation (CSVM)

According to the switching profile of CSVM in 3L-ANPC, the corresponding DFI integral bounds can be derived, as shown in Fig. 3. There are 8 sections for each fundamental cycle. Especially, in sections 1, 2, 5, and 6, there are three integration areas. In sections 3, 4, 7, and 8, there is only one (ignore the area where the pole voltage is 0). For instance, in section 1, the integration segments should be (-π, 1-), (1-, 1+), and (1+, π). The DFI math model (3) for CSVM in a 3L-ANPC is then generated from (2). Substituting different values for m and n, the carrier harmonics and sideband components of the CMV can be calculated and compared to the simulation.

The DFI model for CSVM in a 3L-ANPC is then generated from (6.1). Using (4.5)~(4.6) and substituting different values for m and n, the carrier harmonics and sideband components of the CMV can be calculated and compared to the simulation at f_0=100 Hz, f_s=10 kHz, 200V DC bus voltage, and 0.8 modulation index. As shown in Fig. 4, the calculated carrier harmonics and the sideband component of both phase voltage and CMV are very close to the simulation. Here only the $f_s\pm15f_0$ sideband harmonics around f_s are plotted. Therefore, the proposed analytical approach for CMV can be extended to multilevel scenarios.

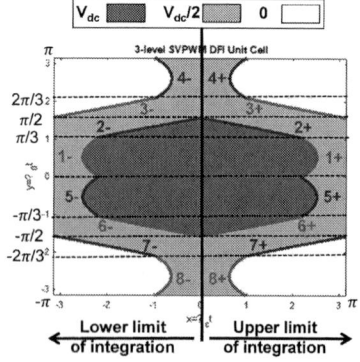

Fig. 3. DFI integral bounds for CSVM in a 3L-ANPC inverter

$$A_{nm} + jB_{nm} = \frac{1}{2\pi} \int_{-\pi}^{\pi} \int_{lowerlimit}^{upperlimit} \frac{insectori}{insectori} U e^{j(mx+ny)} dxdy$$

$$= \int_{-\pi}^{-\frac{2\pi}{3}} \int_{8-}^{8+} \frac{V_{dc}}{2} e^{j(mx+ny)} V_{dc} dxdy + \int_{-\frac{\pi}{3}}^{-\frac{\pi}{3}} \int_{7-}^{7+} \frac{V_{dc}}{2} e^{j(mx+ny)} V_{dc} dxdy$$

$$+ \int_{-\frac{\pi}{2}}^{-\frac{\pi}{3}} \int_{-\pi}^{6-} \frac{V_{dc}}{2} e^{j(mx+ny)} V_{dc} dxdy + \int_{-\frac{\pi}{3}}^{-\frac{\pi}{3}} \int_{6-}^{6+} \frac{V_{dc}}{2} e^{j(mx+ny)} V_{dc} dxdy + \int_{-\frac{\pi}{2}}^{-\frac{\pi}{3}} \int_{6+}^{\pi} \frac{V_{dc}}{2} e^{j(mx+ny)} V_{dc} dxdy$$

$$+ \int_{-\frac{\pi}{2}}^{-\frac{\pi}{3}} \int_{-\pi}^{5-} \frac{V_{dc}}{2} e^{j(mx+ny)} V_{dc} dxdy + \int_{-\frac{\pi}{3}}^{-\frac{\pi}{3}} \int_{5-}^{5+} \frac{V_{dc}}{2} e^{j(mx+ny)} V_{dc} dxdy + \int_{-\frac{\pi}{2}}^{-\frac{\pi}{3}} \int_{5+}^{\pi} \frac{V_{dc}}{2} e^{j(mx+ny)} V_{dc} dxdy$$

$$+ \int_{-\frac{\pi}{2}}^{-\frac{\pi}{3}} \int_{-\pi}^{1-} \frac{V_{dc}}{2} e^{j(mx+ny)} V_{dc} dxdy + \int_{-\frac{\pi}{3}}^{-\frac{\pi}{3}} \int_{1-}^{1+} \frac{V_{dc}}{2} e^{j(mx+ny)} V_{dc} dxdy + \int_{-\frac{\pi}{2}}^{-\frac{\pi}{3}} \int_{1+}^{\pi} \frac{V_{dc}}{2} e^{j(mx+ny)} V_{dc} dxdy$$

$$+ \int_{-\frac{\pi}{2}}^{-\frac{\pi}{3}} \int_{-\pi}^{2-} \frac{V_{dc}}{2} e^{j(mx+ny)} V_{dc} dxdy + \int_{-\frac{\pi}{3}}^{-\frac{\pi}{3}} \int_{12-}^{2+} \frac{V_{dc}}{2} e^{j(mx+ny)} V_{dc} dxdy + \int_{-\frac{\pi}{2}}^{-\frac{\pi}{3}} \int_{2+}^{\pi} \frac{V_{dc}}{2} e^{j(mx+ny)} V_{dc} dxdy$$

$$+ \int_{-\pi}^{\frac{2\pi}{3}} \int_{3-}^{3+} \frac{V_{dc}}{2} e^{j(mx+ny)} V_{dc} dxdy + \int_{-\frac{\pi}{2}}^{\frac{\pi}{2}} \int_{4-}^{4+} \frac{V_{dc}}{2} e^{j(mx+ny)} V_{dc} dxdy$$

(3)

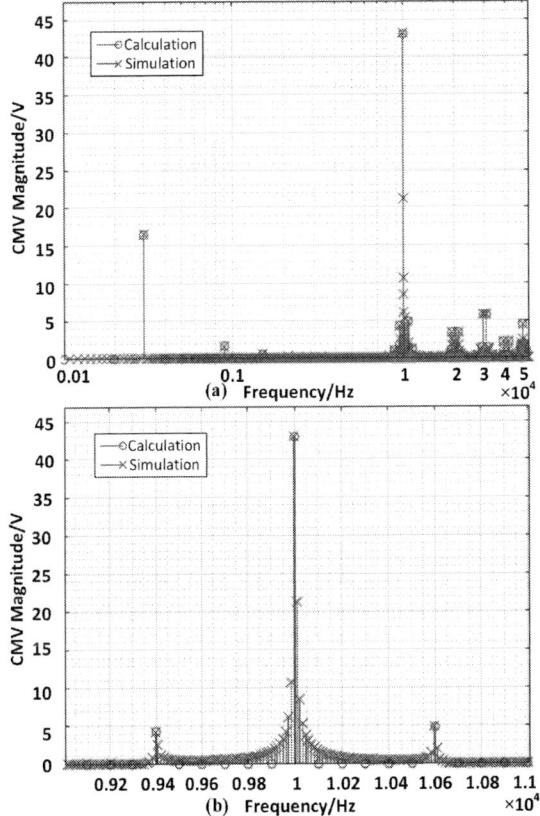

Fig. 4. Comparison of DFI result and simulation for 3L-ANPC inverter.

(a) Wide frequency range, (b) Zoomed-in view @f_s.

B. Nearest Three Space Vectors Modulation (NTSVM)

Direct CSVM modulation does not consider the modulation from the vector assignment point of view. However, determining the switching states, and the corresponding vector duration time is critical to synthesize the reference voltage vector [12]. As the total switching states count of a multilevel converter shows a cubic increase with the converter levels, the computational effort for the CSVM overwhelms quickly and is not preferred to be used for multilevel converters. Instead, using the nearest three vectors to synthesize the reference voltage is arguably the best approach as it saves the computation resources and forms precise vector decomposition. This modulation scheme is called nearest three vector modulation (NTSVM).

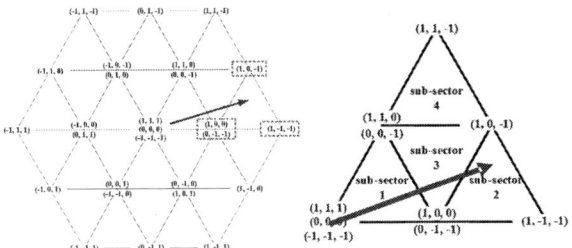

Fig. 5. NTSVM vector plane and reference synthesis.

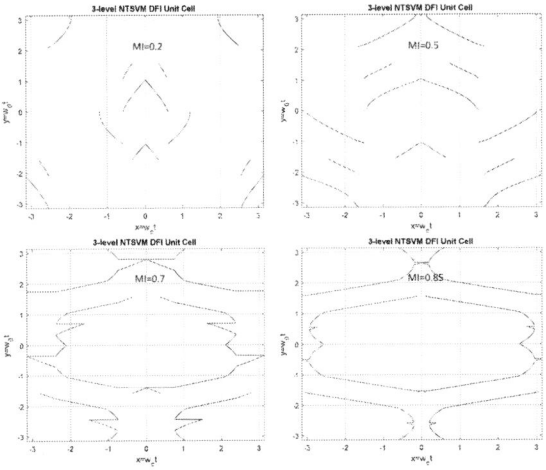

Fig. 6. DFI integral bounds for NTSVM in a 3L-ANPC inverter.

The NTSVM mechanism is illustrated in Fig. 5. Where the reference vector (blue) locates in sector-1 and subsector-2, the NTSVM requires to use the vectors on the three vertices of the subsector-2 triangle. Different from CSVM, which has a simplified modulation waveform that could be directly extended from a 2-level inverter, the modulation waveform of NTSVM is very complicated because of the subsector alternation. Depending on the modulation index, the modulation waveform shows a very different profile. The DFI bounds could be derived and plotted as shown in Fig. 6. Following the same steps that were applied in section 6.2.1, the analytical CMV spectrum vs the simulation result could be obtained. Fig. 7 shows the comparison given DC bus voltage = 800 V, MI=0.8, f_s=10 kHz and f_0=100 Hz. In Fig. 8, the peak CMV at the switching frequency point at different modulation indices is estimated and compared with the simulation. The result shows very accurate CMV prediction performance for NTSVM.

978-1-6654-8901-0/22 $31.00 © 2022 IEEE

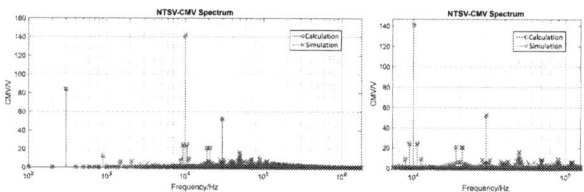

Fig. 7. Comparison of DFI result and simulation for 3L-NTSVM.

Left: Wide frequency range, Right: Zoomed-in around f_s.

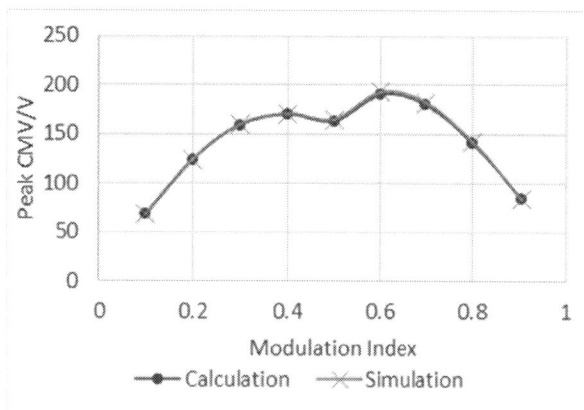

Fig. 8. Comparison of DFI result and simulation for 3L-NTSVM CMV peak @f_s.

C. Reduced-CMV Modulation (RCMVM)

NTSVM alternates the space vectors based on the minimum switching action principle to save the switching loss. However, some of the vectors will result in CMV higher than Vdc/6. Therefore, by avoiding using the 8 voltage vectors that generate $V_{dc}/6$ CMV, the system CMV could be constrained within $V_{dc}/6$. These 8 vectors are marked in Fig. 9. Other than that, the vector synthesis mechanism is the same as NTSVM. Due to the forced forbidden of some vectors, which means the 8 redundant switching states will not be used during modulation, the minimum switching action mechanism could not be realized. For example, in NTSVM sub-sector 2, the vector sequence minimum switching action would be (1, 0, 0), (1, 0, -1), (1, -1, -1) (0, -1, -1). However, in RCMVM sub-sector 2, the (0, -1, -1) vector is banned because it generates $V_{dc}/2$ CMV, the switching sequence becomes (1, 0, 0), (1, 0, -1), (1, -1, -1) (1, 0, 0). The last state transition includes an extra switching. Therefore, compared with NTSVM, the switching loss of RCMVM will be increased.

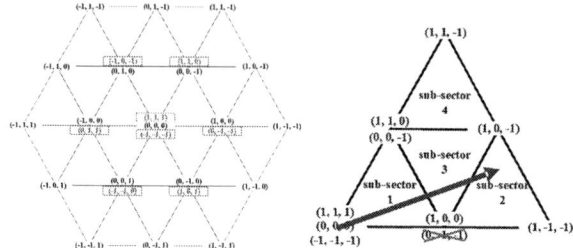

Fig. 9. RCMVM vector plane and reference synthesis.

Similar to NTSVM, the modulation waveform of RCMVM is also complicated compared with CSVM. The DFI bounds could be derived and plotted as shown in Fig. 10. The analytical CMV spectrum vs the simulation result could be obtained through the CMV spectrum estimation model. Fig. 11 shows the comparison given DC bus voltage = 800 V, MI=0.8, f_s=10 kHz and f_0=100 Hz. In Fig. 12, the peak CMV at the switching frequency point at different modulation indices is estimated and compared with the simulation. The result shows very accurate CMV prediction performance for RCMVM.

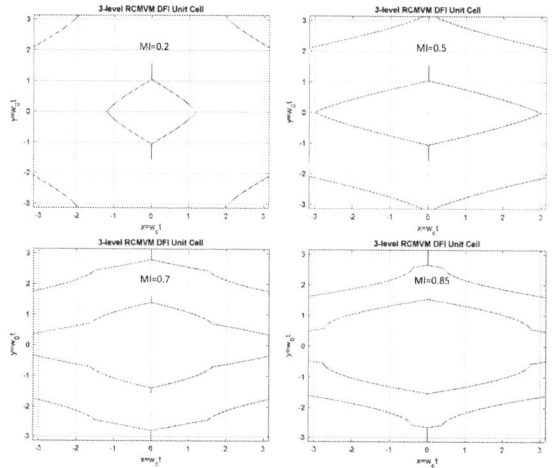

Fig. 10. DFI integral bounds for RCMVM in a 3L-ANPC inverter.

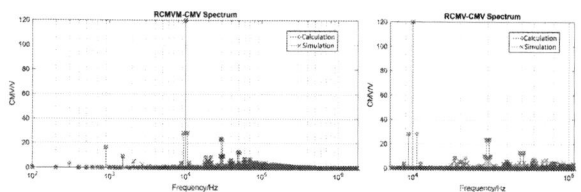

Fig. 11. Comparison of DFI result and simulation for 3L-RCMVM.

Left: Wide frequency range, Right: Zoomed-in around f_s.

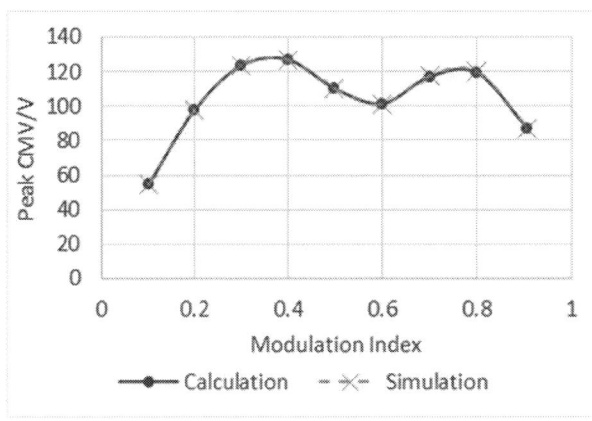

Fig. 12. Comparison of DFI result and simulation for 3L-RCMVM CMV peak @f_s.

III. EXPERIMENTAL VALIDATION

To validate the 3-level modulation performance, a water-cooled 50 kW rated GaN-based 3-level inverter is built. The GaN HEMTs are from GaN System, GS66516T (650 V, 60 A). 3 pieces are paralleled at each switching position. The switching frequency is set to 10 kHz which is controlled by the FPGA, and the DC-link voltage is designed to undertake at most 800 V for high voltage applications. The prototype is shown in Fig. 13. The inverter is a stacked system with the bottom board carrying the power devices and the top board for the DC-link capacitors and input/output interfaces. The whole system sits on a copper grounding plate, where the inverter heatsink and motor case are solidly grounded.

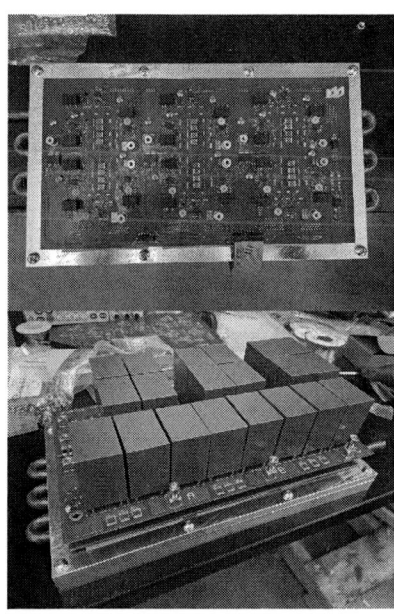

Fig. 13. Three-level GaN-based ANPC inverter prototype.

Top: Top view, Bottom: Final assembly.

Considering the frequency range of the prediction is low, the CMV spectrum components can be considered as linear to the DC input voltage. Therefore, a preliminary 200 V test result scaled up by 4 times can be used to emulate the 800 V performance on the low-frequency spectrum and will be used to compare with the 800 V simulation result to validate its prediction feasibility. Use the controller to implement three different 3L modulation schemes to sample the CMV data spectrum and compare it with the model estimation. Take CSVM CMV spectrum @MI=0.5 as an example, as shown in Fig 14, the spectrum of the prediction result on each harmonic point meets the tested spectrum. The complete prediction accuracy is shown in Fig. 15. For all three PWMs, the prediction results match the experimental results perfectly. This shows a promising accuracy of the potential software integration for an online prediction, which could collaborate with some dynamic modulation scheme switching algorithms. The real 800 V test and comparison will be the priority of the future work, before the prototype passed all voltage and power targets.

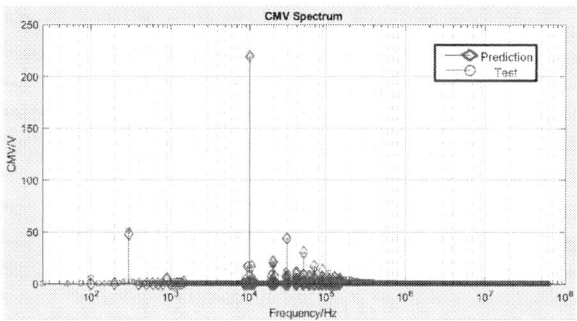

Fig. 14. Comparison of CMV spectrum between experimental result and prediction result for 3L-CSVM @MI=0.5. The "test" spectrum is generated based on a 200 V experiment result with 4-time amplification.

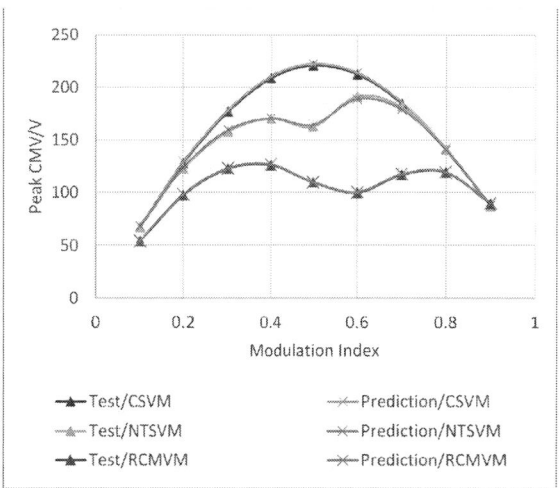

Fig. 15. Comparison of CMV spectrum peak value between experimental result and prediction result. The "test" data points are calculated based on a 200 V experiment result with 4-time amplification.

IV. CONCLUSION

The GaN-based multi-level topology could benefit the EV application under a high-voltage scenario from both power density and efficiency aspects. To overcome the CMV issue that is strengthened by the high voltage and high switching frequency that are introduced by the GaN device implementation, proper CM modeling is needed, which is less studied in past research work. This paper successfully extended the proposed CMV prediction model to the 3-level inverter. With NTSVM and RCMVM, though the modulation waveforms are complex, the suggested CMV prediction procedure with DFI still applies. The computation result matches the simulation result very accurately under current test condition. For the future work, a validation of the model with 800 V test result is planned. With the model being built, another future work is trying to integrate it into an online computation algorithm to dynamically switch modulation schemes according to the system operating point. This also needs differential-mode modeling to characterize the differential-model performance of various PWMs in the three-level inverter, a more comprehensive modeling work is required.

ACKNOWLEDGMENT

Authors would like to thank the support of Mercedes-Benz R&D North America. The experimental validation made use of the Engineering Research Center Shared Facilities supported by the Engineering Research Center Program of the National Science Foundation and DOE and the CURENT Industry Partnership Program.

REFERENCES

[1] Meintz, Andrew, et al. "Enabling fast charging–Vehicle considerations." *Journal of Power Sources* 367 (2017): 216-227.

[2] Ronanki, Deepak, Apoorva Kelkar, and Sheldon S. Williamson. "Extreme fast charging technology—Prospects to enhance sustainable electric transportation." *Energies* 12.19 (2019): 3721.

[3] Tu, Hao, et al. "Extreme fast charging of electric vehicles: A technology overview." *IEEE Transactions on Transportation Electrification* 5.4 (2019): 861-878.

[4] Un, Emre, and Ahmet M. Hava. "A near-state PWM method with reduced switching losses and reduced common-mode voltage for three-phase voltage source inverters." *IEEE Transactions on Industry Applications* 45.2 (2009): 782-793.

[5] Hou, Chung-Chuan, et al. "Common-mode voltage reduction pulsewidth modulation techniques for three-phase grid-connected converters." *IEEE transactions on Power Electronics* 28.4 (2012): 1971-1979..

[6] Hava, Ahmet M., and Emre Ün. "Performance analysis of reduced common-mode voltage PWM methods and comparison with standard PWM methods for three-phase voltage-source inverters." *IEEE Transactions on Power Electronics* 24.1 (2009): 241-252.

[7] Zhang, Haoran, et al. "Multilevel inverter modulation schemes to eliminate common-mode voltages." *IEEE transactions on industry applications* 36.6 (2000): 1645-1653.

[8] Huang, Yang, et al. "Analytical characterization of CM and DM performance of three-phase voltage-source inverters under various PWM patterns." *IEEE Transactions on Power Electronics* 36.4 (2020): 4091-4104.

[9] Zhang, Xuning, et al. "Filter design oriented EMI prediction model for DC-fed motor drive system using double fourier integral transformation method." *Proceedings of The 7th International Power Electronics and Motion Control Conference.* Vol. 2. IEEE, 2012.

[10] Sun, Yaxiu, Ruifeng Sun, and Junying Zhao. "Modeling of PWM drive motor system DM noise source." *2011 International Conference on Electrical and Control Engineering.* IEEE, 2011.

[11] Quan, Zhongyi, and Yun Wei Li. "Impact of PWM schemes on the common-mode voltage of interleaved three-phase two-level voltage source converters." *IEEE Transactions on Industrial Electronics* 66.2 (2018): 852-864.

[12] Celanovic, Nikola, and Dushan Boroyevich. "A fast space-vector modulation algorithm for multilevel three-phase converters." *IEEE transactions on industry applications* 37.2 (2001): 637-641.

Bidirectional High Voltage Conversion Ratio High-Frequency DC/DC Converter with Low Number of Components

Pedram Chavoshipour Heris
Department of Electrical Engineering
University of Arkansas
Fayetteville, Arkansas, United States
pedramc@uark.edu

Zahra Saadatizadeh
Department of Electrical Engineering
University of Arkansas
Fayetteville, Arkansas, United States
zahras @uark.edu

Rahul Biswash
Department of Electrical Engineering
University of Arkansas
Fayetteville, Arkansas, United States
rbiswash@uark.edu

Alan Mantooth
Department of Electrical Engineering
University of Arkansas
Fayetteville, Arkansas, United States
mantooth @uark.edu

Abstract— **In this paper, a coupled-inductor-based bidirectional dc/dc converter is proposed. The proposed converter can achieve a wide conversion ratio range for both step-up and step-down modes by the utilized coupled inductor. A small primary inductor along with the primary capacitor has canceled the input current ripple of the converter on the high current side. This feature would avoid large filters for the terminals. Additionally, the converter has a common-grounded structure between input and output terminals which avoids dv/dt issues. Furthermore, the normalized voltage stress on the switches is lower than the high voltage side of the converter which leads to low switching losses. In this paper, the converter has been analyzed theoretically. The PSCAD/EMTDC simulation results and experimental results of the proposed converter are extracted and presented. The experimental results will be presented in the final version of the paper. It will be implemented for 1MHz frequency operation using GaN to achieve a more compact prototype of the proposed topology.**

Keywords—Bidirectional DC-DC converter, non-isolated, high-voltage-gain, high frequency, input current ripple cancellation, coupled inductor based

I. INTRODUCTION

Bidirectional DC-DC converters have the feature of power flows in both directions which can be applied to electrical vehicles (EV) and versatile industrial applications [1]-[2]. Since the main energy source of EVs which are based on fuel cells is too low to be connected to the DC bus, a bidirectional high voltage gain DC-DC converter is required [3]-[4]. The fuel cells can be packed in series to increase the voltage, however, their voltage would be dropped by the increase of output current, and the voltage of the DC bus will fluctuate with the load change which is common in this application [5]. The proposed converter in [5] has a wide range of conversion ratios which can supply the required voltage to the bus with a wide range of input voltage. Also, its current ripple on the high current side is zero which avoids the shocks to the fuel cell packs and the overall power supply. This application of the converter is shown in Fig. 1. Generally, bidirectional converters are divided into two main groups of isolated and non-isolated topologies. The isolated bidirectional converters are safe and can be used in high-power applications, however, they require a large number of high-power switches due to the high voltage stress on them which may lead to large switching and conduction losses [6]. Non-isolated bidirectional DC-DC converters are low cost

and volume due to not having isolated transformers, also, they can be common grounded which reduces the dv/dt issues [7]. Furthermore, nonisolated bidirectional converters can be coupled inductor based or switched-inductor based structures. The switched-inductor based ones have the features of, lower costs, and no core losses in high-frequency operations. On the other hand, coupled inductor-based converters can provide a wide range of voltage gains in application with variable loads [7]. Some non-isolated bidirectional high voltage conversion ratio dc-dc converters are presented in [8]-[13] which have one coupled inductor in their structures in which have been operated in low frequency condition. In [14], dual active bridge (DAB) isolated bidirectional dc-dc converter is operated for high frequency of 500kHz. In this converter, the current ripple at low voltage side is high. This converter is suitable for high power applications.

In this paper a new non-isolated coupled inductor based power converter is proposed. The structure of the proposed converter is simple. Utilizing a small auxiliary input inductor, the input current ripple is cancelled at low voltage side. The proposed converter has simple switching pattern. Proposed converter comparing the other coupled inductor based converters of the same type has higher ratio of voltage gain over total components number. Moreover, the voltage stress on switches in the proposed converter is in the medium level comparing to other conventional converters of the same type. Finally, the proposed converter is designed for high frequency of 1 MHz and the simulation and experimental results verify the accuracy performance of the converter.

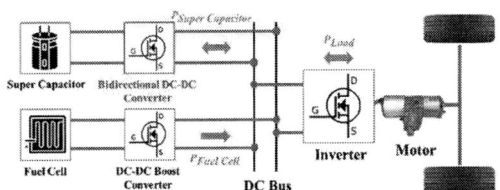

Fig. 1. Application of a bidirectional DC-DC converter with a wide range of voltage gain.

II. OPERATING PRINCIPLES OF THE PROPOSED CONVERTER

The power circuit of the proposed bidirectional converter is shown in Fig. 1(a). From Fig. 1(a), the proposed converter include three switches (S_1 , S_2 and S_3), a two-winding

978-1-6654-8901-0/22 $31.00 © 2022 IEEE

coupled inductor which has n_p and n_s number of primary and secondary windings, the magnetizing inductance (L_m) and leakage inductances (L_k), one capacitor (C_1). In Fig. 1, the turns ratio of transformer is equal to $N = n_s / n_p$. By increasing the turns ratio of transformers the voltage gains can be more increased. The capacitances of the capacitors C_1 , C_ℓ and C_H are assumed to be large enough, therefore, the voltages of the capacitors can be considered constant values. It is obvious that the inductance L_{aux} is applied to eliminate the input current ripple. The power circuit of proposed converter and its equivalent circuits in the step-up and step-down operation are shown in Fig. 2.

In the step-up operation, the power is transferred from low voltage side (V_ℓ) to high voltage side (V_H) and the output load of R_H is placed in the high voltage side which is shown in Fig. 2(b).

In the step-down operation, the power is transferred from high voltage side (V_H) to low voltage side (V_ℓ) and the output load R_ℓ is placed in the low voltage side which is shown in Fig. 2(c).

In Fig. 3, the switching pattern of three existence switches during a switching period (T_s)is shown. In Fig. 4, the equivalent circuits during a switching period (T_s) are shown. Fig. 3 illustrates the voltage and current waveforms of the proposed converter considering ideal coupled inductors.

Analysis of the proposed converter

In this part, the analysis of proposed converter during a switching period is given.

Time Interval of $0 < t < DT_s$: During this time interval, the switches, S_1 and S_2 are conducting, while the switch S_3 is OFF. This mode is shown in Fig. 4(a). Therefore, the voltage v_{Lm} is equal to V_ℓ. As a result, the inductor current can be as follows:

$$i_{Lm} = (V_\ell / L_m)(t - t_0) + i_{Lm}\big|_{t=t_0} \tag{1}$$

Time Interval of $DT_s < t < T_s$: During this time interval, the switches, S_1 and S_2 is turned OFF, while the switch S_3 is ON. This operating mode is indicated in Fig. 4(b). Therefore, the voltage v_{Lm} is calculated as $[(V_\ell - V_H) / (1 + N)] + V_\ell$. Then, the inductor's current can be written as follows:

$$i_{Lm} = \{[(V_\ell - V_H) / (1 + N)] + V_\ell\}(t - t_0) / L_m + i_{Lm}\big|_{t=DT_s} \tag{2}$$

III. STEADY STATE ANALYSIS

A. Voltage Gain

In steady state, the voltage balance law for the inductor L_m can be written as follows:

$$\bar{v}_{Lm} = DV_\ell + (1-D)\{[(V_\ell - V_H) / (1+N)] + V_\ell\} = 0 \tag{3}$$

By simplifying the above equation, the voltage conversion ratio of $G = V_H / V_\ell$ is obtained as follows:

$$G = V_H / V_\ell = (N + 2 - D) / (1 - D) \tag{4}$$

B. Voltage Stresses on Switches

The voltage stress on switch S_1 during $DT_s < t < T_s$ is calculated as follows:

$$V_{S1} = V_\ell / (1 - D_1) = V_H / (2 + N - D) \tag{5}$$

The voltage stress on switch S_2 during $DT_s < t < T_s$ and switch S_3 during $0 < t < DT_s$ is calculated as follows:

$$V_{S2} = V_{S3} = V_H - V_\ell = V_\ell[(N+1) / (1-D)]$$
$$= V_H[(N+1) / (2+N-D)] \tag{6}$$

C. Average Currents of Inductor and Switches

Based on Fig. 2(a), the average currents through the switches during a switching period are calculated as $I_{S3} = I_{S2} = I_H$, $I_{S1} = I_{Lm}$. The average input current are calculated as $I_\ell = I_{Lm} - I_H$.

Considering the power balance law for the proposed converter, the average input current and the average current of the inductor are calculated as follows:

$$I_\ell = -G I_H = \frac{-(2 + N - D)I_H}{(1-D)} \tag{7}$$

$$I_{Lm} = (1-G)I_H = \frac{-(1 + N - D)I_H}{(1-D)} \tag{8}$$

The average current of capacitor during a switching period is calculated as follows:

$$i_C = \begin{cases} -I_H / D & \text{During } DT_s \\ I_H / (1-D) & \text{During } (1-D)T_s \end{cases} \tag{9}$$

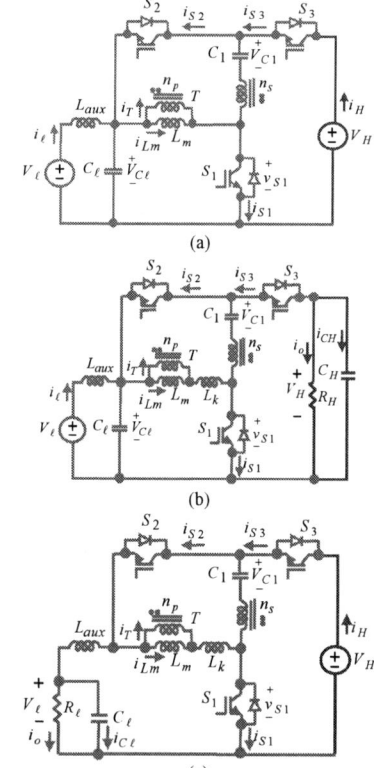

(a)

(b)

(c)

Fig. 2. Proposed bidirectional converter and its equivalent circuits for step-up and step-down operations; (a) proposed converter; (b) equivalent circuit for step-up operation; (c) equivalent circuit for step-down operation.

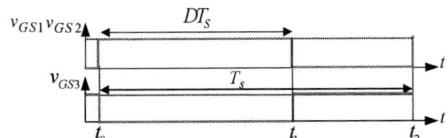

Fig. 3. Switching pattern of switches.

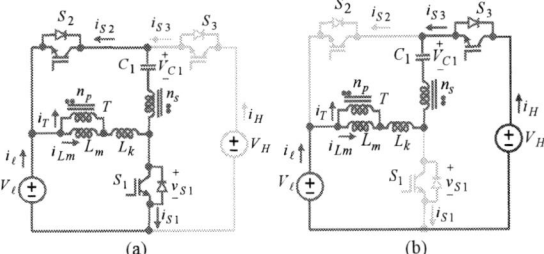

Fig. 4. Equivalent circuits of the proposed converter; (a) Mode 1; (b) Mode 2.

[$V_{D\max}$], number of switches (N_S), diodes (N_D), inductors (N_I), capacitors (N_C), the total components number (N_T). By considering that, the voltage gain is increased by using more components number. As a result, the criteria of total voltage gain over total components number (G_T / N_T) would be a fair factor to be compared as Fig. 5 (a). Considering Fig. 5 (a), the proposed SISO converter has highest value of G_T / N_T which verifies the proposed converter has better performance comparing to other conventional SISO converters. Fig. 5(b) shows the normalized voltage stress ($V_{S\_\max} / V_o$), which the proposed SISO converter has the medium value comparing to other compared converters.

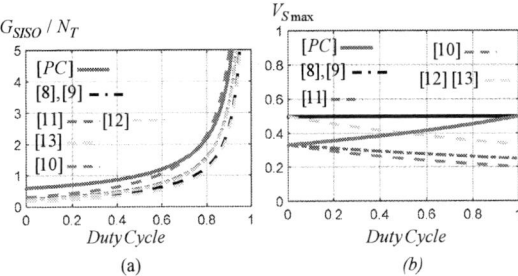

Fig. 5. The comparison results of proposed converter with conventional converters versus duty cycle (D); (a) voltage gain over total components number; (b) the total normalized voltage stress on switches.

IV. PERFORMANCE COMPARISON

The proposed bidirectional DC-DC converter is compared with the other conventional DC-DC converters of the same type in Table I. The compared DC characteristics include the voltage conversion ratio (G_{port}), the maximum voltage stress on switch [$V_{S\max}$], the maximum voltage stress on diodes

TABLE I.   COMPARISON OF PROPOSED COUPLED INDUCTOR BASED DC-DC CONVERTER WITH CONVENTIONAL STRUCTURES.

DC-DC Converters	$G_{port} = \dfrac{V_o}{V_i}$	$V_{S\max}$	$V_{D-\max}$	N_S	N_D	N_I	N_C	N_T
[8], [9]	$\dfrac{n+1}{1-D}$	$V_S = \dfrac{V_o}{1+n}$	$V_S = \dfrac{nV_o}{1+n}$	1	3	1: COUPLED	3	8
[10]	$\dfrac{(n+2)(1+D)-D}{1-D}$	$\dfrac{V_o}{(n+2)(1+D)-}$	$\dfrac{(n+1)V_o}{(n+2)(1+D)-D}$	1	4	1: COUPLED	4	10
[11]	$\dfrac{n(1+D)+2}{1-D}$	$V_S = \dfrac{V_o}{2+n+nD}$	$\dfrac{(n+1)V_o}{n(1+D)+2}$	1	4	1: TYPICAL 1: COUPLED TOTAL: 2	5	12
12	$\dfrac{1+n(1+D)}{1-D}$	$\dfrac{V_o}{n(1+D)+1}$	$\dfrac{nV_o}{n(1+D)+1}$	1	4	1: COUPLED	3	9
13	$\dfrac{1+n(1+D)}{1-D}$	$\dfrac{V_o}{n(1+D)+1}$	$\dfrac{(n+1)V_o}{n(1+D)+1}$	1	4	1: TYPICAL 1: COUPLED TOTAL: 2	5	12
Proposed SISO Cell converter	$\dfrac{n+2-D}{1-D}$	$V_S = \dfrac{V_o}{2+n-D}$	$V_D = \dfrac{(1+n)V_o}{(2+n-D)}$	3	-	1: COUPLED	2	6

V. EXPEDRIMENTAL AND SIMULATION RESULTS

To demonstrate the feasibility of the proposed bidirectional DC-DC converter, it is simulated in PSCAD/EMTDC software and the simulation results are extracted for both step-up and step-down operations of the converter shown in Fig. 6. The simulation parameters are listed in Table II. The voltage conversion ratio of the proposed converter, voltage and current stresses of switches, and required condition of input currents ripple elimination at the low voltage side are summarized in Table III.

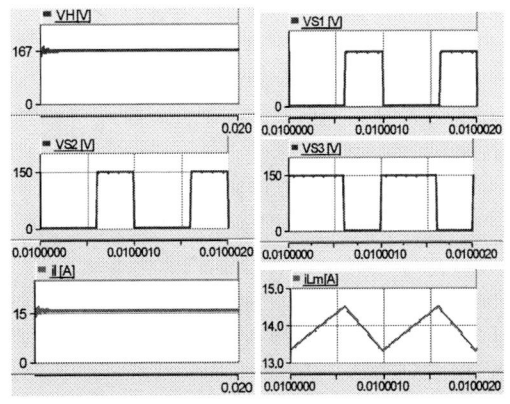

978-1-6654-8901-0/22 $31.00 © 2022 IEEE

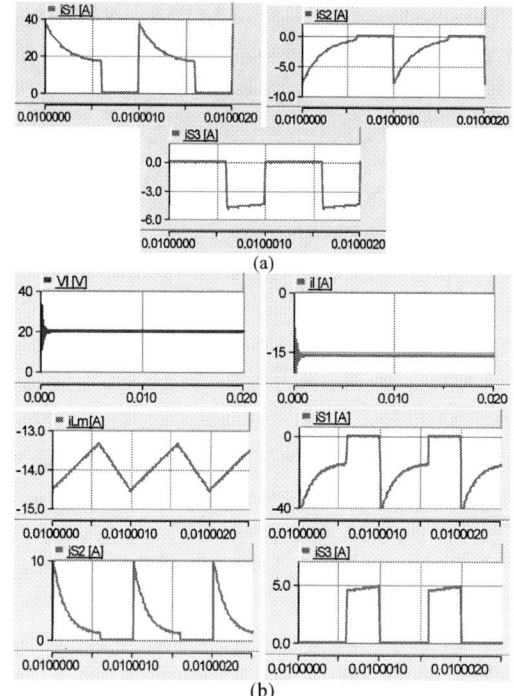

(a)

(b)

Fig. 6. Simulation results; (a) step-up operation; and (b) step-down operation.

TABLE II. SIMULATION PARAMETERS

$L_{aux}=1.5\mu H$	$L_m=10\mu H$	$D=0.6$	$f_s=1MHz$
			$P_{out}=300W$
$R_\ell=1.25\Omega$	$N=2$	$C_1=C_\ell=C_H=10\mu F$	$V_\ell=20V$
$R_H=90\Omega$			$V_H=170V$

The voltage conversion ratio of the proposed converter, voltage and current stresses of switches, and required condition of input currents ripple elimination at the low voltage side are summarized in Table III.

The power circuit of the proposed converter is implemented and operated for 300W with 1MHz frequency operation. The 3D model of the implemented circuit is shown in Fig. 7 for better illustration of the sizes. The implemented components are listed as in Table IV. The extracted experimental results are shown in Fig. 8. The extracted experimental results have good match with the ones calculated and extracted from simulation results.

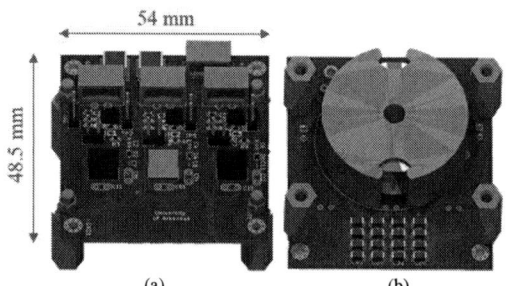

(a)   (b)

Fig. 7. Model of the implemented circuit; (a) front; (b) back.

TABLE III. CHARACTERISTICS OF THE PROPOSED TPC

Voltage conversion ratio;
$$G=V_H/V_\ell=(N+2-D)/(1-D)$$
$$V_H=(2+2-0.6)/(0.4)\times20=170V$$

Voltage stresses of switches v_{s1}, v_{s2} and v_{s3};
$$V_{S1}=V_\ell/(1-D_1)=V_H/(2+N-D)=20/0.4=50V$$
$$V_{S2}=V_{S3}=V_H-V_\ell=V_\ell[(N+1)/(1-D)]$$
$$=V_H[(N+1)/(2+N-D)]=170-20=150V$$

Required condition of input currents ripple elimination at low voltage side;
$$L_{aux}\neq0$$

Current stress on switches, I_ℓ and I_{Lm};
$$I_{S3}=I_{S2}=I_H,\ I_{S1}=I_{Lm},$$
$$I_\ell=I_{Lm}-I_H=-GI_H=\frac{-(2+N-D)I_H}{(1-D)}$$
$$I_{Lm}=(1-G)I_H=\frac{-(1+N-D)I_H}{(1-D)}$$
$$I_{S3}=I_{S2}=I_H=1.88A,\ I_{Lm}=-14.1A,$$
$$I_\ell=15.98A$$

Table IV   Experimental parameters

Components	Specifications
Switches	GS66508B-MR
Gate drivers	LM5114AMF
Gate driver isolation cores	PC95ER9.5/5-Z Material: PC95
Ferrite Cores	P30/19-3F46 Material: 3F46

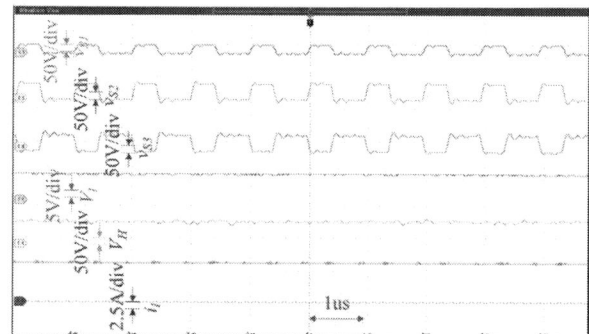

Fig. 8. Experimental results for boost operation; including v_{s1}, v_{s2}, v_{s3}, V_l, V_H and i_l (discharging mode of the battery).

ACKNOWLEDGMENT

Testing for this project was conducted at the National Center for Reliable Electric Power Transmission (NCREPT), the University of Arkansas' High-Power Test Facility.

REFERENCES

[1] Babaei, E. and Saadatizadeh, Z. (2018), High voltage gain dc–dc converters based on coupled inductors. IET Power Electronics, 11: 434-452.

[2] Saadatizadeh, Z, Chavoshipour Heris, P, Sabahi, M, Tarafdar Hagh, M, Maalandish, M. A new non-isolated free ripple input current bidirectional DC-DC converter with capability of zero voltage switching. Int J Circ Theor Appl. 2018; 46: 519– 542.

[3] Babaei, E., Saadatizadeh, Z. and Cecati, C. (2017), High step-up high step-down bidirectional DC/DC converter. IET Power Electronics, 10: 1556-1571.

[4] Z. Saadatizadeh, P. C. Heris, Y. Yang and F. Blaabjerg, "High Step-Up/Down Switched-Capacitor Based Bidirectional DC-DC Converter," 2020 IEEE 21st Workshop on Control and Modeling for Power Electronics (COMPEL), 2020, pp. 1-6.

[5] Y. Zhang, H. Liu, J. Li and M. Sumner, "A Low-Current Ripple and Wide Voltage-Gain Range Bidirectional DC-DC Converter with Coupled Inductor," in IEEE Transactions on Power Electronics, vol. 35, no. 2, pp. 1525-1535, Feb. 2020, doi: 10.1109/TPEL.2019.2921570.

[6] P. He and A. Khaligh, "Comprehensive Analyses and Comparison of 1 kW Isolated DC–DC Converters for Bidirectional EV Charging Systems," in IEEE Transactions on Transportation Electrification, vol. 3, no. 1, pp. 147-156, March 2017, doi: 10.1109/TTE.2016.2630927.

[7] A. R. N. Akhormeh, K. Abbaszadeh, M. Moradzadeh and A. Shahirinia, "High-Gain Bidirectional Quadratic DC–DC Converter Based on Coupled Inductor with Current Ripple Reduction Capability," in IEEE Transactions on Industrial Electronics, vol. 68, no. 9, pp. 7826-7837, Sept. 2021, doi: 10.1109/TIE.2020.3013551.

[8] S. Pourjafar, F. Sedaghati, H. Shayeghi, and M. Maalandish, "High step-up DC–DC converter with coupled inductor suitable for renewable applications," IET Power Electron., vol. 12, no. 1, pp. 92–101, Jan. 2019.

[9] M. Das and V. Agarwal, "Design and analysis of a high-efficiency DC–DC converter with soft switching capability for renewable energy applications requiring high voltage gain," IEEE Trans. Ind. Electron., vol. 63, no. 5, pp. 2936–2944, May 2016.

[10] H. Ardi and A. Ajami, "Study on a high voltage gain SEPIC-based DC–DC converter with continuous input current for sustainable energy applications," IEEE Trans. Power Electron., vol. 33, no. 12, pp. 10403–10409, Dec. 2018.

[11] A New High-Gain, High-Efficiency SEPIC-Based DC–DC Converter for Renewable Energy Applications Sara Hasanpour , Mojtaba Forouzesh, Student Member, IEEE, Yam P. Siwakoti, Senior Member, IEEE, and Frede Blaabjerg , Fellow, IEEE

[12] B. Axelrod, Y. Beck, and Y. Berkovich, "High step-up DC–DC converter based on the switched-coupled-inductor boost converter and diodecapacitor multiplier: Steady state and dynamics," IET Power Electron., vol. 8, no. 8, pp. 1420–1428, Aug. 2015.

[13] M. Forouzesh, K. Yari, A. Baghramian, and S. Hasanpour, "Single-switch high step-up converter based on coupled inductor and switched capacitor techniques with quasi-resonant operation," IET Power Electron, vol. 10, no. 2, pp. 240–250, 2017.

[14] Y. Park, S. Chakraborty and A. Khaligh, "DAB converter for EV onboard chargers using bare-die SiC MOSFETs and leakage-integrated planar transformer," IEEE Transactions on Transportation Electrification, vol. 8, no. 1, pp. 209-224, March 2022.

978-1-6654-8901-0/22 $31.00 © 2022 IEEE

Short Circuit Fault Induced Failure of SiC MOSFETs in DC Solid-State Circuit Breakers

Shuyan Zhao
Department of Electrical and Computer Engineering, College of Engineering
Drexel University
Philadelphia, PA, USA
shuyan.zhao@drexel.edu

Reza Kheirollahi
Department of Electrical and Computer Engineering, College of Engineering
Drexel University
Philadelphia, PA, USA
reza.kheirollahi@drexel.edu

Hua Zhang
Department of Electrical and Computer Engineering, College of Engineering
Rowan University
Glassboro, NJ, USA
zhangh@rowan.edu

Fei Lu
Department of Electrical and Computer Engineering, College of Engineering
Drexel University
Philadelphia, PA, USA
fei.lu@drexel.edu

Abstract—This paper investigates the failure modes of SiC MOSFETs-based passive voltage clamping dc solid-state circuit breakers (SSCBs). There are two major concerns of main SiC switch failure that are analyzed in this paper: 1) gate-source voltage ringing related switch degradation/failure when cutting off heavy fault current with high d*i*/d*t*. 2) thermal runaway directly caused by the short circuit surge current in the main switch before the cut-off and transient power strike during cut-off. Two 10A/86A dc interruption experiments are conducted to demonstrate the failure caused by gate ringing due to the high d*i*/d*t* issue coupled with unavoidable parasitic common source impedance of the device package. The thermal runaway failure due to the short circuit current surge and transient turn-off power strike is also demonstrated by a 631A dc interruption test. At last, an active commutation current injection scheme is discussed as future research trends to address the revealed gate ringing and cut-off power strike issues in SSCBs.

Keywords—DC circuit breaker, failure, thermal runway, gate ringing, power strike, SiC MOSFET

I. INTRODUCTION

Due to recent escalation of global fossil fuel energy crisis, clean renewable energies, and their applications, such as photovoltaic, wind, ocean wave, and hydrogen generations are now gaining more attentions. They are promising alternative energy resources in the near future [1]-[3].

Direct current (dc) systems have been witnessing a great progress recently due to its better renewable energy resources penetration than traditional ac systems. However, absence of reliable fault isolation and protection device is still a shortcoming of dc systems that prevents its further deployment in practical applications. Other than conventional ac systems, there is no periodic natural current zero crossings in dc systems, which imposes serious challenges on dc circuit breakers (DCCBs) [4].

A typical dc system with DCCBs is shown in Fig. 1. Facilitated by the rapid development of wide bandgap semiconductors, now DCCBs are gaining better reliability and faster response speed. Hybrid and solid-state circuit breakers are two categories of semiconductor assisted DCCBs and are recent research hot spots. Hybrid circuit breakers utilize solid state

The information, data, or work presented herein was funded in part by the Advanced Research Projects Agency-Energy (ARPA-E), U.S. Department of Energy, under Award Number DE-AR0001114 in the BREAKERS program monitored by Dr. Isik Kizilyalli.

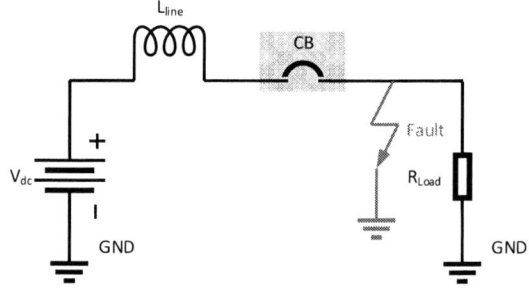

Fig. 1. Topology of a typical dc system, showing dc circuit breakers for fault protection.

switches-based power interrupter, which can achieve high efficiency and fast response speed than traditional mechanical breakers. It has been successfully validated and deployed in real engineering implementations, such as HVDC transmission systems [5], [6]. However, in low- and medium voltage level dc systems where the fault loop impedance is usually low, solid state circuit breakers (SSCBs) are more advantageous for its much faster response speed in the microseconds level [7]-[9].

The state-of-the-art third generation commercial SiC MOSFETs have obtained enhanced voltage and current ratings and high operation temperature as well as reduced on-state resistance, which makes high efficiency dc SSCBs promising. Extensive research has been conducted on SiC MOSFET based SSCBs such as fault current bypass type [10] and parallel voltage clamping type [11].

However, a common concern of those passive clamping based SSCBs is that the extremely high d*i*/d*t* due to the hard turn-off operation, which couples with device package parasitic common source inductances, causes gate source voltage ringing issues [12]. It might cause non-reversible and even permanent damage to gate oxide layer which puts SSCB out of control. Another concern is that the hard turn-off transient power strike on main switch, combined with a continuous normal and fault current conduction power loss before that, put SSCB at serious risk of main semiconductor devices transient thermal runaway [13].

This paper aims to investigate the two failure modes of SiC

978-1-6654-8901-0/22 $31.00 © 2022 IEEE

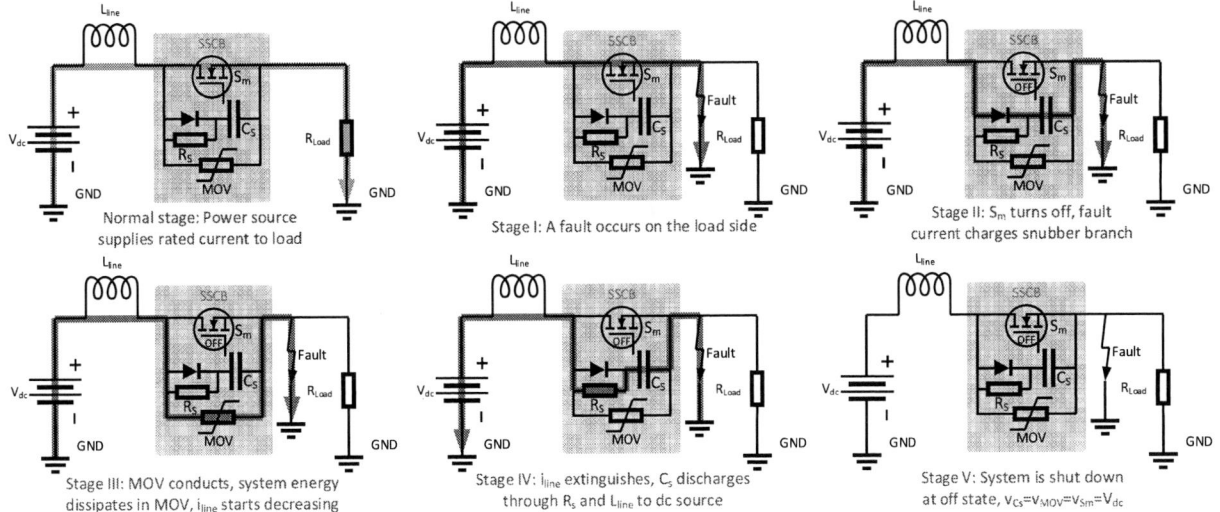

Fig. 2. Working principles of voltage clamping circuit based SSCB, showing stages of normal operation, breaking and off state.

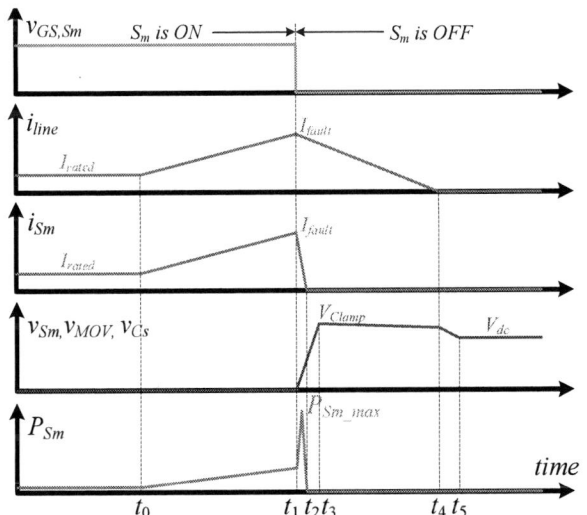

Fig. 3. Critical waveforms regarding working stages of VCC based SSCB.

MOSFET in passive voltage clamping based dc SSCB applications by analytical and experimental demonstrations. First, basic working principles of SSCB is briefly introduced, followed by transient power strike and gate ringing analysis during the turn-off transient. Second, two groups of dc current interruption experiments are conducted to demonstrate the main switch failure concerns. Finally, a discussion is presented to suggest that a zero current switching (ZCS) based soft turn-off solution can potentially mitigate the gate ringing and power strike concerns, thence avoiding SiC MOSFET failure in SSCBs. Besides, associated design guidelines are provided for future implementations.

II. SiC MOSFET Based DC Solid-State Circuit Braekers

A. Working principles of passive voltage clamping circuits

Figs. 2 and 3 show the working stages of voltage clamping circuit (VCC) based SSCBs with critical waveforms, including normal operation, breaking, and off state.

Normal stage ($t < t_0$): In normal operation stage, power source supplies rated current to the load through SSCB continuously.

Stage I ($t_0 < t < t_1$): A fault occurs on the load side. Fault current in i_{Sm} and i_{line} ramps up quickly.

Stage II ($t_1 < t < t_3$): The fault is detected by the SSCB. Then, S_m reacts to turn off at $t = t_1$. i_{Sm} is extinguished to zero at $t = t_2$. The commutated current charges snubber capacitor C_S. $v_{Sm} = v_{MOV} = v_{Cs}$ keeps increasing in this stage to establish a high voltage across SSCB.

Stage III ($t_3 < t < t_4$): $v_{Sm} = v_{MOV} = v_{Cs}$ achieves MOV clamping voltage V_{clamp} at $t = t_3$. Fault current stops charging snubber. Instead, it commutates to the MOV branch. The SSCB voltage $v_{Sm} \approx V_{clamp} > V_{dc}$, thus ensuring i_{line} to start decreasing. It is extinguished to zero at $t = t_4$ when v_{Sm} roughly equals tripping voltage V_{trip} of MOV.

Stage IV ($t_4 < t < t_5$): C_s starts discharging through R_s and L_{line} to dc source at $t = t_4$, causing a slight reverse current in i_{line}.

Stage V ($t_5 < t$): C_s stop discharging at $t = t_5$ and the SSCB hold dc bus voltage, $v_{Sm} = v_{MOV} = v_{Cs} = V_{dc}$. The fault current interruption process completes.

III. Two Failure Modes: Power Strike and Gate Ringing in DC Solid-State Circuit Breakers Application

It is noted that the risk of SiC MOSFET damage at SSCB normal operation stage is low since it constantly conducts rated current at a low voltage drop. Most of device failure happens at the SSCB fault interruption process, which can be categorized into (1) thermal runaway caused by transient power strike due to high fault current magnitude, and (2) gate oxide failure caused by gate source voltage ringing due to high di/dt.

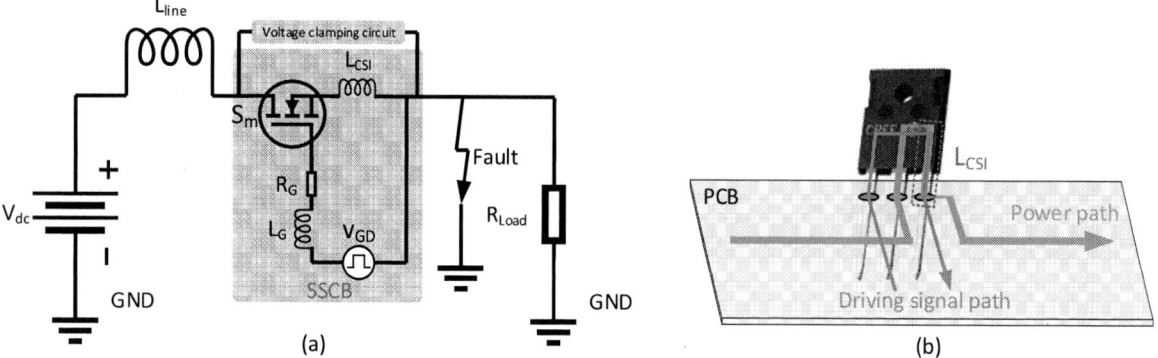

Fig. 4. SiC MOSFET based SSCB in a dc system, showing (a) parasitic components on gate driving loop and (b) common source inductance L_{CSI} on a switch.

A. Thermal runaway caused by transient power strike

Based on the SSCB working principle, the thermal runaway of SiC MOSFET is caused by a three-stage power strike.

1) During the normal operation stage of an SSCB, there is a constant conduction loss on SiC MOSFET which is depicted as follows.

$$P_{loss1} = I_{rated}^2 \times R_{ds(on)} \qquad (1)$$

Where I_{rated} refers to rated load current and $R_{ds(on)}$ means the semiconductor resistance. The continuous power loss P_{loss1} leads to the first stage temperature rise of SiC MOSFET, which depends on cooling system design of the SSCB.

2) During stage I of an SSCB when SiC MOSFET conducts fault current, the transient conduction loss can be inferred as follows.

$$P_{loss2} = I_{fault}^2 \times R_{ds(on)} \qquad (2)$$

Where I_{fault} indicates fault current magnitude, which can achieve up to $10\times I_{rated}$. Although $R_{ds(on)}$ of state-of-the-art SiC MOSFETs is low, the large fault current might still cause a significant power strike on the main switch. This transient power loss P_{loss2} leads to the second stage temperature rise of SiC MOSFET. It is noted that since the duration of this stage is usually much smaller than the time constant of the equivalent thermal model, the cooling design has very limited effect on the second stage temperature increasing.

3) During stage II of an SSCB when SiC MOSFET turns off and i_{Sm} extinguished to zero, the transient switching loss is depicted as follows.

$$P_{loss3} = \frac{I_{fault\_max}^2}{C_t}\left(\frac{t^2}{2 \cdot T_f} - \frac{t^3}{2 \cdot T_f^2}\right), \; P_{loss3\_max} = \frac{2 \cdot I_{fault\_max}^2 \cdot T_f}{27 \cdot C_t} \quad (3)$$

Where I_{fault_max} indicates the maximum fault current magnitude before breaking. C_t means the total output capacitance of S_m including parasitic capacitance C_{oss} and external capacitance C_s. T_f refers to switch current falling time $t_2 - t_1$. (3) indicates

that the overlap of fault current and the voltage of SiC MOSFET leads to a significant transient power strike. This transient switching loss P_{loss3} has much higher magnitude but lower duration time than P_{loss2}. It is noted that P_{loss3} leads to the third stage temperature rise of SiC MOSFET. The external cooling system has almost no influence on the third stage temperature increasing neither.

A good cooling design is effective to limit the first stage temperature rise, in which the risk of thermal runaway is low. However, the risk is high at the second and third stages, especially when a heavy fault occurs, causing I_{fault} (I_{fault_max}) extremely large. It is highly possible that a thermal failure happens during the SSCB working stages I or II, which depends on specific fault magnitude, duration and switch transient thermal impedance.

B. Gate oxide failure caused by gate source voltage ringing

Gate oxide breakdown in SSCBs is another major category of switch failure, which mostly results from gate source voltage ringing during turn-off caused by high di/dt coupled with common source inductance (L_{CSI}) of SiC MOSFETs.

As indicated in Fig. 3, due to the large fault current magnitude as well as ultrafast current extinguishing speed, there is a serious di/dt issue on S_m during the interruption process, whose influence on switch failure is explained as follows.

Fig. 4(a) shows parasitic components on the gate driving signal loop. The following ac analysis from t_1 to t_2 is conducted to specify the gate source voltage ringing.

$$L_t C_{iss} \frac{d^2 v_{gs,CSC}(t)}{dt^2} + R_G C_{iss} \frac{dv_{gs,CSC}(t)}{dt} + v_{gs,CSC}(t) = -L_{CSI} \frac{di_{S_m}(t)}{dt}$$
$$(4)$$

Where $v_{gs,CSC}(t)$ means the common source coupled gate source voltage ringing. L_t is total parasitic inductance of the gate driving loop ($L_t = L_{CSI} + L_G$). L_{CSI} is the SiC MOSFET common source inductance of both driving signal and power paths as indicated in Fig. 4(b), while L_G is parasitic inductance of gate driving circuit layout. C_{iss} indicates S_m input capacitance. R_G refers to gate driving loop resistance. Thence, $v_{gs,CSC}(t)$ can be derived as follows.

$$v_{gs,CSC}(t) = \frac{L_{CSI} \cdot I_{fault\_max}}{T_f} \cdot [1 - \frac{\omega_0}{\omega_d} e^{-\alpha t} \sin(\omega_d t)]$$

$$\alpha = \frac{R_G}{2L_t}, \omega_0 = \frac{1}{\sqrt{L_t C_{iss}}}, \omega_d = \sqrt{\omega_0^2 - \alpha^2} \qquad (5)$$

Equation (5) indicates that the gate source voltage ringing directly relates to L_{CSI} and di/dt (I_{fault_max}/T_f), which mainly depend on switch package and fault magnitude, respectively. When the ringing effect causes v_{gs} exceed threshold value during turn-off, SiC MOSFET might be falsely turned on, causing SSCB out of control. When the ringing effect causes v_{gs} exceed allowable voltage range (-8V ~ +19V), there is a high risk of gate oxide degradation or even breakdown.

IV. EXPERIMENTAL DEMONSTRATION

To demonstrate the abovementioned failure mechanisms in SSCB applications, dc fault interruption experiments are conducted with a 1.2 kV discrete MOSFET (C3M0075120D) based SSCB prototype. Test circuit follows Fig. 2 where the dc source is emulated by pre-charged dc link capacitors. The load side is directly shorted to emulate a fault. During the tests, the SSCB turns on first and dc link capacitors discharge through L_{line}, generating a fast-ramping fault current. V_{DC} and L_{line} are tuned to achieve desired fault current magnitude and duration time for various tests.

A. Transient power strike test

A transient power strike test is first conducted, whose experimental waveforms are shown in Fig. 5. After the short circuit occurs, the fault current in S_m ramps quickly to 631A within 200 μs conduction duration. It shows that the MOSFET current is saturated with a conduction drop v_{Sm} increased up to 740V before turn-off. The transient power strike reaches 467 kW, causing an instant thermal runaway failure. It shows that v_{Sm} suddenly decreases to zero voltage after S_m fails, which cannot hold dc bus voltage V_{DC} anymore after receiving turn-off signal from gate driving circuit, indicating that the MOSFET has lost the voltage blocking capability.

The experimental results successfully demonstrate the second stage transient power strike induced thermal runaway failure of SiC MOSFET, which is consistent with the analysis in Section III A. However, it must be noted that in practical appli-

cations, usually the fault current magnitude and duration are not as extremely high as this test. Therefore, in most of real SSCBs, both this second stage (fault conduction) and the third stage (fault current and voltage overlap) power strikes are combined simultaneous factors leading to ultimate thermal runaway failure of the main MOSFET switch.

B. Gate source voltage ringing test

Another major failure mechanism is the gate source voltage ringing issue caused by di/dt during MOSFET turn-off, which is experimentally demonstrated in Figs. 6 and 7.

Fig. 6 shows v_{Sm}, i_{line} and v_{GS} waveforms of a 10A dc interruption test. In this experiment, the SSCB successfully cuts off a 10A dc within 1.2 μs with a maximum clamping voltage v_{Sm} = 829V. However, it indicates that the di/dt coupled with L_{CSI} of switch package induces a serious gate source voltage ringing which makes the S_m turn-off unreliable. v_{GS} negative voltage of -12.5V exceeds manufacturer recommended minimum value, putting MOSFET gate oxide at high risk of degradation or breakdown.

Fig. 7 shows another 86A dc interruption test results using the same MOSFET device, in which v_{Sm} and v_{GS} waveforms are presented. Due to much higher di/dt in this test, a fatal gate ringing occurs, which makes S_m falsely turn on thence the SSCB is out of control. The gate ringing thereafter deteriorates and v_{GS} lower than -20V exceeds manufacturer recommended minimum value, causing gate oxide breakdown as indicated by digital multimeter measurement in Fig. 7. Finally, the failed MOSFET loses voltage blocking capability.

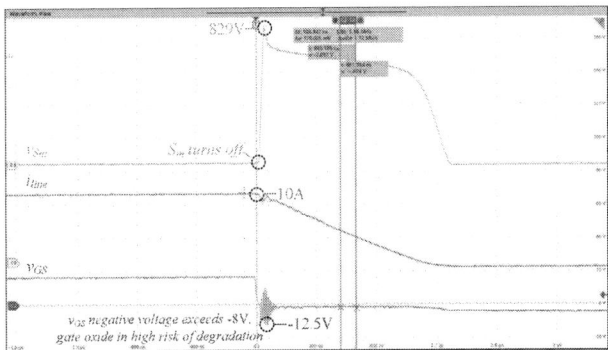

Fig. 6. 10A dc interruption experiment, showing gate voltage ringing issue.

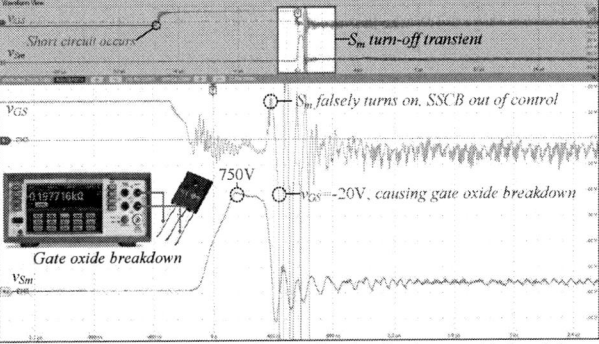

Fig. 7. 86A dc interruption experiment, showing fatal gate oxide breakdown of SiC MOSFET due to gate voltage ringing issue.

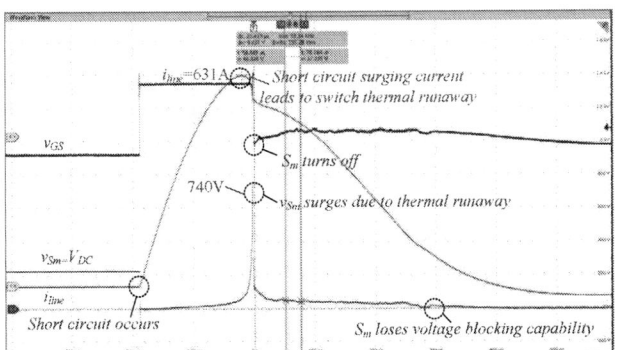

Fig. 5. 631A dc interruption experiment, showing S_m thermal failure due to transient power strike.

978-1-6654-8901-0/22 $31.00 © 2022 IEEE

V. DISCUSSION

Regarding SiC MOSFET failures in passive voltage clamping based SSCBs, obtaining high reliability and fast operation speed are two critical design priorities. As discussed in this paper, main switch failure issue in SSCBs includes two main sources: a) transient power strike during fault conduction and current interruption, and b) false turn-on and gate oxide breakdown due to the gate source voltage ringing. These two factors require extensive efforts in the future.

As a new solution, Fig. 8 shows a novel current commutation based SSCB with soft turn-off capability, which is different from the conventional passive voltage clamping circuit based hard-switching SSCB.

Fig. 8 provides the circuit topologies, state diagrams, and critical waveforms to compare working principles of both solutions. In hard-switching SSCBs, after receiving a trigger signal, S_m turns off. Since voltage and current overlap during interruption, transient power strike is unavoidable. However, in soft-switching SSCBs, there is an extra state between the SSCB triggering stage and S_m turn-off process, named as current commutation. It forces the fault current in the main switch to zero by injecting a resonant pulse current. In this case, S_m turns off when its current is already zero, meaning zero transient power. It is noted that injected pulse current is fast and does not impact the operation speed of soft-switching SSCBs. Meanwhile, since S_m achieves a zero current soft turn-off, there is no evident di/dt stress on S_m during fault interruption process, which effectively gets SiC MOSFET rid of gate ringing issues.

Recently, active commutation current injection (ACCI) technology has been introduced in hybrid circuit breakers [14], [15], but has yet been widely discussed in SSCBs. Achieving soft turn-off using ACCI in SiC MOSFET based SSCBs brings two main advantages: a) it eliminates transient power on the main switch completely; b) reducing current to zero during turn-off removes the induced voltage ringing on the gate of S_m. Although soft-switching SSCBs require more components, they feature higher reliability and lifetime. Therefore, it is meaningful to continue this research to propose novel topologies and optimize circuit components.

VI. CONCLUSIONS

This paper investigates short circuit fault induced failure of SiC MOSFET in passive voltage clamping circuit based SSCBs. There are two major concerns of SiC MOSFET main switch failure identified in this paper: a) second stage (fault current conduction) and third stage (fault current and voltage overlap) power strikes thermal runaway failure, and b) gate oxide failure, caused by high di/dt with device package common source inductance induced gate source voltage ringing. Two failure mechanisms are analyzed and experimentally demonstrated in this paper. At last, to overcome the identified SiC MOSFET failures in passive clamping based SSCBs, an active current commutation solution is discussed, which can realize zero current soft turn-off, providing directions for future research and developments of SSCBs.

ACKNOWLEDGMENT

The information, data, or work presented herein was funded in part by the Advanced Research Projects Agency-Energy (ARPA-E), U.S. Department of Energy, under Award Number DE-AR0001114 in the BREAKERS program monitored by Dr. Isik Kizilyalli. The views and opinions of authors expressed herein do not necessarily state or reflect those of the United States Government or any agency thereof.

REFERENCES

[1] T. Dragičević, X. Lu, J. C. Vasquez and J. M. Guerrero, "DC Microgrids—Part I: A Review of Control Strategies and Stabilization Techniques," in IEEE Transactions on Power Electronics, vol. 31, no. 7, pp. 4876-4891, July 2016, doi: 10.1109/TPEL.2015.2478859.

[2] J. K. Motwani et al., "Asymmetrical DC-DC Modular Multilevel Converter using Switching Cycle Control," 2022 IEEE 23rd Workshop on Control and Modeling for Power Electronics (COMPEL), 2022, pp. 1-8, doi: 10.1109/COMPEL53829.2022.9830011.

[3] J. Ma, M. Zhu, Y. Li and X. Cai, "Monopolar Fault Reconfiguration of Bipolar Half Bridge Converter for Reliable Load Supply in DC Distribution System," in IEEE Transactions on Power Electronics, vol. 37, no. 9, pp. 11305-11318, Sept. 2022.

[4] Huo, Q., Xiong, J., Zhang, N., Guo, X., Wu, L. and Wei, T., "Review of DC circuit breaker application," in Electric Power Systems Research, vol. 209, pp. 107946, 2022.

[5] Z. J. Zhang and M. Saeedifard, "Overvoltage Suppression and Energy Balancing for Sequential Tripping of Hybrid DC Circuit Breakers," in IEEE Transactions on Industrial Electronics, 2022, doi: 10.1109/TIE.2022.3203765.

[6] K. Liu, L. Qi, Z. Dongye, G. Tang and X. Cui, "Low Inductance Design for Symmetrical Submodules in Hybrid HVDC Circuit Breaker," in IEEE Transactions on Power Electronics, vol. 36, no. 11, pp. 12321-12331, Nov. 2021, doi: 10.1109/TPEL.2021.3077884.

[7] S. Rahimpour, O. Husev and D. Vinnikov, "Impedance-Source DC Solid-State Circuit Breakers: An Overview," 2022 International Symposium on Power Electronics, Electrical Drives, Automation and Motion (SPEEDAM), 2022, pp. 186-191.

[8] S. Zhao et al., "A 4 kV/120 A SiC Solid-State DC Circuit Breaker Powered By a Load-Independent IPT System," in IEEE Transactions on Industry Applications, vol. 58, no. 1, pp. 1115-1125, Jan.-Feb. 2022.

[9] X. Song, P. Cairoli, Y. Du and A. Antoniazzi, "A Review of Thyristor Based DC Solid-State Circuit Breakers," in IEEE Open Journal of Power Electronics, vol. 2, pp. 659-672, 2021.

[10] R. Kheirollahi, S. Zhao and F. Lu, "Fault Current Bypass-Based LVDC Solid-State Circuit Breakers," in IEEE Transactions on Power Electronics, vol. 37, no. 1, pp. 7-13, Jan. 2022.

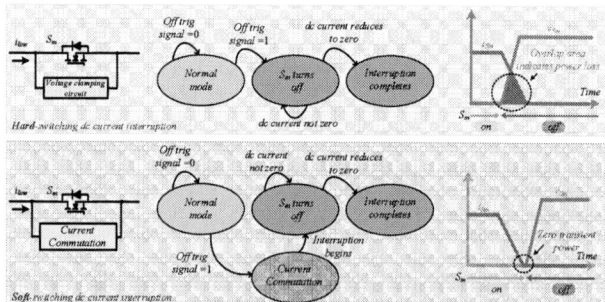

Fig. 8. A novel active current commutation based SSCB with soft turn-off capability, showing topology, state diagrams and critical waveforms compared to hard switching SSCBs.

[11] S. Zhao, R. Kheirollahi, Y. Wang, H. Zhang and F. Lu, "Implementing Symmetrical Structure in MOV-RCD Snubber-Based DC Solid-State Circuit Breakers," in IEEE Transactions on Power Electronics, vol. 37, no. 5, pp. 6051-6061, May 2022, doi: 10.1109/TPEL.2021.3133113.

[12] W. Zhang, X. Wang, M. S. A. Dahidah, G. N. Thompson, V. Pickert and M. A. Elgendy, "An Investigation of Gate Voltage Oscillation and its Suppression for SiC MOSFET," in IEEE Access, vol. 8, pp. 127781-127788, 2020, doi: 10.1109/ACCESS.2020.3008940.

[13] S. Zhao, R. Kheirollahi, Y. Wang, H. Zhang and F. Lu, "Investigation of Limitations in Passive Voltage Clamping-Based Solid-State DC Circuit Breakers," in IEEE Open Journal of Power Electronics, vol. 3, pp. 209-221, 2022, doi: 10.1109/OJPEL.2022.3163072.

[14] Y. Zhou, Y. Feng, N. Shatalov, R. Na and Z. J. Shen, "An Ultraefficient DC Hybrid Circuit Breaker Architecture Based on Transient Commutation Current Injection," in IEEE Journal of Emerging and Selected Topics in Power Electronics, vol. 9, no. 3, pp. 2500-2509, June 2021, doi: 10.1109/JESTPE.2020.2983354.

[15] N. Shatalov, Y. Zhou, R. Na and Z. J. Shen, "Design and Operation of Bi-Directional Hybrid Circuit Breaker Based on Transient Commutation Current Injection," 2021 IEEE Applied Power Electronics Conference and Exposition (APEC), 2021, pp. 1128-1133.

.

Area-Efficient High-Voltage (HV) Lateral MOSFETs for Discrete Device Development and Power IC Integration

Sundar Babu Isukapati[a], Seung Yup Jang[a], Woongje Sung[a]

[a] College of Nanoscale Science and Engineering, State University of New York Polytechnic Institute, Albany, NY, USA, sundar.isukapati@sunypoly.edu

Abstract—**This paper reports the design of area-efficient high voltage lateral MOSFETs for discrete device development and also for integration in power IC development. Utilizing the three metal layered back-end-of-the-line (BEOL) process, the footprint of the devices has been significantly reduced without any deviation from the static electrical performances. The reported devices were fabricated on a six-inch N epi/P epi/ 4H-SiC N+ substrate. The reported HV lateral devices are the best in class with superior breakdown voltage (BV) - specific on-resistance ($R_{on,sp}$) trade-off. The devices demonstrated a BV of 430V at drain-source current (I_{ds}) of 1mA and a $R_{on,sp,active}$ of 6.2 m$\Omega \cdot$cm^2 at a gate-source voltage (V_{gs}) of 25V at 25 °C.**

Keywords—4H-SiC, Power IC, RESURF, Lateral MOSFET, chip size, CMOS, HVIC, PIC

I. INTRODUCTION

Due to its inherent material properties, Silicon has been limited in its power processing and high-temperature operational capabilities [1]. Alternatively, with its superior material properties, 4H-SiC has been the favorite for applications requiring high power and high temperatures. When compared to its silicon counterparts, the 4H-SiC power electronics components are smaller, reliable, and highly efficient [1]. These capabilities pave a way for significant energy savings and clean energy sectors. There has been a significant advancement in technological aspects like the availability of high-quality epi, and manufacturing process maturity over the last decade that lead to the commercialization of 4H-SiC discrete power MOSFETs.

Along with the well-sought-after and extensive discrete power MOSFET applications, SiC has tremendous potential in the development of HV Power ICs for high-power and high-temperature applications. SiC Power ICs have the potential to exceed the current generation Si and SOI technologies in terms of power capability and high-temperature operation [2]. In Power ICs the integration of the HV power MOSFET and the control circuitry into a single IC chip will remarkably reduce the footprint, cost, and also parasitics from the interconnections [3]. This will consequently improve the dynamic performance and system-level efficiency. Unlike the discrete level integration, since all the power and control components are fabricated on a single chip, there will be no external projections

The information, data, or work presented herein was funded by Advanced Research Projects Agency-Energy (ARPA-E) U.S. Department of Energy, under Award Number DE—AR0001028, Program Director: Dr. Isik Kizilyalli.

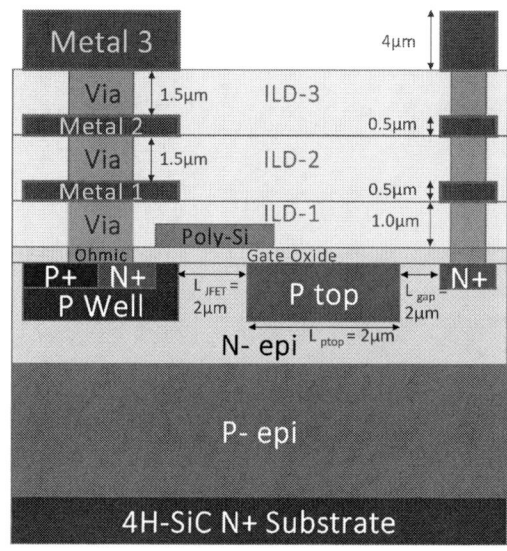

Figure 1: Cross-section of the Lateral HV nMOSFET; device fabricated on 150mm N epi/P epi/ 4H-SiC N+ substrate; device implements a three metal layered BEOL process

rating thereby harnessing the high temperate operational perk of 4H-SiC. To pursue the half-bridge circuit, a major building block for the development of power electronic converters, there is a need for the integration of two power MOSFETs (high-side and low-side) on a single chip. This is feasible with no constraints only with the lateral architectural devices. In vertical architecture, since the devices share the same substrate, multiple power MOSFETs operating at different voltages cannot be integrated on a single chip [4]. But, unlike vertical devices, due to their architecture, lateral devices inherently demand a larger cell pitch. So, any improvement in the cell pitch and chip size significantly impacts the development of high-performance economical devices and ICs. Hence, the prime motivation for this work is to design area-efficient HV lateral MOSFETs for discrete device development as well as for the monolithic integration in the HV Power IC applications. These area-efficient MOSFETs save a significant chip size allowing flexible cell design.

The HV lateral MOSFETs reported in this work are one of the best in class to demonstrate a breakdown voltage (BV) of 430 V with a specific on-resistance of active area ($R_{on,sp,\ active}$)

Design A	Design B	Design C
Layout top view	*Layout top view*	*Layout top view*
	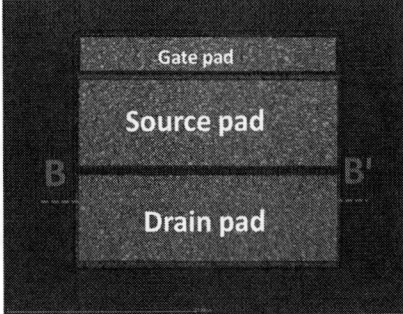	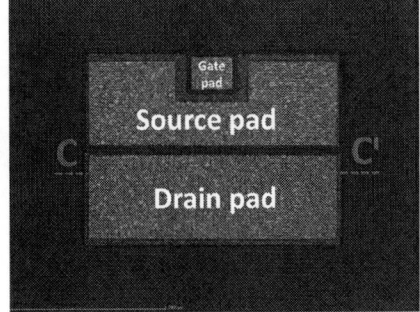
Chip size = **3.32mm²**	Chip size = **2.50mm²**	Chip size = **1.8mm²**
Active Area = **1.96mm²**	Active Area = **1.84mm²**	Active Area = **1.51mm²**

Figure 2: Optical layout top views of Designs A, B, and C; corresponding chip sizes and active areas have been mentioned

of 6.2 mΩ·cm² at a gate-source voltage (V_{gs}) of 25 V at 25 °C. The area-efficient designs were achieved utilizing the three-metal layered BEOL process, a significant requisite in developing robust and compact HV power converter ICs. The fabrication process, design, and electrical characteristics of the reported area efficient devices are discussed in Sections II and III.

II. DESIGN OF THE HV LATERAL MOSFETs

Fig. 1 shows the cross-section of the nominal HV lateral MOSFET. The HV NMOS was fabricated on the N epi/P epi/N+ substrate stack that was optimized for the development of Power ICs [5].

The HV NMOS was designed based on the REduced SURface Field (RESURF) concept. The optimum doping concentrations of the N epi and P epi were optimized for better trade-off relation between BV and $R_{on,sp}$ using the 2-D device simulations. The bottom P-epi not only helps for isolation purposes but also forms a double RESURF with the surface Al implanted P top to uniformly distribute the electric field and thereupon increase the breakdown voltage in the lateral direction. Hence, the doping and thickness of P-epi were optimized to support the highest possible breakdown voltage and simultaneously prevent the depletion layer from reaching the N+ substrate. The reach-through of the depletion layer to the N+ substrate leads to shorting of all the HV power and control circuitry. The length of the P top (L_{ptop}) and the gap between the P top and the N+ drain (L_{gap}) determine the gate-to-drain length (L_{gd}) of the lateral MOSFET which dictates the blocking capability as well as the $R_{on,sp}$ of the device. The dimensions of L_{ptop} and L_{gap} were chosen such that the breakdown doesn't occur on the surface while simultaneously targeting the optimum BV-$R_{on,sp}$ trade-off. To ensure the field plate effect, the P top was placed so that the edge of the polysilicon overlaps the P top layer. The critical dimensions of the HV lateral devices reported in this work are shown in Fig. 1.

Design A

Fig.2 shows the conventional layout view of the HV lateral MOSFET with drain, source, and gate pads. The corresponding 3-D cross-sectional layout is shown in Fig. 3. The design implements interdigitated source and drain patterns. In Design A, all three metal layers are engaged over each of the source/drain fingers. Hence the drain, source, and gate metal pads are placed outside of the active area. This type of layout technique occupies a chip size of 3.32 mm². In this design, the cell pitch of the device is dictated by the critical dimension of the metal 3 (metal 3 - metal 3 spacing) as indicated in the layout crossectional view in Fig. 3. The metal 3 spacing between the source and drain should be mandatorily large (to avoid any electrical short) which consequently increases the cell pitch of the device and hence the $R_{on,sp}$.

Design B and C

Employing the three metal layers BEOL process that was developed for the power IC technology, designs B and C were designed. Designs B and C overcome the shortcomings of the conventional Design A by placing the source and drain pads on the active area as shown in Fig. 2. Although the designs still embrace the interdigitated source and drain patterns, the metal layers are not engaged over each other for every finger. The metal engagement for source and drain fingers is dictated by the top pad region. For example, at the source pad region, metals are not engaged with each other on the drain fingers and vice versa. Fig. 3 shows the 3-D layout cross-sectional view of the devices with Designs B and C. Design B has a chip size of 2.5mm². In Design C, the chip area was further shrunk by incorporating gate into the source pad achieving a chip size of 1.8mm² as shown in Fig. 2. Designs B and C significantly reduce the chip size when compared to Design A. Another significant advantage of Designs B and C when compared to Design A is, unlike Design A, the cell pitch of the devices with designs B and C can be aggressively designed to reduce the cell pitch as there are no metal 3 source and drain interdigitated paths running across the layout.

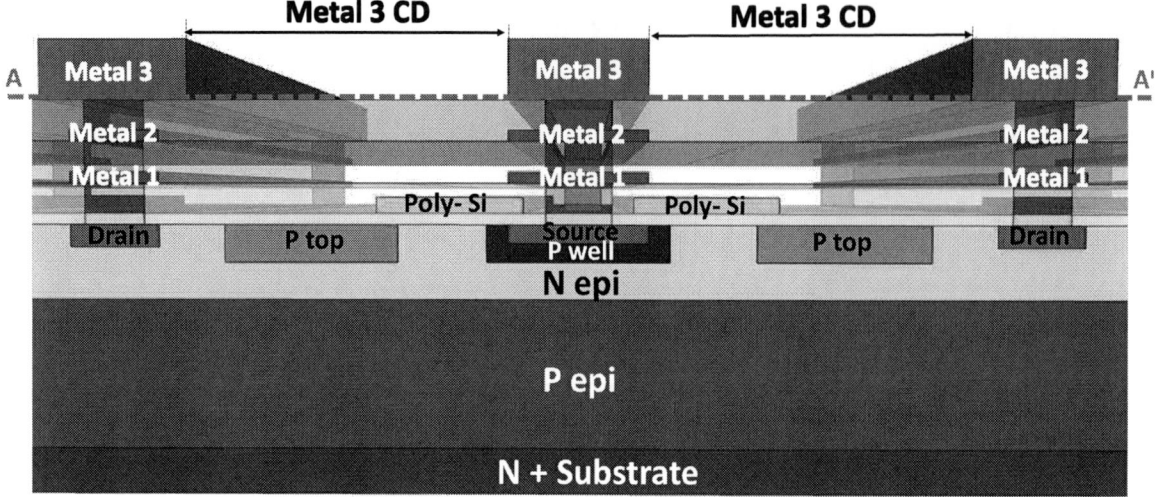

Figure 3: 3-D cross-sectional view of Design A when cut across- A-A'

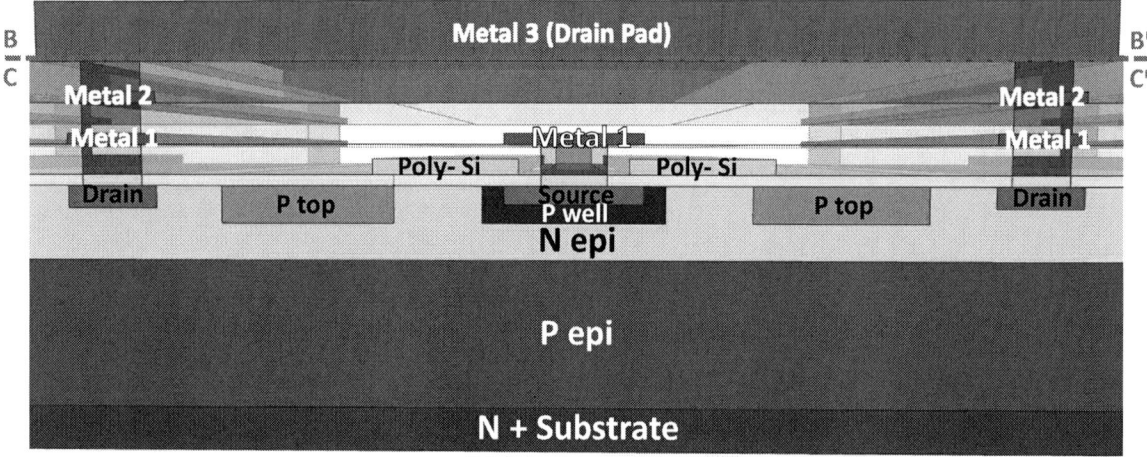

Figure 4: 3-D cross-sectional view of Designs B and C when cut across B-B' and C-C' respectively

III. FABRICATION, RESULTS, AND DISCUSSION

An N epi layer with doping concentration of 6.5×10^{16} cm^{-3} and a thickness of 2.5 μm, and a P epi layer with doping of 2×10^{16} cm^{-3} and thickness of 6 μm on 4H-SiC N$^+$ substrate was the initial material. A bottom P epi was implemented such that the LV devices are completely isolated from the HV devices in a Power IC [5]. The fabrication process followed by the electrical characterizations is discussed in this section.

Ion Implantation

Aluminum and Nitrogen implants were used for the formation of p-type (P well, P top, and P+ source) and n-type (N+ source and drain) regions in the device respectively. Subsequent to the implantation process, the implanted ions were activated by subjecting the wafers to an annealing process at 1650 ℃ for 10 min.

Gate Oxide and Poly-Si formation

A 50nm targeted thickness of gate oxide was formed using high-temperature CVD oxide. The post oxidation annealing was performed in N$_2$O and N$_2$ ambient for 180 minutes. Followed by the gate oxide process, a 0.5μm gate polysilicon was deposited, patterned, and etched.

Ohmic and BEOL process

Once the interlayer dielectric (ILD -1) with a thickness of 1μm was deposited and etched, Nickel was used for forming the ohmic contacts for n type and p type regions. The contacts were annealed at 960 ℃ for 2 min to exhibit ohmic behavior. A 0.5μm thick Aluminum metal was used to form a metal 1 layer. Another ILD (ILD-2) with a thickness of 2μm was deposited and etched to form a via. A 0.5μm thick metal 2 was then formed followed by another 2μm ILD (ILD-3). The ILD-3 was then etched to form a via and the final metal 3 was formed by depositing Aluminum with a thickness of 4μm.

The on-state characteristics of the HV lateral MOSFETs were measured at 25 ℃ and 200 ℃. Fig. 5 shows the typical output characteristics of the HV lateral MOSFETs. Since the cell pitch of all the designs remains the same, the extracted $R_{on,sp, active}$ of the designs A, B, and C stay similar at 6.2 mΩ·cm^2 at V$_{gs}$ of 25V at 25 ℃. The extracted $R_{on,sp,active}$

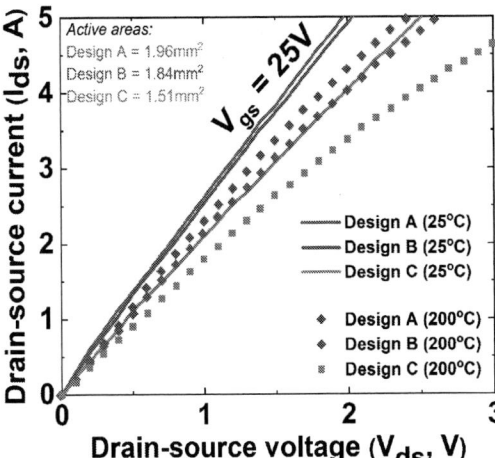

Figure 5: The typical output characteristics of designs A, B, and C at Vgs of 25V at 25 ºC and 200 ºC. Extracted Ron,sp, active of the designs A, B, and C stay similar at 6.2 mΩ·cm² at Vgs of 25V and Vds of 0.1V at 25 ºC

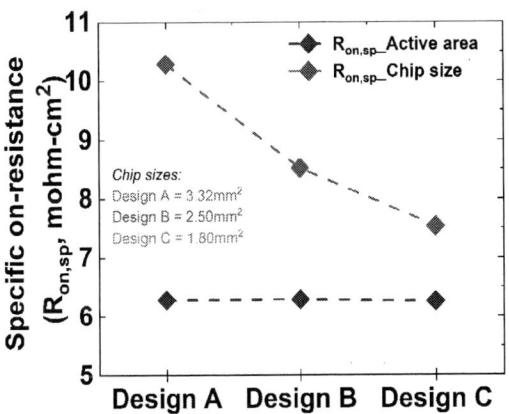

Figure 6: The Ron,sp comparison for all the three types of design variations; Ron, sp extracted for both active area (Ron,sp,active) and chip size (Ron,sp,chip) at Vgs of 25V at 25 ºC; The Ron,sp, active remains same for all the three designs while Ron,sp,chip is significantly higher for design A and design B due to their chip sizes

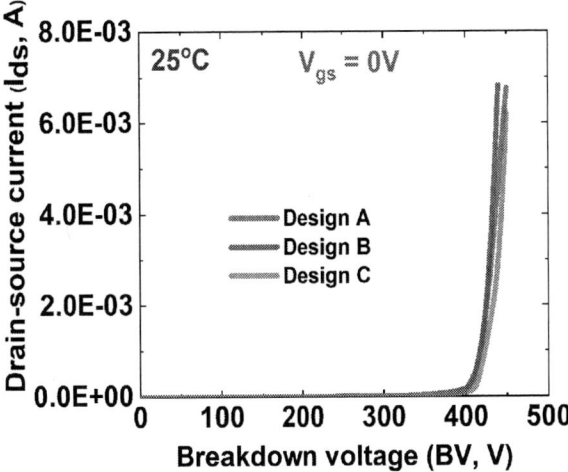

Figure 7: Blocking characteristics of Designs A, B, and C; all three designs demonstrate a breakdown voltage of 430V at a drain-source current of 1mA and a gate-source voltage of 0V at 25 ºC

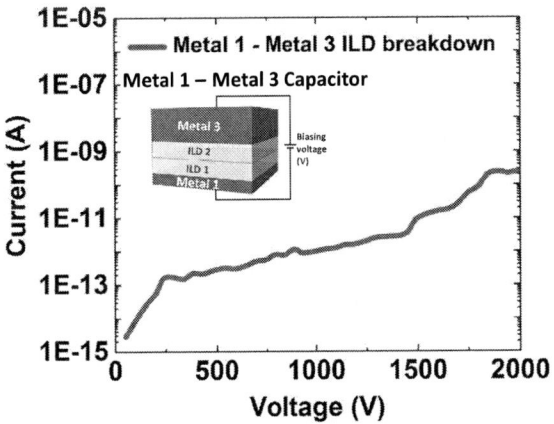

Figure 8: Blocking capability of the 3.5μm ILD between Metal 1 and Metal 3 from a MIM capacitor; inset shows the 3-D cross-section of the MIM capacitor; ILD demonstrates a high blocking capability of about 1800V at 0.1nA of leakage

increases with the increase in temperature due to the increase in the JFET, drift, and metal resistances [6]. When contemplated the designs, the $R_{on, sp, active}$ for all the three designs remain the same but the extracted $R_{on,sp, chip}$ of designs B and C are lower when compared to Design A due to their significant reduction in chip sizes as shown in Fig. 6. The reduction of $R_{on,sp,chip}$ leads to the implementation of designs B and C in developing area efficient Power ICs. In addition, the blocking characteristics have also been evaluated. All three designs demonstrate a blocking capability of 430V in the lateral direction at I_{ds} of 1mA at 25 ºC as shown in Fig. 7.

In designs B and C, it is imperative that the interlayer dielectric (ILD) between metal 1 and metal 3 should withstand a voltage that's much larger than the breakdown voltage of the device since the source and drain metal lines crossover with each other as shown in Fig. 4. Hence, as shown in Fig. 8, when measured a MIM capacitor between metal 1 and metal 3, the 3.5μm ILD demonstrated a blocking capability of 1800V at 0.1nA validating the safe and robust operation of designs B and C.

IV. CONCLUSIONS

The process, design, and electrical performances of the fabricated HV lateral MOSFETs have been reported. The HV lateral MOSFETs not only have the best BV-$R_{on,sp, active}$ trade-off but also offer the smallest footprint. These area-efficient layout techniques will be extremely beneficial for discrete development and also be seamlessly integrated into the Power ICs in SiC.

978-1-6654-8901-0/22 $31.00 © 2022 IEEE

ACKNOWLEDGMENT

The authors would like to thank the Program Director Dr. Isik Kizilyalli and his team at ARPA-E for their support and encouragement. The authors would also like to thank the team at the ADI Hillview facility for the fabrication of the devices.

REFERENCES

[1] A. Elasser and T. P. Chow, "Silicon carbide benefits and advantages for power electronics circuits and systems," *Proceedings of the IEEE*, vol. 90, no. 6, pp. 969–986, Jun. 2002, doi: 10.1109/JPROC.2002.1021562.

[2] D. Disney and Z. J. Shen, "Review of Silicon Power Semiconductor Technologies for Power Supply on Chip and Power Supply in Package Applications," IEEE Trans. Power Electron., vol. 28, no. 9, pp. 4168–4181, 2013, doi: 10.1109/TPEL.2013.2242095.

[3] D. Kinzer, "GaN power IC technology: Past, present, and future," in *2017 29th International Symposium on Power Semiconductor Devices and IC's (ISPSD)*, May 2017, pp. 19–24. doi: 10.23919/ISPSD.2017.7988981.

[4] M. Okamoto, A. Yao, H. Sato, and S. Harada, "First Demonstration of a Monolithic SiC Power IC Integrating a Vertical MOSFET with a CMOS Gate Buffer," in 2021 33rd International Symposium on Power Semiconductor Devices and ICs (ISPSD), May 2021, pp. 71–74. doi: 10.23919/ISPSD50666.2021.9452262.

[5] Sundar Babu Isukapati, Adam J Morgan, Woongje Sung, Hua Zhang, Tianshi Liu, Ayman Fayed, Anant K. Agarwal, Emran Ashik, and Bongmook Lee, "Development of Isolated CMOS and HV MOSFET on an N- epi/P- epi/4H-SiC N+ Substrate for Power IC Applications," in 2021 IEEE 8th Workshop on Wide Bandgap Power Devices and Applications (WiPDA), Nov. 2021, pp. 118–122. doi: 10.1109/WiPDA49284.2021.9645134.

[6] N. Yun, J. Lynch, and W. Sung, "Demonstration and analysis of a 600 V, 10 A, 4H-SiC lateral single RESURF MOSFET for power ICs applications," *Appl. Phys. Lett.*, vol. 114, no. 19, p. 192104, May 2019, doi: 10.1063/1.5094407

978-1-6654-8901-0/22 $31.00 © 2022 IEEE

A New Layout Method for Junction Field Effect Transistors (JFETs) on 4H-SiC that Provides a Significant Reduction in On-Resistance

Justin Lynch[a], Nick Yun[a], Seung Yup Jang[a], Adam J. Morgan[a], Woongje Sung[a]

[a]State University of New York Polytechnic Institute, Albany, NY 12203 USA (lynchjm@sunypoly.edu)

Abstract— **In this work, we demonstrate a new layout technique for 1.2kV-rated lateral-vertical 4H-SiC JFETs that provides a 21% reduction of the specific on-resistance ($R_{on,sp}$) when compared to JFETs using the conventional stripe layout. Both the proposed and conventional layouts were fabricated on the same substrate and achieved a $R_{on,sp}$ of 3.13 mΩ-cm^2 and 3.97 mΩ-cm^2 at a V_{GS} of 0 V and 2.46 mΩ-cm^2 and 3.12 mΩ-cm^2 at a V_{GS} of 2 V during on wafer measurement, respectively. Additionally, the proposed layout approach showed no adverse influence on the blocking characteristics of the device. The demonstration of this proposed layout approach shows the high-performance potential of 4H-SiC JFETs.**

Keywords— *4H-SiC, JFET, Surface-Buried Gate JFET, Lateral-Vertical JFET*

I. Introduction

The use of SiC power transistors for 1.2kV voltage rating applications, including power conversion, electric vehicles, and aeronautics, is becoming increasingly preferred compared to their Si counterparts. SiC as a semiconductor material offers low conduction loss and reliable device operation when compared to Si due to its inert material properties, such as a wide bandgap, high critical electric field, and high thermal conductivity. The SiC power Metal Oxide Semiconductor Field Effect Transistor (MOSFET) and Junction Field Effect Transistor (JFET) are two of the industry used SiC power transistors in 1.2kV applications. While the SiC power MOSFET is a popular device choice for circuit designers, its performance can be limited by the precondition of a metal-oxide-semiconductor interface. The SiC MOS interface, which is used as the gate of the MOSFET, has long lead to several issues still hampering SiC power MOSFET performance today. Due to a large number of interface states found at the oxide-SiC interface, the channel mobility of the SiC MOSFET is typically limited to values below 50 cm^2/V·s, a small fraction of SiC's bulk mobility. This poor mobility leads to high channel resistances, which can thwart the SiC MOSFETs performance at low (< 1.7kV) voltage ratings. In addition to poor channel mobility, issues with the gate oxide reliability can lead to threshold voltage shifts and gate oxide breakdown in harsh environment performance, a key application area of SiC [1-3].

In certain applications, one approach to overcome the poor MOS interface and the problems a MOS structure can lead to is the implementation of devices that do not rely on a MOS gate. The JFET is able to offer the capabilities of a FET switch without the use of a MOS gate. The JFET is a three terminal switch which relies on a p/n junction for the control of current. The SiC JFET still offers the advantage of using SiC as a power device substrate, without the MOS risks. The JFET, if designed properly, is able to exploit its design to offer a normally-on device with competitive on-resistance, high ruggedness due to the lack of a MOS stack, and a simpler fabrication method due to the removal of the gate oxide process [4-6]. Applications of SiC JFETs today include solid state circuit breakers, digital ICs and power converters [7-10]. Additionally, with the use of normally-on JFETs in cascode configurations, the performance of SiC JFETs can be tailored to nearly any application need [11].

Two popular varieties of SiC normally-on JFETs are the vertical JFET or lateral-vertical JFET. Despite the promise of very low $R_{on,sp}$ vertical JFETs struggle with gate control and can have a high pinch-off voltage (V_{pi}), or gate voltage required to close the channel. Lateral-vertical JFETs have slightly larger $R_{on,sp}$ values compared to their vertical counterparts, but with the addition of a second lateral channel, the gate control is improved and thus reasonable V_{pi} can be obtained. Currently, SiC JFETs are commercially available and have been the focus of recent research and development for further applications [12-13].

In this work, the improvement of the lateral-vertical SiC JFETs is investigated using a new device architecture with comparable electrical performance to commercially available JFETs today. The device's structure, fabrication process, electrical characteristics, and potential improvements will be discussed in this paper.

II. Device Architecture and Fabrication

A. Device Architecture

Fig. 1 shows the top-view fully processed 150mm wafer that the devices were fabricated on. The fabrication process will be discussion in the next sub section. Fig. 2 shows the 2-D schematic top and cross-sectional view of the Surface-Buried Gate (SBG) JFET as well as a 3-D view, labeled with important device components. The structure of the device's channels has a significant impact on important device characteristics, $R_{on,sp}$ and V_{pi} and Breakdown Voltage (BV). Device parameters that influence the channel resistances include both lateral and vertical channel dimensions, as well as channel and gate doping profiles. With the proper profiles and

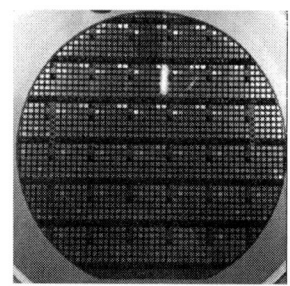

Figure 1: Image of fabricated JFETs on 6-inch 4H-SiC wafer.

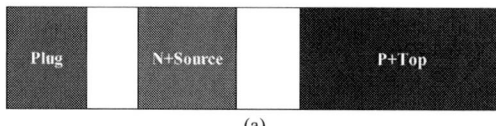

(a)

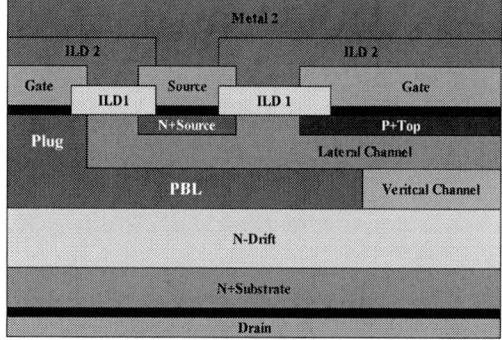

(b)

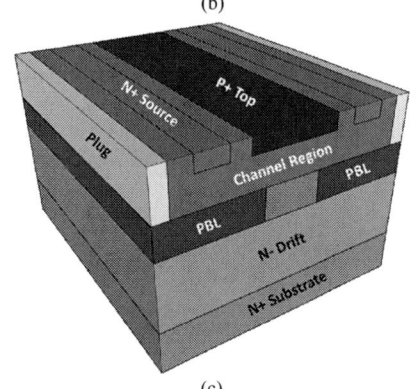

(c)

Figure 2: (a) Schematic top-view layout, (b) half-cell cross-sectional view of the traditional layout, and (c) a 3-D schematic view.

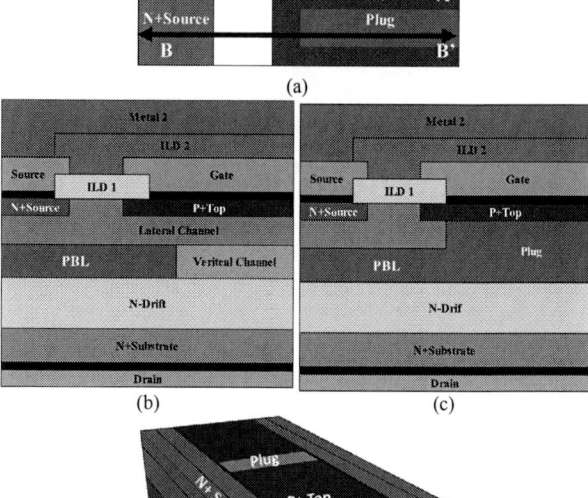

(a)

(b)   (c)

Figure 3: (a) Schematic top-view layout of the new SBG JFET, (b) the cross-sectional view along the A-A' cut, (c) the cross-sectional view along the B-B' cut, and (d) a 3-D schematic view.

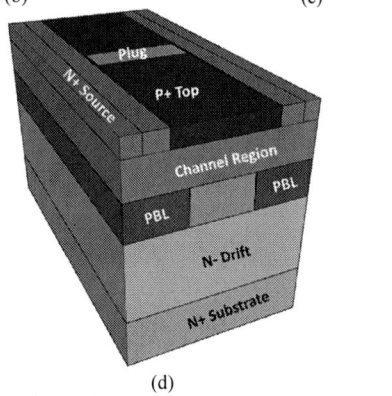

(d)

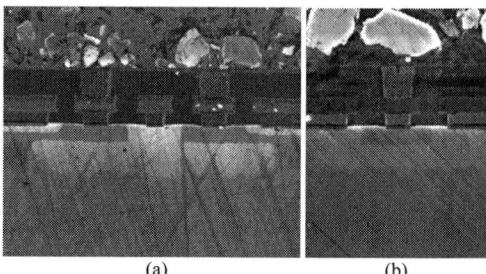

(a)   (b)

Figure 4: Cross-sectional SEM images of (a) the traditional layout JFET and (b) the advanced layout JFET along the B-B' cut.

device design, the device performance can achieve commercially viable performance. Without the correct channel design, the JFET resistance can become undesirably high and even change the device operation from normally-on to normally-off. The P-buried gate (PBL) of the JFET is connected to the gate metal on the surface of the device through a highly doped Plug. The doping of these implants has an effect on the channel resistance as stated above and the gate resistance of the device. In addition to the buried gate, a P+top is included in the device and forms the surface gate. These two gates work in conjunction, giving this JFET the SBG structure. The proposed layout, shown schematically in both 2-D and 3-D in Fig. 3 along with important device components, places the Plug in the orthogonal direction, which allows for a significantly reduced cell pitch. The cell pitch was reduced by 40%, a reduction from 8.4 µm to 5 µm, leading to an increase in channel density and corresponding decrease in $R_{on,sp}$. The placement of the Plug interrupts the channel periodically, but the gain in additional device fingers from the decreased cell pitch leads to an overall channel density increase within the chip. The placement of the Plug in the orthogonal direction was varied in design to investigate the optimal spacing of what was labeled Plug interruptions. The interruption ratio was varied X:1, where for every X µm of channel there is 1µm of Plug interruption. The variations included were 5:1, 10:1, and 20:1. Figure 4 shows the cross-sectional SEM images of the both the conventional and proposed layouts of the JFETs.

978-1-6654-8901-0/22 $31.00 © 2022 IEEE   128

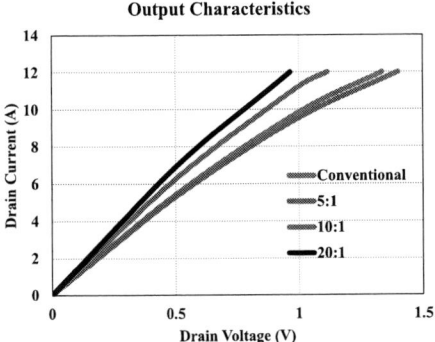

Figure 5: Typical I-V curves for JFETs of interest. Curves taken from the same die.

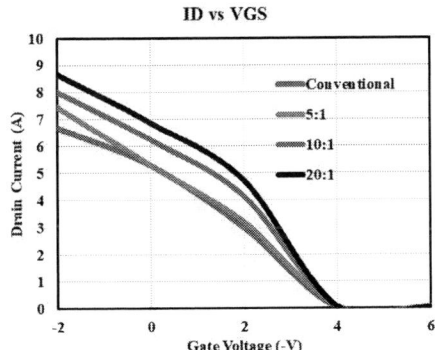

Figure 6: Transfer Characteristics (I_D vs V_{GS}) at a V_D of 0.5 V. All devices show similar behavior.

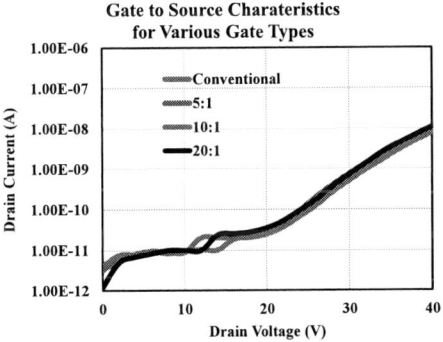

Figure 7: Typical gate to source characteristics of the various JFETs. No substantial difference is seen between the fabricated JFETs.

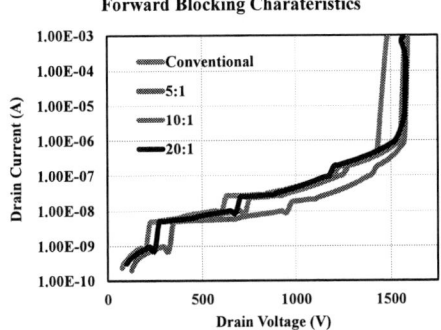

Figure 8: Typical blocking curves for JFETs of interest at $V_{GS} = -25$ V. Curves taken from the same die.

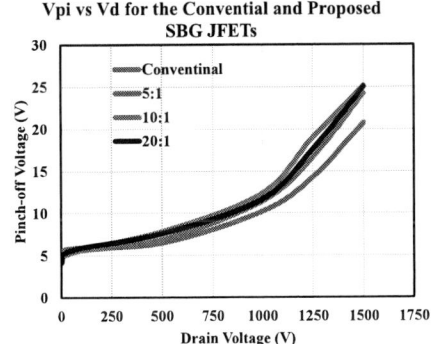

Figure 9: V_{pi} at various V_D values for the SBG JFETs of interest.

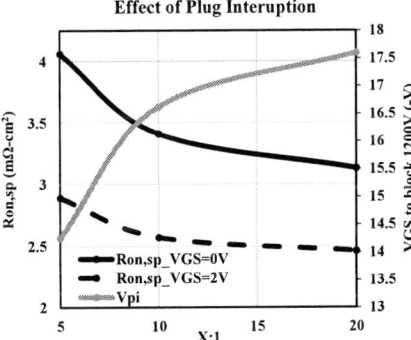

Figure 10: Effect of the frequency of Plug interruption on device a $R_{on,sp}$ and V_{pi}

B. Device Fabrication

The devices were fabricated at a 6-inch Analog Devices Inc. (ADI) fabrication facility in Hillview, San Jose, CA. High temperature implantations of nitrogen (n-type) was used to form the N+ source and channel regions, while aluminum (p-type) was used to form the P+ top, PBL, Plug, and JTE regions of the devices. Additionally, the P+ implant was used in conjunction with the JTE implant for the formation of a hybrid-JTE edge termination to ensure the high BV values of the devices. Once ion implantation was completed, an activation anneal of 1650 °C for 10 minutes was applied. Following the activation anneal, inter layer dielectric (ILD) was deposited and subsequently patterned and etched for the front-side ohmic contacts. Nickel was then deposited for the ohmic contact on the front-side and underwent an rapid thermal anneal (RTA) for silicidation process. After, un-silicided nickel was removed from the front-side. Backside nickel was then deposited and both back and front ohmic contacts underwent an ohmic anneal of approximately 950 °C for 2 minutes. Following the ohmic anneal, the gate and source Ti/Al metal stack was deposited, patterned and etched. A second ILD layer was deposited and patterned to form both the source and gate openings for the device. Once completed, a thick layer of aluminum was deposited and patterned for both gate and source pads. The surface of the devices were then passivated using nitride and polyimide.

978-1-6654-8901-0/22 $31.00 © 2022 IEEE

Device	$V_{GS} = 0V$			$V_{GS} = 2V$			Simulated V_{pi} (-V)	Actual V_{pi} (-V)	ΔV_{pi} (-V)
	Simulated $R_{on,sp}$ (mΩ-cm²)	Actual $R_{on,sp}$ (mΩ-cm²)	$\Delta R_{on,sp}$ (mΩ-cm²)	Simulated $R_{on,sp}$ (mΩ-cm²)	Actual $R_{on,sp}$ (mΩ-cm²)	$\Delta R_{on,sp}$ (mΩ-cm²)			
Conventional SBG	3.39	3.97	0.58	2.89	3.12	.23	8	28.1	20
5:1 SBG	N/A	4.06	N/A	N/A	2.89	N/A	N/A	25	N/A
10:1 SBG	N/A	3.31	N/A	N/A	2.57	N/A	N/A	25	N/A
20:1 SBG	N/A	3.13	N/A	N/A	2.46	N/A	N/A	25	N/A
BG	2.96	3.82	0.86	2.6	3.2	0.6	15	>25	>10
SG	2.89	3.92	1.03	2.85	3.64	0.79	13	>20	>5

Table 1: Summary of simulated and measured data for the fabricated JFETs of various gate types. Simulated data was obtained using Sentaurus 2-D simulation.

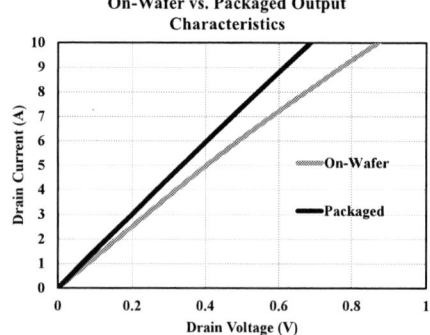

Figure 11: Comparison of on-wafer and packaged output characteristics of a fabricated proposed 20:1 JFET

III. ELECTRICAL RESULTS

A. Eelctrtical Performance

The I-V curves for the fabricated devices at a $V_{GS} = 0$ V are shown in Fig. 5 and showcases the improvement gained with the new layout technique. The lowest $R_{on,sp}$ achieved at a V_{GS} of 0 V using the proposed and conventional layouts were 3.13 mΩ-cm² and 3.97 mΩ-cm², respectively. When the V_{GS} is increased to 2 V, the lowest $R_{on,sp}$ achieved was then 2.46 mΩ-cm² and 3.12 mΩ-cm² for the proposed and conventional layouts. This low $R_{on,sp}$ at a $V_{GS} = 2$ V value highlights the potential of the advanced layout devices. The $R_{on,sp}$ values are calculated using the active area of the device. The transfer characteristics of the devices are shown in Fig. 6, and shows the JFETs ability to close the channel at a low gate voltage. Fig. 7-9 show the Gate-Source behavior and forward blocking behavior of the various SBG JFETs. These figures confirm that the blocking capability and gate function of the conventional SBG JFET is not compromised in the proposed layout. Near ideal BV values were obtained with a V_{GS} of -25 V using the hybrid-JTE edge termination. The design of the hybrid-JTE edge termination was done using Sentaurus TCAD 2-D simulations and following the method described in reference [14]. The summary of the effect of Plug interruption ratio on device performance is shown in Fig. 10. Fewer interruptions (larger X:1 ratio) are desirable for improved device $R_{on,sp}$, and had limited effect on the blocking capability of the device. All fabricated devices were able to achieve the target BV voltage (1.2kV) while using reasonable V_{GS} values below -20 V, a crucial characteristic for SiC JFETs applications and implantation in cascode configurations [4]. From the device performance, a Plug interruption ratio of 10:1 provides the best balance of device performance.

B. Device Improvement

Despite the promising results from the proposed SiC JFET devices, improvements in device layout and performance are attainable. Table 1 shows the fabricated JFETs showed large deviation from the same structure simulated in Sentaurus 2-D simulation. The large variation between simulated and measured $R_{on,sp}$ is believed to be originated from a higher resistance in the channel, and other non-ideal resistances (i.e. contact, measurement probes, etc.). With increased channel doping, along with reduced channel width to keep a reasonable V_{pi}, it is believed the channel resistance can be improved, reducing the difference between simulated and measured $R_{on,sp}$. Table 1 also includes Buried Gate (BG) and Surface Gate (SG) structures which utilize either solely the PBL (BG) or P+Top (SG) as a gate. These devices were fabricated on the same wafer as the SBG JFETs. A large difference in measured and simulated V_{pi} for both the BG and SBG structures, but not the SG, is also seen within Table 1. It is hypothesized that this difference is due to the PBL not fully contributing to device pinch-off, due to its high resistance. By increasing the doping concentration in the PBL the V_{pi} difference should decrease. Another potential area of improvement is the introduction of channel spreading layer (CSL) below the vertical channel of the JFET. As seen before with SiC MOSFETs as in [15], the introduction of high energy nitrogen implants would allow for a tighter cell pitch, as well as reduce the resistance of current flow in the vertical channel.

In addition to the above hypothesized improvements additional areas of improvement were quantified. Significant improvement in the forward operation of the device can be achieved with full current extraction from the devices. After using a few of the fabricated SiC JFETs in traditional TO-247 packaging, significant improvement in $R_{on,sp}$ values were seen. A typical 15-20% reduction was seen due to full current extraction in the devices, indicated an issue with the thick top metal or on-wafer measurement set-up. Fig. 11 shows the IV from a 20:1 plug interruption proposed JFET device before and after packaging. As seen in the image, a significant improvement in performance, 3.55 mΩ-cm² to 2.86 mΩ-cm² (18%), is observed. When converting this $R_{on,sp}$ value to account for total chip area, the number changes to 4.1 mΩ-cm², a value in line with the currently available commercial JFETs in [12]. With further substrate grinding and use of the champion fabricated devices in packaging, the $R_{on,sp}$ could be lowered further. With room for potential improvement on already commercially viable SiC JFETs, the fabricated proposed JFETs show immense potential as a 1.2kV SiC power device.

IV. CONCLUSIONS

This work demonstrates a new layout technique for lateral-vertical 1.2kV SiC JFETs. The design and layout of the proposed new SBG JFET, which allows for a cell pitch reduction of 40% to be achieved, was compared to the conventional SBG JFET. After device structure and fabrication was discussed, the electrical characteristics of the fabricated proposed JFET and conventional JFET were compared. With the proposed structure a 21% reduction of the $R_{on,sp}$ was achieved compared to conventional JFETs. The proposed JFET achieved a $R_{on,sp}$ of 3.13 mΩ-cm^2 at a V_{GS} of 0 V and 2.46 mΩ-cm^2 at a V_{GS} of 2 V during on wafer measurement. In addition to the output characteristics, no deviation can be seen between the proposed and conventional JFETs when comparing both blocking and gate behavior, confirming the viability of the new proposed JFETs. Both structures were able to achieve the desired BV with reasonable V_{GS} values. Additional methods to improve future device performance was also noted. In total, the proposed SiC JFET offers the lateral-vertical SiC JFET to keep its superior blocking capabilities and while achieving commercially viable $R_{on,sp}$ values.

ACKNOWLEDGMENT

The authors would like to acknowledge the contributions of Igal Deckman, Dennis Rossman, Sung Kim, Duy-son Nguyen, and Jin-Ho Seo to this work and Alexander Bialy in the Metrology Department, SUNY Polytechnic Institute, for SEM images presented.

REFERENCES

[1] J. Wang and X. Jiang, "Review and Analysis of SiC MOSFET's Ruggedness and Reliability," IET Power Electronics, vol. 13, no. 3, 1 Feb. 2020, pp. 445–455.

[2] A. J. Lelis, R. Green, and D. Habersat, "SiC MOSFET Threshold-Stability Issues," Materials Science in Semiconductor Processing, vol. 78, May 2018, pp. 32–37.

[3] G. De Martino, F. Pezzimenti, and F. G. Della Corte, "Interface Trap Effects in the Design of a 4H-SiC MOSFET for Low Voltage Applications," 2018 International Semiconductor Conference (CAS), Oct. 2018.

[4] K. Rueschenschmidt, M. Treu, R. Rupp, P. Freidrichs, R. Elpelt, D. Peters, and P. Blaschitz, "SiC JFET: Currently the Best Solution for an Unipolar SiC High Power Switch," Materials Science Forum, vol. 600-603, Sept. 2008, pp. 901–906.

[5] M. Treu, R.Rupp, P. Blaschitz, K. Ruschenschmidt, T. Sekinger, P. Friedrichs, R. Elpelt, and D. Peters, "Strategic Considerations for Unipolar SiC Switch Options: JFET vs. MOSFET," 2007 IEEE Industry Applications Annual Meeting, 23-27 Sept. 2007

[6] W. Sung, E. Van Brunt, B. J. Baliga, and A. Q. Huang, "A Comparative Study of Gate Structures for 9.4-kV 4H-SiC Normally on Vertical JFETs," IEEE Transactions on Electron Devices, vol. 59, no. 9, 3 Aug. 2012, pp. 2417–2423.

[7] D. He, Z. Shuai, Z. Lei, W. Wang, X. Yang, and Z. John Shen, "A SiC JFET-Based Solid State Circuit Breaker with Digitally Controlled Current-Time Profiles," IEEE Journal of Emerging and Selected Topics in Power Electronics, vol. 7, no. 3, 21 Mar. 2019, pp. 1556–1565.

[8] P. G. Neudeck, D. Spry, M. Krasowski, N. Prokop, and L. Chen, "Demonstration of 4H-SiC JFET Digital ICs across 1000°C Temperature Range without Change to Input Voltages," Materials Science Forum, vol. 963, 1 July 2019, pp. 813–817.

[9] U. Mehrotra, B. Ballard, and D. C. Hopkins, "High Current Medium Voltage Bidirectional Solid State Circuit Breaker Using SiC JFET Super Cascode," 2020 IEEE Energy Conversion Congress and Exposition (ECCE), 11-15 Oct. 2020.

[10] S. V. Araújo, B. Sahan, P. Zacharias, R. Rupp, and X. Zhang, "Application of SiC Normally-on JFETs in Photovoltaic Power Converters: Suitable Circuits and Potentials," Materials Science Forum, vol. 645-648, Apr. 2010, pp. 1111–1114.

[11] Z. Li and A. Bhalla, "USCi SiC JFET Cascode and Super Cascode Technologies," PCIM Asia 2018; International Exhibition and Conference for Power Electronics, Intelligent Motion, Renewable Energy and Energy Management, 2018, pp. 1-6.

[12] Untied SiC, "1200V-142mW SiC Normally-on JFET," UF3N120140 datasheet, Feb. 2022.

[13] M. Kong, J. Guo, J. Gao, K. Huang, B. Zhang, and B. Wang., "A High-Performance 4H-SiC JFET with Reverse Recovery Capability and Low Switching Loss," IEEE Transactions on Electron Devices, vol. 68, no. 10, 10 Aug. 2021, pp. 5022–5028.

[14] W. Sung, and B. J. Baliga, "A Comparative Study 4500V Edge Termination Techniques for SiC Devices," IEEE Transactions on Electron Devices, vol. 64, no. 4, pp. 1647-1652, Apr. 2017.

[15] D. Kim, N. Yun, S. Y. Jang, A. J. Morgan and W. Sung, "An Inclusive Structural Analysis on the Design of 1.2kV 4H-SiC Planar MOSFETs," in IEEE Journal of the Electron Devices Society, vol. 9, 02 Sept. 2021, pp.804-812,

Busbar Design and Optimization for High Power Three-phase Inverter with WBG Device

Yuxuan Wu*, Mustafeez-ul-Hassan, and Fang Luo
Department of Electrical and Computer Engineering
Stony Brook University
Stony Brook, USA
yuxuan.wu@stonybrook.edu

Abstract—**The wide-band gap devices can switch at a higher frequency with a higher dV/dt well as improve switching performance. The optimization of busbars can reduce the power loop parasitic which will reduce the overvoltages during the device's state transient. Busbar is not only required to carry and distribute the current to the power module, but also need to provide a path for decouple the high-frequency current ringing and minimize the impact of high-frequency current coupling. This paper introduced a workflow of a high power three-phase inverter design by considering the power converter layout to achieve the maximum power and , and a comparative study between three-different busbar designs is presented.**

Index Terms—**busbar design, DC-AC inverter, WBG device, two-level motor drive, high-power inverter**

I. INTRODUCTION

As the wide band-gap (WBG) devices can achieve higher voltage and current slew rate (dV/dt and di/dt) due to its smaller parasitic inside power semiconductor, the switching performance of power semiconductor have significantly improved in terms of switching losses, power turn-on/off delay, and maximum switching frequency [1]. The application of WBG devices have draw plenty attention on its power loop design.

The WBG devices can switch at a higher frequency with a higher dV/dt well as improved switching performance. The optimization of busbars can reduce the power loop parasitic which will reduce the overvoltages during the device's state transient. When designing the busbar, the busbar is optimized under follow perspective. First, power loop inductance, by optimizing the DC loop inductance, the need for the DC-link decoupling component reduces as well as the electromagnetic interference (EMI) filtering component [2]. Second, current distribution, by avoiding currents accumulated on a spot and increases the reliability of the system [3]. Third, system layout, modules' layout, gate-source loop, and DC power loop can impact the converter volume as well as the power density of the converter [4].

The busbar is the artery of an inverter. When designing busbars for inverters, the busbars need to be designed at a system level. This means before designing a busbar, the inverter layout, electrical design, control method, and mechanical design should all be considered [5]. For example, thermal coupling between power modules can directly impact the module layouts in the inverter; the communication method

between the controller and inverter will impact the gate drive (GD) layout and result in different inverter layouts; particle discharge will result in busbars having poor reliability, optimizing the busbar to avoid particle discharge can improve the overall reliability of inverter; the busbar needs to design according to the inverter limitation. Due to the constrain of the system, the optimization of the busbar needs to be investigated.

[6] [7] discussed the commutation loop optimization of commutation loop inductance for multi-layer busbar. Those two studies show multi-layer busbars are good for parasitic optimization, but the difference in power module layouts will constrain the inverter design and eventually result in different busbar layouts, and multi-layer busbars might not always be the best choice. [8] has considered busbars design at the system level, and it has adjusted the power module design into consideration, but it has failed to optimize the design in volume and resulted in a low power density.

This paper carried out a comparative study between multi-layer busbar for high current inverters and achieved high power density and high reliability. In section II, the design of a high-power inverter is illustrated, the thermal evaluation process is presented for determining power modules layout, and gate-driver design consideration. In section III, three different busbar designs a) multi-layer PCB busbar, b) two-layer aluminum busbar, and c) improved two-layer copper busbar, are designed for the simulation-based evaluation studies. In section, the high power inverter's performance is evaluated with busbar b) and c) in terms of voltage stress test, power converter switching performance, and inverter comparison.

II. DESIGN AND EVALUATION OF INVERTER

The inverter is designed to handle 150 kVA of power and perform DC/AC or AC/DC conversion. The inverter is using two-level three-phase configuration by using three half-bridge power modules which are rated for 1200 V 150 A [9].

A. Inverter Design Consideration

The optimization of busbars can reduce the power loop parasitic which will reduce the overvoltages during the device's state transient. When designing the busbar, the busbar is optimized under follow perspective. First, determined the layout of power modules. The layout of the power module of the inverter can be determined by performing thermal analysis

978-1-6654-8901-0/22 $31.00 © 2022 IEEE

of the inverter, and thermal simulation can be perform to evaluated the design. Second, inverter layout optimization. gate driver optimization and design By optimizing the DC loop inductance, the need for the DC-link decoupling component reduces as well as the EMI filtering component. Second, current distribution, by avoiding currents accumulated on a spot and increased the reliability of the system. Third, system layout, modules' layout, gate-source loop, and DC power loop can reflect the converter volume as well as the power density of the converter. The busbar is the artery of an inverter. When designing busbars for inverters, the busbars need to be designed at a system level. This means that before designing a busbar, the inverter layout, electrical design, control method, and machinal design should all be considered.

B. Evaluation of Thermal Performance

To simulate the thermal losses of the modules, the motor drive is simulated in PLECS with various switching frequencies, the simulation is presented in Fig. 1 The PLECS module is generated based on the datasheet provided by the power module manufacturer, and the simulation shows the minimum efficiency is simulated at 99% when the switching frequency is at 90 kHz. The power loss of the converter is 2 kW from the simulation. The simulation result shows switching losses have contributed 82% of the losses, well as the conduction loss is only contributing 18% of the losses, as shown in Fig. 2.

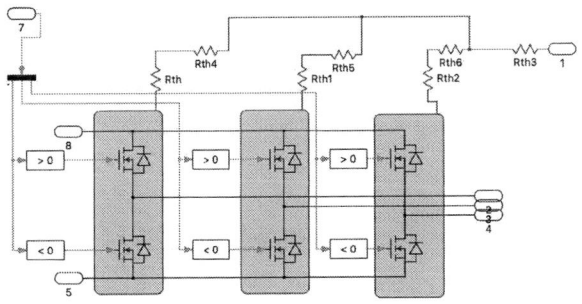

Fig. 1. Thermal simulation schematic of the inverter in PLECS.

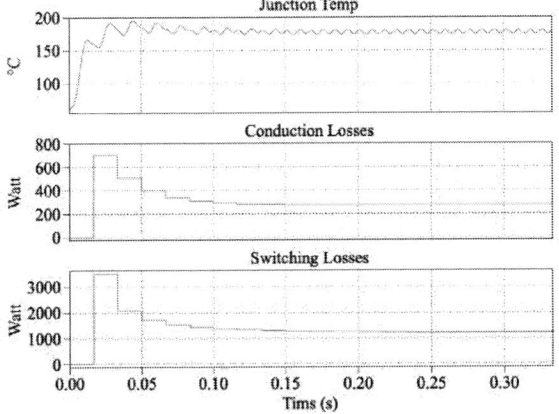

Fig. 2. Thermal simulation and losses result by PLECS.

By plugging the thermal losses into the model of the heat sink and the power modules, and thermal conductivity values into the simulation, the values used for thermal simulation are presented in Table I. The thermal simulation results as shown in Fig. 3. The thermal simulation result shows, the minimum gap allowed between the modules is 3 mm for avoiding any thermal coupling, and the gap between the modules is adjusted to 5 mm to avoid any unexpected thermal coupling. The thermal simulation shows there is no hot spot between modules.

TABLE I
PARASITIC FOR THERMAL EVALUATION

Parasitic	Value
Losses on each die	100 W
$R_{th(Junction-Case)}$	0.11 $°C/W$
$R_{th(Case-HeatSink)}$	0.056 $°C/W$
$R_{th(HeatSink-Liquid)}$	0.018 $°C/W$

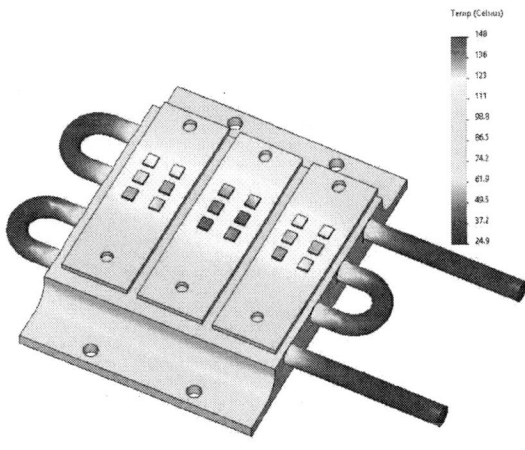

Fig. 3. Thermal simulation result for power module layout.

C. Gate Driver Design and Optimization

After the power modules are layout properly, the GD IC with an active miller clamp (AMC) has been chosen for this inverter. The AMC can prevent the miss triggering when the switching device has high capacitance, which can be found in the high current power module which paralleling multiple devices for high power capacity [10]. The peak current of the GD IC required for the converter is calculated with Eq.(1). In this case, a GD with 4 A peak current capacity is chosen for the inverter.

$$I_{GPeak} = Q_G/t_{SW} \qquad (1)$$

The GD circuit is lying parallel to the power modules, and the gate-source loop is off-setting the power loop, therefore, avoiding coupling between the power and external gate-source loop.

978-1-6654-8901-0/22 $31.00 © 2022 IEEE

D. DC Link Design Consideration

As discussed in the section II-A, this inverter is designed to target high power density well as reliability. The decoupling capacitor is mounted as close to the power module as possible to minimize the parasitic inductance in the power loop. The equivalent circuit of the inverter is presented in Fig.. The DC link capacitor bank for ripple current filtering is placed on the very top of the inverter. There are three different positions for the DC link power input: a) power input from the DC link capacitor bank, b) power input from the decoupling capacitor bank and c) power input directly to the inverter. The ideal for such a high-power inverter design with high power density is expecting the PCB which holds the decoupling capacitors and DC link ripples current filtering component are only required to bypass the high-frequency ripple current, well as the main high current is directly distributed to the power module input. The maximum current rating for the DC link components is listed in Table II.

TABLE II
CURRENT RATING FOR DC LINK COMPONENTS

Components	Current Rating (A)
DC link Busbar	180
Decoupling Cap PCB	20
DC current ripple filer PCB	20

E. Inverter layout

To sum up, the high-power inverter's structure is presented in Fig. 4.. The busbar which carries the large current is placed directly on the power modules, and the busbar is used to conduct the current between the decoupling capacitor bank and DC ripple current filtering capacitors.

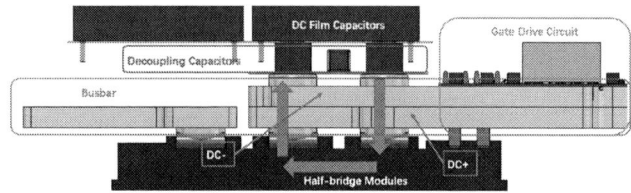

Fig. 4. DC-link busbar design flow

III. BUSBAR DESIGN FOR HIGH-POWER INVERTER

After settling down the layout of the inverter, three different types of busbars a) multilayer PCB busbar, b) two-layer aluminum busbar, and c) improved two-layer copper busbar are presented and compared. The busbars are evaluated from the following perspectives a) parasitic evaluation with finite element analysis (FEA), b) volume, c) cost, d) partial discharge. The busbar design flow is presented in Fig. 5.

A. Introduction of Busbars

1) Multi-layer PCB Busbar: The commutation loop of the multilayer PCB is presented in Fig. 6. The PCB busbar is designed with 8 layers of PCB with 3 Oz/ft^2 copper for

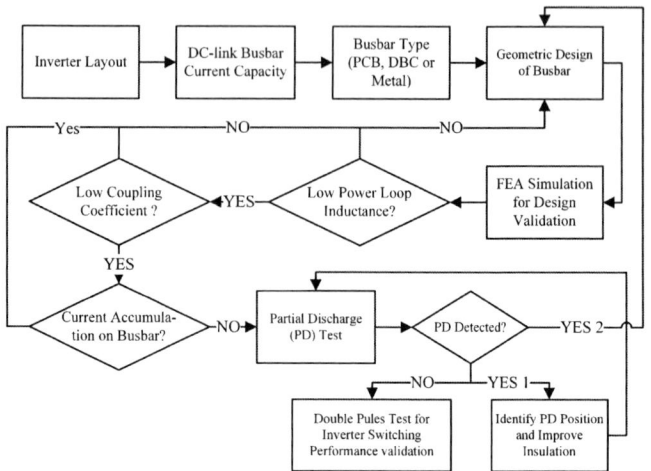

Fig. 5. DC-link busbar design flow

high current capacities. The clearance required for the 800 V DC link is 4 mm to avoid partial discharge happening on the busbar. The thickness of PCB is 2.4 mm for better insulation between the layers.

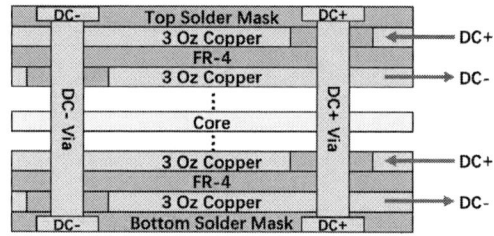

Fig. 6. Layer structure of the PCB Busbar (PCB BB)

2) Two-layer Aluminum Busbar: This busbar was an improved design on the previous one by resolving some issues which were found during testing. The busbar design is shown in Fig. 7. The two-layer structure on top of the power module is eliminated, and the clearance between DC+ and DC- busbar is increased to 3 mm. The clearance between the DC link and the gate driver circuit increases. Most importantly, the loop size of the DC-link has decreased by reducing the thickness of the busbar.

3) Improved Two-layer Copper Busbar : This busbar was an improved design on the previous one by resolved some issues which were found during testing, the busbar design is shown in Fig. 8. The two-layer structure on top of the power modules is eliminated, and the clearance between DC+ and DC- busbar is increased to 3 mm. Most importantly, the clearance between the DC link and the gate driver circuit increases, which reduce the coupling between the DC-link busbar and gate-source by 74%. Lastly, the loop size of DC-link has decreased by reducing the thickness of the busbar, which results lower power loop inductance.

978-1-6654-8901-0/22 $31.00 © 2022 IEEE

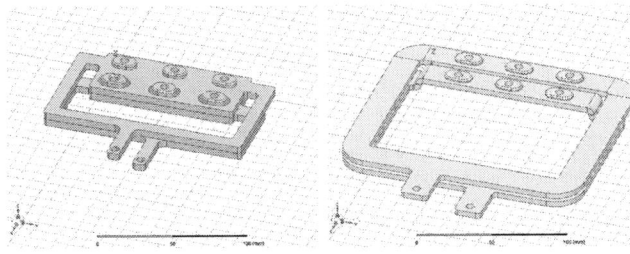

Fig. 7. 3D drawing of two-layer
aluminum busbar(Al BB)

Fig. 8. 3D drawing of improved
two-layer copper Busbar (Cu BB)

B. Dimension and Cost

The DC-link of a high-power inverter needs to handle a large current up to 180 A, the DC link PCB needs to use 3 Oz/ft^2 copper instead of 2 Oz/ft^2 and a multilayer PCB, which is eight-layers in this case, is required. This results in the cost of the PCB busbar increasing from \$65 per piece (with 2 Oz/ft^2 copper and four-layer) to \$280 per piece (with 3 Oz/ft^2 copper and eight-layer). The dimension and cost of the three busbars presented in the previous section are presented in Table III.

C. Optimization of Busbars

1) Design Optimization: The inductance of the power module when the top and bottom devices turn on is 14 nH which is measured with Bode 100 impedance analyzer. The power loop inductance for three busbars are found by using Ansys Q3D, and the power loop inductance of the busbars with a module in short circuit is measured with Bode 100 as well. The results are presented in table III.

The result shows PCB busbar has the smallest DC link power loop, and the inductance from the power input to the inverter is also the smallest due to the flux cancellation design of the PCB busbar. However, the performance of the improved two-layer busbar is not far behind the PCB busbar when considering the high current power module is introducing the majority of power loop inductance in the system.

2) Current Distribution Evaluation: The current density have simulated in Ansys Q3D for observing current distribution on busbars. The simulation result is presented in Fig.9.

The simulation result shows that the current in Al busbar is distributed unevenly compared with Cu busbar. The Cu busbar have less hot spot which is less likely cause insulation failure.

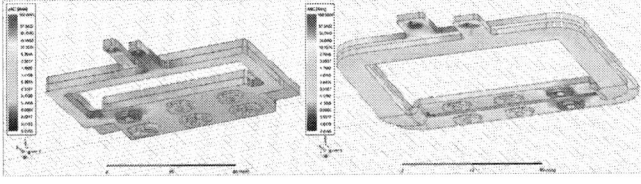

Fig. 9.   Current density simulation for busbars
a) Result of Al busbar b) Result of Cu busbar

3) Partial Discharge Test: One major concern in high voltage power loop design is PD. Once PD initial voltage (PDIV) reached, the performance of the insulation material will decay exponentially. Moreover, since the power loop inductance of the high current module is large and the dV/dt and di/dt of WBG are higher, the overvoltage during the switching will be higher than Si IGBT. Therefore, PD testing for the DC link busbar is important. The PD testing platform is referencing [11]. The test results for three busbars are presented in Table III.

The aluminum busbar has the poorest performance and has failed the PD test at 1 kV, which is necessary for an 800 V DC-link. The PCB busbar passed the test for 800V DC-link but failed the test for 1.1 kV DC-link. The copper busbar has the best result on the PD test, and it is ready for 800 V and 1.1 kV DC-link.

TABLE III
BUSBAR PROPERTY COMPARISON

Property	PCB BB	Al BB	Cu BB
Length (mm)	150	120	185
Width (mm)	180	145	180
Thickness (mm)	2.4	14	6
Volume (mm^3)	64800	243600	199800
Cost (\$/piece)	280[a]	136	184
Simulated L_{Power} (nH) @ 30 MHz[b]	0.753	2.591	1.094
Measured L_{Power} (nH) @ 30 MHz[c]	14.54	17.22	15.13
Power Loop Gate-Source Loop Coupling Coefficient	0.36%	0.82%	0.21%
PDIV (V)	1300	1050	2100
V_{DCMax} (V)	800	650 800 (improved)	1300

[a]Price quoted from ALLPCB.com.
[b]Inductance exclude power module.
[c]Inductance include power module.

By analyzing the failure point of the two busbars. PD of the aluminum busbar occurred at the gap between the terminal block and the top layer. The PD problem was resolved by filling up the gap with Kapton Taps. No failure was observed at the surface of the PCB busbar, so a defect is very likely happening inside the PCB; the electric field of the inner layers might be higher than the surface layers.

Since the defect of the PCB busbar cannot be confirmed, and the cost of the PCB busbar is higher than the two-layer busbar, only the two-layer busbar is optimized and improved.

IV. TWO-LAYER BUSBAR DESIGN VALIDATION

The inverter design is validated with double pulse testing (DPT) and continued testing. DPT can be used to evaluate the switching performance of the inverter and improve the gate drive circuit accordingly. The DPT platform is presented in Fig. 10, and the equipment used for validation is listed in Table IV.

The DPT result for the Al busbar is presented in Fig. 11. When the external turn-on gate resistor is 1 Ω, the gate

978-1-6654-8901-0/22 \$31.00 © 2022 IEEE   135

TABLE IV
EQUIPMENT FOR DPT

Equipment	Model	Specification
DPT Capacitor Bank	Lab made	3300 μF
Load Inductor	Lab made	131 μH
DC Power Supply	Spellman SL1PN600	1 kV 600 mA Max
Current Probe	Tektronix TRCP0300	30 MHz, 300 A Max
Voltage Probe (V_{GS})	Tektronix THDP0200	200 MHz, 1.5 kV Max
Voltage Probe (V_{DS})	Tektronix TPP0850	800 MHz, 1 kV RMS
Oscilloscope	Tektronix MSO56	6-Channel Scope

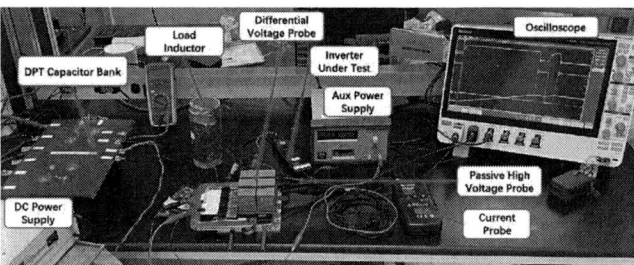

Fig. 10. DPT setup.

signal is distorted, and miss triggering is observed at 650 V. To resolve this problem, the external turn-on gate resistor is adjusted to 4.7 Ω, and the inverter is capable to switch at 1 kV without any miss triggering.

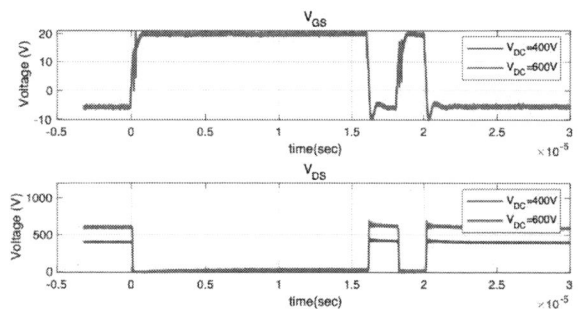

Fig. 11. DPT result of inverter with Al busbar.

Fig. 12 shows the motor drive with Al busbar and Cu busbar is designed for reducing the gate-source loop and power loop. By improving the busbar design, the power loop inductance is reduced from 17 nH to 15 nH.

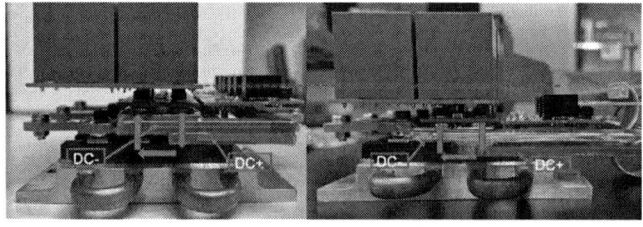

Fig. 12. Inverter with Al busbar (left) and Cu busbar (right).

As the power loop inductance and the coupling coefficient are reduced, the DPT is capable to perform at 1 kV which is even higher than the DC-link rated voltage. The DPT result is presented in Fig. 13. The result shows the switching waveform is clear and normal when the device current is 147 A at 1 kV with improved Cu busbar when the external gate resistor is 1 Ω. From Fig. 14, The dV/dt when $R_{G(ext)}=1\Omega$ is 10.17 V/ns, well as 6.47 V/n when $R_{G(ext)} = 4.7\Omega$.

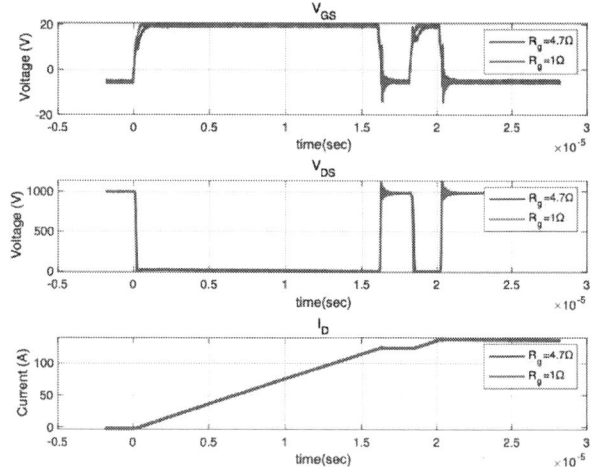

Fig. 13. DPT result of inverter with Cu busbar.

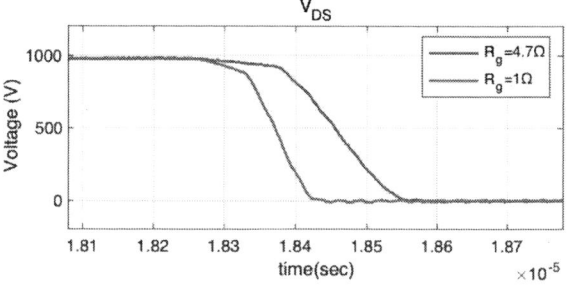

Fig. 14. $V_{DS(on)}$ waveform when $I_L = 147A$ with different $R_{G(ext)}$.

To sum up, by improving the busbar design according to the method proposed in Section III, the switching performance of the inverter during turn-on can be improved by 56%. The reduced coupling coefficient between the gate-source and power loop and power loop size reduce the value of the external gate resistor which allows the power semiconductor to switch at its rated speed.

V. CONCLUSION

The paper presented a workflow on busbar design for high-power inverters. Busbar design of high-power inverters needs to have low power loop inductance, low coupling coefficient, and high partial discharge initial voltage. Moreover, the paper provided a method to design the busbar on the converter level. The experimental validation process allows evaluating the switching performance of the inverter.

REFERENCES

[1] F. Luo, Mustafeez-Ul-Hassan, Z. Yuan and K. Choksi, "High-Density Motor Drive Development for Electric Aircraft Propulsion: Cryogenic

and non-Cryo Solutions," 2022 International Power Electronics Conference (IPEC-Himeji 2022- ECCE Asia), 2022, pp. 2130-2134, doi: 10.23919/IPEC-Himeji2022-ECCE53331.2022.9807228.

[2] H. Bishnoi, P. Mattavelli, R. Burgos and D. Boroyevich, "EMI Behavioral Models of DC-Fed Three-Phase Motor Drive Systems," in IEEE Transactions on Power Electronics, vol. 29, no. 9, pp. 4633-4645, Sept. 2014, doi: 10.1109/TPEL.2013.2284436.

[3] A. B. Mirza, A. I. Emon, S. S. Vala and F. Luo, "A Comprehensive Analysis of Current Spikes in a Split-Phase Inverter," 2022 IEEE Applied Power Electronics Conference and Exposition (APEC), 2022, pp. 1580-1585, doi: 10.1109/APEC43599.2022.9773407.

[4] A. Deshpande and F. Luo, "Multilayer busbar design for a Si IGBT and SiC MOSFET hybrid switch based 100 kW three-level T-type PEBB," 2017 IEEE 5th Workshop on Wide Bandgap Power Devices and Applications (WiPDA), 2017, pp. 20-24, doi: 10.1109/WiPDA.2017.8170496.

[5] K. Choksi, Y. Wu, Mustafeez-ul-Hassan and F. Luo, "Evaluation of Factors Impacting Reflected Wave Phenomenon in WBG Based Motor Drives," 2022 International Power Electronics Conference (IPEC-Himeji 2022- ECCE Asia), 2022, pp. 736-740, doi: 10.23919/IPEC-Himeji2022-ECCE53331.2022.9807042.

[6] R. S. Krishna Moorthy et al., "Estimation, Minimization, and Validation of Commutation Loop Inductance for a 135-kW SiC EV Traction Inverter," in IEEE Journal of Emerging and Selected Topics in Power Electronics, vol. 8, no. 1, pp. 286-297, March 2020, doi: 10.1109/JESTPE.2019.2952884.

[7] C. Chen, X. Pei, Y. Chen and Y. Kang, "Investigation, Evaluation, and Optimization of Stray Inductance in Laminated Busbar," in IEEE Transactions on Power Electronics, vol. 29, no. 7, pp. 3679-3693, July 2014, doi: 10.1109/TPEL.2013.2282621.

[8] YZ. Yuan et al., "Design and Evaluation of Laminated Busbar for Three-Level T-Type NPC Power Electronics Building Block With Enhanced Dynamic Current Sharing," in IEEE Journal of Emerging and Selected Topics in Power Electronics, vol. 8, no. 1, pp. 395-406, March 2020, doi: 10.1109/JESTPE.2019.2947488.

[9] "SIC MOSFET module FCA150AC120" [Online]. https://www.sansha.co.jp [Accessed: 16-Sep-2022].

[10] A. B. Mirza, A. I. Emon, S. S. Vala and F. Luo, "Noise Immune Cascaded Gate Driver Solution for Driving High Speed GaN Power Devices," 2021 IEEE Energy Conversion Congress and Exposition (ECCE), 2021, pp. 5366-5371, doi: 10.1109/ECCE47101.2021.9595515.

[11] Z. Yuan et al., "Design of Partial-discharge-free Busbar for More-electric Aircraft Application with Low Pressure Condition," 2021 IEEE Applied Power Electronics Conference and Exposition (APEC), 2021, pp. 1178-1182, doi: 10.1109/APEC42165.2021.9487285.

Analysis of a Switching Event and its Impact on Gate Drive in Gallium-Nitride Based Bi-Directional Switches

Mustafeez-ul-Hassan
Department of Electrical and Computer
Engineering, Stony Brook University
Stony Brook, NY, USA
mustafeez.hassan@stonybrook.edu

Yuxuan Wu
Department of Electrical and Computer
Engineering, Stony Brook University
Stony Brook, NY, USA
yuxuan.wu@stonybrook.edu

Fang Luo
Department of Electrical and Computer
Engineering, Stony Brook University
Stony Brook, NY, USA
fang.luo@stonybrook.edu

Abstract—**Application of wide band gap (WBG) devices in power electronics converters enable high frequency and high-speed operation, together with reduced conduction losses. Gallium-Nitride (GaN) devices are more promising as they offer low chip area, and stable variation of channel resistance. Such devices can be employed in current source inverters (CSI) where two of them need to be connected in anti-series to make a reverse voltage blocking (RVB) switch. RVB switches are not only important in CSIs, but also find their applications in matrix converters, high voltage DC circuit breakers and T-type voltage source inverters. This paper discusses a switching event comprising of turn-on and turn-off phenomenon in RVB switch. Since the impact of layout parasitic onto gate drivers has been a common issue for GaN based converters, the paper presented a switching events, and discussed all the contributors which help or worsen the driving of bi-directional switches.**

I. INTRODUCTION

Power electronics converters have been of extreme importance in transforming energy from one form to another, whereas high speed and high frequency power conversion is critical to compact and efficient power electronics converters. The energy can be transformed from one DC voltage level to another DC voltage level depending upon the type of application. The converters can also be employed in converting from DC to AC or vice versa. With the advent of wide-band gap devices, i.e. SiC MOSFETS, and GaN devices; the researchers have found it interesting to apply such devices into the power electronics converters.

Amongst the switches, the ones commonly used and easily available in market block voltage in one direction and can conduct current in both the directions. Historically, these switches find their applications in DC-DC converters, and voltage source inverters. However, with the challenge of increased power efficiencies and power densities, converters like matrix converters, Vienna rectifiers, current source inverters (CSIs), DC circuit breakers and neutral point clamped inverters are also finding increased attention and application [1]–[3]. Such converter types either need reverse voltage blocking (RVB) or bi-

directional (BD) current carrying capability. Both the configurations need at-least two switching devices in series or anti-series to make a RVB and/or BD switch. Different combinations of such devices are studied and presented in [4], [5] and are shown below.

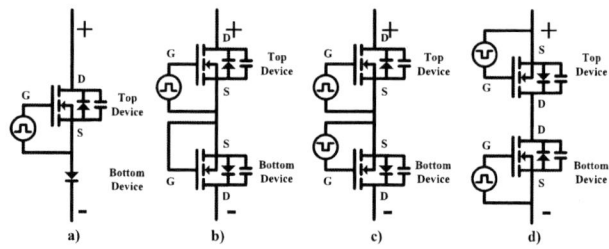

Fig. 1: RVB switch with a) single transistor b) two transistors; BD switch with dual transistor c) common source d) common drain

This paper considers a bi-directional current carrying switch only; as such a combination not only gives current control in both the directions, but also offers reduced conduction losses as compared to reverse voltage blocking switches [5]. Additionally, the bi-directional switches considered are common source configured, as it saves the cost and area complexity associated with gate driver. For further simplification, a dual channel gate driver might be employed to control both the switches independently and having more flexibility in terms of making a switch either to be BD or RVB.

Furthermore, the paper considers a GaN based bi-directional switch with its application for a current source inverter as such devices make it more interesting and challenging because of their intrinsic fast speeds, and capability of higher switching frequencies. Although GaN devices are expected to show improved performance, yet such converters are extremely sensitive to layout, and gate driving requirements. The CSI developed is expected to offer reduced motor side over-voltages [6] and better EMI [7]. The rationale for GaN based switching devices are their reduced on-state resistance, increased switching speeds, and stable breakdown and

978-1-6654-8901-0/22 $31.00 © 2022 IEEE

threshold voltages especially at reduced temperatures [8], [9].

For the analysis, LTspice was used to model the switching cells with all the stray components and converters as a whole. Hardware results are also presented to prove the concept at the end of paper. The paper is organized as follows: Chapter II discusses the system configured and identifies a switching event in a typical three phase converter and explains the gate driving conditions for double pulse tests, whereas chapter III presents the hardware results and summarizes all the influences impacting the gate drive of bidirectional switches.

II. SYSTEM CONFIGURATION

In order to understand the impact of a complete switching event, a 300 V_{LL-RMS} three phase CSI was considered as a potential application of BD switch. For such a voltage, 650 V GaN device was found to be compatible with enough voltage blocking and current carrying capability. Furthermore, space vector pulse width modulation (SVPWM) utilized as the modulation technique to understand the switching interval and commutation loop. Schematics of a three phase CSI has been shown in Fig. 2 where L_{DC} represents the DC link storage inductor and L_{stray} represents the parasitic power loop inductance due to printed circuit board (PCB) layout of any switching branch. $T_1 - T_6$ represent the common source configured bi-directional current carrying and RVB switching devices, and L_{dec} and C_{dec} represent decoupling capacitors. Similarly, $L_{stray-uc}$ represent half of the un-compensated stray inductance between the top and bottom switching devices constituting the same phase leg i.e. T_1T_4, T_3T_6 and T_5T_2. The output filter capacitor is represented with C_F connected through lead inductance of L_{leads}.

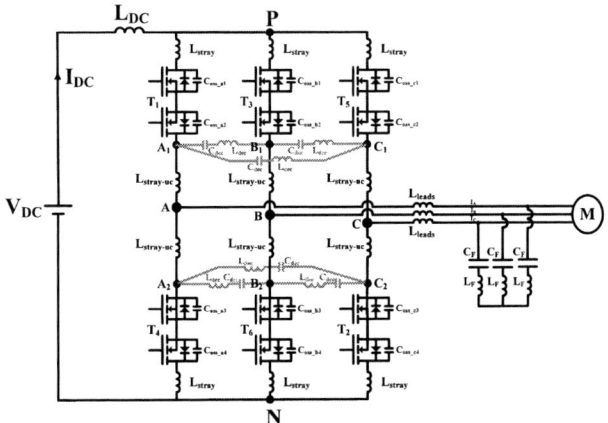

Fig. 2: Three phase CSI with parasitics

A. Identification of a Switching Event involving Bi-Directional Switches

A unified SVPWM strategy was considered for a CSI and has been shown in Fig. 3, where switching transitions in any of the sectors, e.g. sector II can be realized as:

$$T_1T_2 \rightarrow T_2T_3 \rightarrow T_2T_5 \rightarrow T_2T_3 \rightarrow T_1T_2$$

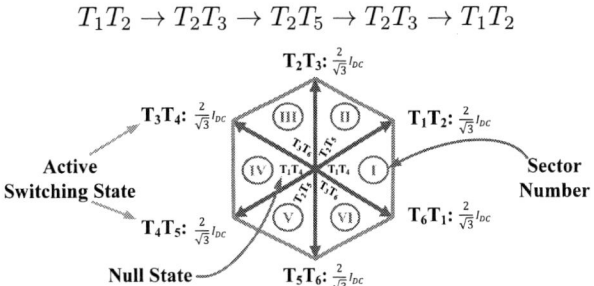

Fig. 3: Unified SVPWM representation of CSI

Considering the above switching sequence, converter is assumed to be in switching state of T_1T_2, and a transition happens while transferring to T_2T_3. As the commutation happens between T_1 and T_3, the residual energy through the initial conducting path must be quenched to ensure a safe transition. During this transition, the switching event can be identified as APB, and such an understanding can be extended to APC and BPC as well. Similar kind of commutation intervals can be identified for bottom half of converter and are presented as ANB, ANC and BNC [10]. Given the decoupling capacitors are selected to be much larger then (C_{oss}) of switching devices, the improved and shortened commutation loops can be identified as A_1PB_1, A_1PC_1, B_1PC_1.

B. Development of Double Pulse Test (DPT) Platform

Considering the transition of voltage happening between the horizontal loops as identified previously, a DPT platform can be configured between the switching nodes. Similar DPT platform has been shown for A_1PB_1 in Fig. 4 where transistors T_1 and T_3 have been decomposed into common source configured T_{xA} and T_{xB}. In the DPT, T_{1A} and T_{3B} are permanently kept in on state, whereas T_{1B} is configured in off state, and a double pulse is applied to T_{3A}. Equivalent DPT circuit during both the states is presented below.

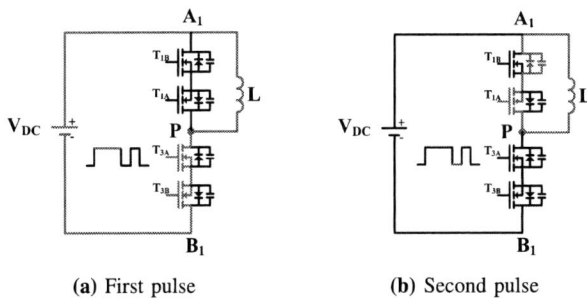

(a) First pulse   (b) Second pulse

Fig. 4: Equivalent circuit during a switching instance in top half of converter

978-1-6654-8901-0/22 $31.00 © 2022 IEEE

III. HARDWARE DEVELOPMENT AND RESULTS

A. Impact on Gate Drive During the Switching Events

During the evaluation of DPT, the impact of numerous quantities were considered and analyzed. It is worth mentioning that successful conduction of DPT for an individual switching cell may not necessarily imply successful DPT in the presence of switching devices of other phases. From the experimentation, it was found that the common source configured device which is always kept on (T_{1A} in Fig. 4b) across the clamping inductor is most vulnerable to switching transients, and can easily be mistriggered therefore leading to immediate interruption of inductor current and electrical failure of the switching device in the switching cell.

Considering the vulnerability of devices ($T_{1A}/T_{3A}/T_{5A}$) to switching instants during a converter operation, this section of paper considers all the parameters tested and summarizes about their influence on performance of gate driver.

1) Switching Devices of Adjacent Phases

For the successful operation of converter, doing DPT is not sufficient to ensure the proper functioning of the converter. Having individually tested all the commutation loops, all six common source configured devices T_1-T_6 were populated and DPT were repeated on individual loops. It was worth mentioning that neighbouring devices were found to have influence on the always kept on (T_{1A}) across the clamping inductor and should be fixed. From experimentation, it was concluded that application of ferrite beads in the gate drive loop helped reduce the possible mistriggering.

2) Applied Gate Drive Voltage on Switching Devices of Adjacent Phases

Having brought the neighbouring switching devices in circuit, it was found that leaving their gates floating results in erroneous behavior. However, application of proper turn-off gate driving voltage helps better achieve the switching performance.

3) Decoupling Capacitors

Placement of decoupling capacitor across the switching cell helps mitigate the high frequency noise and resonances. From the experimentation, it was observed that not only the choice of proper decoupling capacitor was important, but also balanced capacitance between all three phases was found to be critically important. It was observed that choice of decoupling capacitors with lower equivalent series inductance (ESL) results in relatively better switching performance.

4) Optimized Gate Loop

Two type of gate loops were tried and investigated. Gate loop was placed in the same plane as power plane, and perpendicular plane as well. It was found that gate loop placed vertical to power loop had a significant improvement in performance of gate driving wave-forms.

5) Buffer Stage Ahead of Isolated Gate Driver

Having added a buffer stage integrated circuit (IC) before the gate driving IC helps better noise immunity and keeps the gate voltages better as compared to the ones with only single gate driver IC.

6) Higher Sinking Current Gate Driving ICs

From the experimentation, it was also found that utilizing the gate driving ICs with higher current sourcing/sinking capabilities results in much better and cleaner gate wave-forms and hence appropriate behavior of the switching devices.

B. Depiction of Mistriggering

In order to verify the impact of all the quantities presented above on the actual converter level performance, a power board for three phase inverter was developed, and tests were conducted sequentially. A typical event of mistriggering is also captured and presented in where it can be shown that always kept on device's gate voltage bleeds to lower voltages leading to potential failure of devices and whole converter eventually.

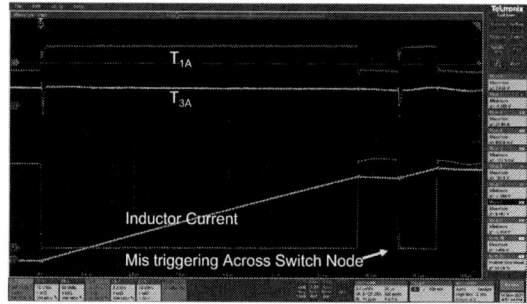

Fig. 5: Mistriggering in always on device

C. Development of a Fully Functional Three Phase Converter

In order to understand the operation of a bi-directional switch, a three phase CSI was developed using GaN devices. Each switching device T_1-T_6 was driven by a dual channel gate driver which can precisely be utilized to insert and control the overlap time. For the DC link inductor, powdered core having 250 uH inductance was selected, whereas a "Δ" connected output filter capacitor bank was placed. Double pulse tests were carried onto individual phases at first, and totally integrated three phase system at last. For the measurements, oscilloscope of 1 GHz bandwidth by Tektronix was utilized, whereas 200 MHz active and 800 MHz bandwidth passive probes labelled as THDP0200 and TPP0850, were employed for gate-source and drain-source voltages respectively. The whole converter setup is presented in Fig. 6a where a three phase resistive load is connected at the output. Having properly controlled the mistriggering across the always on devices, successful DPT at rated voltages were conducted as seen from Fig. 6b. Power loop ringing frequency of 138 MHz was observed which shows the successful operation of a bidirectional switch at such a high speed.

(a) CSI test setup

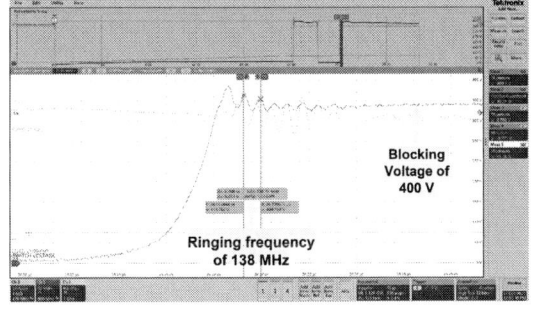

(b) DPT confirming the optimized power loop design
Fig. 6: Three phase CSI setup and key wave-forms

IV. CONCLUSION

This paper signifies the importance of bi-directional current carrying switches in different converter using GaN devices. Configuring two devices in series/anti-series possess challenges to the gate driving circuitry which have been explored. In the paper, causes of mis-triggering and their remedies were proposed to make sure a converter level bi-directional switch is fully operational and performs well. The switches developed were used to develop a three phase CSI and was successfully operated under loaded conditions.

ACKNOWLEDGEMENT

The authors would like to acknowledge the financial support lended by NASA to carry out this research under ULI: Development of the Cryogenic Hydrogen-Energy Electric Transport Aircraft (CHEETA) Design Concept under award number 80NSSC19M0125. Furthermore, NSF CAREER award numbered 1846917 titled Semiconductor-Based EMI Mitigation Architecture for Future Power Electronics Systems is also acknowledged.

REFERENCES

[1] P. Wheeler, J. Rodriguez, J. Clare, L. Empringham, and A. Weinstein, "Matrix converters: a technology review," *IEEE Transactions on Industrial Electronics*, vol. 49, no. 2, pp. 276–288, 2002.

[2] J. Kolar, U. Drofenik, and F. Zach, "Vienna rectifier ii-a novel single-stage high-frequency isolated three-phase pwm rectifier system," *IEEE Transactions on Industrial Electronics*, vol. 46, no. 4, pp. 674–691, 1999.

[3] H. Dai and T. M. Jahns, "Comparative investigation of pwm current-source inverters for future machine drives using high-frequency wide-bandgap power switches," in *2018 IEEE Applied Power Electronics Conference and Exposition (APEC)*, 2018, pp. 2601–2608.

[4] H. Dai, R. A. Torres, T. M. Jahns, and B. Sarlioglu, "Characterization and implementation of hybrid reverse-voltage-blocking and bidirectional switches using wbg devices in emerging motor drive applications," in *2019 IEEE Applied Power Electronics Conference and Exposition (APEC)*, 2019, pp. 297–304.

[5] M. ul Hassan, A. I. Emon, Z. Yuan, H. Peng, and F. Luo, "Performance comparison and modelling of instantaneous current sharing amongst gan hemt switch configurations for current source inverters," in *2022 IEEE Applied Power Electronics Conference and Exposition (APEC)*, 2022, pp. 2014–2020.

[6] Y. Wu, K. Choksi, M. ul Hassan, and F. Luo, "An extendable and accurate high- frequency modelling of three-phase cable for prediction of reflected wave phenomenon," in *2022 IEEE Applied Power Electronics Conference and Exposition (APEC)*, 2022, pp. 944–950.

[7] R. A. Torres, H. Dai, T. M. Jahns, and B. Sarlioglu, "Operation and analysis of current-source inverters using dual-gate four-quadrant wide-bandgap power switches," in *2019 IEEE Energy Conversion Congress and Exposition (ECCE)*. IEEE, 2019, pp. 2353–2360.

[8] H. Gui, R. Chen, J. Niu, Z. Zhang, L. M. Tolbert, F. F. Wang, B. J. Blalock, D. Costinett, and B. B. Choi, "Review of power electronics components at cryogenic temperatures," *IEEE transactions on power electronics*, vol. 35, no. 5, pp. 5144–5156, 2019.

[9] M. ul Hassan, A. I. Emon, F. Luo, and V. Solovyov, "Design and validation of a 20 kva, fully cryogenic, 2-level gan based current source inverter for full electric aircrafts," *IEEE Transactions on Transportation Electrification*, pp. 1–1, 2022.

[10] L. G. A. Rodrigues, G. Lefèvre, J. Martin, and J.-P. Ferrieux, "Switching cell design optimization of sic-based power modules for current source inverter applications," in *2017 19th European Conference on Power Electronics and Applications (EPE'17 ECCE Europe)*. IEEE, 2017, pp. P–1.

In-Situ Ultrafast Sensing Techniques for Prognostics and Protection of SiC Devices

Ali Parsa Sirat, Chondon Roy, Daniel Evans, James Gafford, and Babak Parkhideh
Department of Electrical and Computer Engineering
Energy Production & Infrastructure Center (EPIC)
University of North Carolina at Charlotte
Charlotte, North Carolina, USA
Email: {aparsasi, jgafford, bparkhideh}@uncc.edu

Abstract—**Embedding diagnostic and prognostic into power electronics has the potential to increase the reliability and resiliency of and increasing automated adaptability in variable operating conditions has increased the reliability and resiliency of these systems, especially with the transition to wide bandgap semiconductors where higher voltages, lower on-state resistances, and faster switching speed can lead to rapid failures under conditions such as shoot through. This value proposition requires minimally invasive sensing elements that provide real-time monitoring of online system operations such that parasitic values are not introduced, which erode performance. Presented is an in-situ current sensing circuit with integration to the controller to provide enhanced operational capabilities such as sub-microsecond short circuit protection and power semiconductor device on-state resistance measurement. Techniques developed for measurement and protection are not limited to the tested SiC devices and may be extended to numerous types of critical components. These techniques can provide detailed, real-time state of health estimation for critical components and system capabilities, thus, enhancing system reliability.**

Keywords—*Diagnostics, Prognostics, SiC, Ultrafast Sensor, On-state Resistance, Shoot-through Detection*

I. INTRODUCTION

Advances in wide bandgap (WBG) semiconductor devices have expanded operating envelopes for power electronics systems [1-3]. A DC-AC power electronic inverter or "Intelligent Power Stages (IPSs)", can be noted as an essential subsystem of a new grid management architecture [4-6] using WBG devices for higher power quality and faster operation. These increasingly capable systems provide avenues of expansion for commercially relevant power and energy conditioning systems. However, the faster they function, the faster sensing mechanisms and operation algorithms are needed. The interoperable SiC DC-AC inverter power stage can be enhanced with integrated advanced sensing, which would be protecting and monitor the state at the power semiconductor device and intelligent gate driver level [7].

In recent reliability improvement trends, degradation is determined by continuously monitoring the aging characteristics of the elements, such as the on-resistance of semiconductor switches [22-26]. Also, with more optimized-layout power converter technologies, a need for faster fault detection is inevitable. Due to the very low self-inductance and

on-resistance of WBG, protecting them against short circuit faults is extremely crucial. In [19], below one-microsecond fault clearance for hundreds of amps currents had been achieved using either the de-saturation method or high bandwidth current monitoring.

The two-level three-phase topology features silicon carbide (SiC) power modules, with integrated online health monitoring and protection sensors. The commercially available SiC device outcomes are incorporated in a low schedule risk method to achieve high efficiency and power density. The IPS can also include subsystem interoperability, embedded intelligence (or high-speed current and voltage sensors), advanced EMI mitigation, and embedded and supervisory controls supporting multiple applications. Ultrafast in-situ sensors are employed for monitoring and protecting the health conditions of SiC devices [8].

In this paper, in-situ ultrafast protection and health monitoring schemes based on high-frequency current sensing, and device health online using on-state resistance (R_{DSON}) measurement are proposed, implemented, and tested. In section II, the schemes and signal routing mechanisms are discussed. In section III, sensors' design, implementation, and properties are discussed. In section IV the experimental result, the experimental results are provided, and the conclusion is finally stated in section V.

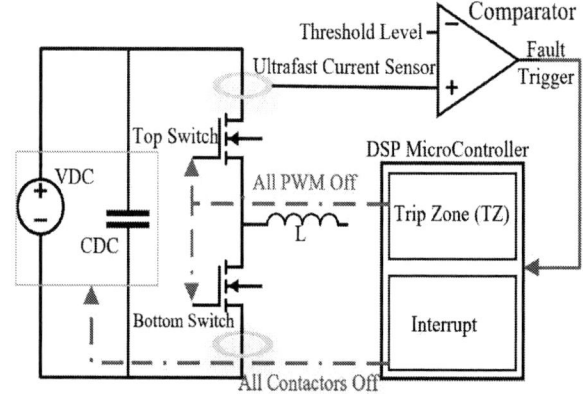

Fig. 1: Signal routing of the proposed shoot-through protection scheme.

II. PROTECTION AND RELIABILITY SCHEMES

A. Shoot-Through Protection

One of the key features for diagnostics of wide bandgap semiconductor power devices is the shoot-through protection. High-speed sensors can be employed for monitoring and protecting the health conditions of wide bandgap semiconductor (such as SiC) power modules. Sub-microsecond fault detection and tripping signal routes are deployed in this paper by utilizing an ultrafast shoot-through protection scheme based on high-frequency current sensing thru PCB-embedded Rogowski coil and signal routing via the built-in EPWM trip zone of the C2000 series microcontrollers. Fig. 1 shows the scheme of the shoot-through protection that can be applied to each power module in power converters. This method to clear the fault current with a sub-microsecond protection time is implemented by retrofitting an existing control board with a single digital input signal and without the need of modifying the gate driver design to add fault functionality there. This approach makes integration easier with existing designs without the need to redesign gate drivers. An ultrafast and non-intrusive current sensing is needed for this protection scheme. In-situ Rogowski-based current sensor and their combination are among the best techniques to have a bandwidth over 10 MHz [9-12]. While there is practically no appropriate commercial non-invasive Rogowski coil sensor with high-current/precision in the market, PCB-embedded Rogowski coil sensors can be modified and integrated into the layout of power converters, via a PCB-embedded differential coil and an analog integrator circuit. In the next section, more details of the current sensor are provided.

B. Monitoring On-State Resistance for Aging

The key feature for prognostics is resistance change and characterization of SiC semiconductor power modules [13]. An isolated drain-to-source using the on-state resistance (R_{DSON}) of the SiC device sensor and an algorithm to monitor the on-state resistance of SiC power modules are developed for that purpose. Instantaneous V_{DSON} and current through the device are required for the R_{DSON} calculation. Fig. 2 presents the V_{DSON} current sensor configuration on the three-phase inverter

[27]. The instantaneous current is measured using the Rogowski coil switch current sensor at the drain terminal for the high-side device and the source terminal for the low-side device. During the on-state, the device current is the same as the phase inductor current [28]. The phase-inductor current was used for the R_{DSON} calculation in [27]. The use of the proposed switch current sensors for the R_{DSON} calculation and control could potentially remove the need for the phase-current sensors.

III. IN-SITU ULTRAFAST CURRENT AND VOLTAGE SENSOR

Ultrafast current and voltage sensing are needed for the protection, reliability, and control of WBG power converters [14]. The hardware for this paper is employed in an inverter with a switching speed of 30 kHz. This low switching frequency, applicable for high voltage and power inverters, means that no high-frequency sensor is needed for control, yet for sufficient online protection and monitoring, ultrafast sensors are required to be deployed with fast edge rates. On the other hand, super-fast sensing is highly challenging in WBG power electronics, due to their layout, geometry, and physics [15]. In this work, an ultrafast non-invasive current sensor has been considered and put into operation for protection, and switch current measurement, as well as a V_{DSON} sensor for health monitoring of the devices.

A. Ultrafast Switch Current Sensor

The inductive sensing method has been always one of the best isolated current detecting solutions in many applications. To be specific about the Rogowski coil (RC), the coil's raw output (up to the self-integrating pole) needs to be integrated. Opamp-based analog integrators are among the best options for integrating the coil raw output [16-22]. In Fig. 3 (a) the terminated coil (by R_t) is going through an inverting integrator. In Fig 3 (b) the coil and integrator gains compose a unique gain starting from the frequency of F_a to F_d. All practical analog integrators begin to integrate after a specific frequency (F_a), due to the internal input resistance in their differential mode (R_f) to avoid DC saturation of the opamp.

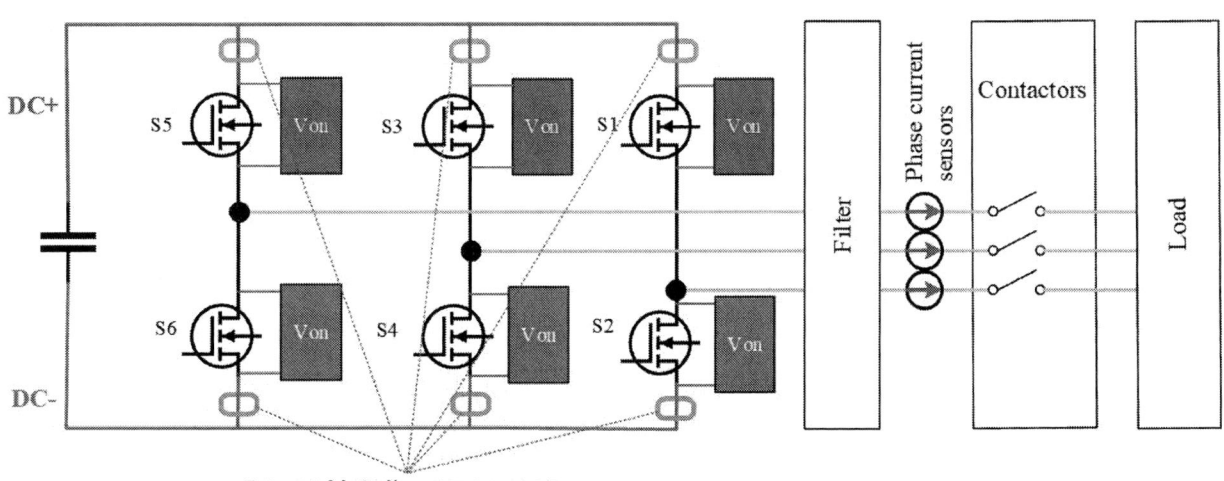

Fig. 2: On-state resistance measurement scheme. V_{DSON} and device current sensor configuration [27].

978-1-6654-8901-0/22 $31.00 © 2022 IEEE   143

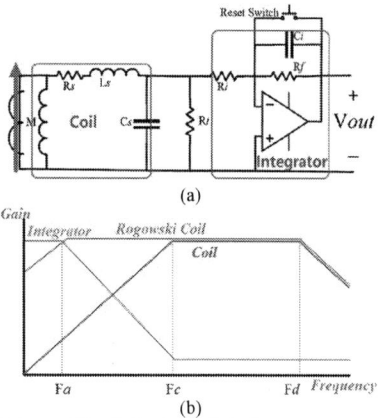

(a)

(b)

Fig. 3: Rogowski coil circuit (a), Gain frequency response (b).

However, the DC component is missing in simple -$M.di/dt$ detectors, (in which M is the mutual inductance between the sensor's coil and the actual conductor that carries the current, and di/dt is the current change in a time) because there is no current change in DC component and consequently, the raw output of the inductive-based current sensors are acting as AC-coupled current probes. By principle, a conventional RC switch current sensor needs to be reset periodically.

Technically, one can terminate the coil with a resistor of R_t (the coil damping/termination resistor) to achieve coil self-integration in higher frequencies. Moreover, F_c is the frequency in which the coil starts to integrate itself via R_t. The terminated coil can operate up to F_d which is essentially higher than the coil resonant frequency of F_r:

$$F_r = 1/2\pi.sqrt(L_sC_s) \ (Hz \ ) \ (1)$$

In the case of the switch current, although the RC cannot sense the DC, resetting C_i during the off state of the switch, can force the integration of the coil output from zero in each cycle. In better words, this explains how resetting the integrating capacitor periodically can lead to maintaining the DC component of the switch current.

The value of the components that are shown in Fig. 1 is needed to be considered for the calculation of the sensor's sensitivity and gain frequency response. M, Rs, Ls, and Cs, which are the mutual inductance, coil resistance, coil self-inductance, and coil stray capacitance, respectively, can be calculated from the coil geometry. Except for the mutual inductance (M) that senses the change in observed current regarding time, the other RLC components are parasitic elements, and the bigger they are, the lower bandwidth the coil has. The coil termination resistor is called the damping resistance of R_t. If the R_t is open, the coil will have a resonant frequency of F_r (1). The double pole at F_r can be shifted as two single poles to the lower and higher frequencies of F_c and F_d, correspondingly. In frequencies between these two single poles, the coil will have a self-integrated output voltage.

The analog integrator is needed to integrate the coil differential +20 dB slope over frequencies below F_c. The integrator factor can be considered as $1/R_iC_i$. Also, the final sensitivity of the sensor is roughly:

$$Sensitivity = M/R_iC_i \quad (V/A) \quad (2)$$

TABLE I.   PROTOTYPED CURRENT SENSOR PROPERTIES

Property	Value
Edge Side Length (mm)	30
Width (mm)	3
Height (mm)	1.6
Relative Permittivity	4 (FR4)
Turn Numbers	132
Mutual Inductance (M)	2.1 nH
Self-Inductance (L_s)	500 nH
Stray Capacitance (C_s)	17 pF
Winding Resistance (R_s)	0.9 Ω
Termination Resistor (R_t)	24 Ω
Integrator Capacitance (C_i)	1 nF
Integrator Resistance (R_i)	45 Ω
Integrator Low-band (F_a)	5 kHz
Self-integration Starting Frequency (F_c)	10 MHz
Undamped Coil Resonant Frequency (F_r)	54 MHz
Final Sensitivity	44 mV/A

A square-shaped Z-axis PC-embedded coil as in Fig. 4 has been designed to build up the custom-designed Rogowski coil switch current sensor [18-21]. This geometry allows us to have a high enough number of turns (to achieve a mutual inductance of 2-3 nH). The square edge side length of 30 mm along with the 3 mm coil traces' width and the PCB height of 1.6 mm are good fits for a high-current commercial SiC power module such as CAS300M17BM2 (Fig. 4(b)). Table I provides the sensor measured properties' values schematic in Fig. 3. The design is meant to operate over a range frequency of 5 kHz up to over 50 MHz. A high slew-rate voltage-feedback dual opamp IC has been used in this design. The performance of the sensor has been validated in the double pulse test with high currents in the next section.

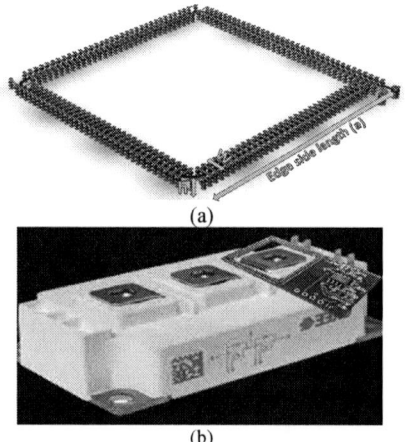

(a)

(b)

Fig. 4: The coil 3D geometry (a), Sensor vs. SiC module size (b).

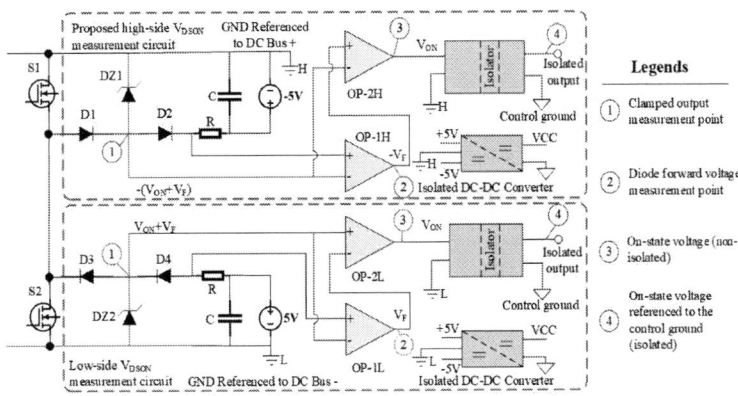

Fig. 5: On-state voltage sensor schematic [27].

B. Isolated V_{DSON} Sensor

SiC (or most other FET types) device degradation mechanisms lead to an increase of V_{DSON} at a certain current in long-term operation [23-26]. Real-time monitoring of the device degradation could predict possible device failure enabling users to act to prevent system interruption, thus increasing system reliability. For the accurate measurement of R_{DSON}, a fast and accurate isolated V_{DSON} measurement sensor is essential. Fast-settling isolated V_{DSON} sensors have been utilized with a high voltage blocking capability. The schematics of the high-side and low-side V_{DSON} measurement sensors are presented in Fig. 5. Detail design and evaluation of each of the sensors have been presented in [27], where they used the low-frequency control current from a LEM current sensor for R_{DSON} estimation.

IV. EXPERIMENTAL RESULTS

The proposed monitoring and protection schemes via ultrafast in-situ current and voltage sensors can be implemented within the sensing capabilities of an interoperable DC-AC inverter power stage to perceive the state of health at the level of the power semiconductor devices and intelligent gate drivers [29, 30].

A double pulse test setup based on a SiC power module (CAS300M17BM2) and a TI microcontroller (F28379D) has been provided to investigate the proposed protection scheme in Fig. 1. The results of the 650 V, double pulse test with a trip

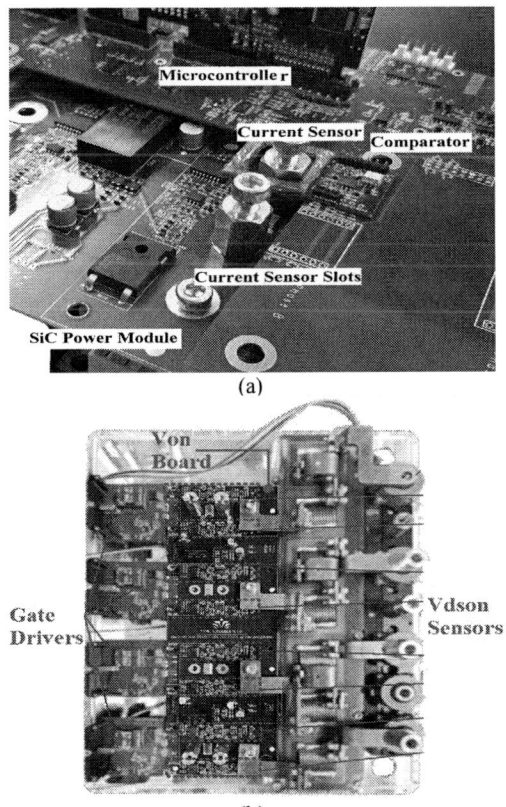

(a)

(b)

Fig. 6: Shoot-through protection components (a), R-on monitoring setup specification (b).

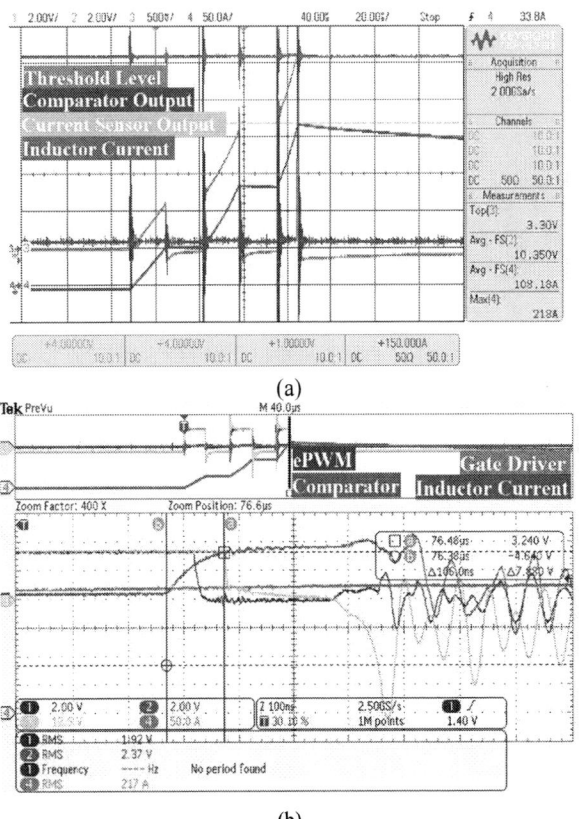

(a)

(b)

Fig. 7: Shoot-through protection in double pulse mode (a), Zoom-in of the fault clearing moment(b).

978-1-6654-8901-0/22 $31.00 © 2022 IEEE

limit of approximately 220 A have been presented in Fig. 7 and the timings recorded for each step in the signal chain in table II. the total fault clearance time of 521 ns has been achieved in this scenario. Although fault clearing through the gate driver fault pin can occur in less than 300 ns [19], note that the current sensor output had a lower latency response than the Tektronix current sensor used for comparison, so it was given a delay time of 0 ns. The timing of the signal chain was measured with low voltage, non-isolated probes for accurate timing measurement before putting the circuit into the double pulse test. During the turning-off of the measured device, and the transients on the Tektronix differential probes obscure the device turn-off time so the datasheet value is used. In addition, fault clearing enables the pin of the gate driver would need extra circuitry.

TABLE II.   SHOOT-THROUGH PROTECTION TIMING

The Component	Delay Time
Comparator Propagation Delay	162 ns
GPIO Input – RC Pullup Time	22 ns
ePWM Trip Latency	20 ns
Gate Driver Delay	106 ns
SiC Module Turn-Off time	211 ns
Shoot-through protection delay	521 ns

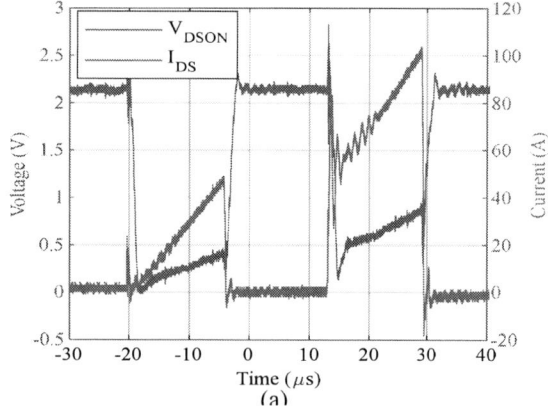

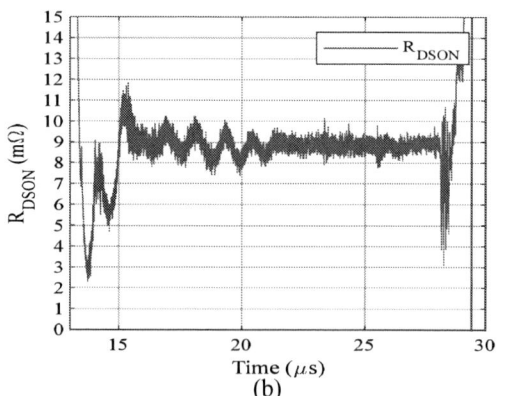

Fig. 8: R_{DSON} measurement using isolated V_{DSON} measurement and current from RC current sensor: Measured V_{DSON} and current (scaled to Amperes) (a), and calculated R_{DSON} using the isolated V_{DSON} and current from RC current sensor (b).

The R_{DSON} measurement was calculated using the ultrafast current sensor simultaneously used for shoot-through protection and the V_{ON} sensor's output voltage, presented in section 3, with the result as seen in Fig. 8. These results were obtained under a double pulse test using 400 V with a peak current target of 100 A. These integrated sensors yielded an R_{DSON} measurement of roughly 9 mΩ. This is within the range listed in the datasheet with the exact value not being measured using precision equipment before the tests. The ultrafast current sensor and VON sensor have also not been calibration before the double pulse test to increase the accuracy.

V. CONCLUSION

The presented ultrafast current sensor has been utilized for a sub-microsecond shoot-through protection scheme with results being much faster than existing desaturation techniques. Utilizing the existing trip zone faults, the sensors can be easily added to the controllers without the need to modify the existing gate drive circuitry. The ultrafast current sensor can also be used in conjunction with V_{DSON} sensors for accurately measuring and monitoring the on-state resistance. Future work will look to improve the accuracy of the sensor, reduce protection latencies, and explore novel ways to improve the ultrafast current sensor.

ACKNOWLEDGMENT

This project was supported by Oak Ridge National Laboratory (ORNL) funded through the Department of Energy (DOE) - Office of Electricity's (OE), Transformer Resilience and Advanced Components (TRAC) program led by the program manager Andre Pereira. The authors would also like to thank the Energy Production and Infrastructure Center (EPIC) and Electrical and Computer Engineering Department at the University of North Carolina at Charlotte.

REFERENCES

[1]   Bindra, A., 2015. Wide-bandgap-based power devices: Reshaping the power electronics landscape. IEEE Power Electronics Magazine, 2(1), pp.42-47.

[2]   Habibi, S., Rahimi, R., Ferdowsi, M. and Shamsi, P., 2021. DC Bus Voltage Selection for a Grid-Connected Low-Voltage DC Residential Nanogrid Using Real Data with Modified Load Profiles. Energies, 14(21), p.7001.

[3]   Nematirad, R. and Pahwa, A., 2022, April. Solar Radiation Forecasting Using Artificial Neural Networks Considering Feature Selection. In 2022 IEEE Kansas Power and Energy Conference (KPEC) (pp. 1-4). IEEE.

[4]   Heydari-Doostabad, H., Hosseini, S.H., Ghazi, R. and O'Donnell, T., 2021. Pseudo dc-link EV home charger with a high semiconductor device utilization factor. IEEE Transactions on Industrial Electronics, 69(3), pp.2459-2469.

[5]   Hosseini, S.H., Ghazi, R. and Heydari-Doostabad, H., 2020. An extendable quadratic bidirectional DC–DC converter for V2G and G2V applications. IEEE Transactions on Industrial Electronics, 68(6), pp.4859-4869.

[6]   Chinthavali, M., Moorthy, R.S.K. and Adib, A., 2021, June. Standard Modular Architecture for Consumer End Plug and Play Interfaces. In 2021 IEEE Applied Power Electronics Conference and Exposition (APEC) (pp. 2105-2112). IEEE.

[7]   Henn, J., Lüdecke, C., Laumen, M., Beushausen, S., Kalker, S., van der Broeck, C.H., Engelmann, G. and de Doncker, R.W., 2021. Intelligent Gate Drivers for Future Power Converters. IEEE Transactions on Power Electronics, 37(3), pp.3484-3503.

978-1-6654-8901-0/22 $31.00 © 2022 IEEE

[8] Naghibi, J., Mehran, K. and Foster, M.P., 2021. High-Frequency Non-Invasive Magnetic Field-Based Condition Monitoring of SiC Power MOSFET Modules. Energies, 14(20), p.6720.

[9] Hewson, C. and Aberdeen, J., 2018, March. An improved Rogowski coil configuration for a high speed, a compact current sensor with high immunity to voltage transients. In 2018 IEEE Applied Power Electronics Conference and Exposition (APEC) (pp. 571-578). Ieee.

[10] Ziegler, S., Woodward, R.C., Iu, H.H.C. and Borle, L.J., 2009. Current sensing techniques: A review. IEEE Sensors Journal, 9(4), pp.354-376.

[11] Hewson, C.R., Ray, W.F. and Davis, R.M., 2006, March. Verification of Rogowski current transducer's ability to measure fast switching transients. In Twenty-First Annual IEEE Applied Power Electronics Conference and Exposition, 2006. APEC'06. (pp. 7-pp). IEEE.

[12] Yin, S., Wu, Y., Dong, M., Lin, J. and Li, H., 2022. Design of Current Transformer for In Situ Switching Current Measurement of Discrete SiC Power Devices. IEEE Journal of Emerging and Selected Topics in Industrial Electronics.

[13] Marzoughi, A., Burgos, R. and Boroyevich, D., 2018. Characterization and performance evaluation of the state-of-the-art 3.3 kV 30 A full-SiC MOSFETs. IEEE Transactions on Industry Applications, 55(1), pp.575-583.

[14] Beheshtaein, S., Cuzner, R.M., Forouzesh, M., Savaghebi, M. and Guerrero, J.M., 2019. DC microgrid protection: A comprehensive review. IEEE Journal of Emerging and Selected Topics in Power Electronics.

[15] Belous, A., 2021. Power Electronics Devices Based on Wide-Gap Semiconductors. In Handbook of Microwave and Radar Engineering (pp. 389-435). Springer, Cham.

[16] Rafiq, A., Pramanick, S.K. and Maheshwari, R., 2020. Design of PCB coil based high bandwidth current sensor with power-loop stray inductance characterization. IEEE Transactions on Industrial Electronics, 68(12), pp.12791-12801.

[17] Ibrahim, M.E. and Abd-Elhady, A.M., 2020. Design and modeling of a two-winding Rogowski coil sensor for measuring three-phase currents of a motor fed through a three-core cable. IEEE Sensors Journal, 21(6), pp.8289-8296.

[18] Wang, J., Shen, Z., Burgos, R. and Boroyevich, D., 2015, November. Design of a high-bandwidth Rogowski current sensor for gate-drive shortcircuit protection of 1.7 kV SiC MOSFET power modules. In 2015 IEEE 3rd Workshop on Wide Bandgap Power Devices and Applications (WiPDA) (pp. 104-107). IEEE.

[19] Mocevic, S., Wang, J., Burgos, R., Boroyevich, D., Jaksic, M., Stancu, C. and Peaslee, B., 2020. Comparison and discussion on shortcircuit protections for silicon-carbide MOSFET modules: Desaturation versus Rogowski switch-current sensor. IEEE Transactions on Industry Applications, 56(3), pp.2880-2893.

[20] Mocevic, S., Wang, J., Burgos, R. and Boroyevich, D., 2021, May. Rogowski switch-current sensor self-calibration on enhanced gate driver for 10 kV SiC MOSFETs. In 2021 IEEE 12th Energy Conversion Congress & Exposition-Asia (ECCE-Asia) (pp. 1420-1425). IEEE.

[21] Wang, J., Mocevic, S., Burgos, R. and Boroyevich, D., 2020. High-scalability enhanced gate drivers for SiC MOSFET modules with transient immunity beyond 100 V/ns. IEEE Transactions on Power Electronics, 35(10), pp.10180-10199.

[22] Moreno, M.V.R., Robles, G., Albarracín, R., Rey, J.A. and Tarifa, J.M.M., 2017. Study on the self-integration of a Rogowski coil used in the measurement of partial discharge pulses. Electrical Engineering, 99(3), pp.817-826.

[23] Wei, J., Liu, S., Li, S., Fang, J., Li, T. and Sun, W., 2018. Comprehensive investigations on degradations of dynamic characteristics for SiC power MOSFET s under repetitive avalanche shocks. IEEE Transactions on Power Electronics, 34(3), pp.2748-2757.

[24] Pu, S., Yang, F., Vankayalapati, B.T. and Akin, B., 2021. Aging mechanisms and accelerated lifetime tests for SiC MOSFETs: an overview. IEEE Journal of Emerging and Selected Topics in Power Electronics, 10(1), pp.1232-1254.

[25] Ugur, E., Xu, C., Yang, F., Pu, S. and Akin, B., 2020. A new complete condition monitoring method for SiC power MOSFETs. IEEE Transactions on Industrial Electronics, 68(2), pp.1654-1664.

[26] Farhadi, M., Yang, F., Pu, S., Vankayalapati, B.T. and Akin, B., 2021. Temperature-independent gate-oxide degradation monitoring of SiC MOSFETs based on junction capacitances. IEEE Transactions on Power Electronics, 36(7), pp.8308-8324.

[27] C. Roy, N. Kim, J. Gafford and B. Parkhideh, "On-State Voltage Measurement of High-Side Power Transistors in Three-Phase Four-Leg Inverter for In-Situ Prognostics," in Proc. IEEE Energy Conversion Congress and Exposition (ECCE), Oct. 2021, pp. 2770-2776.

[28] C. Roy and B. Parkhideh. "Design consideration for characterization and study of dynamic on-state resistance of GaN devices." In *Proc. IEEE 7th Workshop on Wide Bandgap Power Devices and Applications (WiPDA)*, Oct. 2019, pp. 181-186.

[29] Rezaei, Omid, Omid Mirzapour, Mohammad Panahazari, and Hassan Gholami. 2022. "Hybrid AC/DC Provisional Microgrid Planning Model Considering Converter Aging" Electricity 3, no. 2: 236-250.

[30] O. Mirzapour and M. Sahraei-Ardakani, "Environmental Impacts of Power Flow Control with Variable-Impedance FACTS," 2020 52nd North American Power Symposium (NAPS), 2021, pp. 1-6, doi: 10.1109/NAPS50074.2021.9449793.

A Flux Balancing Strategy for 10-kV SiC-Based Dual-Active-Bridge Converter

Zihan Gao[1], Pengfei Yao[2], Haiguo Li[1], Fred Wang[1,3]

[1]Min H. Kao Department of Electrical Engineering and Computer Science, the University of Tennessee, Knoxville, TN, USA
[2]China Huaneng Group Co., Ltd., Beijing, China
[3]Oak Ridge National Laboratory, Oak Ridge, TN, USA
Email: {zgao15, hli96}prb@vols.utk.edu , yaopf19921210@163.com, fred.wang@utk.edu

Abstract— The transformer flux unbalance in dual-active-bridge (DAB) converters is a critical issue due to the electrical parameter and modulation mismatch, and load or control transients. Compared to low voltage DAB converters, the unbalance problem in medium voltage DAB converters may cause more severe problem, and become more difficult to deal with, because of the high operating voltage and insulation requirement. In this paper, a flux balancing strategy for a medium voltage (MV) DAB converter is proposed, including the transformer design with ferrite gap, current harmonic-based unbalance detection, as well as the flux balancing control scheme. With the proposed ferrite gap, nonlinear magnetizing current and its modeling are induced. Then, by analyzing the current harmonics, the flux unbalance level can be detected, and hence controlled. Test results have verified the proposed method in a MV DAB converter.

Keywords—Dual-active-bridge converters; flux balancing; medium voltage

I. INTRODUCTION

Medium voltage (MV) dc/dc converters have increasingly been implemented widely as the dc high power and voltage conversion are needed. The dual-active-bridge (DAB) converters with medium frequency MV transformer are commonly selected as the topology which can transmit power bidirectionally. As the great progress has been made on the wide-bandgap devices, higher and higher MV devices, especially SiC MOSFETs, are available, enabling MV power electronic converters. Along with the technical progress on MV SiC devices and applications, the MV transformers are also under development. Due to the high voltage requirement, the insulation performances of these transformers have to be carefully considered, which also affects the power density and efficiency of the transformers. To improve the performance efforts have been made on analyzing and designing insulation thermal and magnetics.

The MV converters are having higher and higher power densities, leaving less material margin on devices and transformers. However, with converter parameter mismatch,

This work was supported by the Advanced Manufacturing Office (AMO), United States Department of Energy, under Award no. DE-EE0008410, and made use of the Engineering Research Center Shared Facilities supported by the Engineering Research Center Program of the National Science Foundation and DOE under NSF Award Number EEC-1041877 and the CURENT Industry Partnership Program.

such as channel resistances, switching transition times, or modulation and control transients, volt-second unbalance can be imposed on the transformer impedance, causing dc component in the transformer magnetizing current, and hence biased flux density [1, 2]. The bias in flux density may cause transformer saturation, increase losses, or even damage the MV SiC devices.

To tackle this issue, many approaches on estimation, detection and control have been proposed in the literature, and the methods can be categorized as the methods of modeling or estimating the flux unbalance, the methods of detecting the flux unbalance and methods to compensate the flux unbalance. First, [3-5] introduced the prediction method of potential dc bias in the transformer. Studies also noticed that by using zero-voltage-switching transient [6], the dc bias of the flux can be limited. To detect the flux unbalance, hall-effect and coil-based sensors can be used to measure the magnetizing current, or the core flux density [1]. To improve the sensitivity, other indirect methods can also be adopted, e.g., ferrite "magnetic ear" [7, 8], magnetostriction based sensor [9, 10], or passive impedance-based measurement [11]. For transient flux balancing during load and control transients, the transient modulation schemes have also been well discussed [12, 13]. Nevertheless, due to the high operating voltage and the use of nanocrystalline cores, not all can be implemented in MV DAB converters. New methods with non-contact detection and robust control are desirable for MV DAB converters [14].

In this paper, a second-order harmonic-based indirect flux sensing and balancing scheme for medium-frequency MV transformer and DAB converter is introduced. A magnetic gap is used to create partial saturation and harmonics in the magnetizing current, with which the dc bias of the flux unbalance can be detected and then controlled. The analysis and implementation have been performed for the MV DAB converter. The analysis and tests reveal that the proposed method is feasible in MV DAB converters to prevent transformer saturation caused by flux unbalance.

In Section II, the method of detecting flux unbalance with a non-linear magnetic gap is introduced and validated. In Section III, the control scheme to balance the core flux is then discussed, followed by the test results verifying the scheme in Section IV. Finally, Section V summarizes and concludes the proposed scheme.

978-1-6654-8901-0/22 $31.00 © 2022 IEEE

II. HARMONIC-BASED FLUX UNBALANCE DETECTION

A. The Magnetic Gap and Non-Linear Magnetizing Current

As has been discussed in the introduction, several approaches for flux unbalance detection have been introduced. However, for MV transformers, due to the implementation complexity, the current sensing methods is still preferable. Due to the low level of magnetizing current compared to the load current, a strategy must be taken to improve the measurement accuracy. Otherwise, the flux unbalance, which is represented by the dc bias in the magnetizing current will be submerged into the dc offset and sensor noise.

In the utility transformers, discussion and modeling have been made on transformer saturation and magnetizing current harmonics. In [15], current harmonics in the magnetizing current during dc biased flux has been studied. The results of the saturation test of the 50 Hz transformer can also be extended into medium frequency MV transformers. Due to the high operating point, partial saturation may be observed due to the non-linearity of the core, resulting in nonlinear magnetizing current. If the flux DC bias is different, the saturation level within one switching cycle will be different, too. Then, the saturation may be used as the information to sense to extract the flux unbalance level, by which no DC bias measurement will be used for the flux balancing feedback.

As some of the MV transformers tend to have an air gap to prevent saturation due to flux unbalance, one of the ways to introduce non-linearity in the transformer core is to replace the air gap with a magnetic gap. The magnetic gap can be a thin ferrite sheet with a lower saturation flux density, in contrast to the nanocrystalline and amorphous cores. The amorphous material has high permeability and saturating flux density, while the ferrite can have relatively lower permeability and saturation level. If the saturation level of the two materials can be mismatched, a non-linear current curve with a partial saturation can be realized in the magnetizing current.

To verify the introduced non-linearity, a transformer with an amorphous core and ferrite gap has been made and tested as in Fig. 1. The estimated saturation current of the transformer ferrite gap is 2.5 A. The peak-to-peak value of the magnetizing current is around 5 A. In Fig. 1 (a), the flux is symmetrical. The positive and negative half-cycles are also asymmetric, which means different saturation level and the variation of the

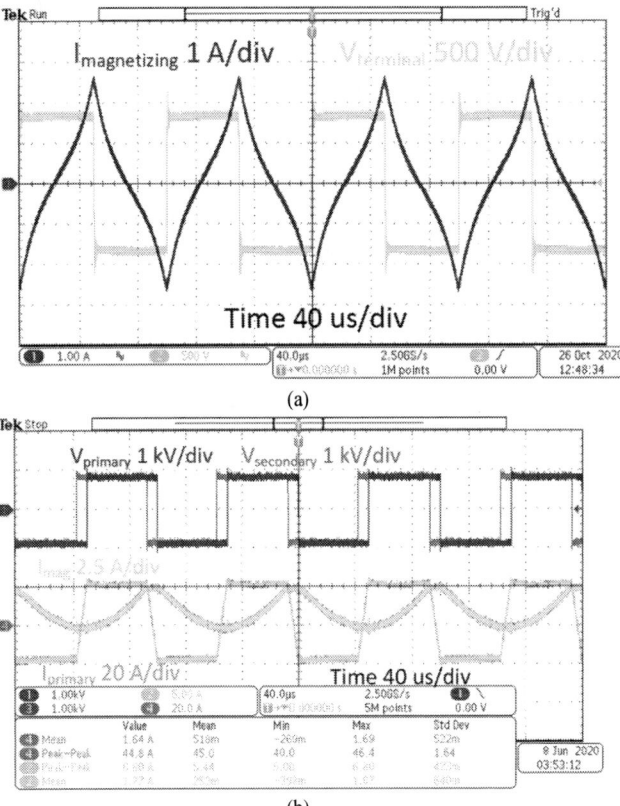

Fig. 1. The current waveform with ferrite gap: (a) balanced flux, (b) unbalanced flux.

magnetizing current, the higher the current, the smaller the inductance, and hence the faster current slope. In Fig. 1 (b), the transformer is working in full load mode, and the DC bias of 1.7 A in magnetizing current is inserted. Similar to the waveforms in Fig. 1 (a), the transformer current is with large distortion, along with the significant DC bias. The peak-to-peak value is 6.8 A. The positive and negative half-cycles are also asymmetric, which means different saturation level and the variation of the magnetizing current, the higher the current, the smaller the inductance.

B. Modeling of the Magnetic Gap in DAB Converters

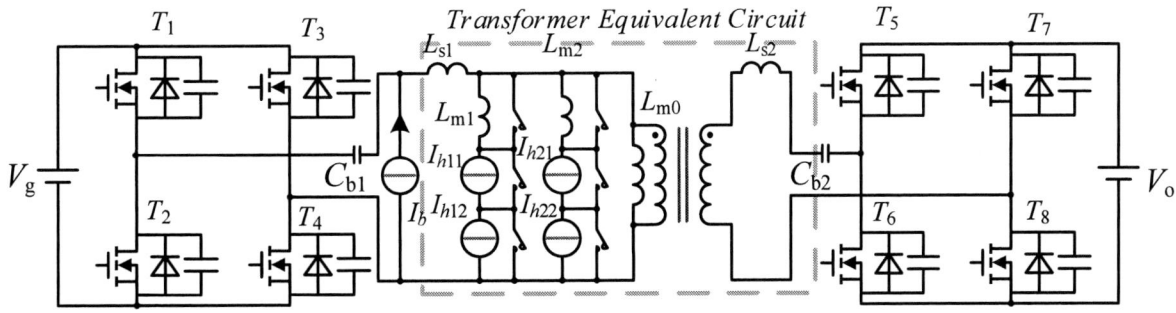

Fig. 2. The model of DAB with ferrite gaped transformer.

978-1-6654-8901-0/22 $31.00 © 2022 IEEE

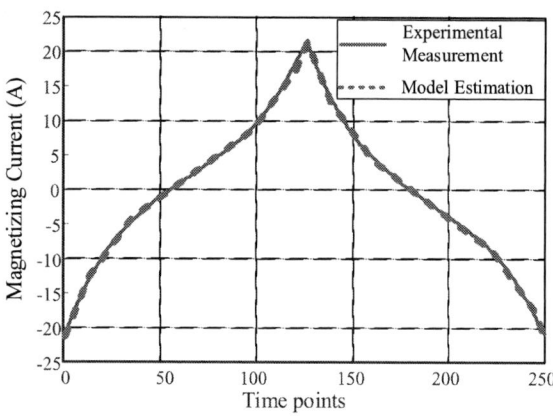

Fig. 3. The curve fitting of transformer magnetizing current with ferrite gap.

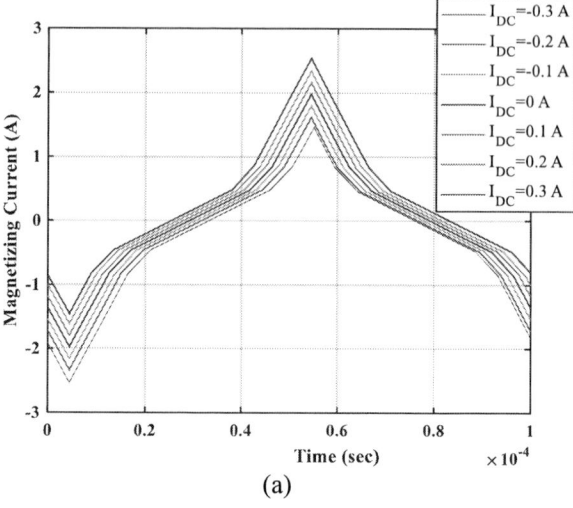

(a)

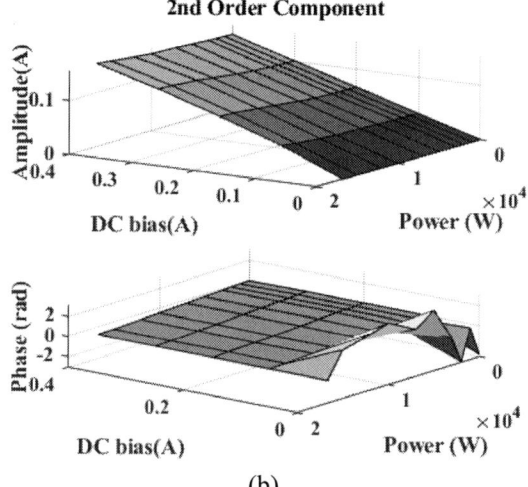

(b)

Fig. 4. The modeled transformer current: (a) at different load condition, (b) with different flux unbalance at no load.

To avoid high complexities of analysis, a piecewise linear model can be used to model the proposed transformer with a magnetic gap without considering the detailed permeability change of the combined transformer core. In order to model the partial transformer saturation, an inductive network has to be adopted to have a variant magnetic inductance. The topology of the model is shown in Fig. 2.

In the proposed DAB model, virtual DC blocking capacitors C_{b1}, C_{b2}, are inserted in series with the transformer windings, to avoid DC errors in calculation from the DAB, and paralleled DC source Ib is added to introduce the DC bias current of the magnetizing current. To simplify the analysis, only one side of the transformer has the DC bias. L_{m0} serves as the unsaturated magnetizing inductance, while Lm1 and L_{m2} are the paralleled inductance to mimic the inductance reduction during saturation. Two stages are used to have a smoother transition from the unsaturated state to the saturated state, which will be verified with the curve fitting. Current sources I_{h11}, I_{h12} I_{h21} and I_{h22} with by passing virtual switches are inserted to model the behavior of B-H hysteresis of the saturable ferrite sheets. If the magnetizing inductance of the unsaturated core is L_{nonsat}, ferrite partially saturated L_{pasat}, and ferrite gap fully saturated L_{sat}, the values of L_{m0}, L_{m1} and L_{m2} can be determined as

$$
\begin{cases}
L_{m0} = L_{unsat} \\
\dfrac{L_{m0}L_{Lm1}}{L_{m0}+L_{m1}} = L_{pasat} \\
\dfrac{L_{m0}L_{m1}L_{m2}}{L_{m0}L_{m1}+L_{m1}L_{m2}+L_{m0}L_{m2}} = L_{sat}
\end{cases}
\tag{1}
$$

As only the steady-state flux unbalance is of interest here, the state-space method has been used to analyze the non-linear behavior of the magnetic gap in the DAB converter. Due to the complexity of the adopted model, the computation of the state space has been performed in the combined platform of Simulink/MATLAB and PLECS, so that the topology can be easily implemented in Simulink and all the state matrices can be extracted automatically in PLECS. Then, the initial state can be found at the steady state by using the equation of

$$
\mathbf{x}_0 = \prod_{i=1}^{k} e^{\tilde{\mathbf{A}}_i t_i} \mathbf{x}_0
\tag{2}
$$

Where $\mathbf{x}_0$ stands for the initial vector of state variables in one switching cycle, and $\tilde{\mathbf{A}}_i$ the state matrix at the subinterval i as well as t_i the time duration. The matrix exponential computes the state after one switching interval, in which several intervals model both the saturated and non-saturated transformer states, and finally at the end of the switching cycle the state variables should coincide with the initial condition. Then, with the particle swarm algorithm, both t_i and $\mathbf{x}_0$ can be found with negligible computation error, with the controlled preset saturation level in each branch.

The method has been verified with the open-circuit test of the transformer, and meanwhile the current waveform is also used to curve fitting parameters of the model. In Fig. 3, both test waveform and curve-fitted modeling result have been

978-1-6654-8901-0/22 $31.00 © 2022 IEEE

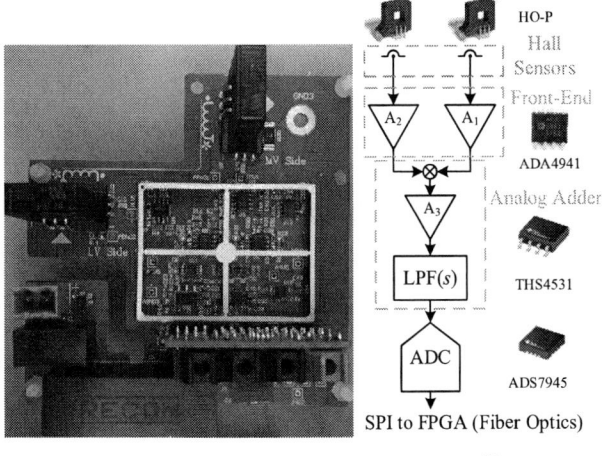

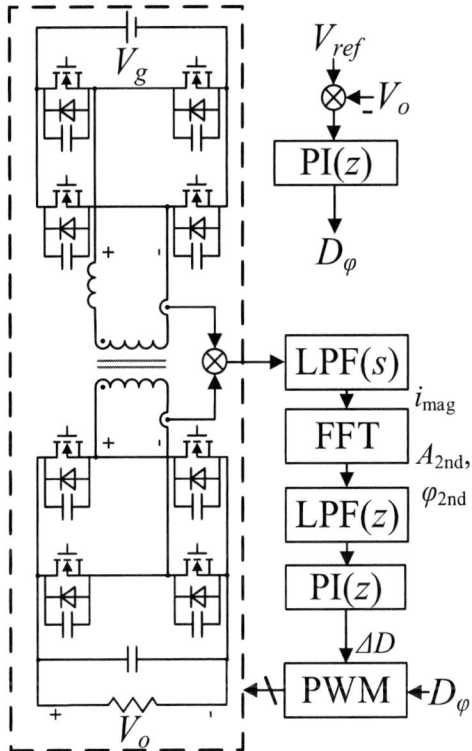

Fig. 5. (a) photo and (b) diagram of Transformer magnetizing current sensor for MV DAB converter.

shown. To achieve high precision, one core with 4 turns winding has been used to enlarge the current for more accurate measurement. With the test result, the non-saturated inductance is 22 mH, partial transitional saturated 10.46 mH, and ferrite fully saturated 5.85 mH. The threshold of saturation on each branch is 0.44 A and 0.98 A respectively, with a hysteresis of 0.12 A. The curve fitting error is 2.4%.

With the curve-fitted model, the DAB operating waveform can be derived. And the waveform with ferrite gap can be seen in Fig. 4. Then the current harmonic can be extracted from the modeled current at a different level of dc flux unbalance [14].

C. Magnetizing Current Sensing

The proposed scheme of magnetizing current sensor is shown in Fig. 5. To meet the insulation requirement, the two winding currents are sensed separately, with through-hole Hall effect sensors. Then, the two analog signals representing two currents are summed together with the same or different preset gain. The gain is tuned so that the output of the summation circuit should be the magnetizing current, which means the gains of the two scalers should be

$$\frac{A_1}{A_2}\frac{A_{Hall,1}}{A_{Hall,2}} = \frac{N_1}{N_2} \tag{3}$$

where A_1, A_2 are the gains of the scalers, and $A_{Hall,1}$, $A_{Hall,2}$ the Hall sensor gains, N_1, N_2 the winding turns for both sides.

III. FLUX BALANCING CONTROL

As the detection scheme has been proposed, analyzed, and verified, the corresponding controller can be designed to control both the power conversion of the DAB and also the DC flux of the transformer. The diagram showing both the DAB controller and flux balancing scheme is depicted in Fig. 6. From the control diagram, the DC-link voltage of the output side is sampled, and then a notch filter in the digital controller is used to filter out the second order power harmonic induced

Fig. 6. Control scheme of DAB including flux balancing control.

by the downstream single-phase DC/AC converter. The output voltage is controlled by the PI controller, and then the phase shift reference D is transferred into the pulse-width modulation (PWM) generator.

The sampling frequency of the transformer sensor is of importance, as using a high sampling rate leads to a high cost on board and communication design, while a too low rate may result in poor precision and unstable control response. Since the sampling rate of the flux balancing scheme has to be several times higher than the switching frequency, a reasonable sampling rate is needed. Since only the second order harmonic is needed, 8-point FFT within a single switching can be used. As the switching frequency is 10 kHz, the sampling rate is then 80 kHz, and Nyquist frequency 40 kHz. As high noise from the Hall sensors has been found, digital filters have to be used in the digital controller, after the FFT calculation has been conducted. The signal processing considering noise and polarity of flux unbalance is illustrated in Fig. 7 (a).

The dc offset unbalance and compensation can be injected independently on one side. The duty-cycle compensation ΔD can be added on either the low voltage side, or the high voltage side. In the converters discussed, the compensation is solely added on the duty-cycle of secondary side, or the MV side, as has been found that the MV SiC device has more deviation. The duty cycle compensation is shown in Fig. 7 (b). The corresponding change of the duty-cycle on For each step of the duty cycle change, the volt-second change can be

$$\Delta\lambda = V_{DC} \cdot \Delta t_{PWM} \tag{4}$$

978-1-6654-8901-0/22 $31.00 © 2022 IEEE   151

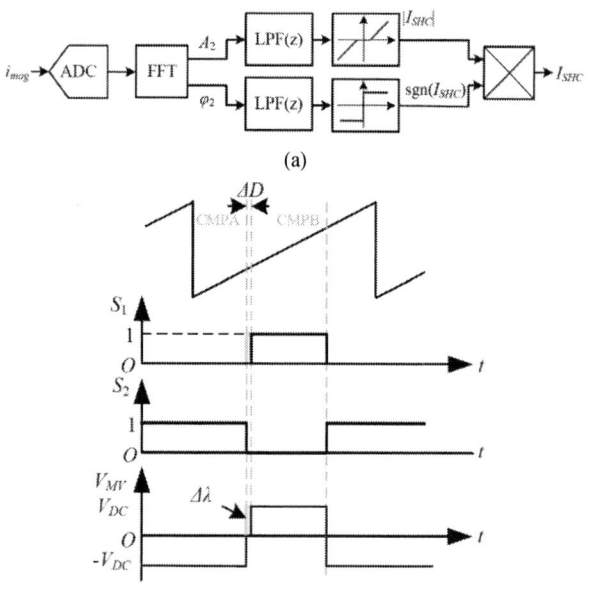

(a)

(b)

Fig. 7. The diagrams of the (a) second order harmonic signal processing, (b) duty-cycle compensation.

Where $\Delta\lambda$ the change of volt-second product, V_{DC} the DC-link voltage on the compensating side, and Δt_{PWM} the resolution of the PWM modulator.

IV. TEST VERIFICATION

With the measurement, control, and modulation schemes aforementioned, the complete scheme has been verified in an 850/6700-V DAB converter. The photograph of the MV DAB converter under test is shown in Fig. 8. For the MV DAB converter, the flux balancing detection and control are operating during the test. With some injected volt-second unbalance on MV side, a disturbance can be injected into the magnetizing current, so that the stability of the flux balancing method can be tested.

From Fig. 9, when the volt-second unbalance of 1.5% was injected at the no-load condition, an evident current o set step-up can be found on the magnetizing current, and after around 1.02 second, the envelope of the magnetizing current resumed to the value before the disturbance is added. Unipolar distortion can be found in the magnetizing currents in the lower

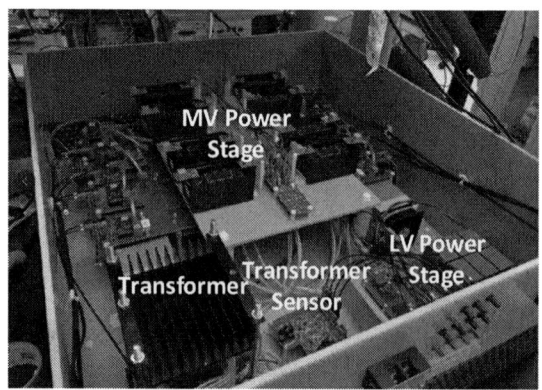

Fig. 8. The photograph of the MV DAB under test.

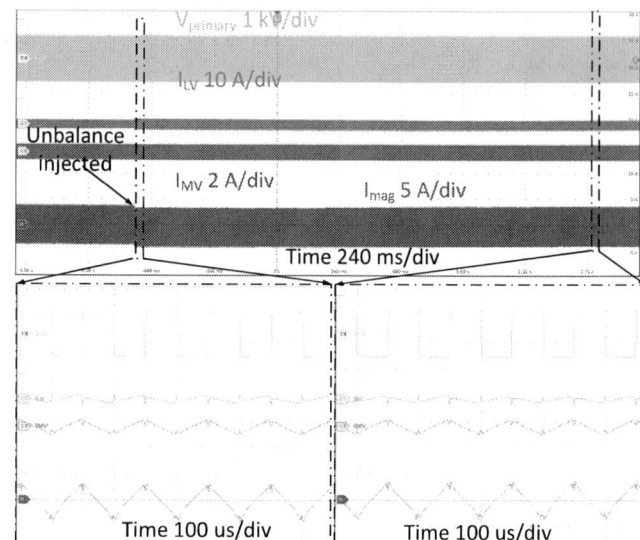

Fig. 9. Test waveform of MV DAB at no load and flux disturbance injection.

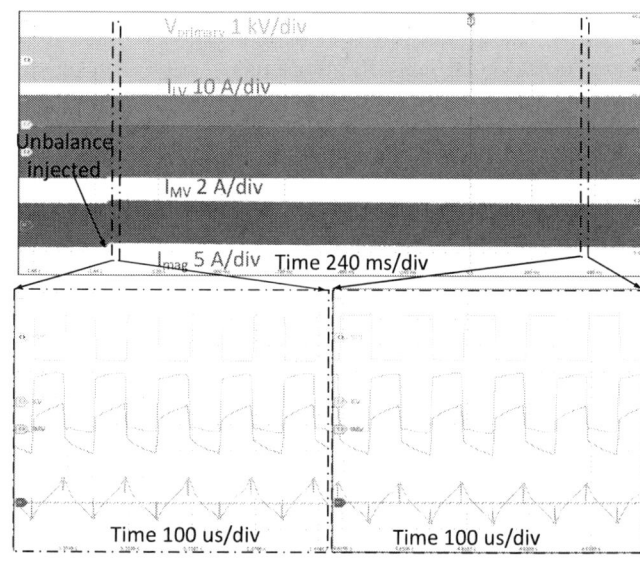

Fig. 10. Test waveform of MV DAB at full load and flux disturbance injection.

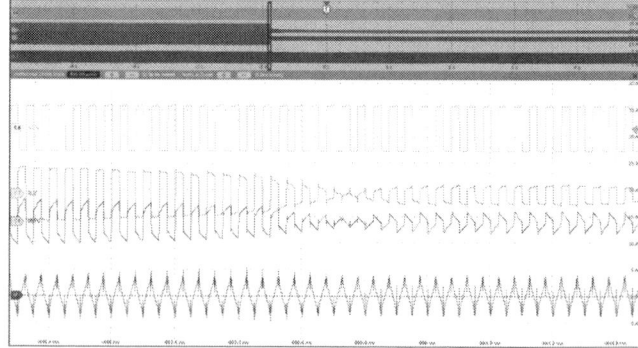

Fig. 11. Test waveform of MV DAB at load transient.

left waveform, the DC bias in the magnetizing current was approximately -1.3 A (reflected on the LV side). After the flux balanced again, in the lower right waveform, the magnetizing

978-1-6654-8901-0/22 $31.00 © 2022 IEEE   152

current is more symmetric with control compensation, and the dc component is negligible.

From Fig. 10, at full-load condition, when the volt-second unbalance of 1% was injected, a current offset step-up can be found on the magnetizing current, too, and it took around 0.9 second for the envelope of the magnetizing current to resume the state before the disturbance was injected. The distortion can be found in the magnetizing currents in the lower left zoomed-in curve, the dc bias in the magnetizing current is approximately 2 A. After the flux balanced by the control, in the lower right zoomed-in waveform, the magnetizing current can be balanced, and the DC component is also negligible.

To verify the stability during the load or control transient, a test has been conducted during the load transient from full-load to no-load, shown in Fig. 11. After the full-load from the DAB serving as the load, the current transition can be found on the transformer winding currents, as well as the magnetizing current. The load transition is set after 5 ms.

V. Conclusion

In this paper, a harmonic-based MV transformer flux balancing scheme, considering both the sensing and control, has been introduced and validated through analyses and tests. First, the non-linearity and magnetic gaps of the transformer core have been introduced. By utilizing the non-linearity of ferrite sheet saturation, the flux unbalance can be indirectly measured by detecting the second-order harmonic current. Then, the balancing control has been discussed with considerations on sampling and modulation. Tests have been performed on a MV DAB converter and verified the proposed method. The proposed flux detection and balancing strategy based on current harmonics are feasible with transformers and DAB converters with medium voltage ratings.

Acknowledgment

The authors want to thank Powerex and Southern Company for providing help on this work.

References

[1] P. Yao, X. Jiang, P. Xue, S. Ji, and F. Wang, "Flux Balancing Control of Ungapped Nanocrystalline Core-Based Transformer in Dual Active Bridge Converters," *IEEE Transactions on Power Electronics,* vol. 35, no. 11, pp. 11463-11474, 2020.

[2] P. Yao, X. Jiang, P. Xue, S. Li, S. Lu, and F. Wang, "Design Optimization of Medium Frequency Transformer for DAB Converters with DC Bias Capacity," *IEEE Journal of Emerging and Selected Topics in Power Electronics,* pp. 1-1, 2020.

[3] D. Yuan, J. Yang, Y. Liu, Y. Jia, R. Liu, and Z. Song, "A Prediction Method of DC Bias for DC-DC Dual-active-bridge Converter with MOSFETs," in *2019 14th IEEE Conference on Industrial Electronics and Applications (ICIEA),* 19-21 June 2019, pp. 1228-1232.

[4] L. Shu, W. Chen, Z. Lin, D. Ma, X. He, and W. A. Syed, "DC Bias Study for DC-DC Dual-Active-Bridge Converter," in *2018 IEEE 4th Southern Power Electronics Conference (SPEC),* 10-13 Dec. 2018, pp. 1-5.

[5] L. Shu, W. Chen, and Z. Song, "Prediction method of DC bias in DC-DC dual-active-bridge converter," *CPSS Transactions on Power Electronics and Applications,* vol. 4, no. 2, pp. 152-162, 2019.

[6] D. Costinett, D. Seltzer, D. Maksimovic, and R. Zane, "Inherent volt-second balancing of magnetic devices in zero-voltage switched power converters," in *2013 Twenty-Eighth Annual IEEE Applied Power Electronics Conference and Exposition (APEC),* 17-21 March 2013 2013, pp. 9-15.

[7] G. Ortiz, L. Fässler, J. W. Kolar, and O. Apeldoorn, "Application of the magnetic ear for flux balancing of a 160kW/20kHz DC-DC converter transformer," in *2013 Twenty-Eighth Annual IEEE Applied Power Electronics Conference and Exposition (APEC),* 17-21 March 2013, pp. 2118-2124.

[8] G. Ortiz, L. Fässler, J. W. Kolar, and O. Apeldoorn, "Flux Balancing of Isolation Transformers and Application of "The Magnetic Ear" for Closed-Loop Volt–Second Compensation," *IEEE Transactions on Power Electronics,* vol. 29, no. 8, pp. 4078-4090, 2014.

[9] L. Schrittwieser, M. Mauerer, D. Bortis, G. Ortiz, and J. W. Kolar, "Novel principle for flux sensing in the application of a DC + AC current sensor," in *2014 International Power Electronics Conference (IPEC-Hiroshima 2014 - ECCE ASIA),* 18-21 May 2014 2014, pp. 1291-1298.

[10] L. Schrittwieser, M. Mauerer, D. Bortis, G. Ortiz, and J. W. Kolar, "Novel Principle for Flux Sensing in the Application of a DC + AC Current Sensor," *IEEE Transactions on Industry Applications,* vol. 51, no. 5, pp. 4100-4110, 2015.

[11] L. Zhu, H. Bai, A. Brown, and M. McAmmond, "Design a 400 V–12 V 6 kW Bidirectional Auxiliary Power Module for Electric or Autonomous Vehicles With Fast Precharge Dynamics and Zero DC-Bias Current," *IEEE Transactions on Power Electronics,* vol. 36, no. 5, pp. 5323-5335, 2021.

[12] B. Zhao, Q. Song, W. Liu, and Y. Zhao, "Transient DC Bias and Current Impact Effects of High-Frequency-Isolated Bidirectional DC–DC Converter in Practice," *IEEE Transactions on Power Electronics,* vol. 31, no. 4, pp. 3203-3216, 2016.

[13] M. Stojadinović, E. Kalkounis, F. Jauch, and J. Biela, "Generalized PWM generator with transformer flux balancing for dual active bridge converter," in *2017 19th European Conference on Power Electronics and Applications (EPE'17 ECCE Europe),* 11-14 Sept. 2017 2017, pp. P.1-P.10.

[14] Z. Gao *et al.*, "A Transformer Flux Balancing Scheme Based on Magnetizing Current Harmonic in Dual-Active-Bridge Converters," in *2021 IEEE Applied Power Electronics Conference and Exposition (APEC),* 14-17 June 2021 2021, pp. 1894-1899.

[15] Z. Xiaoxin, L. Huiqi, L. Yang, Z. Xiaojun, and C. Zhiguang, "Analysis of the DC bias phenomenon by the harmonic balance method based on the electromagnetic coupling model," in *2013 IEEE INTERNATIONAL CONFERENCE ON MICROWAVE TECHNOLOGY & COMPUTATIONAL ELECTROMAGNETICS,* 25-28 Aug. 2013, pp. 387-390.

978-1-6654-8901-0/22 $31.00 © 2022 IEEE

Short-Circuit Ruggedness and Partial Discharge Evaluation of a 3.3 kV SiC MOSFET Power Module

Ke Wang, Yizhou Cong, Pengyu Fu, Xiao Li, Qianyi Cheng,
Boxue Hu, Jin Wang
The Ohio State University
Columbus, OH (USA)
wang.10302@osu.edu

Ashish Kumar, Kraig Olejniczak, Daniel Pelletier, Zach Cole,
Amol Deshpande, Amit Goyal
Wolfspeed Inc.
Fayetteville, AR (USA)
Kraig.Olejniczak@wolfspeed.com

Abstract—**This paper presents the initial evaluation of a 3.3-kV, SiC MOSFET power module (not fully populated) from Wolfspeed, including static, short circuit (SC), and partial discharge (PD) test results. Static characterization results are presented first. The module performance change after repetitive pulse SC tests is provided by comparing the original static characterization test results to the one performed after the SC stress tests. More electric aircrafts (MEA) require an increased electric system voltage level (e.g., 230 VAC constant or variable frequency, ± 270 VDC or 540 VDC) and make PD a very relevant and urgent issue. Thus, PD evaluation tests for the module package are included for different air pressures from Sea Level (i.e., 760 Torr) to 50,000 feet (i.e., 87 Torr) altitude. The PD inception voltage (PDIV) and extinction voltage (PDEV) obtained via experimentation are presented in this paper.**

Keywords—*3.3 kV power module, static parameters, short circuit evaluation, partial discharge, low air pressure*

I. Introduction

To avoid complex voltage stacking (i.e., multilevel) structures and achieve a simplified control strategy, medium voltage (MV) semiconductors greater than 3000 V are attractive for power conditioning and conversion applications. SiC power MOSFETs have great potential due to their faster switching speed, higher maximum operating temperature, and higher breakdown electric field strength when compared to silicon [1-2]. With the maturing of SiC semiconductor fabrication technology, recent years have seen more MV SiC power modules [3] released to the market and a dramatic decrease in their price. However, many of these SiC devices are still in an early development stage and limited device information is available. These devices need to be evaluated for performance and reliability to enable market pull in the near future.

Due to the quality of the SiC/SiO₂ interface and the high electrical field in the gate oxide, SiC power MOSFETs have suffered from reliability issues especially under extreme conditions [4]. Among all reliability issues, failures and degradation by SC stress are the most serious. During SC events, since a high drain–source voltage (V_{ds}) and saturation current are impacting the device simultaneously, an extremely high instantaneous power dissipation can be observed [5]. Investigations on the mechanisms of transient failures caused by single-pulse SC stress are widely reported. In order to extract the non-instantaneous degradation trend, several papers have already reported on repetitive SC stress and its influence upon SiC power MOSFET static parameters [6-8].

Another robustness challenge in MV and future high-voltage power modules is the existence of partial discharge (PD). PD is a precursor phenomenon of insulation breakdown. As the number of hybrid and all-electric aircraft applications accelerate, the device PD performance at sea level and normal air pressure and high altitude and low air pressure will continue to gain greater attention [9]. Recent studies on PD inside silicone gel-filled power modules are more focused on DC and AC sinusoidal voltage excitations [10-11]. However, PD testing of fully constructed power modules provides little to no specific insight to the module designer or about the power module package design. The test plan must include numerous tests, following the manufacturing build process, in order to systematically decouple the PD impact of each additional part and/or assembly introduced inside the module cavity. Also, the number of studies related to low pressure pulse PD testing is currently limited [12].

In this paper, extensive characterization and evaluation of a 3.3-kV SiC power module from Wolfspeed are presented. The device static parameters and their shift in performance after being stressed by repetitive SC tests are shown in Section II. Threshold voltage, I-V curves, drain-to-source and gate-to-source leakage currents are compared and analyzed. In Section III, PD evaluation of the device package at low pressures are presented. Tests have been conducted with several different low air pressure conditions, from sea level to 50,000 feet altitude. Observations and conclusions related to PD behavior and appropriate qualification test conditions are put forward.

II. Short-Circuit Ruggedness Evaluation of the 3.3 kV SiC Power Module

A. Short-circuit experimental setup

An Agilent B1506 curve tracer was used to measure the device static characteristics, and a customized SC test setup was built as shown in Fig. 1. The devices under test (DUT) is the lower devices of the 3.3-kV half-bridge power module. The drain, gate, and source of the upper devices were shorted together to protect them from potential damage. A 10-kΩ current-limiting resistor was implemented in the setup. This current-limiting resistor could protect the power source if the DUT can't be turned off during the short circuit event. Four identical 50-μF film capacitors were connected in parallel to form the dc-link capacitor bank. Due to the power module size and the expected high short-circuit current, there was no PCB busbar or decoupling capacitor close to the power module. A copper busbar was directly connected to the power module test setup as shown in Fig. 2. The power loop stray inductance (L_s)

978-1-6654-8901-0/22 $31.00 © 2022 IEEE

includes the equivalent series inductance (ESL) of the dc-link capacitor bank and the stray inductance introduced by the copper busbar. The loop stray inductance L_s induced an overshoot on the drain-to-source voltage during the turn-off transient. Thus, the dc-link capacitors should be properly selected with low ESL and the busbar should be designed carefully to reduce stray inductance. Finite-element simulations were performed on the copper busbar and the total inductance of the power loop was 19.8 nH based on the simulation results. As shown in Fig. 2, the copper busbar was wrapped with Kapton tape and there was a 3-D printed polyamide insulation layer between the positive and negative busbars.

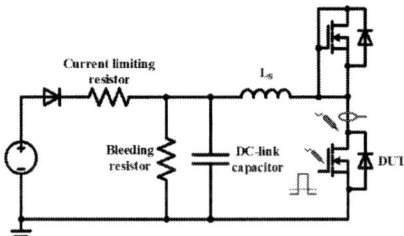

Figure 1. Circuit diagram of the repetitive SC test setup

Figure 2. Test setup for the short circuit characterization of a 3.3 kV SiC MOSFET module

B. Test procedure and results

The preliminary single-pulse SC test was conducted with a 2-kV dc bus voltage and a 1-µs pulse width. Fig. 3 shows the SC test waveforms, including the gate voltage and the drain-to-source voltage and current. The overshoot voltage was 500 V, and the device turn-off speed is 19.0 V/ns with a 50-Ω turn-off gate resistor, $R_{g,off}$. Therefore, the equivalent inductance is calculated to be approximately 26 nH, which is consistent with the sum of the calculated bussing inductance above and the module's power loop inductance of ~10 nH).

The device static parameters, including threshold voltage, V_{th}, gate leakage current, I_{gss}, drain-to-source leakage current, I_{dss}, and the I-V curve are obtained before and after the repetitive-pulse SC test. For the repetitive-pulse SC tests, the dc bus voltage was 2 kV, and the pulse width is 1 µs with a 1-second time interval between adjacent pulses to make sure the junction can be cooled down before the next short circuit event. The SC current magnitude of 2.64 kA peak could be observed during tests. The static parameters were measured after 100 pulses respectively. On gate drive side, according to the preliminary test results, the $R_{g,off}$ was kept to be 50 Ω to limit the drain-to-source overshoot voltage.

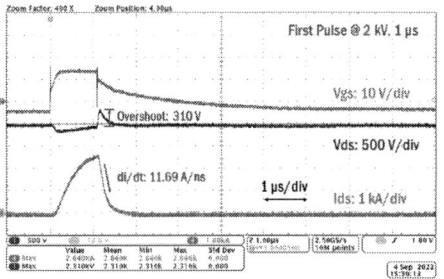

Figure 3. Experimental SC Test waveforms of the 3.3 kV SiC MOSFET module at 2 kV DC and 15 V gate bias

Typical module electrical performance was acquired before and after applying repetitive-pulse SC stresses. Fig. 4 shows threshold voltage, gate leakage current, drain-to-source leakage current, and I-V characteristic shifts. It could be observed that gate threshold voltage, drain-to-source leakage current and on-resistance, R_{Dson}, (extracted at V_{gs} = 5 V and 15 V) shifted slightly from Fig. 4 (a) and (c) – (e). The gate leakage current did not show a significant shift as shown in Fig. 4(b).

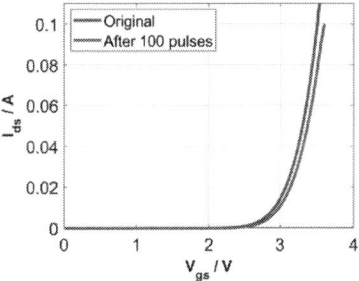

(a) Threshold voltage, V_{th}, via transconductance characteristics pre- and post-stress

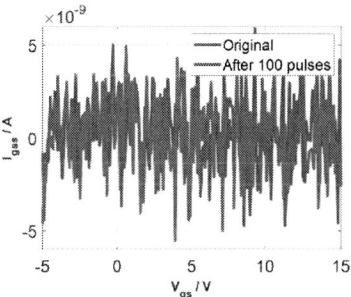

(b) I_{gss} versus V_{gs} pre- and post-stress

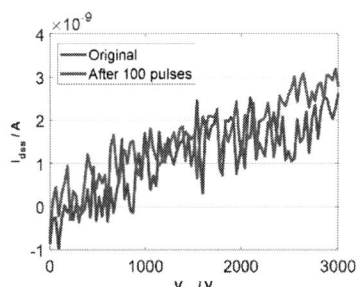

(c) I_{dss} versus V_{ds} pre- and post-stress

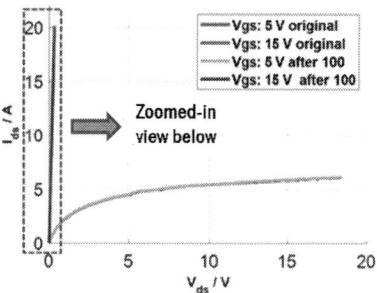

(d) I-V characteristic at 5 V and 15 V pre- and post-stress

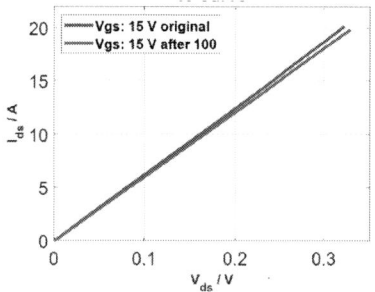

(e) Zoomed I-V characteristic at 15 V pre- and post-stress

Figure. 4. DUT static characteristics before and after repetitive-pulse SC tests

The threshold voltage was measured when the device drain and gate shorted and drain-to-source current was 100 mA at room temperature. The pre-stress V_{th} was 3.54 mV and the post-stress value was 3.60 mV after 100 repetitive-pulse SCs. For leakage currents, the gate leakage current before and after tests did not show much difference between the two conditions; meanwhile, the drain-to-source leakage current after 100 pulses was slightly higher than the original value. The I-V characteristics were also compared before and after the SC stress. In Fig. 4(d), conditions with both 5-V and 15-V gate voltages are presented, and the zoomed-in figure with 15-V gate voltage (recommended gate voltage for turning on) is presented as Fig. 4(e). An increase in R_{DSon} could be observed. The original R_{DSon} with 15-V gate voltage was 16.0 mΩ and the value was 16.5 mΩ after 100 repetitive pulses. To conclude, the repetitive-pulse SC stresses bring positive shifts of V_{th}, R_{DSon} and I_{dss} of the 3.3 kV SiC MOSFET.

Table I – Shift in device parameters after repetitive-pulse SC test

Parameters	Pre-Stress Measurement	Post-Stress Measurement	Shift Percentage
V_{TH}	3.54 V	3.60 V	1.7 %
I_{gss}, average value	0.26 nA	0.28 nA	N/A
I_{dss}, fitting value at 3-kV V_{ds}	2.46 nA	2.98 nA	N/A
R_{DSon}	14.7 mΩ	14.9 mΩ	1.4 %

III. AC PARTIAL DISCHARGE EVALUATION OF THE 3.3 kV SiC POWER MODULE

A. Experimental setup

The test circuit for a 60-Hz AC PD Test is shown in Fig. 5. The test circuit is similar to the circuit that is identified in IEC 60270 [13]. The PD pulse current is monitored via a high-frequency current transformer (HFCT), which has a bandwidth of 85 MHz. The voltage measurement is acquired with a Fluke Electronics 80K-40 voltage probe with a 10-foot cable. The current and voltage sensors are connected to the Techimp PDBaseII™ acquisition unit. A custom designed chamber has been used to achieve the desired low-pressure conditions. The air flow system can connect to the external environment or to a pressure-regulated source of dry air. In all the tests reported herein, bottled dry air is used to control the air humidity. The test sample is a 3.3 kV-rated LM3 power module package from Wolfspeed. No DBC or dies are placed in the test sample. The test sample connection circuit diagram is shown in Fig 6. In this test, the DC- terminal and AC terminals are shorted to evaluate the partial discharge inception voltage (PDIV) and partial discharge extinction voltage (PDEV) across +DC terminal and -DC terminal.

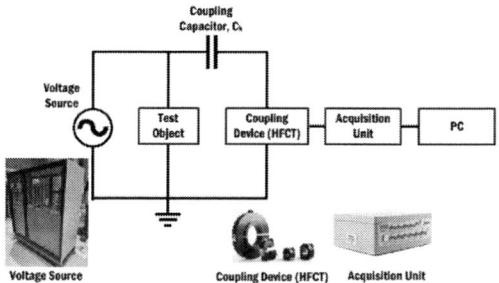

Figure 5. 60 Hz AC PD test circuit

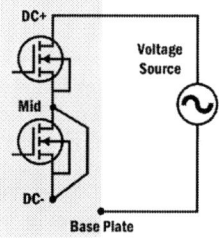

Figure 6. Test sample connection circuit diagram for partial discharge measurement of the 3.3 kV SiC MOSFET module; the PD is measured between DC+ and the baseplate

B. Test procedures and configurations at different air pressure

The 60 Hz AC excitations were applied to the test samples under various low air pressure conditions. The pressure levels involved are: 87 Torr (50,000 ft.), 141 Torr (40,000 ft.), 226 Torr (30,000 ft.), 350 Torr (20,000 ft.), 523 Torr (10,000 ft.), 633 Torr (5,000 ft.) and 760 Torr (Sea Level). Please note that no temperature corrections were made resulting in altitude equivalents being approximate.

978-1-6654-8901-0/22 $31.00 © 2022 IEEE

First, the dry air from a tank was allowed into the chamber, and then the air inside the chamber was pumped out to reach the target air pressure. The air pressure was monitored by both a meter on the inlet valve and an electronic environment meter placed inside the chamber. Every time a new pressure level was reached, a 10-minute waiting time was established before any tests were conducted in order to achieve a stable pressure. For the 60 Hz PD tests, in each test configuration, five sets of PDIV/PDEV data were taken for each pressure level with three-minute time intervals between every two consecutive data sets. For each PDIV/PDEV data set, the test procedures were: (a) raising the voltage to approximately 80 – 90 % of the expected PDIV; (b) raising the voltage by steps with the size of 2 - 4 % of the expected PDIV until PD events could be observed; (c) lower the voltage in steps of approximately 2 - 4 % until PDEV is reached; and, (d) monitoring for one minute after the PDEV is reached to guarantee no additional PD events occurred.

The measuring standard used in this paper follows IEC 60270. Here, PDIV is defined as the lowest voltage where "repetitive" PD pulses are observed, and PDEV is defined as the highest voltage where "repetitive" PD pulses disappear after the applied voltage exceeding the PDIV is reduced back down. The definition of "repetitive" is "subjective." In this paper, "repetitive" is described as when "PD pulses are observed for at least 30 consecutive seconds in a one-minute observation window."

C. Test results and analysis

For all air pressure levels, the environment temperature was 22 ℃ to 23 ℃ and the humidity was controlled between 38.9% to 41.0% to present a normal environment condition. The test setup, including electric parts and the low-pressure chamber, was tested and verified to be PD free up to 5 kV rms with the 60 Hz AC excitations. PDIV and PDEV were measured at 87 Torr, 141 Torr, 226 Torr, 350 Torr, 523 Torr, 633 Torr and 760 Torr. Each test configuration included five test points.

PDIV and PDEV test data for different air pressures are shown in Fig. 7. As seen in Fig. 7, both PDIV and PDEV values decrease with decreasing pressure as expected. The maximum error in each group is less than 11%, which shows good repeatability. PDEV values are normally 6 – 21 % lower than the PDIV values.

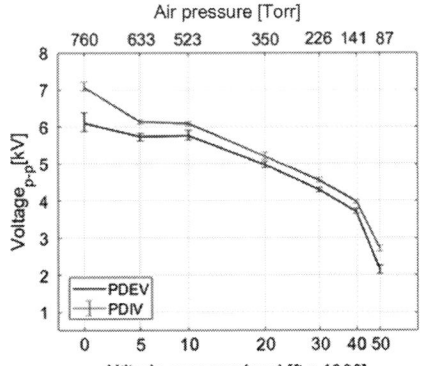

Figure 7. Partial discharge test results of 3.3 kV SiC MOSFET module with 60 Hz sinusoidal excitation source at varying pressure levels (altitude level)

Fig. 8 (a) shows a typical current waveform detected by the HFCT during the PD events, and the PD pattern (b) at 633 Torr. The current triggering level was set as 0.21 mA. Referring to Fig. 8 (b), PD events mostly happened between 0 to 90 degrees and 180 to 270 degrees in phase. The number of PD events occurring in the positive voltage half cycle is less than the negative voltage half cycle. This is due to more electrons being readily available from the electrode under a negative-applied voltage resulting in more electron avalanches. Based on the PD pattern and the shape of the test sample, it's highly possible that these PD events are surface discharges [14]. When the electric field at the surface of the electrode exceeds the breakdown strength of gas, ionizations of air near the surface of electrode occur. This is due to the tangential field on the insulation surface being high enough to cause PDs along the surface of the material. The surface discharges keep occurring until the transient activity stops.

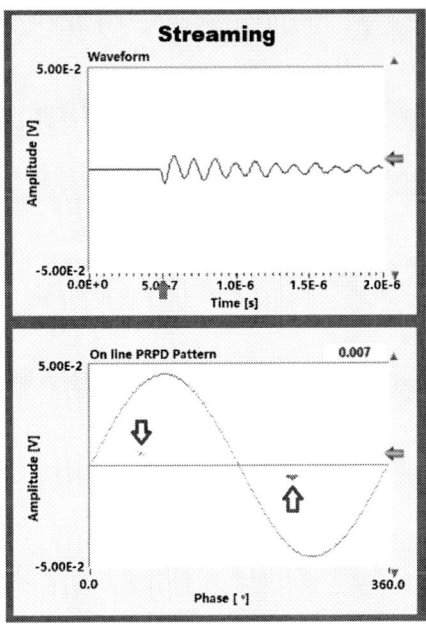

Figure 8. Typical PD events under 632 Torr (equivalent to 5,000 feet) (a) current waveforms (b) pattern

IV. Conclusions

In this paper, a pre- and post-stress evaluation was conducted to determine shift in the threshold voltage (V_{th}), gate leakage current (I_{gss}), drain-to-source leakage current (I_{dss}), and I-V characteristic after a 100 repetitive-pulse SC Test. In this test, the dc bus voltage was 2 kV, and the pulse width is 1 μs with a 1-second time interval between two pulses to make sure the junction can be sufficiently cooled before the next short circuit event. A positive shift of the device static parameters can be confirmed after the SC stress cycles. The device static parameter value shift analysis is planned for 1000 and 10,000 repetitive pulses and will be reported in the future.

Preliminary PD tests of the module package were conducted with 60 Hz AC under different low air pressures (representing the condition from sea level to 50,000 feet altitude). Since the power module is an industry-standard 3.3-kV footprint, not specifically designed for aerospace applications, PD events could be observed during the PD Test at low air pressure. PDIV and PDEV were recorded and analyzed. The PD evaluation reported in this paper is part of the first step for a more comprehensive investigation providing insights for the most appropriate PD test procedures for aviation applications stresses.

REFERENCES

[1] S. Madhusoodhanan *et al.*, "Comparison study of 12kV n-type SiC IGBT with 10kV SiC MOSFET and 6.5kV Si IGBT based on 3L-NPC VSC applications," 2012 IEEE Energy Conversion Congress and Exposition (ECCE), Raleigh, NC, 2012, pp. 310-317, doi: 10.1109/ECCE.2012.6342807.

[2] L. Schrittwieser, J. W. Kolar and T. B. Soeiro, "99% Efficient three-phase buck-type SiC MOSFET PFC rectifier minimizing life cycle cost in DC data centers," in CPSS Transactions on Power Electronics and Applications, vol. 2, no. 1, pp. 47-58, 2017, doi: 10.24295/CPSSTPEA.2017.00006.

[3] A. Marzoughi, J. Wang, R. Burgos and D. Boroyevich, "Characterization and Evaluation of the State-of-the-Art 3.3-kV 400-A SiC MOSFETs," in IEEE Transactions on Industrial Electronics, vol. 64, no. 10, pp. 8247-8257, Oct. 2017.

[4] J. Wei, S. Liu, L, Yang, J. Fang, T. Li, S. Li and W. Sun, "Comprehensive Analysis of Electrical Parameters Degradations for SiC Power MOSFETs Under Repetitive Short-Circuit Stress," in IEEE Transactions on Electron Devices, vol. 65, no. 12, pp. 5440-5447, Dec. 2018, doi: 10.1109/TED.2018.2873672.

[5] C. Chen, D. Labrousse, S. Lefebvre, M. Petit, C. Buttay, and H. Morel, "Study of short-circuit robustness of SiC MOSFETs, analysis of the failure modes and comparison with BJTs," Microelectron. Rel., vol. 55, pp. 1708–1713, Aug. 2015, doi: 10.1016/j.microrel.2015.06.097.

[6] A. Castellazzi, A. Fayyaz, L. Yang, M. Riccio, and A. Irace, "Short circuit robustness of SiC power MOSFETs: Experimental analysis," in

Proc. ISPSD, Waikoloa, Hawaii, Jun. 2014, pp. 71–74, doi:10.1109/ISPSD.2014.6855978.

[7] S. Mbarek, F. Fouquet, P. Dherbecourt, M. Masmoudi, and O. Latry, "Gate oxide degradation of SiC MOSFET under short-circuit aging tests," Microelectron. Rel., vol. 64, pp. 415–418, Sep. 2016, doi: 10.1016/j.microrel.2016.07.132.

[8] X. Zhou, H. Su, Y. Wang, R. Yue, G. Dai, and J. Li, "Investigations on the degradation of 1.2-kV 4H-SiC MOSFETs under repetitive short-circuit tests," IEEE Trans. Electron Devices, vol. 63, no. 11, pp. 4346–4351, Nov. 2016, doi: 10.1109/TED.2016. 2606882.

[9] Z. Wei, D. Grosiean, D. Schweickart, P. Fu and J. Wang, "Low Pressure Partial Discharge Tests with Ultra High dv/dt PWM Voltages for Aircraft Motor Windings," 2021 IEEE Electrical Insulation Conference (EIC), 2021, pp. 256-259, doi: 10.1109/EIC49891.2021.9612319.

[10] T. M. Do, J-L. Augé, and O. Lesaint. "A study of parameters influencing streamer inception in silicone gel," IEEE Transactions on Dielectrics and Electrical Insulation, vol. 16, 2009, pp. 893-899.

[11] M. Sato, A. Kumada, K. Hidaka, "Surface discharges in silicone gel on AlN substrate," IEEE Transactions on Dielectrics and Electrical Insulation, vol. 23, 2016, pp. 494-500.

[12] Z. Wei, D. Kasten, B. Hu, Y. Abdullah, J. Wang, D. Grosjean and D. Schweickart, "Study of Partial Discharge Behavior at Flight-Altitude Pressures under 60 Hz and Impulse Voltages for Samples Related to Aircraft Motors," 2018 IEEE International Power Modulator and High Voltage Conference (IPMHVC), 2018, pp. 180-185, doi: 10.1109/IPMHVC.2018.8936742.

[13] High Voltage Test Techniques – Partial Discahrge Measurements, IEC Std. 60270, 2000.

[14] H. Illias, Teo Soon Yuan, A. H. A. Bakar, H. Mokhlis, G. Chen and P. L. Lewin, "Partial discharge patterns in high voltage insulation," 2012 IEEE International Conference on Power and Energy (PECon), 2012, pp. 750-755, doi: 10.1109/PECon.2012.6450316.

Advantages of SiC-Based Devices on the Design of Dual-Active Bridge DC/DC Converter for DC faults

Shrivatsal Sharma*, Yos Prabowo*, Subhransu Satpathy*, Subhashish Bhattacharya*
*NC State University, Raleigh, NC, USA.
Email: *ssharm39@ncsu.edu, yprabow@ncsu.edu, ssatpat2@ncsu.edu, sbhatta4@ncsu.edu*

Abstract—DC short circuit fault ride-through is a critical feature for the reliability and performance of DC microgrids. This paper presents the advantage that SiC-based devices offer for designing a Dual-Active Bridge (DAB) DC-DC converter while considering DC short circuit events. It is known that SiC-MOSFET devices have a higher transient current carrying capability than Si-IGBT devices due to their superior thermal conductivity. This characteristic of SiC-MOSFET devices is utilized to improve the design of DAB for DC short circuit fault ride-through applications. An analytical model is developed to understand the performance of a DAB during DC short circuit faults. Switching and thermal simulations are used to validate the analytical model and compare DAB designs based on SiC-MOSFET and Si-IGBT. The advantages of SiC-MOSFET enabled DAB compared to Si-IGBT enabled DAB are also quantified for a particular application of DAB in a DC microgrid. It is shown that for fault ride-through applications, DAB enabled with SiC-MOSFET can be designed for lower phase shifts compared to Si-IGBT enabled DAB, which inherently reduces the inductor size and circulating current in a DAB.

Index terms - DC fault, DC microgrid, Dual-Active Bridge, SiC, Si

I. INTRODUCTION

In recent years, DC microgrid (MG) systems have gained much traction due to their advantages over traditional AC MG systems [1]–[3]. In a DC MG system, distributed energy resources (DERs) are typically integrated with the grid using DC-DC converters. A typical structure of a DC MG with DERs and power converters is shown in Fig. 1. Protection and faults detection in DC MGs are quite challenging and an active area of research [4]. A robust design of DC-DC converters is needed to ensure a safe fault-through and enhance the reliability and performance of DC MG systems. Dual-active bridge (DAB) DC-DC converter is an attractive solution to integrate DERs with a DC MG. A DAB is known for its high power density, buck/boost operation, bidirectional power flow, zero voltage switching capability, galvanic isolation, and inherent fault isolation capability [5]. The analysis of the fault response characteristics of a DAB has been shown in [6], [7]. However, the advantages that SiC-MOSFET offers in the design of DAB for fault ride-through applications have not been reported in the literature.

For fault ride-through applications, the magnitude of the transient current that devices can carry is critical [7]. The transient current is limited by the junction temperature of the devices, which is a function of their thermal conductivity [7]. For a given cooling architecture of a system, high

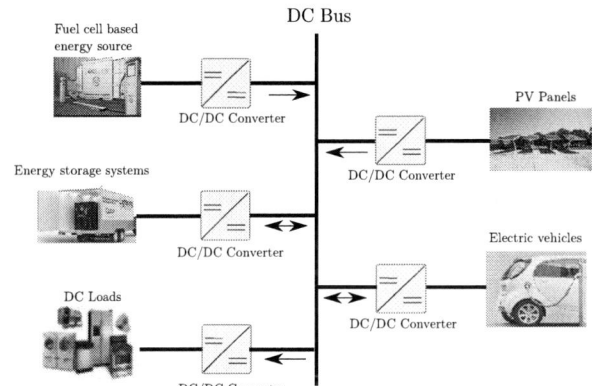

Fig. 1: Typical structure of a DC MG

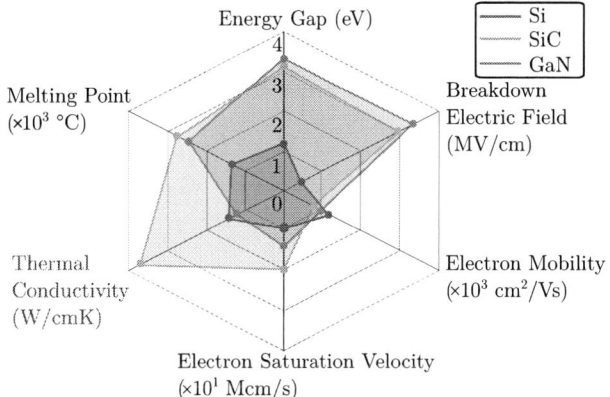

Fig. 2: Key material properties of Si, 4H-SiC and GaN at 25°C and atmospheric pressure [9]–[11]

thermal conductivity allows a better heat transfer from the semiconductor device junction to its case. It thus enables high transient current carrying capability [8]. Fig. 2 shows the comparison of the key material properties of Si, SiC and GaN [9]–[11]. It can be seen that SiC-based devices have much better thermal conductivity than Si and GaN-based devices. Thus, SiC-based devices can carry a higher transient current than Si and GaN-based devices. In this work, the potential advantage of higher thermal conductivity of SiC-MOSFET devices than Si-IGBT devices is explored and quantified for the design of a DAB for fault ride-through applications. GaN-based devices are not considered for comparison due to their lower thermal conductivity than Si-based ones.

978-1-6654-8901-0/22 $31.00 © 2022 IEEE

This paper is organized as follows. In section II, an analytical model is derived to evaluate the performance of a DAB under DC short circuit fault conditions. The analytical model is validated with switching simulations and is used to show the potential advantage in the design of DAB with SiC-MOSFET compared to the DAB enabled by the conventional Si-IGBT for fault ride-through applications. In section III, thermal simulations incorporating the electrical and thermal characteristics of devices are performed to further validate the analytical model. Using the analytical model and simulations, section III also highlights and quantifies the advantages of SiC-MOSFET compared to Si-IGBT for a design of DAB that integrates DERs with a DC MG. Finally, section IV summarizes the key ideas of the paper.

II. ANALYSIS OF DAB CONVERTER DURING DC FAULT

In this section, the performance of a DAB is analyzed during a DC short circuit fault, focusing on fault-ride through applications. Thus, the protection algorithms such as desaturation detection during a DC short circuit fault are not considered. The converter is also assumed to be operated in the steady-state condition before the DC short circuit fault.

Fig. 3 shows the schematic of a DAB with a short-circuit fault on the secondary side. In this section, first, the operation of a DAB during normal operation with simple phase shift (SPS) modulation is discussed, and then the DC fault condition is analyzed. SPS modulation is considered as it is commonly used due to its ease of implementation. In Fig. 3, V_s, V_o, v_p, v_s, i_p, n, and L_s are the input voltage, output voltage, primary side ac voltage, secondary side ac voltage, inductor current, turns ratio of transformer, and power transfer inductance, respectively. In a DAB, the power transfer with SPS modulation is controlled using the phase-shift, ϕ, between the primary and secondary ac voltages. For the system shown in Fig. 3, the design of L_s at rated power, P, is given by (1).

$$L_s = \frac{M V_s^2 D(1-D) T_s}{2P} \quad (1)$$

where T_s is the switching period, and M and D are given by (2).

$$M = \frac{V_o}{n V_s}; \quad D = \frac{\phi}{\pi} \quad (2)$$

Fig. 4 shows a typical inductor waveform of a DAB controlled using SPS modulation with I_{p1} and I_{p2} are the peak inductor currents at the switching instants. I_{p1} and I_{p2} are given by (3) and (4), respectively [12].

$$I_{p1} = \frac{((2D-1) + M)V_s T_s}{4L_s} \quad (3)$$

$$I_{p2} = \frac{(1 + M(2D-1))V_s T_s}{4L_s} \quad (4)$$

The inductor rms current can be obtained using I_{p1} and I_{p2} and is given by (5).

$$i_{p-rms} = \sqrt{\frac{I_{p1}^2 + I_{p2}^2 + I_{p1} I_{p2}(1-2D)}{3}} \quad (5)$$

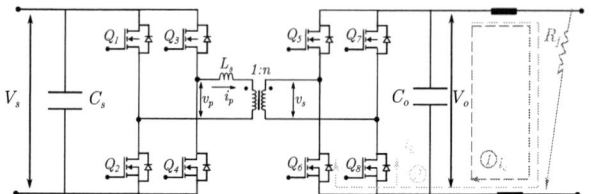

Fig. 3: Schematic of a DAB with short-circuit fault on the output side.

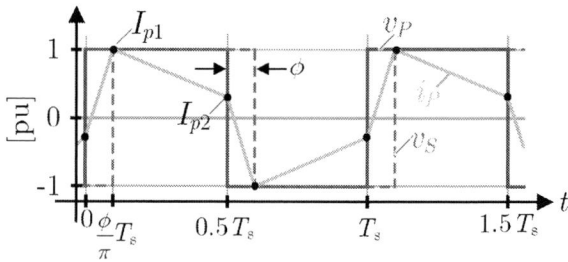

Fig. 4: A typical inductor waveform of a DAB with SPS modulation.

A DC short circuit fault can be analyzed by either shorting the primary side or the secondary side of a DAB. Fig. 3 shows a DAB with a DC fault on the secondary side. Similar analysis can also be done for a DC short circuit fault at the primary side. A transient short-circuit event consists of two stages, a capacitor discharge stage and a freewheeling stage. In the freewheeling stage, the natural transient response of the uncontrolled bridge defines the performance until the converter reaches its steady-state value, and the fault current in this stage passes through two legs of the bridge freewheeling diodes [6]. The inductor currents at switching events during a DC fault at the secondary side, I_{p1f} and I_{p2f}, can be obtained by substituting V_o with 0 V and ϕ as $\pi/2$ rad in (2). I_{p1f} and I_{p2f} are given by (6).

$$I_{p1f} = 0; \quad I_{p2f} = \frac{V_s T_s}{4L_s} \quad (6)$$

It can also be seen from (6) that during a DC short circuit fault, the inductor current will be triangular as only input voltage gets applied across the inductor. The inductor rms current, i_{p-rmsf}, during a DC fault can also be obtained by substituting I_{p1}, I_{p2}, and D in (5) by I_{p1f}, I_{p2f}, and 0.5, respectively. To compare the inductor current before and after the fault, as a worst-case scenario, it is assumed the DAB operated at the rated load condition before the fault. The ratio between the steady-state inductor rms current after a DC short circuit fault, and the steady-state inductor rms current at the rated load condition is given in (7).

$$\frac{i_{p-rmsf}}{i_{p-rms}} = \frac{V_s n}{\sqrt{\begin{array}{c} -8 D^3 V_o V_s n + 12 D^2 V_o V_s n + \\ V_o^2 - 2 V_o V_s n + V_s^2 n^2 \end{array}}} \quad (7)$$

The ratio of inductor rms currents given in (7) can be simplified by substituting n with $\frac{V_o}{V_s}$. The simplified ratio is given

978-1-6654-8901-0/22 $31.00 © 2022 IEEE

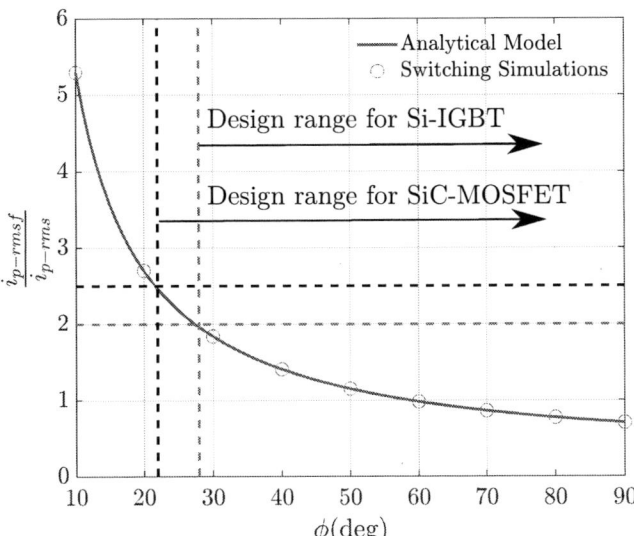

Fig. 5: Ratio of rms inductor currents.

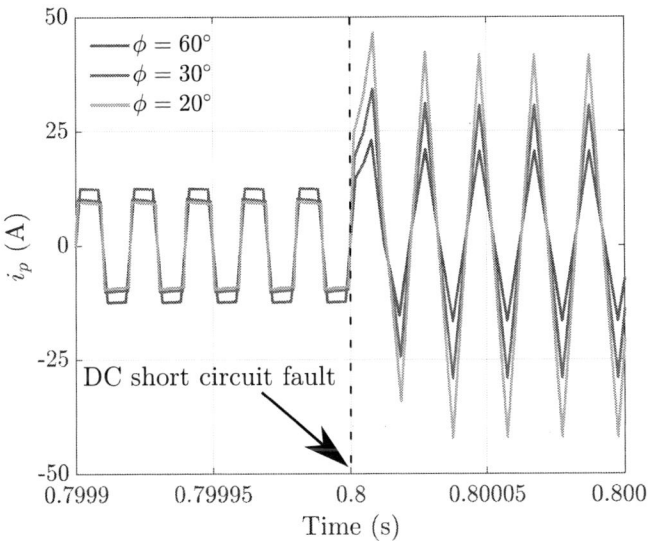

Fig. 6: Inductor current during DC short circuit fault for different ϕ.

in (8).

$$\frac{i_{p-rmsf}}{i_{p-rms}} = \frac{1}{\sqrt{-8\,D^3 + 12\,D^2}} \qquad (8)$$

After substituting D from (2) in (8), the ratio of rms inductor currents as a function of ϕ is shown in Fig. 5. The ratios of rms inductor currents for different phase shifts are also obtained using switching simulations for a system with parameters given in Table I. For each ϕ, the value of L_s is obtained using (1) at the rated load condition. The inductor waveforms for a few phase shifts are shown in Fig. 6. In these results, a DC short circuit fault occurs at 0.8 s, before which the DAB is operated at the rated load condition. The ratios of inductor rms currents for different phase shifts from simulations are also overlaid in Fig. 5, and it can be seen that the analytical model is in good agreement with the switching simulations.

Table I: System Parameters

Parameters	Value
Nominal ESS voltage (v_{nom})	400 V
DC MG voltage (v^*)	400 V
Power rating (P)	3.3 kW
Considered phase shifts (ϕ)	20 °, 30 °, 45 °, 60 °, 90 °
L_s for different ϕ	48 μH, 67 μH, 91 μH, 108 μH, 121 μH
Transformer turns ratio (1 : n)	1:1
Switching frequency (f_{sw})	50 kHz
Heat sink thermal resistance (R_{th-H})	0.5 K/W
Junction to case thermal resistance of FF450R12ME4B11 ($R_{thjc-Si}$)	0.066 K/W
Junction to case thermal resistance of CAB450M12XM3 ($R_{thjc-SiC}$)	0.11 K/W

Fig. 5 shows that the ratio of the inductor rms currents is high for the DAB designed with low phase shifts at the rated load condition. The low phase shift reduces the size of the power transfer inductor and also reduces the reactive current in the DAB. For Si-IGBT enabled DAB, this ratio is typically limited to 2, as Si-IGBTs have a transient current carrying capability of twice the rated current [6], [7]. However, this limitation is due to the junction temperature of the power switches. The junction temperature depends on the thermal conductivity of the devices. SiC-MOSFETs have better thermal conductivity than Si-IGBTs, thus having a higher transient current carrying capacity. This material property of SiC enables the design of DAB with low phase shifts. Considering the data given in data sheets [13]–[15], the ratio of the inductor fault current to steady-state current is considered 2.5. The increased ratio reduces the minimum phase shift to around 20 ° for SiC-MOSFET based DAB from around 30 ° for Si-IGBT based DAB. The DAB designs enabled with Si-IGBT and SiC-MOSFET are compared in the next section using thermal simulations.

III. COMPARISON OF DAB DESIGNS ENABLED WITH SI-IGBT AND SIC-MOSFET

In this section, a DAB integrating DER to a DC MG is considered to compare the designs with Si-IGBT and SiC-MOSFET for a DC short circuit fault ride-through application. The system parameters are shown in Table I. For SiC-MOSFET, a CREE module (CAB450M12XM3) is selected [15], while for Si-IGBT, an Infineon module (FF450R12ME4B11) is considered [16]. Both the devices are of 1200 V voltage rating and 450 A current rating. The junction to case thermal resistance of both devices is also given in Table I.

Switching simulations using PLECS [17] while considering the thermal and electrical characteristics of both the devices are done for all the five combinations of ϕ and L_s given in Table I. Following the analytical model results and pulsed current carrying capability of SiC-MOSFET devices, the minimum ϕ of 20 ° is chosen. Thermal models of both devices

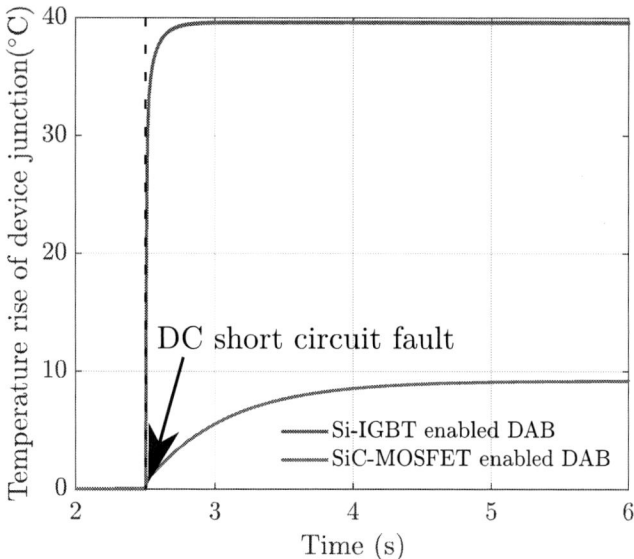

Fig. 7: Comparison of junction temperature due to a DC short circuit fault for a ϕ of 20 ° at the rated load condition (SiC-MOSFET device: CAB450M12XM3 [15], Si-IGBT device: FF450R12ME4EB11 [16]).

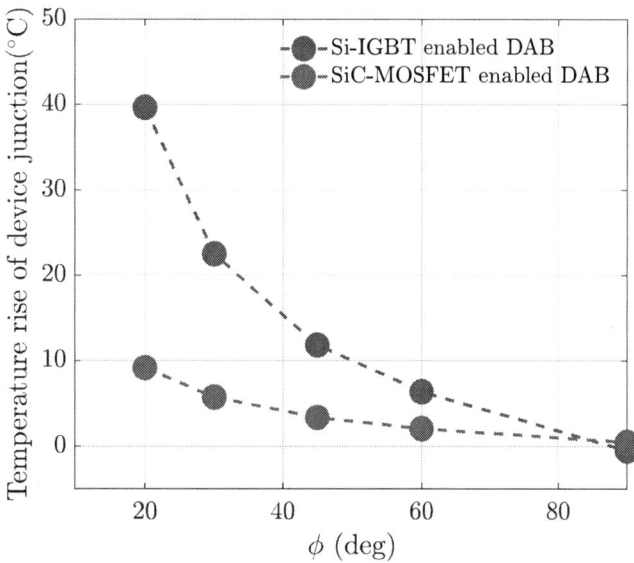

Fig. 8: Comparison of junction temperature rise due to DC short circuit fault (SiC-MOSFET device: CAB450M12XM3 [15], Si-IGBT device: FF450R12ME4EB11 [16]).

are obtained from their websites [18]. For a direct comparison of the performance of both the devices, the same heat sink whose thermal resistance value is given in Table I is used in the simulations for both devices.

Fig. 7 shows the time-domain results of the temperature rise of the junctions of both devices due to a DC short circuit fault. In this result, ϕ is considered 20 °, and the DC short circuit occurs at 2.5 s. Before 2.5 s, the DAB is operated at the rated load condition. The temperature rise of Si-IGBT is significantly more than SiC-MOSFET due to its low thermal conductivity. The temperature rise is also obtained for different

combinations of ϕ and L_s, and the results are shown in Fig. 8. It can be seen that the temperature rise of the device junction increases with the decrease in the phase shifts at the rated load condition. This result is in good correlation with the results in Fig. 5, where it is shown that the ratio between the steady-state inductor rms currents after the DC short circuit fault and before the fault increases with the decrease in the phase shifts. The high ratio of rms currents leads to a significant increase in the device losses in the DC short circuit fault condition compared to the device losses in the rated load condition. This inherently leads to a high temperature rise of the device junction after the DC short circuit fault at low phase shifts.

Fig. 8 also shows that due to the low thermal conductivity of Si-IGBT devices compared to SiC-MOSFET devices, the junction temperature rise for Si-IGBT is much more than that of SiC-MOSFET. Thus, for DC short circuit fault ride-through applications, Si-IGBT devices will reach the allowed junction temperature at a higher phase shift than SiC-MOSFET devices.

From the analytical model results shown in section II and the thermal simulation results in this section, it can be concluded that SiC-MOSFET devices enable the design for fault ride-through applications of DAB at low phase shifts compared to the Si-IGBT devices. A DAB design with a low-phase shift reduces the size of L_s and also reduces the reactive current within the DAB. For the system parameters shown in Table I, a DAB enabled by a SiC-MOSFET can be designed for a ϕ of 20° with L_s of 48 μH. In comparison, a DAB enabled by a Si-IGBT can be designed for a ϕ of 30° with L_s of 67 μH. Thus, there is a reduction of around 25% in the value of L_s.

IV. Conclusions

In this paper, the advantage of SiC-MOSFETs is highlighted for the design of DAB while considering the DC short-circuit events. An analytical model is developed and validated using switching simulations to understand the performance of a DAB during short circuit faults. The analytical model and the thermal simulations incorporating the electrical and thermal characteristics of devices are used to compare the performance of DAB designed with SiC-MOSFET and Si-IGBT. It is shown that for fault ride-through applications, DAB enabled with SiC-MOSFET can be designed for lower phase shifts compared to Si-IGBT enabled DAB, which inherently reduces the inductor size and reactive current in a DAB.

References

[1] M. Saeedifard, M. Graovac, R. Dias, and R. Iravani, "Dc power systems: Challenges and opportunities," in *IEEE PES General Meeting*, 2010, pp. 1–7.

[2] J. Justo, F. Mwasilu, J. Lee, and J. Jung, "Ac-microgrids versus dc-microgrids with distributed energy resources: A review," *"Renewable and Sustainable Energy Reviews"*.

[3] S. Sharma, V. M. Iyer, P. P. Das, and S. Bhattacharya, "A modified droop control algorithm for dc microgrids to achieve accurate current sharing and improved voltage regulation," in *2021 IEEE Applied Power Electronics Conference and Exposition (APEC)*, 2021, pp. 119–125.

978-1-6654-8901-0/22 $31.00 © 2022 IEEE

[4] D. Salomonsson, L. Soder, and A. Sannino, "Protection of low-voltage dc microgrids," *IEEE Transactions on Power Delivery*, vol. 24, no. 3, pp. 1045–1053, 2009.

[5] M. Kheraluwala, R. Gascoigne, D. Divan, and E. Baumann, "Performance characterization of a high-power dual active bridge dc-to-dc converter," *IEEE Transactions on Industry Applications*, vol. 28, no. 6, pp. 1294–1301, 1992.

[6] M. I. Rahman, K. H. Ahmed, and D. Jovcic, "Analysis of dc fault for dual-active bridge dc/dc converter including prototype verification," *IEEE Journal of Emerging and Selected Topics in Power Electronics*, vol. 7, no. 2, pp. 1107–1115, 2019.

[7] S. Beheshtaein, R. M. Cuzner, M. Forouzesh, M. Savaghebi, and J. M. Guerrero, "Dc microgrid protection: A comprehensive review," *IEEE Journal of Emerging and Selected Topics in Power Electronics*, pp. 1–1, 2019.

[8] D. Cittanti, E. Vico, and I. R. Bojoi, "New fom-based performance evaluation of 600/650 v sic and gan semiconductors for next-generation ev drives," *IEEE Access*, vol. 10, pp. 51 693–51 707, 2022.

[9] B. Ozpineci and L. M. Tolbert, *Comparison of wide-bandgap semiconductors for power electronics applications*. United States. Department of Energy, 2004.

[10] J. Hornberger, A. Lostetter, K. Olejniczak, T. McNutt, S. Lal, and A. Mantooth, "Silicon-carbide (sic) semiconductor power electronics for extreme high-temperature environments," in *2004 IEEE Aerospace Conference Proceedings (IEEE Cat. No.04TH8720)*, vol. 4, 2004, pp. 2538–2555 Vol.4.

[11] B. J. Baliga, *Gallium Nitride and Silicon Carbide Power Devices*. World Scientific Publishing Company, 2017.

[12] S. S. Shah, V. M. Iyer, and S. Bhattacharya, "Exact solution of zvs boundaries and ac-port currents in dual active bridge type dc–dc converters," *IEEE Transactions on Power Electronics*, vol. 34, no. 6, pp. 5043–5047, 2019.

[13] Datasheet of c3m0021120d. (accessed: 09.12.2022). [Online]. Available: https://assets.wolfspeed.com/uploads/2020/12/C3M0021120D.pdf

[14] Datasheet of c2m0025120d. (accessed: 09.12.2022). [Online]. Available: https://assets.wolfspeed.com/uploads/2020/12/C2M0025120D.pdf

[15] Datasheet of cab450m12xm3. (accessed: 09.12.2022). Available: https://assets.wolfspeed.com/uploads/2020/12/CAB450M12XM3.pdf

[16] Datasheet of ff450r12me4eb11. (accessed: 09.12.2022). [Online]. Available: https://www.infineon.com/dgdl/Infineon-FF450R12ME4_B11-DS-v02_00-EN.pdf?fileId=db3a3043353fdc16013543e3dc143207

[17] Plexim: Electrical engineering software. (accessed: 09.12.2022). [Online]. Available: https://www.plexim.com/

[18] Thermal semiconductor models for plecs. (accessed: 09.12.2022). [Online]. Available: https://www.plexim.com/download/thermal_models/

978-1-6654-8901-0/22 $31.00 © 2022 IEEE

Active Gate Driving of Cascoded SiC JFETs

Arijit Sengupta
Electrical & Computer Engineering
Vanderbilt University
Nashville, TN, USA
arijit.sengupta@vanderbilt.edu

Sima Azizi Aghdam
Electrical & Computer Engineering
University at Albany, SUNY
Albany, NY, USA
saziziaghdam@albany.edu

Mohammed Agamy
Electrical & Computer Engineering
University at Albany, SUNY
Albany, NY, USA
magamy@albany.edu

Abstract— **In this paper, the gate driving characteristics of a SiC cascoded Junction Field Effect Transistor (JFET) is explored. Firstly, a Conventional Gate Drive (CGD) circuit is analyzed and implemented, followed by the proposal and implementation of two Active Gate Driving (AGD) methods for the cascoded JFET. Two approaches of gate control are explored: i) using the low-voltage -MOSFET gate & (ii) using the JFET gate. Switching characteristics during turn-on and turn-off for the CGD and both AGD approaches are compared, and it is observed that the proposed AGD circuits provide controllability of both switching edges and thus allowing an application optimized switching operation. Analytical, simulation and experimental results are shown to verify the proposed methods.**

Keywords— *Wide bandgap (WBG) Semiconductors, Silicon Carbide (SiC), Active Gate Driver, Cascoded JFET, Switching Characteristics*

I. INTRODUCTION

Commercially available wide bandgap (WBG) power devices will revolutionize next-generation power electronics converters. In comparison with Si power devices, WBG power devices feature higher breakdown electric field, lower specific on-resistance, faster switching speed, and higher junction temperature capability as shown in Fig. 1. These material properties are favorable for the efficiency, power density, specific power, and reliability of power converters [1].

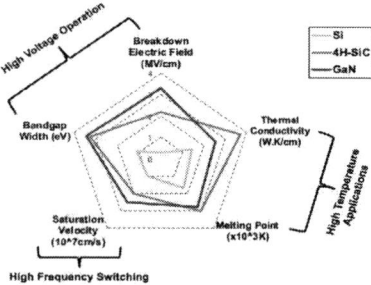

Fig. 1. Summary of Si, SiC, and GaN relevant material properties [2], [3]

Various WBG vertical and lateral, discrete, or integrable device structures have been fabricated and investigated for more than two decades now [2-5]. A wide range of such devices are available commercially now. The SiC power device structures and processes strongly resemble those of vertical, discrete silicon power devices but are capable much superior performance [2-3]. Hence for the purpose of this research, SiC devices have been chosen.

However, to characterize the SiC devices to their fullest potential, it is not sufficient to test and compare them with

respect to a Conventional Gate Driver (CGD) circuit. It has been observed that the fast-switching transition of SiC MOSFETs results in a high dv/dt across the power semiconductor device which also reflects across the gate driver. Therefore, it is important that the gate drivers have a proper design to withstand the high dv/dt across it without affecting the external circuitry [6]. The Active Gate Driver (AGD) design for the cascoded JFET thus becomes essential to ensure proper switching and conduction characteristics. It will enable control of the switching rate by selection of the gate resistance and ensuring reduced on-state resistance based on the turn-on voltage of the gate terminal [7]. A CGD circuit encounters over-voltage during switching transitions. This induces sharp spike of current and electromagnetic induction (EMI) in the circuit, mainly due to the high switching frequency of SiC devices and presence of parasitic inductance in the circuit [8]. A CGD circuit operates with fixed voltage and resistance. A larger external gate resistance can minimize the switching stress and suppress EMI. However, this will increase the switching losses and will slow down the device switching making one of the key features of SiC devices which is very fast switching frequency redundant. One of the approaches to overcome the shortcomings of the CGD topology is by enhancing the performance of the gate driver circuit itself, that is, by using an AGD to replace the CGD as shown in Fig.2. Over the past few years different AGD techniques have been proposed for SiC MOSFETs [8-11].

The circuit proposed in [8] has a two-stage implementation for active gate driving. The first stage is a Field Programmable Gate Array (FPGA) controller circuit which controls the next stage consisting of four transistors to generate corresponding gate drive voltage. This circuit increases the switching speed while minimizing the switching stress. The circuit in [9] is a simpler design compared to the previous one. It uses a parallel capacitor and an auxiliary switch in the external gate resistance branch. This circuit is intended to reduce crosstalk by improving the gate loop impedance and power loop impedance design with the parallel capacitor so designed that it only operates during turn-off and thereby reduces the crosstalk voltage effectively. The circuit proposed in [10] utilizes a two-switch implementation coupled with high-speed comparators to minimize over-current and over-voltage. The gate driver can control the di/dt turn-on and the dv/dt turn-off individually along with lowering the switching losses. The design in [11] uses a gate boost circuit consisting of a delay circuit, resistor, and switch. This gate driver can reduce switching loss and switching time without increasing switching noise. In this paper, two AGD methods of SiC JFETs are presented. The first controls the low voltage MOSFET gate resistance and the second directly controls the

978-1-6654-8901-0/22 $31.00 © 2022 IEEE

gate resistance of the JFET. A two simple transistor design of an active driver on either gate line is proposed to dynamically vary the gate resistance during switching to provide optimal transition speeds based on trade-offs between switching energy loss and circuit component stresses due to slew rates of device voltage and current.

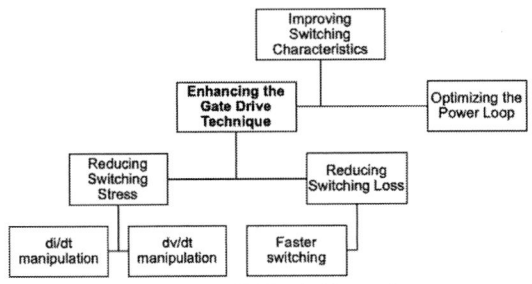

Fig. 2. Methods for improvement of SiC switching performance

II. SIC CASCODED JFET

The cascoded JFET utilizes an N-channel depletion-mode Vertical Trench JFET (VTJFET) with very low $R_{DS(on)}$ and a specialized custom LV MOSFET cascoded in series connection as shown in Fig. 3 (a). The connection schematic of the cascoded JFET is shown in Fig. 3 (b). It simplifies the FET operation and enables gate drive compatibility with all existing SiC MOS and Si IGBT and MOS switches [12-13]. The device is normally-on depending on the thickness of the vertical channel and the doping levels of the structure [14].

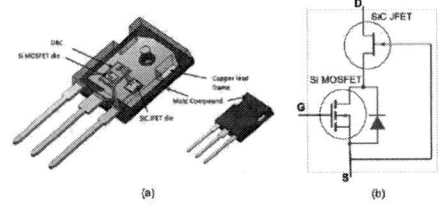

Fig. 3. (a) Illustration of the internal connection of a cascoded JFET in a TO247-3 package, after [12] and (b) schematic showing actual connection between the Si MOSFET and SiC JFET in cascode configuration.

Enhancement-mode JFETs have natural advantages in safe operation as they are normally-off whereas the depletion-mode JFET have advantages with almost twice the saturation current, lower on-resistance and no gate current requirement in the on state [15]. The normally-on JFET remains in the on-state when no negative voltage is fed to the gate, whereas the gate-source junction of the normally-off JFET needs to be forward biased by supplying a positive voltage for it to conduct. Also, to achieve a low on-state resistance, a large gate current should be supplied at steady state for the normally-off JFET which is not required for that of the normally-on JFET [16-17]. Hence, to utilize these advantages, a normally-on JFET is used for the cascode design.

On the other hand, high voltage SiC MOSFETs have been suffering from various reliability and performance issues for years [18]. Manufacturers have given an elaborate effort on improving the performance of the gate-oxide layer of SiC MOSFETs which is one of the major drawback factors especially in high temperature operations. Even though the

structures of MOSFETs have been significantly improved by manipulating the gate-oxide layer and the specific on-state resistance also been reduced due to the increased channel mobilities [16], a cascoded JFET still has better performance than the latest generation of SiC MOSFETs. A cascoded JFET can operate at a larger range of V_{gs}, higher pulsed currents and exhibit significantly lower turn-on and turn-off losses under similar test conditions as described in [13]. Moreover, vertical JFETs, in general, exhibit very low switching losses making them an excellent candidate for high frequency, high-efficiency power conversion applications [19].

III. PREPARE YOUR PAPER BEFORE STYLING

A. Conventional Gate Drive

The conventional gate drive circuit setup is shown in Fig. 4.

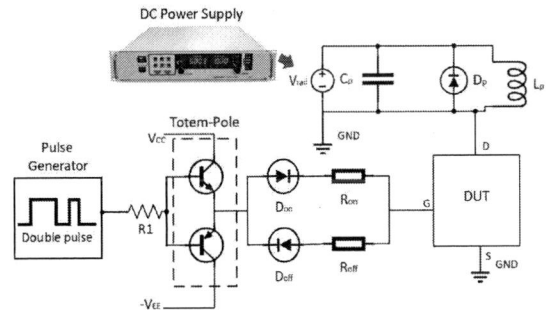

Fig. 4. CGD setup to test switching characteristics

The capacitance between the JFET terminals as seen in Fig. 4 can be expressed in terms of the device parameters as follows:

$$C_{iss} = C_{gd} + C_{gs} \tag{1}$$
$$C_{oss} = C_{gd} + C_{ds} \tag{2}$$
$$C_{rss} = C_{gd} \tag{3}$$

where, C_{iss} is the input capacitance, C_{oss} is the output capacitance, C_{rss} is the reverse transfer capacitance, C_{gd} is the gate to drain capacitance, C_{gs} is the gate to source capacitance and C_{ds} is the drain to source capacitance. These capacitances play an important role in the device performance and will be analyzed in the following sections.

B. Circuit Diagrams of AGD

The circuit diagram of the proposed AGD method as applied to the low voltage MOSFET gate is shown in Fig. 5. The added paths during turn-on and turn-off transitions have been highlighted in red boxes. Various existing models [8-11] were tested before finalizing this design aimed not only at reducing the switching stress and switching losses in the circuit but also to keep the modifications simple and easy to implement. The changes in this AGD setup are very minimal from the CGD setup described above. An alternating ON transition path is added with a small resistor followed by an NMOS. Similarly, an alternating OFF transition path is added with a small resistor followed by a PMOS. The values of the resistors on the main ON and OFF paths are changed to make the overall equivalent gate resistor of the DUT close enough to the CGD analysis for a fair comparison of the switching losses using the two different methods. The main goal here is to dynamically vary the external gate resistor, $R_{G,ext}$ to make the

978-1-6654-8901-0/22 $31.00 © 2022 IEEE

switching transitions faster or slower as needed and thereby improve the switching characteristics.

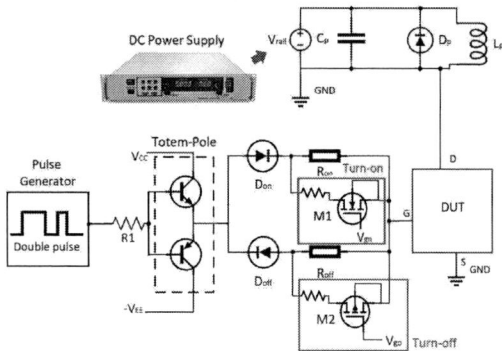

Fig. 5. Proposed AGD setup with turn-on and turn-off path highlighted

C. Analysis of AGD

A detailed circuit analysis of the AGD setup is done using the simplified equivalent small signal model [20] at the external dynamic gate resistor of the DUT during ON and OFF transitions is shown in Fig. 6. The input current, I_{in} is given by:

$$I_{in} = I_{D,on} = I_S e^{(\frac{V_D}{nV_T})}$$
$$or, \ I_{in} = I_S e^{(\frac{V_{in}}{nV_T})}$$
$$or, \ I_{in} = I_S e^{(\frac{V_{CC}}{nV_T})} \tag{4}$$

where V_T is the thermal voltage and $V_T = kT/q$ where k is Boltzmann's constant, T is the absolute temperature in K and q is the magnitude of electronic charge, I_S is the saturation current or scale current, V_D is the forward bias voltage and n is a constant, usually between 1 and 2.

Let the current through Ron be I_1 which can be written as:

$$I_1 = \frac{V_{CC} - V_T - V_{gs,DUT}}{R_{on}} \tag{5}$$

and the current through the lower impedance path be I_2 which can be written as:

$$I_2 = \frac{V_{CC} - V_T - V_{dn}}{R_{on}} + \frac{V_{dn} - V_{gs,DUT}}{r_0} + g_m(V_{gn} - V_{gs,DUT}) \tag{6}$$

where V_{dn} and V_{gn} are the voltages at the drain and gate of the added NMOS, respectively.

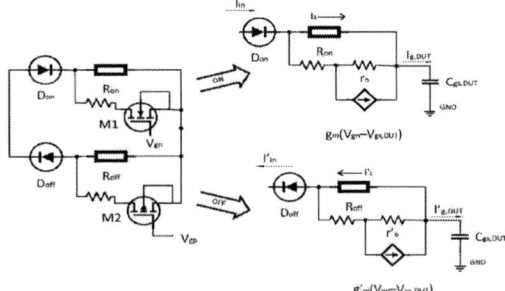

Fig. 6. Small-signal model of the active gate drive (AGD) circuit

The current at the gate of the DUT during the ON transition can thus be written as:

$$I_{G,DUT} = I_{in} + I_1 + I_2 \tag{7}$$

which can be substituted from the previous equations.

Similarly for the OFF transition, the input current, I'_{in} is given by:

$$I'_{in} = I_{D,off} = I_S e^{(\frac{V_D}{nV_T})}$$
$$or, \ I'_{in} = I_S e^{(\frac{V_{in}}{nV_T})}$$
$$or, \ I_{in} = I_S e^{(\frac{V_{EE}}{nV_T})} \tag{8}$$

Let the current through R_{off} be I'_1 which can be written as:

$$I'_1 = \frac{V_{EE}}{R_{off}} \tag{9}$$

and the current through the lower impedance path be I'_2 which can be written as:

$$I_2 = \frac{V_{EE} - V_{dp}}{R_2} + \frac{V_{dp} - V_{gs,DUT}}{r'_0} + g'_m(V_{gp} - V_{gs,DUT}) \tag{10}$$

where V_{dp} and V_{gp} are the voltages at the drain and gate of the added PMOS respectively.

The current at the gate of the DUT during the OFF transition can thus be written as:

$$I'_{G,DUT} = I'_{in} + I'_1 + I'_2 \tag{11}$$

which can be substituted from the previous equations. Since the DUT is a transistor, it can be considered as a capacitive switch as shown in Fig. 6. Thus, the change in current between the ON and OFF states, i.e. $I = I_{G,DUT} - I'_{G,DUT}$. Therefore,

$$\delta V_{gs,DUT} = \frac{\delta I}{C_{gs,DUT}} \tag{12}$$

Integrating this equation over the entire duration of the double pulse generation, which is the pulse width of the two pulses and the OFF state separating the two pulses will yield the corresponding capacitor values of the DUT since the V and I will be known. To do simulation testing and circuit comparison, the small signal analysis is enough and can depict what can be expected out of the circuit model when simulated in LTSpice which is discussed in the next section

D. Simulation Results

The comparison of the CGD and AGD circuits are done side by side as shown in Figs. 7-8 and it is seen that the AGD circuits does minimize the switching losses in both the transistors. For the best set of comparison, the rail voltage is chosen to be 800V and the drain current as 60A which yields a significant amount of switching losses in both turn-on and turn-off of the device and any small differences also will be visible in this test scenario. It is observed from the figures that the miller voltage as well as the turn-on and turn-off losses reduce significantly. The switching transition is faster or slower depending on the gate-source voltage of the NMOS and PMOS at the turn-on and turn-off paths respectively.

It is observed that the PMOS is partially on when $V_{gs,p}$ varies between -5V and -4V and the NMOS is partially on when the $V_{gs,n}$ varies between 5V and 6V respectively. The partially on mode is important as we see the dynamic variance of the external gate resistance during this range as seen in Fig. 9(a) and (b).

978-1-6654-8901-0/22 $31.00 © 2022 IEEE

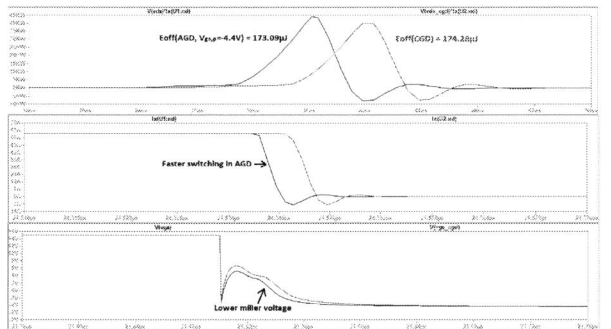

Fig. 7. Simulation results: AGD vs CGD comparison of turn-off characteristics in cascoded JFET

The dynamic variance of the external gate resistance can be clearly observed when the PMOS and NMOS are partially on. The equivalent resistance at the gate of the DUT changes thereby changing the turn-on (E_{on}) and turn-off (E_{off}) switching losses as shown in Fig. 10(a) and (b).

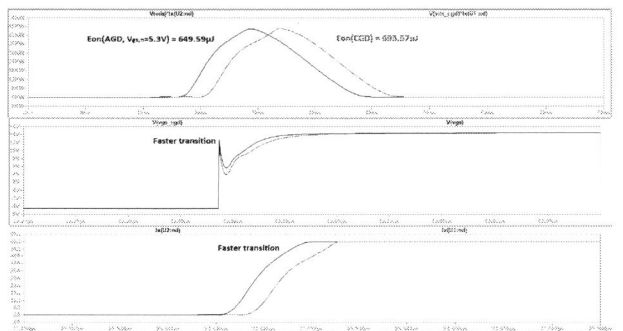

Fig. 8. Simulation results: AGD vs CGD comparison of turn-on characteristics in cascoded JFET

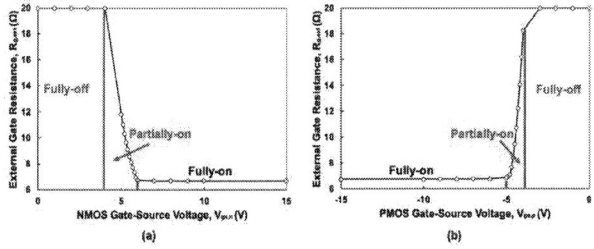

Fig. 9. Simulation results: (a) NMOS Sweep Waveform to calculate dynamic R_g and (b) PMOS Sweep Waveform to calculate dynamic R_g

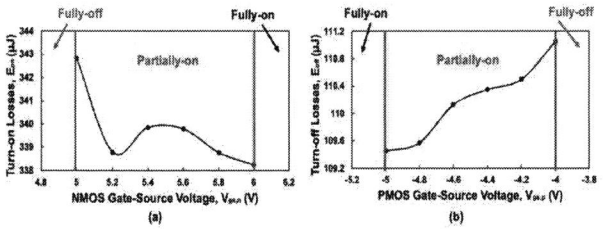

Fig. 10. Simulation results: (a) JFET partially on state E_{on} variance in terms of NMOS' V_{gs} and (b) JFET partially on state E_{off} variance in terms of PMOS' V_{gs}

E. JFET Gate Control

An alternative method to shape the switching transitions is to control the JFET gate current. A similar approach to that

described in the previous section for MOSFET control is applicable to control the resistance of the JFET gate path. Fig. 11 and Fig. 12 show the impact of gate control on the turn on and turn off transitions, respectively. While both gate paths impact the switching transition speed, JFET gate control requires a narrower range of resistance change to impact the voltage and current slew rates more broadly at turn off. This makes this approach attractive for limiting large transient at turn off as in the case of breaker applications or for device protection purposes.

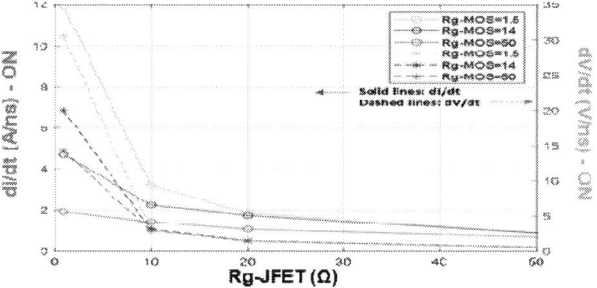

Fig. 11. Impact of AGD control of JFET gate resistance and MOSFET gate resistance on dI_d/dt and dV_{ds}/dt at $V_{ds} = 800V$, $I_d = 75A$ at turn ON

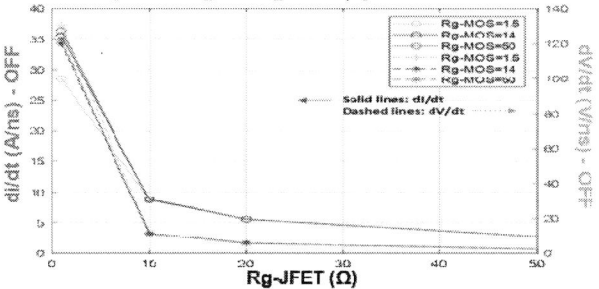

Fig. 12. Impact of AGD control of JFET gate resistance and MOSFET gate resistance on dI_d/dt and dV_{ds}/dt at $V_{ds} = 800V$, $I_d = 75A$ at turn OFF

IV. EXPERIMENTAL RESULTS

A. Experimental Setup

The experimental circuit setup is shown in Fig.13.

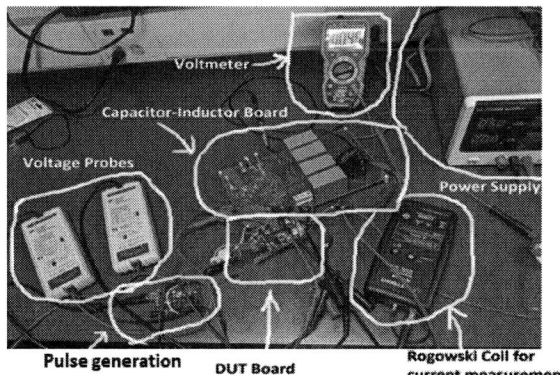

Fig. 13. Circuit setup for double pulse testing

B. Double Pulse Test

To measure the switching parameters of a power device, the standard method used is a double pulse test [21]. The CGD

978-1-6654-8901-0/22 $31.00 © 2022 IEEE

circuit is run at rail voltages 200V, 400V, 600V and 800V. The current is controlled using the 60 μH inductor and is kept within the current limit of the transistors. Fig. 14 shows switching transitions using CGD circuit; where voltages V_{gs} and V_{ds} are measured using differential probes, and the current I_d is shown, measured using a Rogowski coil. The product of V_{ds} and I_d is calculated using the MATH function from oscilloscope and then it is integrated to get the switching losses, i.e., the area under the curve of the product is calculated.

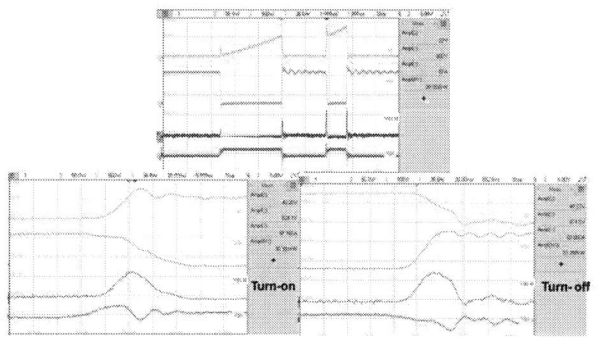

Fig. 14. Experimental results: JFET CGD V_{gs} (pink, 50V/div), V_{ds} (yellow, 500V/div) and I_d (light green, 36A/div) measurement using oscilloscope at V_{rail}=800V, I_L=40A with Power Calculation $V_{ds}.I_d$ (deep green, 20kW/div)

A similar double pulse test is done to measure the switching losses for the AGD circuit. Firstly, the NMOS on the turn-on branch is populated along with the series resistance to measure the turn-on losses. Then, the NMOS along with the series resistor is removed, and the PMOS on the turn-off branch is populated along with its series resistor to calculate the turn-off losses. For the AGD experiment, the rail voltages are chosen to range from 200V to 800V to provide an understanding of the functionality at low and high voltages. Switching characteristics at different NMOS and PMOS gate voltages and hence different switching speeds are shown in Fig. 15.

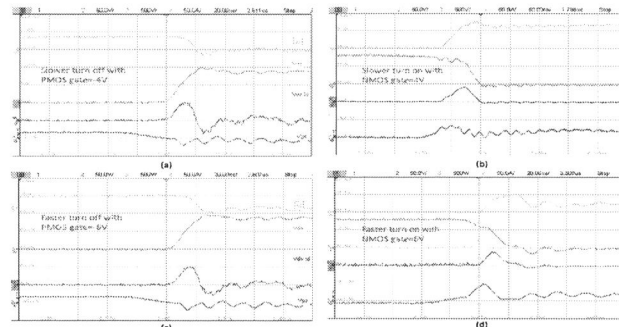

Fig. 15. Experimental results: JFET AGD V_{gs} (pink, 50V/div), V_{ds} (yellow, 500V/div) and I_d (light green, 50A/div) measurement using oscilloscope at V_{rail}=800V, I_L=40A with Power calculation $V_{ds}.I_d$ (deep green)

C. Results & Discussion

Analyzing the experimental results, it is observed that the AGD circuit provides controllability of switching transitions of the DUT. For faster transitions, the switching losses are comparatively lower than the CGD circuit which was also observed in the simulation results. In Figs. 16-17, a comparative plot between the CGD losses and AGD losses are

shown. The turn-on losses are compared when the V_{gs} of the NMOS is at 4V and 6V, whereas the turn-off losses are compared when the V_{gs} of the PMOS is at -4V and -6V. These voltages are chosen since at these values the NMOS and PMOS are at the junction of being fully on and fully off. In between these voltage values, they are partially on. It was observed in simulation that these devices being partially on provides the best comparative results, with the external gate resistance of the DUT being almost equivalent and thus the comparison makes sense. The key points to note from these figures are that:

1) The turn-on losses for the AGD circuit with NMOS' V_{gs} at 6V are lower than the turn-on losses for the CGD upto a certain drain current value of approx. 45A.

2) The turn-off losses for the AGD circuit with PMOS' V_{gs} at -6V are lower than the turn-off losses for the CGD until the full scale of 60A.

Thus, it can be stated that the AGD circuit regulate the switching performance of the devices. The dI/dt and dV/dt comparison of the device are shown in Figs. 18-19. The results show visible changes in either or both of dI/dt and dV/dt when the gate-source voltage of the NMOS is between 5V and 6V and for the PMOS it's between -4V and -5V to make them operate at partially-on mode. The switching losses shown in Fig. 20(a) and (b) was calculated at these instances too. Comparing the switching transitions, visible improvements are noticed include lower under-voltage, lower over-voltage in V_{ds} and lower current spikes in I_d. Higher switching speeds cause more off-state ringing and a higher over-voltage in the V_{gs}, therefore, a programmable gate control voltage can provide the optimal operating condition between switching loss and device stresses and ringing.

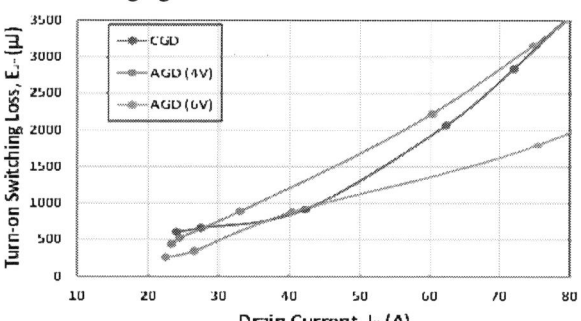

Fig. 16. Experimental results: JFET turn-on switching losses comparison at V_{ds}= 800V for CGD and AGD at NMOS' V_{gs}= 4V, 6V

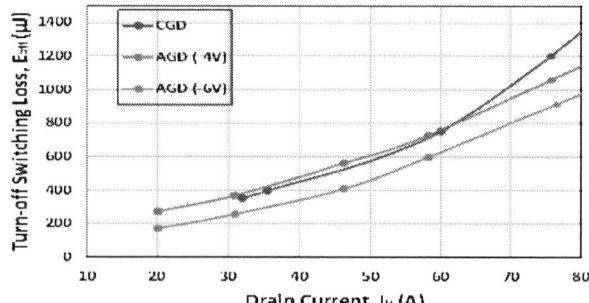

Fig. 17. Experimental results: JFET turn-off switching losses comparison at V_{ds}= 800V for CGD and AGD at PMOS' V_{gs}= -4V, -6V

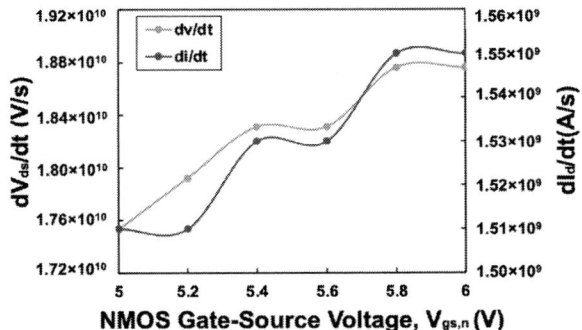

Fig. 18. Experimental Results: JFET dI_d/dt and dV_{ds}/dt at V_{ds}=800V, I_d=50A for various NMOS' V_{gs} during turn-on

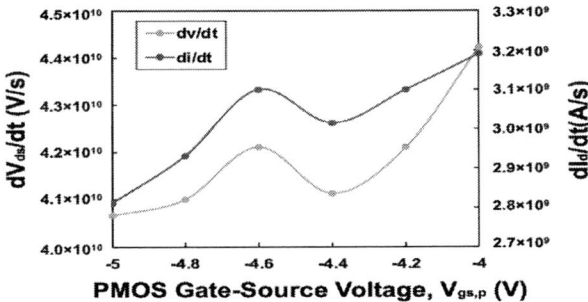

Fig. 19. Experimental Results: JFET dI_d/dt and dV_{ds}/dt at V_{ds}=800V, I_d=50A for various PMOS' V_{gs} during turn-off

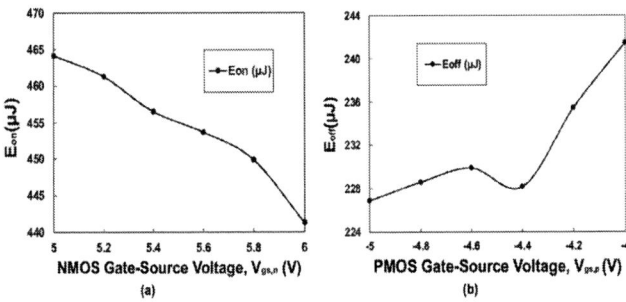

Fig. 20. Experimental results: (a) JFET turn-on losses at V_{ds}=800V, I_d=50A for various NMOS' V_{gs} and (b) JFET turn-off losses at V_{ds}=800V, I_d=50A for various PMOS V_{gs}

V. CONCLUSION

In this paper, an active gate driver circuit is proposed for improving the switching performance of a cascoded JFET. The proposed design is intended to be used to dynamically vary the external gate resistance of the DUT to either speed up or to slow down the switching transitions. The trade-off between using MOSFET or JFET gate paths include voltage stress distribution on devices and amount of energy consumption in clamping circuits used for MOSFET protection. Further, the two device gates have different impacts on switching dynamics: MOSFET gate control having more impact on turn on making it more suitable for switch mode power supplies or inverter operation & JFET having

more impact on the turn off process making it more suitable for breaker operation.

REFERENCES

[1] F. Wang, Z. Zhang, E.A. Jones, "Characterization of Wide Bandgap Power Semiconductor Devices" *IET Energy Engineering*, 2018.

[2] A. Sengupta and M. Agamy, "Comparative Study of SiC MOSFET and JFET using an Active Gate Driver," *Proc. IEEE GPECOM Conf.*, pp. 63-68, 2022.

[3] T. P. Chow, "Wide bandgap semiconductor power devices for energy efficient systems," *Proc. IEEE WiPDA*, pp. 402-405, 2015.

[4] J. Millán, P. Godignon, X. Perpiñà, A. Pérez-Tomás and J. Rebollo, "A Survey of Wide Bandgap Power Semiconductor Devices," *IEEE Trans. Power Elec.*, vol. 29, no. 5, pp. 2155-2163, May 2014.

[5] E. Santi, K. Peng, H. A. Mantooth and J. L. Hudgins, "Modeling of Wide-Bandgap Power Semiconductor Devices—Part II," *IEEE Trans. Electron Devices*, vol. 62, no. 2, pp. 434-442, Feb. 2015.

[6] A. Anurag, S. Acharya and S. Bhattacharya, "Gate Drivers for High-Frequency Application of Silicon-Carbide MOSFETs: Design considerations for faster growth of LV and MV applications," *IEEE Mag. Power Elec.*, vol. 6, no. 3, pp. 18-31, Sept. 2019.

[7] S. Bhattacharya, "Gate drivers for wide bandgap power devices," *Woodhead Publishing*, p.p. 249-300, 2019.

[8] Y. Yang, Y. Wang and Y. Wen, "An active gate driver for improving switching performance of SiC MOSFET," *Proc. IEEE ISNE Symp.*, pp. 1-4, 2018.

[9] Z. Lu, C. Li, H. Wu, W. Li, X. He and S. Li, "Design of Active SiC MOSFET Gate Driver for Crosstalk Suppression Considering Impedance Coordination between Gate Loop and Power Loop," *Proc. IEEE APEC*, pp. 986-990, 2019.

[10] A. Paredes, H. Ghorbani, V. Sala, E. Fernandez and L. Romeral, "A new active gate driver for improving the switching performance of SiC MOSFET," *Proc. IEEE APEC*, pp. 3557-3563, 2017.

[11] K. Yamaguchi, Y. Sasaki and T. Imakubo, "Low loss and low noise gate driver for SiC-MOSFET with gate boost circuit," *Proc. IEEE IECON*, pp. 1594-1598, 2014.

[12] Z. Li and A. Bhalla, "USCi SiC JFET Cascode and Super Cascode Technologies," *Proc. PCIM Asia Conf.*, pp. 1-6, 2018.

[13] United Silicon Carbide, Inc., "1200V-35m SiC Cascode", Rev. A, Jan 2019 [Online]. Accessed: 15 July 2020, Available: https://unitedsic.com/datasheets/DS_UF3C120040K4S.pdf.

[14] M. H. Rashid, "Power electronics: Circuits, Devices & Applications", *Pearson Education Inc.*, 4th ed., 2018.

[15] J. B. Casady, D. C. Sheridan, R. L. Kelley, V. Bondarenko, and A. Ritenour. "A Comparison of 1200 V Normally- OFF & Normally-on Vertical Trench SiC Power JFET Devices," *Mat. Sci. Forum*, vol 679–680, p.p. 641–644, Mar. 2021.

[16] D. Peftitsis and J. Rabkowski, "Gate and Base Drivers for Silicon Carbide Power Transistors: An Overview," *IEEE Trans. Power Elec.*, vol. 31, no. 10, pp. 7194-7213, Oct. 2016.

[17] B. Wrzecionko, D. Bortis, J. Biela and J. W. Kolar, "Novel AC-Coupled Gate Driver for Ultrafast Switching of Normally Off SiC JFETs," *IEEE Trans. Power Elec.*, vol. 27, no. 7, pp. 3452-3463, Jul. 2012.

[18] M. S. Chinthavali, B. Ozpineci and L. M. Tolbert, "High-temperature and high-frequency performance evaluation of 4H-SiC unipolar power devices," *Proc. IEEE APEC*, vol. 1, pp. 322-328, 2005.

[19] A. Ritenour, D. C. Sheridan, V. Bondarenko, and J. B. Casady. "Performance of 15 mm² 1200 V Normally-Off SiC VJFETs with 120 A Saturation Current," *Mat. Sci. Forum*, vol 645–648, p.p. 937–940, Apr. 2010.

[20] A. S. Sedra and K. C. Smith. "Microelectronic Circuits", *Oxford University Press*, 5th ed., 2004.

[21] R. O. C. Lyra, B. J. Cardoso Filho, V. John and T. A. Lipo, "Coaxial current transformer for test and characterization of high-power semiconductor devices under hard and soft switching," *IEEE Trans. Ind. Appl.*, vol. 36, no. 4, pp. 1181-1188, Jul-Aug. 2000.

978-1-6654-8901-0/22 $31.00 © 2022 IEEE

High Frequency High Power Integrated Transformer Design for Resonant Converters with SiC Devices

Tianlong Yuan
Center for Power Electronics Systems (CPES)
Virginia Tech
Blacksburg, USA
tianlong@vt.edu

Feng Jin
Center for Power Electronics Systems (CPES)
Virginia Tech
Blacksburg, USA
fengjin@vt.edu

Zheqing Li
Center for Power Electronics Systems (CPES)
Virginia Tech
Blacksburg, USA
zheqing@vt.edu

Qiang Li
Center for Power Electronics Systems (CPES)
Virginia Tech
Blacksburg, USA
lqvt@vt.edu

Abstract—**Resonant converters are used extensively for voltage regulation and isolation purposes in electric vehicles. The power level and frequency of the converters is increasing with the ongoing development of the wide-band-gap devices, which brings more challenges to the design of magnetic components. The first challenge is that such high power and frequency make it hard to make proper Litz wire. In this paper, a new current sharing method for paralleled Litz wire transformers is introduced. Using this method, the transformer can stand much higher current without increasing the cost or volume. And the power level can be further increased by integrating more matrix transformers. The second challenge is the resonant inductance integration. Different resonant inductance integration methods are summarized and compared. And controlling the leakage inductance by properly breaking the interleaving structure is proposed in this paper. Using this method, the desired leakage inductance can be reached without sacrificing too much winding loss. Comprehensive transformer design and optimization procedure is also presented in this paper. A 30 kW 100 kHz Serial Half Bridge DC-DC converter prototype is designed to verify the methods at last.**

Keywords—Integrated transformer, high power, Silicon carbide (SiC) devices

I. INTRODUCTION

Resonant converters are widely used for voltage regulation and isolation purposes in the ongoing development of electric vehicles. Increasing power and frequency brings more challenges to the design of magnetic components. Serial half bridge topology is used to reduce voltage stress for primary side devices [1]. Previous transformers designed at this power level usually operate at around 50kHz, due to the limitations imposed by wire selection or winding loss [2]-[3]. To reduce the current

stress, an input-parallel output-series transformer structure is adopted, and the parallel windings for both primary and secondary sides are used for each transformer to further reduce the wire size.

Another design challenge of such transformer involves the accurate resonant inductance integration. There are two main solutions according to previous research [4-16]. An additional flux path (made of core or shunt) is used to introduce leakage flux in [4-11]. In [10], leakage inductance is integrated into the center leg as shown in Fig. 2. If the winding structure is balanced as shown in Fig. 2(a), the flux is cancelled in the center leg. In Fig. 2(b), the winding structure is imbalanced; the leakage flux goes through the center leg and the magnetizing flux goes through the outer legs.

Changing the leakage inductance by adjusting the distance between the primary and secondary side windings is used in [12-16]. In [13], the leakage inductance is mainly integrated into the air and the magnetizing flux goes through the core. This method suffers from the larger winding losses at high frequency range compared with perfect interleaved winding structure, due to stronger H field.

Both methods are discussed and compared in this paper and the second method is adopted due to the lower losses and easier

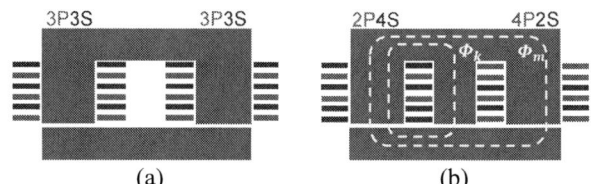

Fig. 2. Integrate leakage inductance in core: (a) balanced structure: $L_k = 0.1uH$; (b) unbalanced structure with center leg: $L_k = 1.4uH$.

structure.

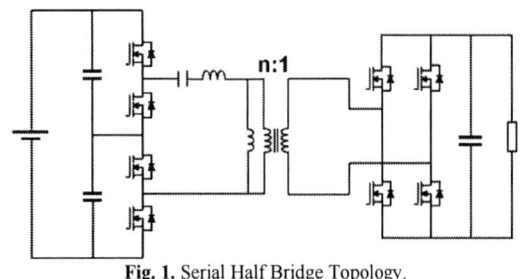

Fig. 1. Serial Half Bridge Topology.

TABLE I
CONVERTER PARAMETERS

Parameters	Value
P_{rated}	30 kW
f_s	100 kHz
V_{in}	550-900V
n	0.58
V_o	600-700V

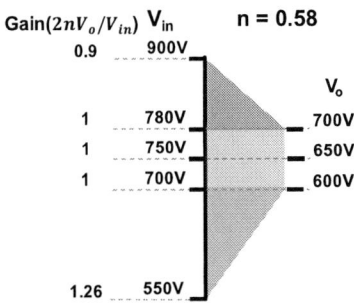

Fig. 3. Voltage regulation requirements.

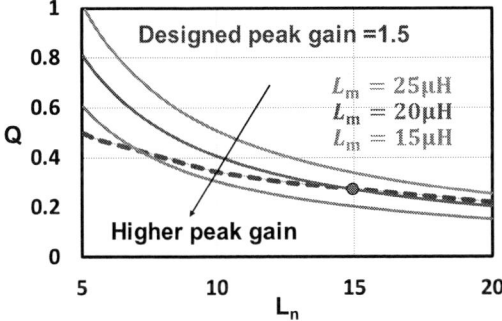

Fig. 4. $L_n Q$ impact on peak gain.

II. TRANSFORMER PARAMETERS AND MATERIAL SELECTION

A. Transformer Parameters Selection

The critical transformer parameters are the turns ratio n, the magnetizing inductance L_m and the leakage inductance L_k. This section mainly focuses on the selection of these parameters based on loss and gain requirements [17]. The voltage regulation requirements for different input voltage levels are shown in Fig. 3. To make the converter work at DCX (DC transformers) mode most of the time, the nominal input voltage is designed to be in the middle of the DCX range. So, the turns ratio is selected to be 0.58.

The peak gain of the converter is controlled by the L_n and Q values as shown in Fig. 4. The required peak gain is 1.26. However, to guarantee that the peak gain is achieved, peak gain is designed to be 1.5 in this case. The peak gain values for different L_m with various L_n and Q are displayed. Although larger L_n makes it easier to reach higher gain requirements with smaller L_k, it will result in wider frequency range for voltage regulation. If the L_n is too small, more leakage inductance is needed which may increase the volume or loss of the transformer. So, L_n is selected to be between 10 to 20. And the designed curve intersects with the curve when L_m equals to 20 µH. At this point, L_n is 15.

$$L_n = L_m/L_k \tag{1}$$

$$Q = \sqrt{L_k/C_k}/(n^2 R_{load}) \tag{2}$$

B. Transformer Material Selection

Soft ferrite is used as the core material because of the smaller P_v and electric conductivity, which leads to smaller core loss at high frequencies [18]. The core loss density of different materials is shown in Fig. 5. DMR96A is selected, finally, because of the smaller core loss density.

Based on the working frequency of 100 kHz, strand AWG is selected to be 40. The wire size is selected to be equivalent AWG 6 based on [19]. However, AWG 6 wire needs 2600 strands and is hard and expensive to manufacture in a real application. So parallel windings are covered in Section III.

III. TRANSFORMER STRUCTURE SELECTION AND CURRENT SHARING CONCEPT

As stated previously, parallel windings will be discussed in this section. First, the analysis of current sharing is conducted based on the example of two windings in parallel. Then, the impact of asymmetry on current sharing is analyzed. At last,

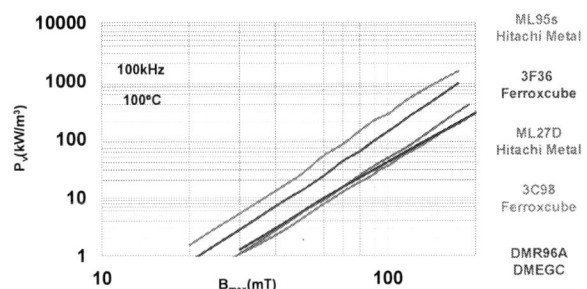

Fig. 5. Core loss density for different materials.

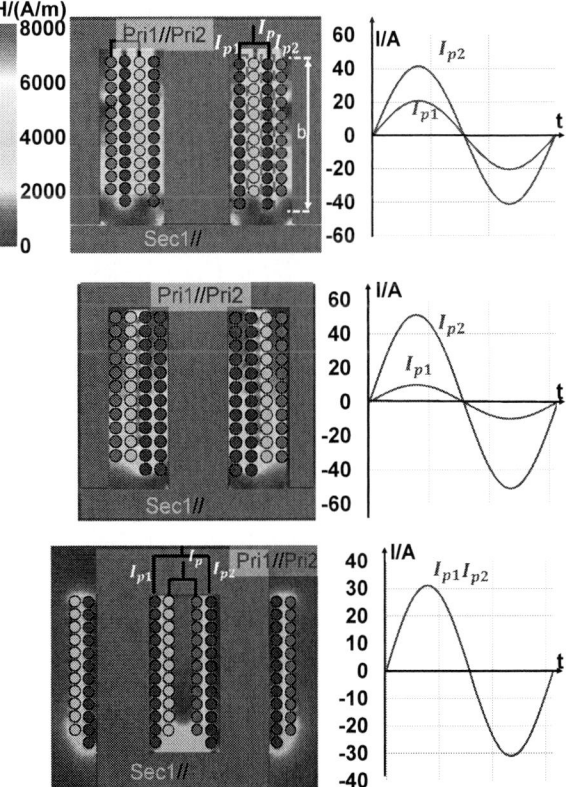

Fig. 6. Current distribution for different winding structures.

978-1-6654-8901-0/22 $31.00 © 2022 IEEE

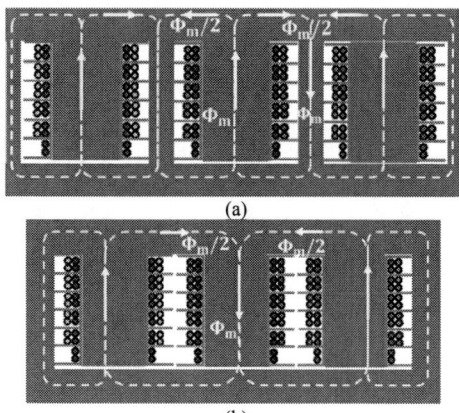

(a)

(b)

Fig. 7. Transformer structure for 3 AWG12 wires in parallel: (a) the flux goes in the same direction for all three legs; (b) the flux in the center leg goes in the opposite direction from outer legs

TABLE II
LOSS AND VOLUME COMPARISON OF DIFFERENT NUMBER OF PARALLEL WINDING GROUPS

	Core Loss	Winding Loss	Volume	Litz Wire Price
AWG 6	30W	24W	0.7L	5$/feet
2 AWG 9	30W	23W	0.67L	1.3$/feet
3 AWG 12	34W	20W	0.87L	1.3$/feet

different numbers of parallel windings are compared from the standpoints of loss, volume, and cost.

A. Current Sharing for Parallel Windings

Finite element analysis (FEA) simulations for different structures are conducted. The H field and current distribution results are shown in Fig. 6. The blue and purple windings are the two primary AWG 9 windings in parallel. The red and orange windings are the two secondary AWG 9 windings in parallel. From the H field, the inner side of the purple winding is lower compared to the field between the purple winding and the orange winding because of the core, but both sides of the blue winding have a stronger H field. So (3)(4) are got, and the current in the purple winding is found to be half of that of the blue winding. From Fig. 6, the current distribution for the UI core is the best because of the symmetry between the two legs. Then the H field around the parallel windings are the same, which results in good current sharing. So, the UI core is selected at last.

$$H \cdot b = I_{p1} \tag{3}$$

$$H \cdot 2b = I_{p2} \tag{4}$$

B. Comparison and Selection of Parallel Winding Groups

To further reduce the wire size, a three parallel windings' structure is also discussed in this section. If the flux in all the legs goes in the same direction (Φ_m in every leg), additional legs must be added flowing $3\Phi_m$ in total as shown in Fig. 7. However, if the flux in the center leg goes in the opposite direction, the flux flowing in the additional legs reduces to Φ_m because of the flux cancellation. So, both the number of legs and the volume is reduced.

The loss and volume data are shown in Table II. The loss is

basically the same. The cost of AWG 9 in parallel is significantly reduced compared to AWG 6. Besides, the volume is reduced as well. But the cost increases if there are three AWG 12 windings in parallel because of the longer winding with more parallel windings. And the volume also increases. So, after comparison, the structure with two AWG 9 windings in parallel is the best with lower loss and smaller volume. But more parallel windings could be used if the current stress is higher.

IV. RESONANT INDUCTANCE INTEGRATION

As previously stated in the introduction, realizing leakage inductance integration as well as maintain good current sharing is another big challenge. In this section, two resonant inductance integration methods are introduced and compared for different resonant inductance. Also, the transformer design process is also discussed. Since there are two transformers with parallel primary side, L_m equals to 40 µH and L_k equals to 2.67 µH for each transformer.

A. Leakage Inductance Integration Methods

The first method confines the leakage inductance in the air as shown in Fig. 8. The leakage inductance is controlled by the distance between the primary side and the secondary side. In Fig. 8, a method of properly breaking the interleaving is proposed. In this way, the leakage inductance is increased without sacrificing the volume and can also be controlled in

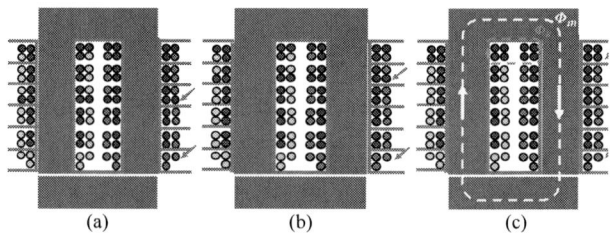

(a)   (b)   (c)

Fig. 8. Integrating the leakage inductance in the air. (a) $L_k = 2\,\mu H$; (b)$L_k = 2.5\,\mu H$; (c)$L_k = 3\,\mu H$

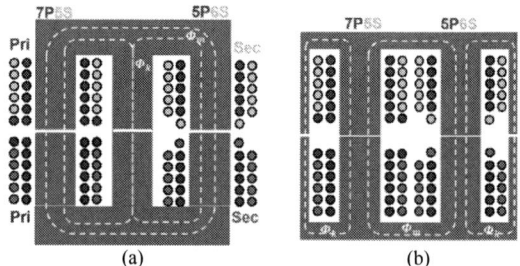

(a)   (b)

Fig. 9. Integrating the leakage inductance in the r. (a) $L_k = 2.4\,\mu H$; (b)$L_k = 2.4\,\mu H$;

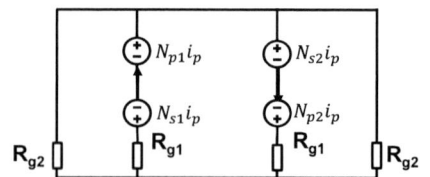

Fig. 10. Reluctance model of the transformer in Fig. 9(b)

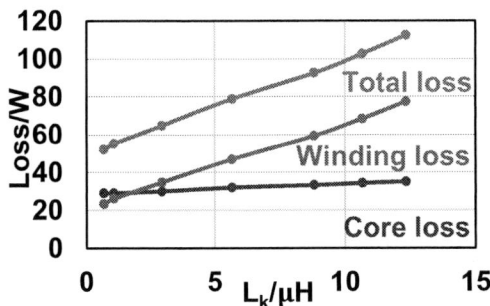

Fig. 11. L_k vs loss while integrating the leakage inductance in the air.

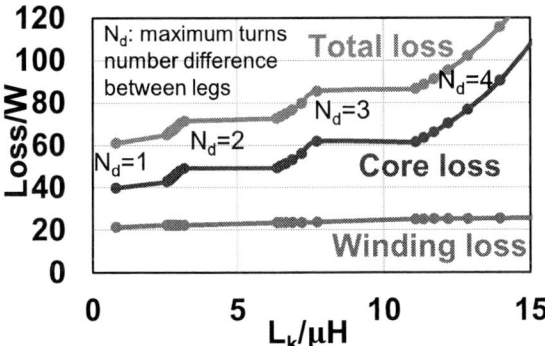

Fig. 12. L_k vs loss while integrating the leakage inductance in the core.

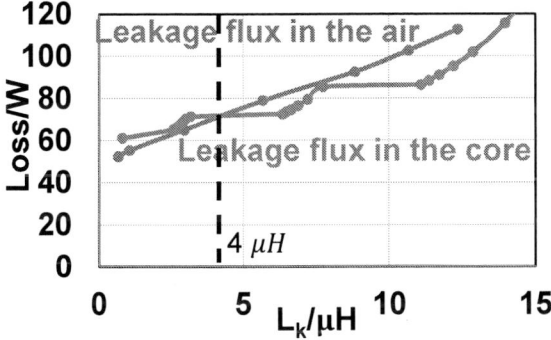

Fig. 13. Total loss comparison for different leakage inductance control

small and steady steps. However, the magnetic field in the air is stronger with increased leakage inductance and thus incurs larger winding loss which may cause thermal issues.

The second method is to integrate leakage inductance in the core, shown in Fig. 9. In Fig. 9(a), the magnetizing flux goes through the outer legs and the leakage flux goes through the center leg. The flux going through the center leg is decided by the MMF difference between the left leg and the right leg and the reluctance of the center leg. So, the leakage inductance is mainly controlled by the airgap in the center leg. And the magnetizing inductance is mainly controlled by the airgap of outer legs. The core loss of the transformer is 35W, winding loss is 24W, total volume is 0.85L.

In Fig. 9(b), the center leg is split into two outer legs. There are two main benefits: the first one is that the flux distribution is better since the leakage flux goes through the outer legs and the magnetizing flux goes through the center legs. So, the core loss is reduced by 13%. Also, the total volume is reduced by 6% because of the thinner top and bottom plates. The second benefit of this approach is that the stray field is smaller with the outer legs. The reluctance model is as shown in Fig. 10. The leakage inductance can then be calculated from the reluctance model as (5). From the equation, we can find that the leakage inductance is determined by the turns number differences and the airgaps. Larger leakage inductance can be got by larger turns number differences and smaller airgaps.

$$L_k = \frac{\left(N_{1p}N_{2s}-N_{2p}N_{1s}\right)^2}{\left(N_{1s}^2+N_{2s}^2\right)R_{g1}+\left(N_{1s}+N_{2s}\right)^2 R_{g2}/2} \tag{5}$$

B. Comparison of the Integration Methods

Both methods have its advantages and disadvantages. So, more discussions at different leakage inductances need to be done.

The loss distribution for different leakage inductances integrated in the air is shown in Fig. 11. From the curve, the core loss does not change much with the leakage inductance. The winding loss increases linearly with the leakage inductance. When the inductance is higher, the total loss is dominated by the winding loss. From Fig. 8, we can see that the leakage inductance can change in steady steps as small as 0.5μH, which can satisfy most of the requirements. To get a fair comparison result, the volume of the transformers for all the points is 0.67L.

The loss distribution for different leakage inductances integrated in the core is shown in Fig. 12. The winding loss does not change much with the leakage inductance and core loss is dominant. Different from the UI core with a steady step of leakage inductance, the leakage inductance can only be changed constantly in certain regions. The steps of leakage inductance are caused by the steps of turns number. For each turns number difference, there is a boundary with the maximum and minimum leakage inductance. The volume of the transformers for all the points is also 0.67L.

The total loss comparison is as shown in Fig. 13. After comparison, for leakage inductance lower than 4 μH, UI core is a better solution because it has lower total loss and is easier to design and manufacture. But with higher leakage inductance, integrating leakage inductance in the core is a better solution

Fig. 14. 30 kW LLC resonant converter using input parallel output series transformers.

978-1-6654-8901-0/22 $31.00 © 2022 IEEE

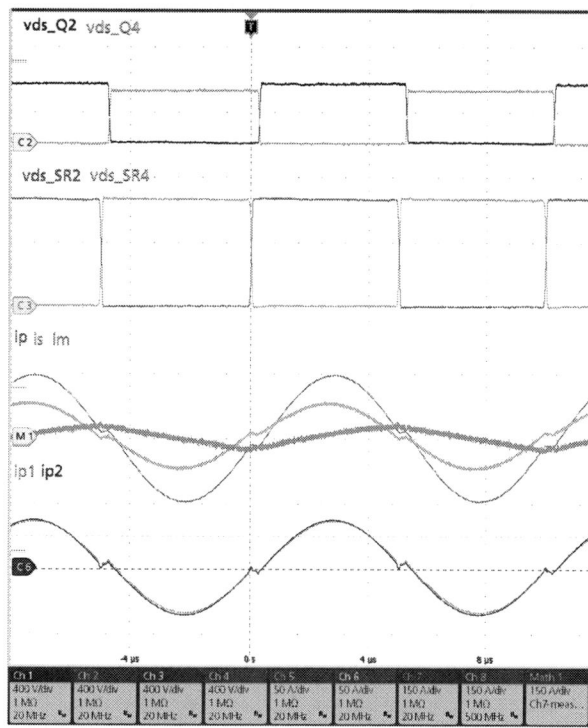

Fig. 15. Waveform tested at 100kHz 30 kW.

because of smaller loss with the same transformer volume. Since the required leakage inductance for each transformer is $2.67 \mu H$, UI core structure shown in Fig. 8(b) is selected at last.

V. EXPERIMENTS AND RESULTS

A resonant converter using 1200 V SiC devices on both sides are designed to verify the concepts as shown in Fig. 14. Both transformers are tested using the impedance analyzer. $L_{m1} = L_{m2} = 40 \ \mu H$, $L_{r1} = L_{r2} = 2.7 \ \mu H$. The desired leakage inductance is met. The testing waveforms at 30 kW 100 kHz are as shown in Fig. 15. From the figure, ZVS is realized. And perfect current sharing is realized for the parallel windings from the bottom of the figure. The peak value for the current in primary side is 150A. And the peak value for each winding is around 38A, which is only one fourth of the total current.

VI. SUMMARY

With the proposed current sharing method, the current stress for the windings can be reduced significantly and would not be a big issue for magnetic design. For small leakage inductance, the proposed leakage inductance control method has smaller total loss and easier core structure compared to integrating the leakage inductance in the core. However, the proposed method has higher loss if the required leakage inductance is higher. One advantage of the proposed method is that the leakage can be changed linearly which can fulfill most of the required leakage inductance. Integrating leakage inductance in the core can only control the inductance constantly in certain regions as shown in Fig. 12. The proposed methods have been implemented into a 100kHz 30 kW LLC resonant converter design. The experimental results also verify the theoretical analysis.

REFERENCES

[1] F. Jin, F. Liu, X. Ruan and X. Meng, "Multi-phase multi-level LLC resonant converter with low voltage stress on the primary-side switches," 2014 IEEE Energy Conversion Congress and Exposition (ECCE), 2014, pp. 4704-4710, doi: 10.1109/ECCE.2014.6954045.

[2] D. Aggeler, J. Biela and J. W. Kolar, "A compact, high voltage 25 kW, 50 kHz DC-DC converter based on SiC JFETs," 2008 Twenty-Third Annual IEEE Applied Power Electronics Conference and Exposition, 2008, pp. 801-807

[3] D. Rothmund, T. Guillod, D. Bortis and J. W. Kolar, "99% Efficient 10 kV SiC-Based 7 kV/400 V DC Transformer for Future Data Centers," in IEEE Journal of Emerging and Selected Topics in Power Electronics, vol. 7, no. 2, pp. 753-767, June 2019, doi: 10.1109/JESTPE.2018.2886139

[4] A. Kats, G. Ivensky and S. Ben-Yaakov, "Application of integrated magnetics in resonant converters," Proceedings of APEC 97 - Applied Power Electronics Conference, 1997, pp. 925-930 vol.2, doi: 10.1109/APEC.1997.575756.

[5] Y. -D. Kim, C. -E. Kim, K. -M. Cho, K. -B. Park and G. -W. Moon, "ZVS phase shift full bridge converter with controlled leakage inductance of transformer," INTELEC 2009 - 31st International Telecommunications Energy Conference, 2009, pp. 1-5, doi: 10.1109/INTLEC.2009.5351846.

[6] B. Sun, R. Burgos, D. Boroyevich, S. Bala and J. Xu, "10 kW High Efficiency Compact GaN-Based DC/DC Converter Design," 2018 IEEE Energy Conversion Congress and Exposition (ECCE), 2018, pp. 6307-6313, doi: 10.1109/ECCE.2018.8557541.

[7] M. Mu, L. Xue, D. Boroyevich, B. Hughes and P. Mattavelli, "Design of integrated transformer and inductor for high frequency dual active bridge GaN Charger for PHEV," 2015 IEEE Applied Power Electronics Conference and Exposition (APEC), 2015, pp. 579-585, doi: 10.1109/APEC.2015.7104407.

[8] S. A. Ansari, J. N. Davidson and M. P. Foster, "Analysis, Design and Modelling of Two Fully- Integrated Transformers with Segmental Magnetic Shunt for LLC Resonant Converters," IECON 2020 The 46th Annual Conference of the IEEE Industrial Electronics Society, 2020, pp. 1273-1278, doi: 10.1109/IECON43393.2020.9254721.

[9] M. A. Bakar and K. Bertilsson, "An improved modelling and construction of power transformer for controlled leakage inductance," 2016 IEEE 16th International Conference on Environment and Electrical Engineering (EEEIC), 2016, pp. 1-5, doi: 10.1109/EEEIC.2016.7555605.

[10] B. Li, Q. Li and F. C. Lee, "High-Frequency PCB Winding Transformer With Integrated Inductors for a Bi-Directional Resonant Converter," in IEEE Transactions on Power Electronics, vol. 34, no. 7, pp. 6123-6135, July 2019, doi: 10.1109/TPEL.2018.2874806.

[11] F. Jin, A. Nabih, C. Chen, X. Chen, Q. Li and F. C. Lee, "A High Efficiency High Density DC/DC Converter for Battery Charger Applications," 2021 IEEE Applied Power Electronics Conference and Exposition (APEC), 2021, pp. 1767-1774, doi: 10.1109/APEC42165.2021.9487108.

[12] Z. Li, C. Zhao, Y. -H. Hsieh and Q. Li, "Partial Fluctuating Power Control of Resonant Converter for Solid-State Transformer," 2022 IEEE Applied Power Electronics Conference and Exposition (APEC), 2022, pp. 770-776, doi: 10.1109/APEC43599.2022.9773455.

[13] S. Zou, J. Lu, A. Mallik and A. Khaligh, "Modeling and Optimization of an Integrated Transformer for Electric Vehicle On-Board Charger Applications," in IEEE Transactions on Transportation Electrification, vol. 4, no. 2, pp. 355-363, June 2018, doi: 10.1109/TTE.2018.2804328.

[14] J. Yang, X. Wu, G. Liu, D. Ping and Z. Deng, "Modeling and Design of Integrated Inductor and Transformer Considering Superposed Flux Density in On-Board-Charger," 2020 IEEE Applied Power Electronics Conference and Exposition (APEC), 2020, pp. 879-884, doi: 10.1109/APEC39645.2020.9124228.

[15] Z. Li, Y. -H. Hsieh, Q. Li, F. C. Lee and M. H. Ahmed, "High-Frequency Transformer Design with High-Voltage Insulation for Modular Power Conversion from Medium-Voltage AC to 400-V DC," 2020 IEEE Energy Conversion Congress and Exposition (ECCE), 2020, pp. 5053-5060, doi: 10.1109/ECCE44975.2020.9236384.

[16] J. Jung, "Bifilar Winding of a Center-Tapped Transformer Including Integrated Resonant Inductance for LLC Resonant Converters," in IEEE

Transactions on Power Electronics, vol. 28, no. 2, pp. 615-620, Feb. 2013, doi: 10.1109/TPEL.2012.2213097.

[17] Bing Lu, Wenduo Liu, Yan Liang, F. C. Lee and J. D. van Wyk, "Optimal design methodology for LLC resonant converter," Twenty-First Annual IEEE Applied Power Electronics Conference and Exposition, 2006. APEC '06., 2006, pp. 6 pp.-, doi: 10.1109/APEC.2006.1620590.

[18] C. Tu, K. Ngo and R. Chen, "A Fast Non-Iterative Design Approach of One-Turn Inductor with Significant AC Flux using Commercially Available Components," 2022 IEEE Applied Power Electronics Conference and Exposition (APEC), 2022, pp. 1486-1491, doi: 10.1109/APEC43599.2022.9773655.

[19] NFPA, "070 National Electrical Code,", Quincy, MA, USA: 2017

A Highly Integrated Sensorless Field Oriented Control BLDC / PMSM Inverter with 99% Efficiency Enabled by an All-in-one System Integrated Full SiC Intelligent Power Module (sIPM®)

Fu-Jen Hsu[1,2], Cheng-Tyng Yen[1], Hsiang-Ting Hung[1], Guan-Wei Lin[3], Chih-Feng Huang[3], Lung-Sheng Lin[3], I-Chi Lin[3], Chih-Fang Huang[2], Ta-Yung Yang[3]

[1] Fast SiC Semiconductor Inc., Hsinchu, Taiwan, E-mail: Frane.Hsu@fastsic.com,
[2] National Tsing Hua University, Hsinchu, Taiwan
[3] Richtek Technology Corp., Hsinchu, Taiwan

Abstract—**SiC MOSFETs are a beneficial solution in motor inverter applications because of their remarkable intrinsic diode reverse recovery behavior. SiC MOSFETs have successfully emerged as the best choice as devices for realizing high performance and high power motor drive in, for example, electric vehicles, but have rarely been used in low power motor drive for applications. In this work, an all-in-one system integrated SiC intelligent power module, which integrates the MCU, memory, SRAM, motor control library, ADC, DAC, gate drivers, and SiC MOSFETs was proposed to explore the benefits and potential of SiC MOSFETs in highly integrated BLDC/PMSM motor drive. As a result, SiC solution generally saves over 80% of space, improves the thermal performance by 27.7°C, and enhances 3% of efficiency compared to the silicon counterparts.**

Keywords- BLDC, PMSM, SiC MOSFET, IPM, motor drive

I. INTRODUCTION

With the advance in technologies, consumers are continuously demanding more compact, lightweight products which require high efficiency and high density power systems. SiC MOSFETs have successfully emerged as the best choice as devices for realizing high performance and high power motor drive in, for example, electric vehicles, but have rarely been used in low power brushless DC (BLDC) and permanent-magnet synchronous (PMSM) motor drive for applications such as medical equipment, home appliance, building controls, and industrial automation.

The trend of BLDC/PMSM motor drives is moving to increase the level of integration to save space, reduce peripheral components, shrink system size, and decrease maintenance. For BLDC/PMSM motor drives with high-level integration, one of the most stringent challenges is thermal management in an increasingly limited space.

In this work, an All-in-one system integrated full SiC intelligent power module (abbreviated as Full SiC sIPM®), which integrates an ARM-based 32-bit Cortex-M0 CPU with 60MHz of frequency, 16KB of memory, 4KB of SRAM, internal ROM with embedded motor control library, 7 channels of 10-bit ADC, 1 channel of 8-bit DAC, high/low-side gate drivers with UVLO and deadtime interlock algorithm, propagation delay matching strategy, and 6 SiC MOSFETs was proposed to explore the benefits and potential of SiC MOSFETs in highly integrated BLDC/PMSM motor drive.

II. EXPERIMENT

In this work, a 650V/4A SiC MOSFET with $R_{DS(on)}$= 420mΩ@V_{GS}=15V was designed to fulfill the system requirements for the proposed sIPM®. These include low $R_{DS(on)}$, low reverse recovery charge Q_{rr}, low forward voltage of body diode, low switching loss, low EMI, high immunity to the noise, good short-circuit capability, and avalanche ruggedness [1] – [5]. Table I includes some characteristics of the devices-under-test. The typical *I-V* curves are shown in Fig. 1(a) to (c). Fig. 1(d) shows the reverse recovery characteristic, with Q_{rr} less than 10nC subtracting output charge Q_{oss}. Fig. 1(e) shows the Q_{oss} characteristic related to the V_{DS} voltage. Fig. 1(f) shows the clamped inductive switching test, known as the "double pulse test", results of the proposed SiC MOSFET. The ultra-low turn-off energy helps the sIPM® system gain more efficiency.

TABLE I
CHARACTERISTICS BETWEEN SiC AND Si SOLUTIONS IN 25°C

Item	Value		Unit
	SiC	Si	
Chip size	1.0	5.1	p.U.
DC characteristics			
BV_{DSS}	650	650	V
I_D	4	4	A
$V_{GS(th)}$	1.6	3.0	V
E_{AS}	140	109	mJ
T_J	175	150	°C
AC characteristics			
C_{iss}	300	550	pF
C_{oss}	22	60	pF
C_{rss}	3.5	5	pF
Body diode characteristics			
V_{SD}	2.45	1.4	V
Q_{rr}	12.3	141	nC

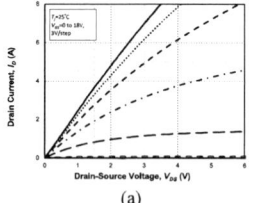 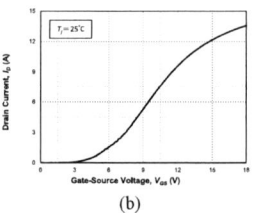

(a)   (b)

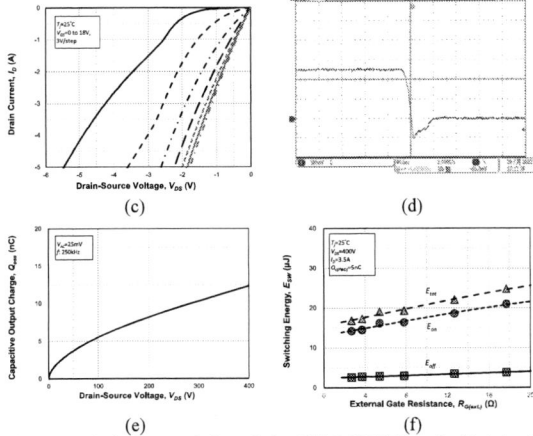

(c)　　　　　　　　　　(d)

(e)　　　　　　　　　　(f)

Fig. 1 Typical characteristics of the SiC MOSFET embedded in the proposed sIPM®. (a) I_D-V_D, (b) I_D-V_G, (c) 3rd quandrant I_D-V_D (d) reverse recovery waveform of the body diode at V_{DS}=400V, di/dt=300A/μs, I_{SD}=2A, the Q_{rr} (incl. Q_{oss}) =12.3nC, T_{rr}=30.5ns, I_{rm}=0.73A (e) Q_{oss}, (f) Test results of clamped inductive switching with proposed SiC MOSFET

Fig. 2(a) shows the short-circuit withstanding time with V_{DS} =310V (same as the rated voltage of the sIPM® systems) voltage. Although the peak current is approximately 34A and simultaneously the V_{DS} value maintains at 310V high level, the proposed SiC MOSFET still works normally. This means that the proposed SiC MOSFET could endure the short circuit time with over 4.7μs at V_{DS} =310V, which is far more than the protection triggering delay time, which is typically lower than 3μs, of the gate driver units. Fig. 2(b) shows avalanche ruggedness test waveforms of the proposed SiC MOSFET, in this test, the proposed SiC MOSFET could endure over 140mJ of avalanche energy without failure.

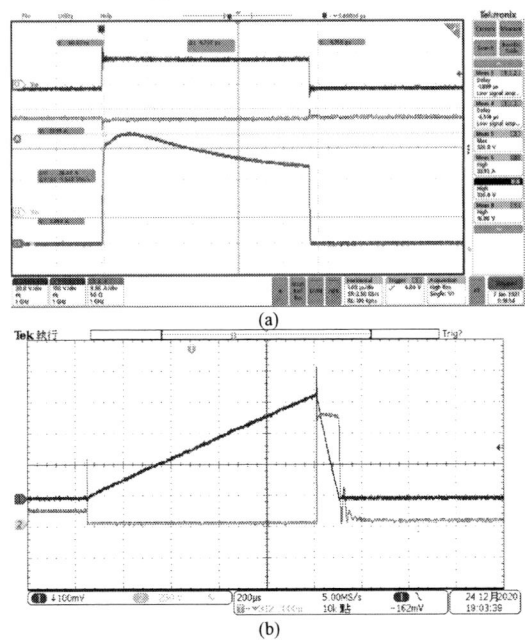

(a)

(b)

Fig. 2 Ruggedness performances of the proposed SiC MOSFET embedded in the proposed sIPM®. (a) short-circuit test waveform at V_{DS}=310V, and (b) avalanche waveform at L=25mH, as a result, E_{AS} >140mJ.

Meanwhile, the proposed sIPM® has been implemented in a motor for air-cleaner with 100W rated power. Fig. 3(a), 3(b), and 3(c) show the outlook of the motor and the neat motor drive PCB enabled by the proposed sIPM®. Fig 2(d) shows the tiny SiC MOSFETs just occupy a small space, suggesting that a larger chip can be used for higher power levels. Table II shows the major parameters of the driving motor.

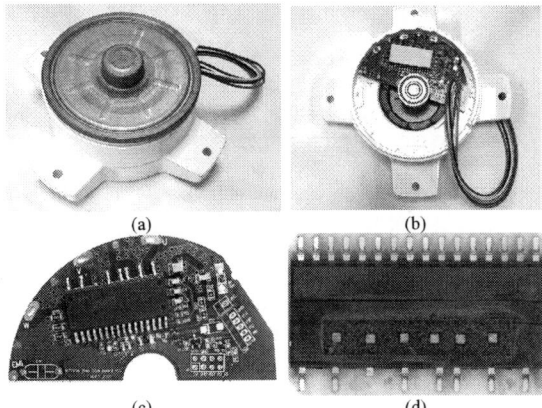

(a)　　　　　　　　　　(b)

(c)　　　　　　　　　　(d)

Fig. 3 (a) Overveiw of the all-in-one motor solution (b) The inner view of the all-in-one motor module, the proposed sIPM® is mounted to the outer metal case by the thermal paste. (c) The peripheral components and PCB of the driving and controlling circuitries mounted on the tested motor. Noted that there are only less than 29 pieces of components required on the PCB. (d) The proposed full SiC sIPM® with integrated MCU, drivers, bootstrap circuits, and 6 of SiC MOSFETs as the three-phase-bridge power stage in a single 23-Lead SOP plastic package body. The typical dimension is around 29mm*12mm*3.3mm (L*W*H).

TABLE II
MAJOR SPECIFICATIONS OF THE OPERATED MOTOR

Item	Value	Unit
Rotational speed	300 to 1,700	rpm
Rated voltage	310	V
Line inductance	18	mH
Line resistance	12.5	Ω

Fig. 4 plots a general sensorless Field Oriented Control (FOC) algorithm [6], which can be compiled into the embedded MCU in the sIPM® simply by the prepared motor control library. The embedded motor control library could greatly help users to implement the sensorless control method quickly. Also, the internal structures are demonstrated in Fig. 5, illustrating the high level of integration of the proposed sIPM® which integrate a 60MHz ARM 32-bit Cortex-M0 CPU, LDOs, gate drivers, bootstrap diodes and SiC MOSFETs in a single SOP-23 package.

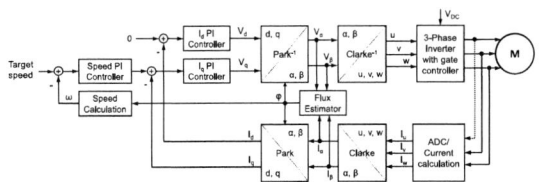

Fig. 4 The block diagram of a general sensorless Field Oriented Control.

978-1-6654-8901-0/22 $31.00 © 2022 IEEE

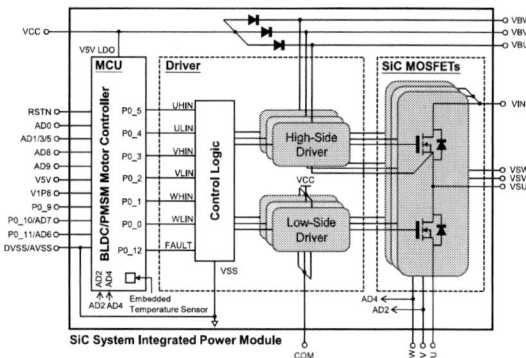

Fig. 5 The function block of the whole sIPM®.

III. RESULT AND CONCLUSION

Fig. 6 shows the waveforms of the proposed motor driving system, whereas Fig. 6(a) shows the waveforms of phase currents used to drive the motor with the proposed sIPM®. Fig. 6(b) shows the waveforms of phase voltages of the motor with the proposed sIPM®. The highly integrated intelligent power module could minimize the stray inductance between the SiC MOSFETs and the gate driver units. As a result, the voltage overshoot could be suppressed at lower than 5.5%

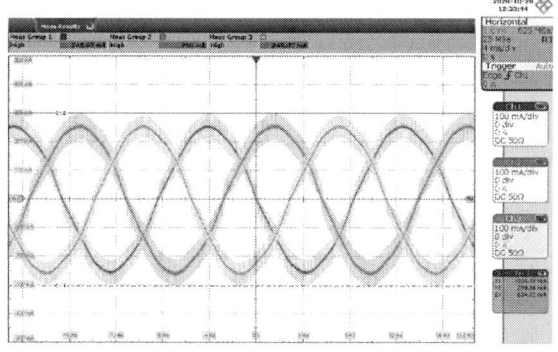

(a)

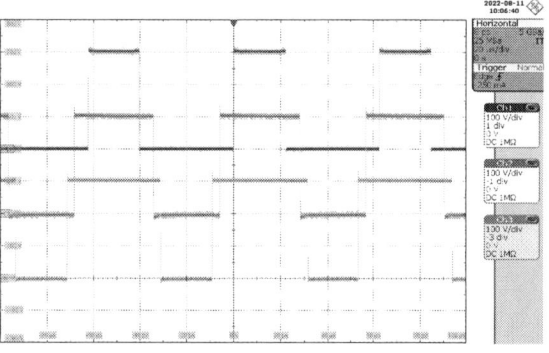

(b)

Fig. 6 (a) Phase current and (b) phase voltage of the tested motor

Fig. 7 shows the sensed temperature with increasing phase current I_q. In Fig. 7, with I_q=0.7A, the corresponding output power is 100W, and the 14.4°C of sensed temperature increase suggested that the overall power loss is less than 1%, meaning the proposed full SiC MOSFET sIPM® delivers an overall

efficiency of more than 99%. In comparison, the estimated efficiency of the original solution with Si MOSFET is only about 96%. The reduced heat and 3% improvement of efficiency are realized by the soft reverse recovery, low Q_{rr} and relatively low forward voltage drop, low switching loss, and low $R_{DS(on)}$ of this SiC MOSFET.

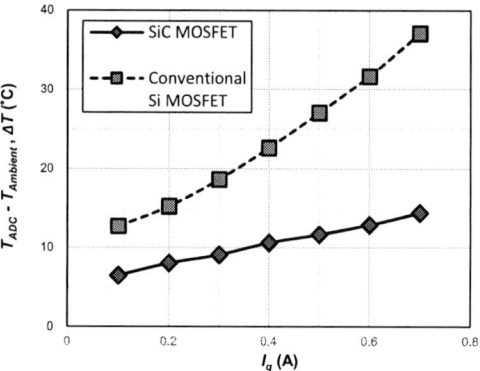

Fig. 7 The steady-state thermal difference between a sIPM® with a conventional Si-based MOSFET and proposed SiC MOSFET. The value Iq is the maximum phase current of the tested motor and ΔT is the difference of the built-in temperature sensor which is nearby the power transistors from the ambient temperature.

TABLE III
COMPARISON OF THE KEY SWITCHING TIMES BETWEEN THE
PROPOSED SIC AND SI SOLUTIONS

Item	Value		Unit
	SiC	Si	
T_f (low side FET conduction)			
U-Phase	25.52	30.56	ns
V-Phase	23.90	31.59	ns
W-Phase	23.06	33.72	ns
T_f (low side body diode conduction)			
U-Phase	101.89	61.59	ns
V-Phase	97.70	63.93	ns
W-Phase	97.46	64.04	ns
T_r (high side FET conduction)			
U-Phase	24.26	31.57	ns
V-Phase	23.24	30.15	ns
W-Phase	23.00	28.63	ns
T_r (high side body diode conduction)			
U-Phase	102.65	61.67	ns
V-Phase	101.22	64.09	ns
W-Phase	99.33	65.24	ns

Table III and Fig. 8 show the related switching behaviors and the comparison of the SiC system and silicon system. To eliminate the switching losses, when the high-side or low-side transistor turn-on/off, the transition time should be as fast as possible. In contrast, when the body diodes transfer from blocking mode to the free-wheeling mode, the transition time should be longer to prevent the EMI issue and the I_{rm} spikes. From Table II, we could clearly find out that the SiC MOSFET solution could simultaneously meet these two criteria – which means it could perform a shorter transistors' on/off transition time and meanwhile provides a smooth and gentle free-wheeling transition behavior. These advantages support the system to enhance its efficiency and suppress the ringing phenomenon and the discontinuity of phase current.

978-1-6654-8901-0/22 $31.00 © 2022 IEEE

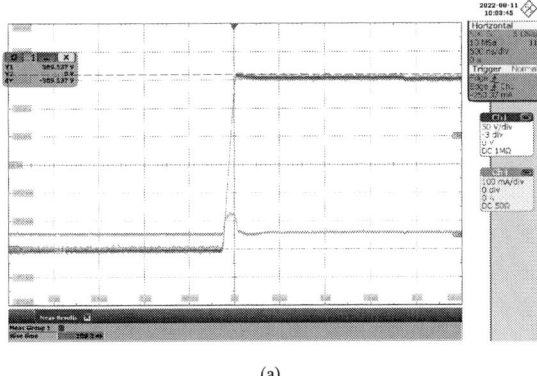

(a)

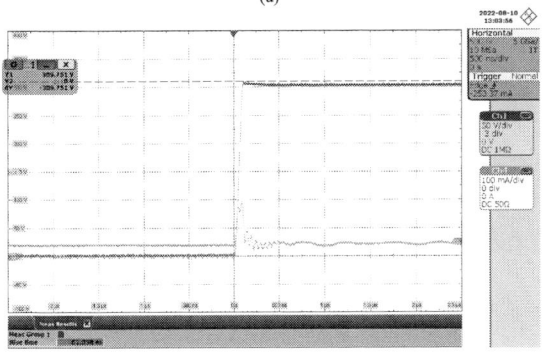

(b)

Fig. 8 The switching waveform in the body diode free-wheeling mode. (blue: phase voltage, brown: reverse recovery current) (a) proposed SiC MOSFET solution, (b) conventional silicon MOSFET solution.

Fig. 9 presents the conduction mode EMI result of the proposed SiC sIPM® solution. Without any EMI filters, this intelligent power module could still pass the CISPR criteria under ~4dB. This result strongly proves the benefits of the usage of SiC MOSFET in BLDC motor drives.

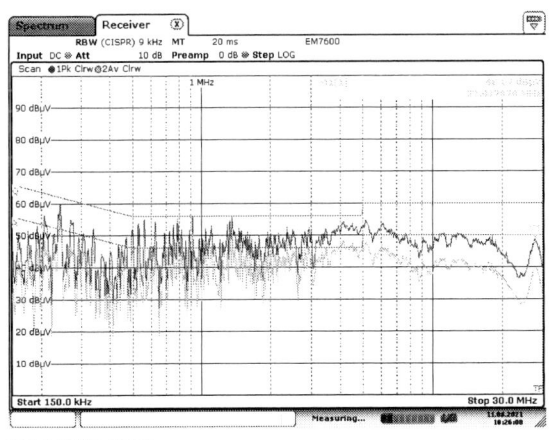

Fig. 9 The conduction EMI results of the purposed all-in-one SiC sIPM® motor driver in Fig. 2. This result is measured without any EMI filters. The result of SiC is perfectly pass the criteria.

IV. CONCLUSION

In conclusion, a SiC MOSFET based all-in-one sIPM® BLDC/PMSM solution is proposed in this work. The depressed noise, good short-circuit capability, and ruggedness avalanche of this SiC MOSFET ensures the reliable and stable operation of the motor drive. The chip area of the proposed SiC MOSFET saves over 80% of space compared to the original Si MOSFET. The benefits of SiC MOSFETs provided in efficiency and thermal management will allow a higher power level in the same form-factor package or the same power level in even smaller packages. On the other hand, when providing better efficiency, the highly integrated SiC solution could still maintain a good EMI performance. These results could push the system further compact and more efficient.

REFERENCES

[1]   D Kastha *et al.*, "Investigation of fault modes of voltage-fed inverter system for induction motor drive", *IEEE Transactions on Industry Applications,* Aug. 1994

[2]   R. S. Chokhawala *et al.*, "A discussion on IGBT short-circuit behavior and fault protection schemes", *IEEE Transactions on Industry Applications,* Apr. 1995.

[3]   F.-J. Hsu *et al.*, "Short-circuit ruggedness analysis of SiC JMOS and DMOS", *IEEE International Symposium on Power Semiconductor Devices & IC's,* May. 2019.

[4]   A. Fayyaz *et al.*, "UIS failure mechanism of SiC power MOSFETs", *IEEE Workshop on Wide Bandgap Power Devices and Applications,* 2016.

[5]   C.-T. Yen *et al.*, "Avalanche ruggedness and reverse-bias reliability of SiC MOSFET with integrated junction barrier controlled Schottky rectifier", *IEEE International Symposium on Power Semiconductor Devices & IC's,* May. 2018.

[6]   B. Akin *et al.*, "Sensorless Field Oriented Control of 3-Phase PMSM", *Application Note, Texas Instruments* 2013.

Noise Analysis of Current Sensor for Medium Voltage Power Converter Enabled by Silicon-Carbide MOSFETs

Morten Rahr Nielsen
AAU Energy
Aalborg University
Aalborg, 9220, Denmark
mrni@energy.aau.dk

Mathias Kirkeby
AAU Energy
Aalborg University
Aalborg, 9220, Denmark
kirkeby.mathias@gmail.com

Hongbo Zhao
AAU Energy
Aalborg University
Aalborg, 9220, Denmark
hzh@energy.aau.dk

Dipen Narendra Dalal
AAU Energy
Aalborg University
Aalborg, 9220, Denmark
dnd@energy.aau.dk

Michael Møller Bech
AAU Energy
Aalborg University
Aalborg, 9220, Denmark
mmb@energy.aau.dk

Stig Munk-Nielsen
AAU Energy
Aalborg University
Aalborg, 9220, Denmark
smn@energy.aau.dk

Abstract—New semiconductor devices based on wide bandgap materials are emerging in medium voltage power electronic converter applications, presenting new opportunities to the industry relying on semiconductor devices. SiC MOSFETs with blocking voltages of 10 kV (and above) is a promising technology, however, their fast switching transitions result in increased output voltage slew rate (dv/dt), which poses challenges to the applicability of the SiC MOSFET technology. This paper examines the impact of the increased dv/dt on the applicability of an off-the-shelf closed loop hall-effect current sensor when utilized in a medium voltage SiC MOSFET based power electronics converter. Analysis of the capacitive couplings in the current sensor is presented along with an experimental determination of the parasitic capacitance between its primary conductor and secondary winding. Experimental measurements have identified two distinct noise components in the current sensor measurement path due to: 1) capacitive coupling between the primary conductor and secondary winding, and 2) an inferred capacitive coupling into the active circuitry of the current sensor.

Index Terms—SiC, MOSFET, Medium Voltage, Current Sensor, Capacitive Coupling, Parasitics

I. INTRODUCTION

At present days, extensive research is being carried out to exploit the emerging silicon-carbide (SiC) MOSFETs as an enabling technology in future medium voltage (MV) power electronic converter (PEC) applications [1], [2]. The SiC technology offers higher breakdown voltage, lower switching and conduction losses, and increased operating temperature compared to the traditional and mature silicon technology [3], [4]. However, the recent technological progress uncovers new challenges, for instance, new requirements for the design of power electronic devices due to the increased electrical field stresses, higher switching frequencies, and increased voltage slew rates (dv/dt) [2].

The increased dv/dt of the SiC MOSFET output voltage can cause capacitive displacement currents within the system, which leads to higher switching losses and EMI issues [5], [6], therefore, research in this field is currently focused on reducing the parasitic capacitive couplings within the system, e.g. in the power module packaging [7], the gate drivers [8], [9], and the filter inductors [6], [10], [11]. A critical parasitic capacitance within the system is the output capacitance as it highly influences the switching performance of the SiC MOSFET due to the charging or discharging of the output capacitance during each switching transition [12], [13]. The output capacitance comprises the internal output capacitance of the power module and the parasitic capacitance of whichever equipment being connected to the output terminal of the power module that experiences the dv/dt (i.e. filter inductors, sensors, etc.).

In [14], the measured high frequency current component overlaying the fundamental current component is reported to be caused by the parasitic capacitance of the filter inductor. The high frequency current component has a significant magnitude compared to the fundamental component, approaching a signal-to-noise ratio (SNR) of unity, which could prove critical for the integrity of the current measurement when utilized for control and protective purposes.

Research pertaining to the challenges arising within current sensing and sampling for closed loop feedback of MV PEC applications is lacking. Therefore, this paper aims at analyzing the impact of the parasitic capacitive couplings in the current sensor and the integrity of current measurements taken with an off-the-shelf closed loop hall-effect current sensor.

II. EVALUATED CURRENT SENSOR

Current sensing is required in a wide range of applications covering many different voltage and current levels, and there-

978-1-6654-8901-0/22 $31.00 © 2022 IEEE

fore, each application has specific requirements for the current sensor regarding size, cost, accuracy, range, bandwidth, and insulation [15]. A presumably current sensing technology for MV PEC applications to meet the above requirements is the mature closed loop hall-effect current sensor technology based on Faraday's law of induction and magnetic field sensing. Closed loop hall-effect sensors have been widely used in low voltage PEC applications, however, their applicability in MV PEC applications has not yet been explored. In this paper, the closed loop hall-effect current sensor LA 55-P from LEM [16] has been chosen as a candidate solution to evaluate its performance when utilized in a MV PEC application.

When the current sensor is placed directly at the output terminal of the half-bridge power module, the voltage at the primary conductor of the current sensor experiences the same dv/dt as the output terminal of the half-bridge power module during each switching transition, which gives rise to concerns regarding parasitic couplings from the primary conductor to the current sensor. Parasitic couplings can, in principle, occur to all parts of the current sensor, however, the parasitic capacitance from its primary conductor to secondary winding is of special concern in this analysis, as the presence of the parasitic capacitance could cause noise current to couple directly into the measurement path of the current sensor. The analytically expected peak magnitude of the noise current due to the parasitic capacitance is directly proportional to the parasitic capacitance, C, and the voltage slew rate across the capacitor, dv/dt, given by the general equation in (1).

$$i = C \frac{dv}{dt} \tag{1}$$

The slew rate of the power module output voltage has been measured to be as high as 30 kV/μs [17], mainly limited by the MOSFET gate resistor and the output capacitance. Even a small parasitic capacitance in the range of a few picofarads could cause hundreds of milliamperes of noise current coupling into the secondary winding of the current sensor, and therefore, directly into its measurement path.

Another important aspect that needs to be considered when working with current sensing of MV PEC applications is the turns ratio of the current sensor. Even a small displacement current in the secondary winding of the current sensor may correspond to a relatively large current in the primary conductor when taking the turns ratio of the current sensor into account.

III. PARASITIC CAPACITANCE EXTRACTION

The need for determining the parasitic capacitance has been emphasized in the last section, and therefore, this section covers an experimental determination of the parasitic capacitance of this certain current sensor.

A. Low Voltage, High Frequency Sweep

The experimental determination of the parasitic capacitance has been done by performing a low voltage, high frequency sweep of the voltage at the primary conductor and measuring the resulting output voltage across the measuring resistor, v_R.

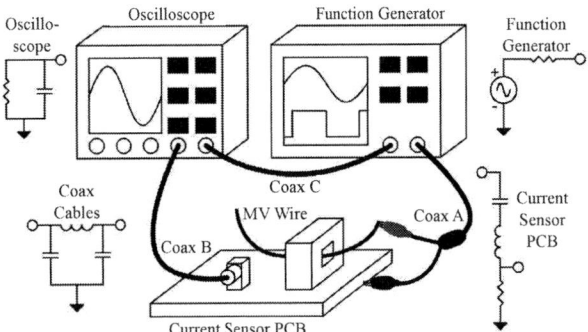

Fig. 1: Configuration of the laboratory setup for experimental determination of parasitic capacitance together with equivalent circuit representations.

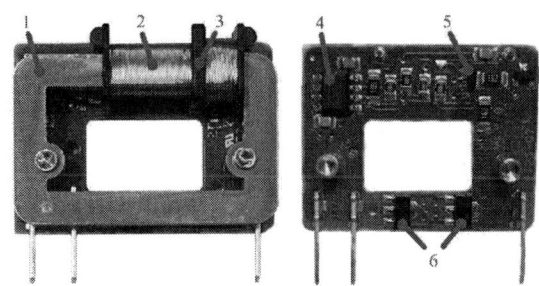

Fig. 2: Internal structure of the LEM LA 55-P with its outer casing removed. 1) Iron core, 2) Secondary winding which spans the air gap, 3) Air gap, 4) Op Amp, 5) Hall-effect element, and 6) Push-pull transistors.

The frequency sweep is performed with a sinusoidal input voltage of ± 10 V in the frequency range from 1 to 50 MHz using a Tektronix AFG31000 arbitrary function generator. Due to the nature of the high frequency extraction method, the importance of considering the parasitic circuit elements introduced by cables as well as input and output impedance of connected laboratory equipment is needed. The laboratory configuration is illustrated in Fig. 1 along with schematics of the equivalent circuit representations used to model the experimental setup in LTspice.

From the low voltage, high frequency sweep, the parasitic capacitance is found to be highly dependent on the placement of the primary conductor inside the current sensor opening due to its internal structure and layout, and a parasitic capacitance of 0.31 pF is taken as the worst-case with the primary conductor placed in close proximity to the secondary winding of the current sensor (upper right corner as shown in Fig. 2).

The experimental determination of the parasitic capacitance is further carried out for three primary turns, defined as the practical upper limit considering the cross-sectional area of the MV wire and the opening of the current sensor, to investigate the influence of adding additional primary turns on the parasitic capacitance. The measurements show that an increase in primary turns with a factor of three increases the parasitic capacitance to 0.65 pF, which is slightly above a factor of two. Hence, in cases where the rated current range of the current sensor allows for additional primary turns, it could prove possible to achieve a comparative lower parasitic capacitance and a lower SNR by increasing the number of primary turns.

IV. EQUIVALENT IMPEDANCE NETWORK FOR NOISE MODELING

With the parasitic capacitance from the primary conductor to secondary winding of the current sensor experimentally determined, an equivalent impedance network has been developed in LTspice. The equivalent impedance network aims to model the behavior of the noise current coupling into the measurement path of the current sensor and to predict its frequency as well as magnitude during a switching transition of the SiC MOSFET. The equivalent impedance network is shown in Fig. 3 in which the equivalent circuit representations indicated on Fig. 1 are used in the modeling.

The analysis of the parasitic capacitance between the primary conductor and secondary winding of the current sensor in Section II suggested that the noise current coupling directly into the measurement path of the current sensor could be modeled as a single capacitance. This simplification of utilizing a simple single-stage model implies that only the first harmonic component is represented by the equivalent impedance network, which needs to be considered when utilizing the model to predict the behavior of the noise current.

Even though the capacitive coupling is relatively small compared to other capacitive couplings within the system, it can cause displacement currents of more than 9 mA in the secondary winding, corresponding to 9 A in the primary conductor when taking into account the turns ratio of the current sensor. The magnitude of the displacement currents has been calculated using (1) at a $\mathrm{d}v/\mathrm{d}t$ of 30 kV/μs for a 10 kV SiC MOSFET half-bridge power module [17].

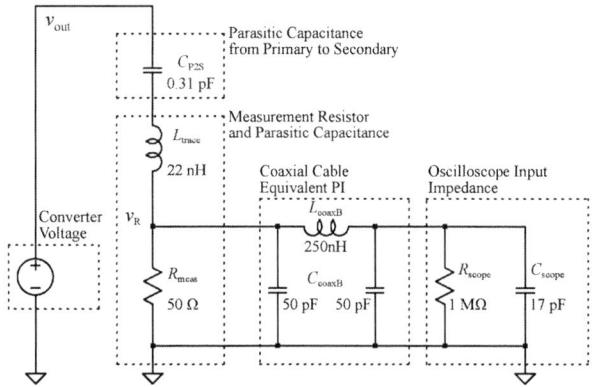

Fig. 3: Equivalent impedance network for noise modeling in LTspice. The converter voltage used as input is the experimentally measured output voltage, v_{out}, from the SiC MOSFET half-bridge power module.

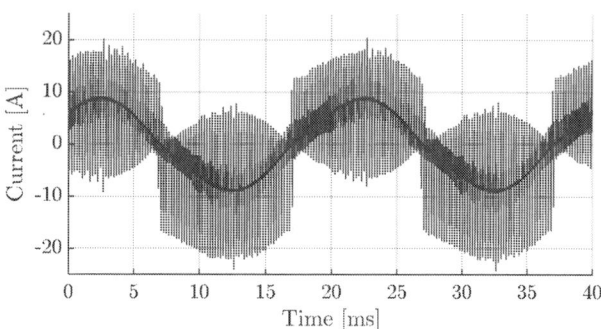

Fig. 4: Continuous current measurement with current sensor, LEM LA 55-P, at 6 kV DC-link voltage and a RMS load current of approx. 7 A.

V. NOISE CURRENT EVALUATION

The actual noise present in the current sensor has been measured during open loop operation of a PEC comprising 10 kV, 20 A, 350 mΩ SiC MOSFET 3rd gen. dies populated in a single-chip half-bridge power modules engineered and manufactured in-house at AAU Energy. A continuous current measurement is shown in Fig. 4, which shows a similar high frequency current component overlaying the fundamental component as the one reported in [14]. These tests have been performed on a single phase test setup shown in Fig. 5 with two half-bridge 10 kV SiC MOSFET power modules connected back-to-back through a shared DC-link and with the half-bridge outputs connected through two series connected 30 mH MV filter inductors to circulate the power. The filter inductors used are designed with a relatively low capacitive coupling of 50 pF terminal to terminal [10].

The experimental measurements have identified two distinct noise current components instead of only one as the initial analysis suggested, and therefore, only included in the equivalent impedance network. The first noise component is caused by the direct capacitive coupling, C_{P2S}, into the measurement path of the current sensor as indicated with a red current path in Fig. 6. The second one is inferred to be caused by capacitive coupling, C_{P2H}, into the hall-effect element resulting in noise currents disturbing the active push-pull amplifier circuit of the current sensor as indicated with a purple current path in Fig. 6.

A. First Noise Component

The measurement of the first noise component was made without an auxiliary power supply to the current sensor and the results are shown in Fig. 7 for a turn-on switching transition of the high-side (HS) MOSFET for varying DC-link voltages. The measurements are performed at zero load current corresponding to the highest $\mathrm{d}v/\mathrm{d}t$ for a switching transition. The upper plot shows the measured output voltage, v_{out}, of power module A. In the lower plot, the solid lines indicate the measured noise current, whereas the dashed lines indicate the corresponding results from the equivalent impedance network in LTspice with the measured output voltage used as converter

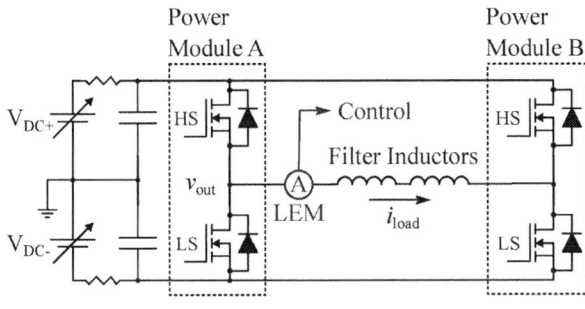

(a) Electrical schematic.

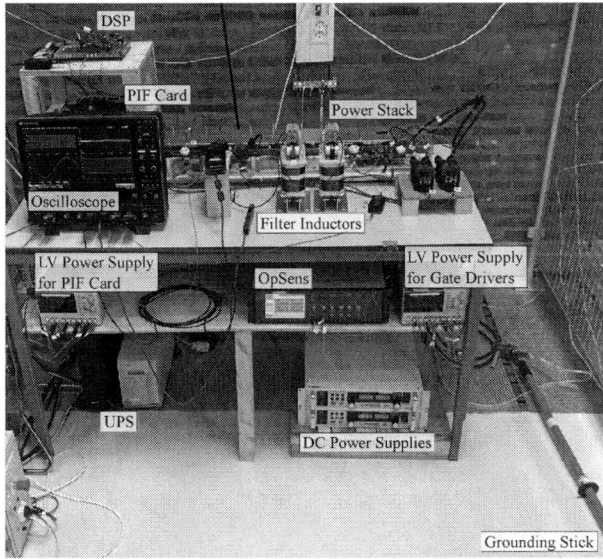

(b) Photograph from the laboratory.

Fig. 5: Single phase test setup used to analyze the impact of switching noise in the current sensor when utilized in a medium voltage power conversion application.

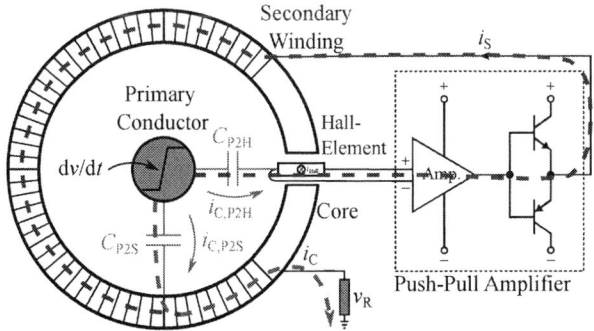

Fig. 6: Schematic of the closed loop hall-effect current sensor with its key parasitic capacitances indicated. The red current path indicates the first noise component and the purple current path indicates the second noise component.

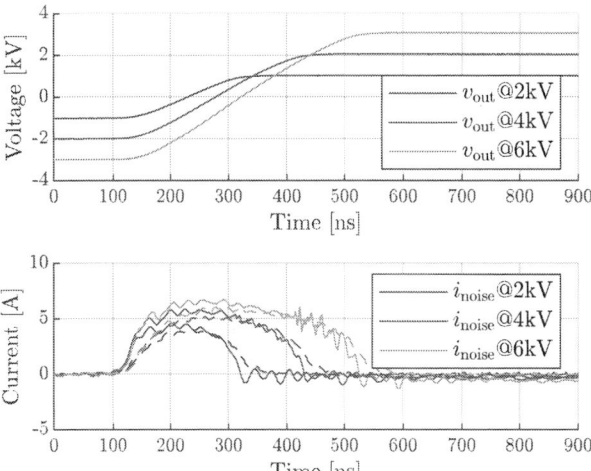

Fig. 7: Measurement of first noise component without an auxiliary power supply to the current sensor. Solid lines are measured values and dashed lines are simulated values from the equivalent impedance network in LTspice shown in Fig. 3 using the measured output voltages as input.

voltage, v_{out}. A high degree of agreement in both magnitude and frequency can be seen between the measured and simulated noise current in Fig. 7 when the current sensor has no auxiliary power supply.

B. Second Noise Component

The measurement of the second noise component was made with an auxiliary power supply to the current sensor and the results are shown in Fig. 8 for a turn-on switching transition of the HS MOSFET at zero load current for varying DC-link voltages. It can be noticed that turning on the current sensor reveals some slower acting response in the noise picked up by the current sensor. The first noise component is presented in a nanosecond timescale, and even though still largest in magnitude, now only covers a fraction of the microsecond timescale associated with the additional noise introduced when the auxiliary power supply is on.

Even though the second noise component is smaller in magnitude compared to the first noise component, it is much more critical for the current sensing and sampling due to its longer duration as will be discussed in the subsequent section.

VI. PROPOSED SAMPLING SCHEME

The identified noise components can negatively impact the integrity of the current sampling if a conventional sampling scheme is adopted as the second noise component takes more than 20 μs to decay to a sufficient level. Therefore, an alternative sampling scheme is proposed to reduce this negative impact by ensuring that the switching noise has had time to decay. The proposed sampling scheme is based on a double sample, single update principle such that samples are taken on the ZERO and PERIOD counter compare events.

978-1-6654-8901-0/22 $31.00 © 2022 IEEE

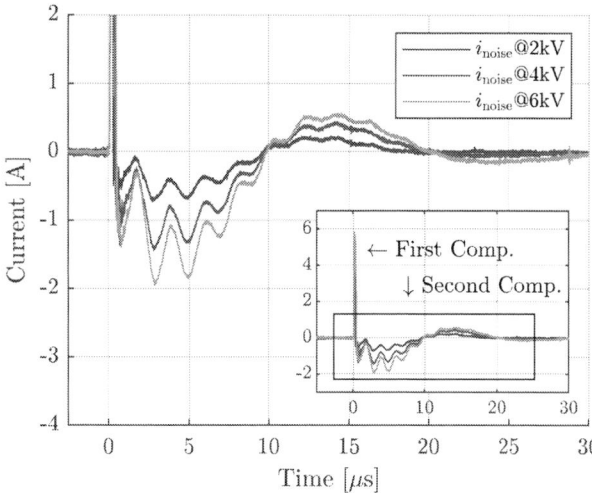

Fig. 8: Measurement of second noise component with an auxiliary power supply to the current sensor. All values are measured experimentally.

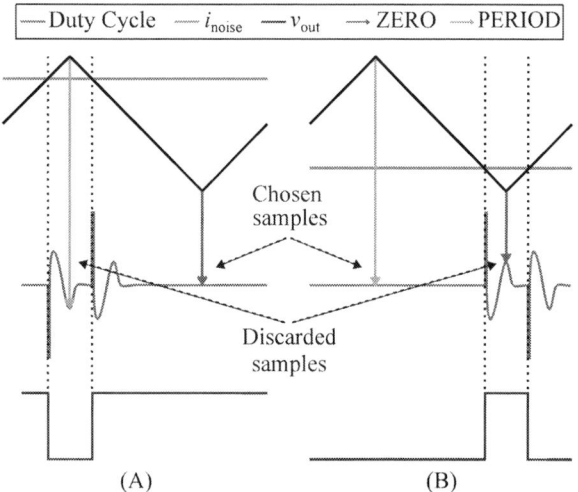

Fig. 9: Illustration of proposed sampling scheme at a (A) high duty cycle and (B) low duty cycle.

Based on the previously calculated duty cycle, one of the current samples is stored/chosen for further computation and the other discarded as shown in Fig. 9. Currents sampled on the ZERO counter compare event are stored for duty cycles above 50%, whereas currents sampled on the PERIOD counter compare event are stored for duty cycles below 50% to ensure that the switching noise has had time to decay.

The proposed sampling scheme has been experimentally evaluated in the setup shown in Fig. 5 at a DC-link voltage of 6 kV and a RMS load current of 7 A. The evaluation compares the sampled currents (ZERO, PERIOD, and chosen) with the voltage across the measuring resistor, v_R, of the current sensor converted into its corresponding noise current as shown in

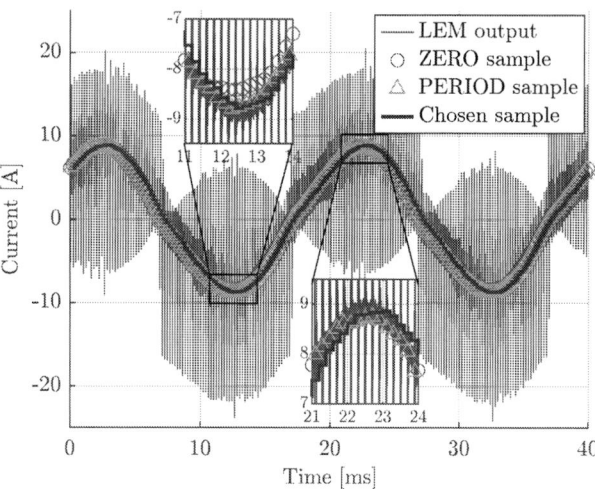

Fig. 10: Sampled currents compared to the measurement output of the current sensor, LEM LA 55-P, converted into its corresponding current.

Fig. 10. The effectiveness of the sampling scheme is most distinct at low or high duty cycles as the noise components are disturbing the sampled current.

It should be noticed that since the converter is operated in open loop without an active dead time compensation, the current waveforms are to be expected to have some distortion around its zero crossings due to converter non-linearities such as dead time voltage error and system parasitics.

VII. CONCLUSION

This paper analyzes the impact of the increased dv/dt of the MV SiC MOSFETs on the applicability of an off-the-shelf current sensor when utilized in a MV PEC application. The paper presents an analysis of the capacitive couplings introduced by the current sensor and an experimental determination of the parasitic capacitance between its primary conductor and secondary winding of 0.31 pF.

An equivalent impedance network has been proposed to predict the behavior of the noise current coupling into the measurement path of the current sensor during a switching transition of the SiC MOSFETs. An experimental comparison is carried out to validate the equivalent impedance network with a high degree of accuracy.

The experimental measurements have identified two distinct noise components in the measurement path of the current sensor, which negatively impacts the integrity of the current sampling if a conventional sampling scheme is adopted.

Therefore, an alternative sampling scheme is proposed and experimentally validated to reduce the negative impact of the switching noise on the integrity of the current sampling.

ACKNOWLEDGMENT

The author would like to thank the research projects MVolt and ComElCo at AAU Energy, Aalborg University, for their economic funding.

REFERENCES

[1] J. B. Casady et al., "New Generation 10kV SiC Power MOSFET and Diodes for Industrial Applications," Proceedings of PCIM Europe 2015; International Exhibition and Conference for Power Electronics, Intelligent Motion, Renewable Energy and Energy Management, 2015, pp. 1-8.

[2] J. Thoma, B. Volzer, D. Kranzer, D. Derix and A. Hensel, "Design and Commissioning of a 10 kV Three-Phase Transformerless Inverter with 15 kV Silicon Carbide MOSFETs," 2018 20th European Conference on Power Electronics and Applications (EPE'18 ECCE Europe), 2018, pp. P.1-P.7.

[3] E. Van Brunt, D. Grider, V. Pala, S. -H. Ryu, J. Casady and J. Palmour, "Development of medium voltage SiC power technology for next generation power electronics," 2015 IEEE International Workshop on Integrated Power Packaging (IWIPP), 2015, pp. 72-74, doi: 10.1109/IWIPP.2015.7295981.

[4] P. Gammon, "Silicon and the wide bandgap semiconductors, shaping the future power electronic device market," 2013 14th International Conference on Ultimate Integration on Silicon (ULIS), 2013, pp. 9-13, doi: 10.1109/ULIS.2013.6523479.

[5] D. N. Dalal et al., "Impact of Power Module Parasitic Capacitances on Medium-Voltage SiC MOSFETs Switching Transients," in IEEE Journal of Emerging and Selected Topics in Power Electronics, vol. 8, no. 1, pp. 298-310, March 2020, doi: 10.1109/JESTPE.2019.2939644.

[6] V. U. Pawaskar, V. T. Nguyen, G. Gohil and P. T. Balsara, "Harmonic filter with low coupling capacitance for Medium Voltage, high dv/dt PWM converters," 2020 IEEE Energy Conversion Congress and Exposition (ECCE), 2020, pp. 3245-3252, doi: 10.1109/ECCE44975.2020.9235838.

[7] A. B. Jørgensen et al., "Reduction of parasitic capacitance in 10 kV SiC MOSFET power modules using 3D FEM," 2017 19th European Conference on Power Electronics and Applications (EPE'17 ECCE Europe), 2017, pp. P.1-P.8, doi: 10.23919/EPE17ECCEEurope.2017.8098962.

[8] D. N. Dalal et al., "Gate driver with high common mode rejection and self turn-on mitigation for a 10 kV SiC MOSFET enabled MV converter," 2017 19th European Conference on Power Electronics and Applications (EPE'17 ECCE Europe), 2017, pp. P.1-P.10, doi: 10.23919/EPE17ECCEEurope.2017.8099274.

[9] A. Anurag, S. Acharya, N. Kolli and S. Bhattacharya, "Gate Drivers for Medium-Voltage SiC Devices," in IEEE Journal of Emerging and Selected Topics in Industrial Electronics, vol. 2, no. 1, pp. 1-12, Jan. 2021, doi: 10.1109/JESTIE.2020.3039108.

[10] H. Zhao et al., "Physics-Based Modeling of Parasitic Capacitance in Medium-Voltage Filter Inductors," in IEEE Transactions on Power Electronics, vol. 36, no. 1, pp. 829-843, Jan. 2021, doi: 10.1109/TPEL.2020.3003157.

[11] R. A. G. Jimenez, G. G. Oggier, D. P. Fernandez and J. C. Balda, "Analysis of Current Resonances due to Winding Parasitic Capacitances in Medium-Voltage Medium-Frequency Transformers," 2022 IEEE Applied Power Electronics Conference and Exposition (APEC), 2022, pp. 939-943, doi: 10.1109/APEC43599.2022.9773427.

[12] X. Song, A. Q. Huang, M. Lee and G. Wang, "A dynamic measurement method for parasitic capacitances of high voltage SiC MOSFETs," 2015 IEEE Energy Conversion Congress and Exposition (ECCE), 2015, pp. 935-941, doi: 10.1109/ECCE.2015.7309788.

[13] X. Huang, S. Ji, J. Palmer, L. Zhang, L. M. Tolbert and F. Wang, "Parasitic Capacitors' Impact on Switching Performance in a 10 kV SiC MOSFET Based Converter," 2018 IEEE 6th Workshop on Wide Bandgap Power Devices and Applications (WiPDA), 2018, pp. 311-318, doi: 10.1109/WiPDA.2018.8569080.

[14] A. Kumar, S. Parashar and S. Bhtattacharya, "Continuous Heat Run Test of Latest Generation Power Modules for 10 kV 4H-SiC MOSFETs in Medium Voltage Power Converters," 2018 IEEE Energy Conversion Congress and Exposition (ECCE), 2018, pp. 1949-1955, doi: 10.1109/ECCE.2018.8557700.

[15] S. Ziegler, R. C. Woodward, H. H. -C. Iu and L. J. Borle, "Current Sensing Techniques: A Review," in IEEE Sensors Journal, vol. 9, no. 4, pp. 354-376, April 2009, doi: 10.1109/JSEN.2009.2013914.

[16] LEM, "Isolated current and voltage transducers," 2018, url: https://www.lem.com/en/file/3139/download

[17] D. N. Dalal et al., "Demonstration of a 10 kV SiC MOSFET based Medium Voltage Power Stack," 2020 IEEE Applied Power Electronics Conference and Exposition (APEC), 2020, pp. 2751-2757, doi: 10.1109/APEC39645.2020.9124441.

GaN Power Converter Applied to Electrocaloric Heat Pump Prototype and Carnot Cycle

Stefan Moench*, Richard Reiner*, Kareem Mansour*, Michael Basler*,
Patrick Waltereit*, Rüdiger Quay*, Kilian Bartholomé[†]

*Fraunhofer Institute for Applied Solid State Physics IAF, Freiburg, Germany (stefan.moench@iaf.fraunhofer.de)
[†]Fraunhofer Institute for Physical Measurement Techniques IPM, Freiburg, Germany (kilian.bartholome@ipm.fraunhofer.de)

Abstract—A 99% efficient power converter is applied to an electrocaloric heat pump prototype for the first time, enabling cyclic electrical field variation in electrocaloric capacitors and high heat pump performance. The electrocaloric effect is an almost fully reversible temperature change in special dielectrics caused by changed electrical field. The GaN power converter achieves high efficiency by zero-voltage switching hysteretic current control. Blocks of commercial multi-layer ceramic capacitors which exhibit an electrocaloric effect are efficiently charged and discharged, synchronized to contacting to a heat sink and source by actuators, which forms a heat pump prototype. Brayton cycles are caused by trapezoidal voltage. Then, arbitrary voltage (E-field) variation for Carnot-like cycles with quasi-isothermal heat transfer is demonstrated and verified by the measured almost constant heat flux during contact to the heat sink. The performance of electrocaloric heat pump prototypes in literature was limited by losses of LC resonant circuits, and is improved by an up to 15.5-fold decrease by this work. The work demonstrates electrocalorics as an emerging power electronics application and contributes to realize future efficient and emission-free, solid state, and electrocaloric heat pump systems.

Keywords—gallium nitride, half-bridges, electrocalorics, heat pumps, energy efficiency.

I. INTRODUCTION

An emerging emission-free heat pump (HP) technology is based on the electrocaloric (EC) effect, which is an almost fully reversible temperature change of some dielectrics as a result of electrical field change [1]. Fig. 1 illustrates a simplified EC heat pump (EC-HP) system [2]. One part is an electrical charging circuit, a GaN half-bridge dc-dc converter in this work, which charges and discharges EC capacitors (EC-C). These EC-Cs act also as heat transfer elements. Synchronized contacting to a heat sink or source pumps heat Q to one side, (cooling or heating). The HP system coefficient of performance ($COP = \dot{Q}/P$) is limited by the electrical loss W per cycle. Beside dielectric loss of the EC-C, and the work input for a thermodynamic cycle, the charging circuit loss mainly limits the COP. High charging efficiency η is required for high COPs [2]–[4]. For resistive loads, an improvement of the converter efficiency η, e.g., from 99% to 99.1%, seems irrelevant. However, for the EC capacitive-load application, the performance is determined by the loss $(1 - \eta)$ of the converter instead, which makes any efficiency improvement worthwhile. Large-scale HPs will have a high reactive charging power cycling in the circuit, up to 100 times higher than the

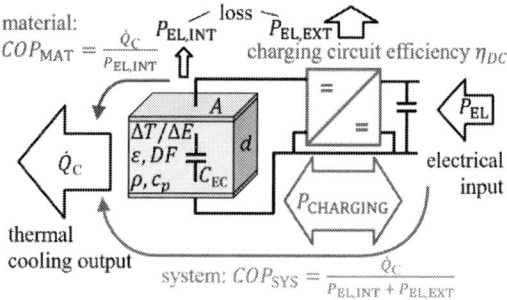

Fig. 1. An electrocaloric capacitor is charged and discharged to cause the electrocaloric effect (a reversible temperature change). High charging efficiency enables a high heat pump coefficient of performance [2].

electrical power loss. This power is equal to the electrical input power for the capacitive load. Wide bandgap semiconductors in converters enable at the same time a compact and highly-efficient system. To the best of the author's knowledge, for efficient charging of electrocaloric components so far only two approaches by E. Defay [3] (LIST, 80% efficient) and Y. Meng [5] (UCLA, 70% efficient) using resonant circuits were applied to prototypes. A Marx modulator adapted for electroactive polymers in [6] (88% efficient) could also be applied. The COP of EC-HP prototypes was however limited by LC resonant circuit loss. The authors recently proposed to use an over 99% efficient power converter [2] (and control [7]), which this work applies to a EC-HP prototype for the first time. Also, Carnot-like cycles are realized by field variation.

II. RESULTS

A. GaN power converter

EC-Cs are charged by a GaN half-bridge converter ($190\,\text{m}\Omega$, $600\,\text{V}$ GaN HEMTs, $600\,\mu\text{H}$ power inductor, as in [2]). Over 99% charging efficiency is achieved by using hysteretic current control with zero-voltage switching (implemented as described in [7]). For the selected operation points in this work, the resulting variable switching frequency is below $70\,\text{kHz}$. Two groups of the same number of EC-Cs ($C_{\text{EC},1}$ and $C_{\text{EC},2}$) are arranged between DC-terminals (input and ground) and output of the half-bridge converter as shown in Fig. 2. This arrangement does not require an additional large (dc-link) buffer capacitor at the input of the converter, in contrast to a setup where only one EC-C is connected to the converter (as

978-1-6654-8901-0/22 $31.00 © 2022 IEEE

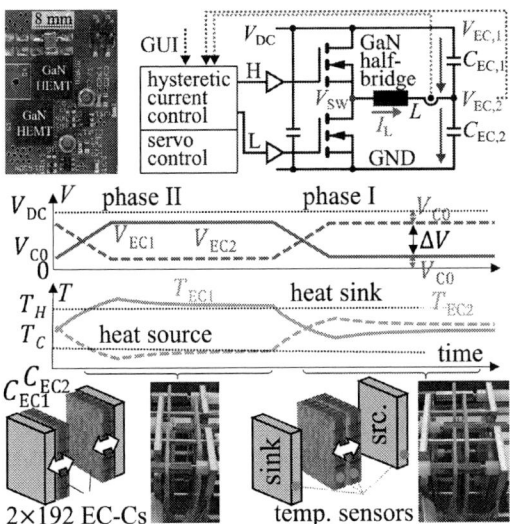

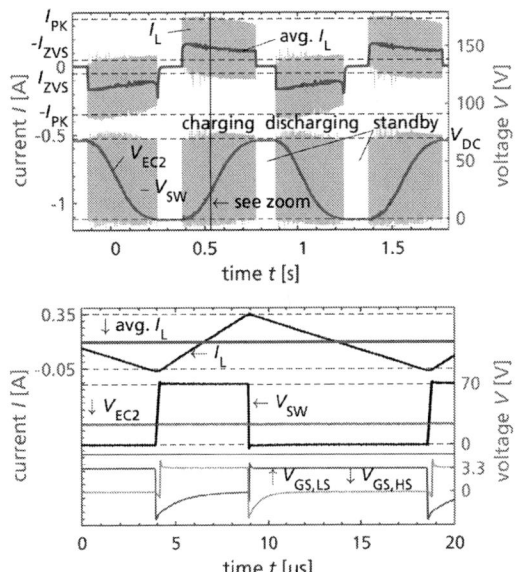

Fig. 2. A GaN half-bridge converter changes the E-field in EC capacitors to cause the electrocaloric effect (two phases) for electrocaloric heat pumping.

in [2]), since the energy from one EC-C is directly transferred to the other instead of to an external buffer capacitor. This strategy, using two capacitors, was also used with LC resonant circuits for electrocalorics in [3], [5] and is also known from efficient drivers for piezoelectric actuators [8]. From this arrangement it also follows that one capacitor is charged while the other is discharged (resulting in reversible temperature increase or decrease under adiabatic conditions), and vice versa. Fig. 2 shows the electrical circuit and the typical electrical waveforms with the typical thermal response (two-stage cascaded thermal configuration with Brayton cycles). The timing between electrical charging and mechanical movement (two phases, either 0° or 180°-phase shift between electrical signals and actuator, see Fig. 2) selects between heating or cooling.

Fig. 3 (top) shows measured switching waveforms (inductor current and averaged charging current; switch-node voltage and output-voltage equal to the microscopic average value of the switch-node voltage) at 1 Hz cycle frequency, inductor peak/valley setpoints of 0.05/0.35 A (which are swapped twice per cycle for charging/discharging). The average charging current is around 125 mA and the switching frequency below 70 kHz. Fig. 3 (bottom, zoom) shows the triangular inductor current resulting from the hysteretic current control, the switch-node voltage and the gate voltages. Two different optimal dead times were selected to achieve optimal zero-voltage switching at the peak and valley.

B. Heat pump prototype

Fig. 4 shows the EC-HP prototype built by this work, which is similar to [3] but with two series-connected thermal stages (instead of one) consisting of EC elements between the heat sink and source, and using a higher number of EC-Cs. Since the capacitors in the two thermal stages are oppositely charged and discharged, the thermal series connection allows to double

Fig. 3. Measured switch-node voltage, EC-C output voltage, inductor current (hysteretic current mode 0.05/0.35 A) and averaged charging current. The non-linear voltage change despite almost constant charging current is due to the non-linear capacitance of the electrocaloric loads.

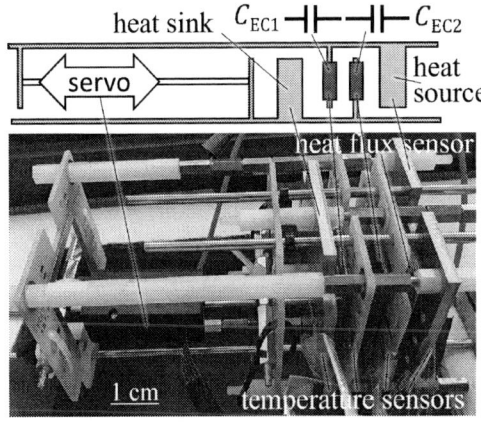

Fig. 4. Electrocaloric heat pump prototype: A servo moves two $3 \times 3 \times 0.3\,\text{cm}^3$ blocks of $16 \times 12 = 192$ paralleled electrocaloric capacitors between a solid brass heat sink/source. The EC-Cs are charged/discharged (synchronized to the movement) by a GaN-based dc-dc converter for high efficiency.

the temperature span of the HP. It is pointed out that an arbitrary high number of thermally series-connected stages can be realized with the same prototype approach, where every second stage is electrically paralleled such that the electrical circuit still only sees two groups of electrical loads [9]–[11]. Multi-stage cascades to increase the temperature span is also known from magnetocalorics [12], [13]. This work uses commercially available multi-layer ceramic capacitors (Y5V, MC1210F476Z6R3CT, Zr-doped BaTiO3 ceramic according to [14]) which exhibit an EC effect. These capacitors are intended for electrical applications, but show a reversible temperature change as side effect for voltages exceeding the rated datasheet value. Even though the EC effect of these

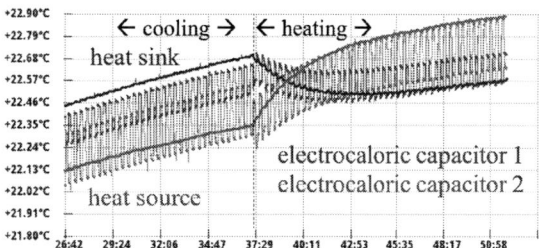

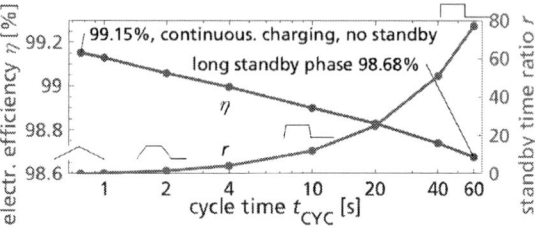

Fig. 5. Measured temperature span build-up after cooling-to-heating phase reversal. (The slow drift is from unstable environment temperature)

Fig. 6. Effect of increasing cycle time (standby phase) on charging efficiency.

MLCCs is low and the permittivity high, resulting in a poor EC figure-of-merit and thus low HP COP [2], they are still suitable to demonstrate EC system prototypes which require availability of a large number of components. Improved EC materials (lead-based and lead-free ceramics, polymers and liquid crystals [15]) are part of ongoing research. For example EC PMN-based ceramics [16]–[18], PST-based ceramics [19] and polymers [20] were already demonstrated which could reach high system performance.

The prototype is first operated using the efficient converter for continuous alternating charging/discharging of the EC-Cs with a trapezoidal voltage and synchronized actuation of the EC-Cs between the sink and source. This mechanical approach [3], [9]–[11], [21]–[24], or electrostatic actuation [5], [25], is often used for electrocaloric prototypes since it enables a high performance thermal switch which is otherwise difficult to realize. The temperatures of the heat sink, source and both EC-Cs are measured and shown in Fig. 5 (0.05 Hz cycles) at cooling-to-heating reversal (by changing the phase shift) with a temperature span build-up. This measurement demonstrates the electrical, mechanical and thermal prototype functionality.

C. Measurement of electrical charging efficiency

For efficiency measurements, the EC-Cs are operated only up to 70 V to avoid device failure even though a higher EC effect might be triggered by applying beyond 200 V (well above the 6.3 V rating from the datasheet). The power converter thus is also only operated well below the initially designed 600 V-class.

The same operation points as in literature [3] were selected, and since the same EC components were used in this work, the efficiency improvement compared to LC resonant circuits is directly quantifiable. From the measured electrical power loss and capacitor charging power the power loss per cycle and capacitor is derived using a power analyzer.

The charging efficiency is derived as one minus the measured (total) electrical power input divided by the measured average charging power (from the rectified value of a low-pass filtered inductor current signal to consider the bidirectional operation mode change within a cycle). The measured power loss and thus the stated electrical efficiency includes both the losses of the power converter (power converter efficiency $\eta_{DC} < 100\%$) and the dielectric loss (dissipation factor $DF > 0\%$) of the EC-Cs. Since the EC figure-of-merit of

the used components is very low, the additional input work for a useful thermodynamic cycle is negligible compared to the input work due to the losses of the converter. In this case, it does not matter if the electrical power loss is measured during adiabatic conditions without HP operation (servos not moving), or during actual useful thermodynamic cycles. Even though the cycle frequency of the prototype is limited due to the thermal time constants of the physically large EC components, electrically significantly higher cycle frequencies than the critical thermal cycle frequency can be characterized to quantify a best-case performance.

a) Standby-phase: First, the effect of standby-phase ratio on efficiency is investigated. No standby phase is achieved if the load capacitance is continuously charged and discharged. The maximum cycle frequency thus is limited by the average charging current, effective load capacitance and the voltage change ΔV. Here, the maximum cycle frequency is 1.4 Hz (standby-ratio $r = 0$). The measurement in Fig. 3 at 1 Hz is for a small standby-ratio of $r = 0.3$. For a cycle time of 60 s (0.0167 Hz), the standby ratio is as high as $r = 77$. This long cycle time allows the heat transfer between the EC elements and the heat sink/source to almost finish, which is limited by the geometry of the components. However, even though the long cycle time allows an almost complete heat transfer, maximizing the heat transferred per cycle, the long standby phases at the same time reduce the efficiency of the converter due to unavoidable static leakage losses of the used power transistors, the capacitive load, and also partly by the measurement equipment (the loss of the voltage probes was subtracted from the measurements based on the measured RMS-values from the power analyzer).

Fig. 6 shows the effect of increasing cycle time on the electrical efficiency. The maximum electrical charging efficiency of 99.15% (which includes the loss from the EC-Cs) is achieved for $r = 0$ (1.4 Hz), and slightly reduces to 98.68% at $r = 77$ (0.0167 Hz). To better understand this, absolute powers are also stated here: At $r = 0$ (99.15% efficiency), the absolute input power of the converter is 39.1 mW (energy of 27.9 mJ per cycle, which is 73 µJ per capacitor and cycle), and the average charging power 4.52 W. The converter and used transistors operate already significantly below the typical power (≈ 1 kW) for these kinds of components. At $r = 77$ (98.68% efficient), the total input power (loss) is just 0.61 mW (energy of 36.6 mJ per cycle, which is 95 µJ per capacitor and cycle) after subtracting the losses of the voltage probes not required for final products, and the average charging power 0.049 W. For this sub-1 mW power loss of the power stage the achieved 98.68% efficiency is considered to be still very high. Care was taken to eliminate any unnecessary leakage current paths in the circuit. The additional loss at the low cycle frequency is equivalent to a static resistance of $34\,\text{M}\Omega$ or $2\,\mu\text{A}$ leakage current at 70 V, and its origin subject to future investigations. Even though this work uses already 384 commercial capacitors to form a suitable minimum load for efficient converter operation, theoretically, the power converter is capable of charging a significantly higher number ($> 100 - 1000$ times) of EC elements with high efficiency. A significant higher number of components will also be required for actual HPs, thermally distributed in series to increase the temperature span, and in parallel to increase the absolute cooling or heating power. Interestingly, in 2014 it was stated, that: "the existence of necessary electronic components is much questionable (high voltages, low frequency, low currents, ultra low leakage)" [26]. This was identified as one unsolved hurdle to achieve high EC-HP COPs. This work now demonstrates that even at the low cycle frequencies efficient electronics for electrocalorics are actually possible, partly by using low leakage and efficient GaN power transistors in a highly optimized circuit with optimized control, and a sufficiently high number of EC-Cs to limit the loss in the standby phases, which paves the way towards highly efficient electrocaloric heat pump systems.

b) Efficiency Improvement: Now, the total input work (electrical loss) using this work's converter is measured and shown in Fig. 7, normalized to 12 individual capacitors (12 Cs) for direct comparison to the measured input work using a LC resonant circuit reported in [3] (with 60 s cycle time and actual heat pump prototype operation). The input work is additionally measured for a fast cycle time (below 1 s with zero standby) which gives the highest electrical efficiency as previously discussed, and also for a cycle time of 60 s, as in [3], required to fully extract the heat from the large components with a large thermal time constant [27] for actual HP prototype operation. It should be mentioned that optimized (thin) geometries [28] for EC MLCCs are subject of ongoing research, aiming for cycle frequencies above 20 Hz using different system approaches [29], since the absolute cooling or heating power scales with increasing cycle frequency despite requiring only the same amount of EC material (improving the thermal power density, similar to electrical power converters where the power density also scales with switching frequency).

Fig. 7 shows that at minimum offset voltage (discussed later), the measured input power is improved by a 11.9-fold decrease compared to [3] at a comparable cycle time of 60 s, and by a 15.5-fold decrease at a fast cycle time.

As observed in [3] and further investigated in [2], offset fields (instead of starting the charging at 0 V) can be used to avoid the high permittivity range of the non-linear EC-C, which significantly reduces the reactive charging power required and thus electrical loss, while at the same time the thermal EC effect is only slightly reduced, improving the HP system COP. As visible in Fig. 2, an offset voltage V_{C0}

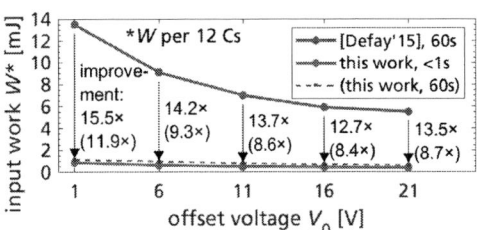

Fig. 7. Improvement by up to 15.5-fold decrease of electrical input work.

TABLE I
COMPARISON OF ELECTRICAL AND THERMAL RESULTS.

	Ref. [3]	this work	
topology	LC resonant	GaN half-bridge	
$V_{C0} + \Delta V$	$0 + 70\,\text{V}$	$0 + 70\,\text{V}$	
cycle frequency f_{CYC}	0.0167 Hz	0.0167 Hz	1.4 Hz
charging efficiency η	80%	98.68%	99.15%
no. of Cs	2×12	2×192	
el. loss W	$24 \times 1125\,\mu\text{J}$	$384 \times 95\,\mu\text{J}$	$384 \times 73\,\mu\text{J}$
heat Q	$24 \times 9.58\,\text{mJ}$	$(384 \times 9.58\,\text{mJ [3]})$	
ΔT per stage	0.175 K	(0.175 K [3])	
$COP = Q/W$	8.5	(100.8)	(131.2)
improvement		$11.9\times$	$15.5\times$

can easily be introduced, but increases the required dc input voltage to $V_{\text{DC}} = \Delta V + 2V_{C0}$. With an offset voltage, the input voltage range of the converter is not fully utilized anymore since an increasing offset voltage increases the reactive charging power by $1 + 2\frac{V_{C0}}{V_{\text{DC}}}$ (assuming a linear ideal and non-electrocaloric capacitive load). Topology variation can be derived where the converter only has ΔV input voltage and the offset is provided externally, for example Fig. 9 from [2] extended to two mid-point connected EC loads. These improvements are not yet implemented in this work, which partly explains why the improvement slightly decreases with increasing offset voltage from a 15.5-fold decrease to an around 13.5-fold decrease, which is still a significant improvement compared to LC resonant circuits.

Tab. I compares the electrical, thermal and mechanical results. From the electrical improvement of $15.5\times$ ($11.9\times$) it can be projected, that, if the same thermal performance (temperature span per stage ΔT and heat absorbed Q) of the components is assumed as measured in [3], that the HP system COP will also be improved likewise by $15.5\times$ ($11.9\times$). This projection of COP improvement is based on the measured heat from literature data for comparability as shown in Tab. I. The thereby assumed heat (3.7 J) per cycle is substantiated by heat flux measurements carried out (later) in this work, which however amounts to only around 2 J (for comparability two times the 1 J per stage due to the two-stage thermal cascade in this work). The measured heat pumped in this work is only around half of the values reported in [3]. This might be explained by the higher inactive thermal mass in this work (additional 1.5 mm thick thermal interface layer, and thermal leakage path from the blocks of EC components to the mechanical PCB-based fixture around them), and by the fact that the commercial ceramic capacitors used in this work

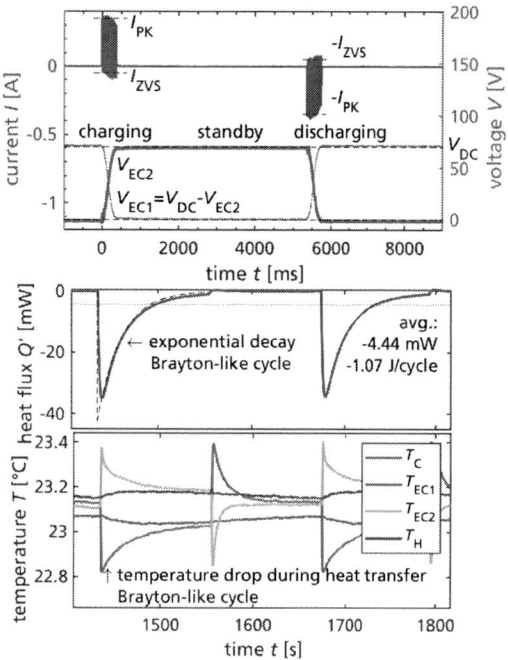

Fig. 8. Electrical step voltage (top) for Brayton thermal cycle (bottom).

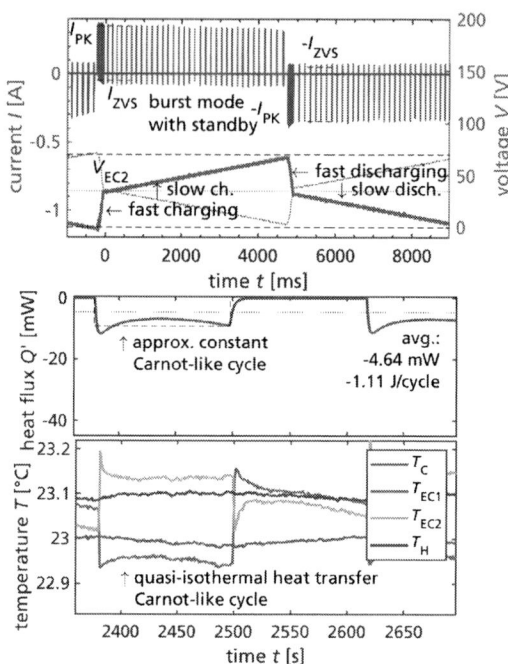

Fig. 9. Electrical field variation (top) for Carnot-like thermal cycle (bottom).

show the EC effect only as a parasitic effect and the exact material composition and manufacturing might have changed in the meantime affecting the EC effect. Furthermore, a high variation of adiabatic temperature change was observed during measurement of individual components, and the components were not selected for the highest EC effect.

D. Field Variation for thermal cycle optimization

The power converter approach also allows arbitrary field (voltage) variation with high efficiency. This can be used to optimize the thermal cycles [30] and to realize Brayton, Carnot, Ericsson (if additional temperature regeneration is implemented), Otto (by turning both half-bridge transistors off) or other thermodynamic cycles [4] depending on the HP prototype system used. The benefit of field variation is demonstrated in this work by Carnot-like cycles in addition to the typically employed Brayton cycles for electrocaloric prototypes. Typ-K thermocouples are routed through a switching card and additional wires and connectors were used to measure the temperature of the heat sink, source ($T_{H/C}$) and both EC elements ($T_{EC1/2}$). No cold junction offset correction was done prior to the measurement, and thus the measured $T_{EC1/2}$ are shown adjusted by +40/+90 mK (for the Brayton measurement) and by +65/+180 mK (for the Carnot-like measurement) to center them in a physically meaningful way between $T_{H/C}$ (which are shown unmodified as measured).

For Brayton cycles, Fig. 8 (top) shows the typical electrical waveforms, where a fast voltage step results in an adiabatic temperature change of the full thermal amplitude. Then the component, which is now hotter (or cooler) than the heat sink (or source) is mechanically connected to the sink (or source).

Heat is now transferred with an initial high temperature drop. The distributed RC-line (thermal resistance and thermal capacitance of the EC component and heat sink) results in almost exponential decay of the heat flux (Fig. 8 (bottom)).

For Carnot-like cycles, Fig. 9 (top) shows the typical electrical field variation waveform: First, a fast voltage step of only half of the total voltage change ($\frac{1}{2}\Delta V$) results in an adiabatic temperature change of only half the theoretically possible full thermal amplitude. This aims to match the resulting temperature change of the EC component to the temperature span between the stages. Now, the EC component is connected to the heatsink with almost same temperature. Then, the electric field (voltage) is slowly increased (maintaining high electrical efficiency by burst-mode operation) by the remaining half of the maximum amplitude. During this time, a quasi-isothermal heat transfer occurs between the EC element and the sink/source. The terms Carnot-like and quasi-isothermal are used here, since only for infinitely slow processes and zero dielectric loss in the EC component a true isothermal and true Carnot cycle is theoretically possible. For the finite-time cyclic operation in the prototype however there still exist a small temperature gradient between the EC elements and sink/source to establish heat flow. The heat flux is measured by a heat flux sensor (greenTEG gSKIN®-XI), attached to the heat sink and partly replacing the thermal interface material (matched thermal conductivity). The heat flux measurement thus allows only to measure heat flux from one EC component during two legs of the thermodynamic cycle (the full cycle could be measured by placing heat flux sensors on both sides of the electrocaloric component). From the measured heat flux, the change of entropy Δs is calculated by time integration.

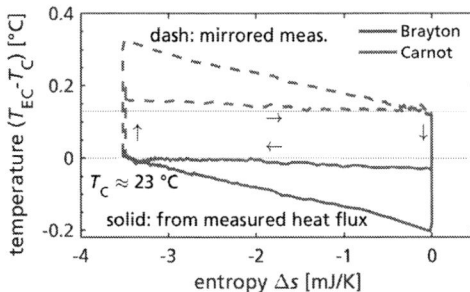

Fig. 10. Thermodynamic cycle of one EC element into the heat sink shows the effectiveness of electric field variation to achieve quasi-isothermal (Carnot-like) heat transfer, in contrast to the Brayton cycle.

Assuming symmetry (similar temperature and heat flux over time during contact to the heat sink/source as during contact of both EC elements when they are disconnected from the sink/source), the other two legs are reconstructed by mirroring.

Fig. 10 shows the thermodynamic cycle based on the measured heat flux and temperatures. For the Brayton cycle a characteristic parallelogram and for the Carnot-like cycle almost a rectangle is observed, while the absolute heat absorbed per cycle is similar ≈ 1 J. This measurement, for the first time, demonstrates Carnot-like operation of an EC-HP prototype. It should be mentioned that the performance can be further improved by additional temperature regeneration as in Ericsson cycles. Then, the full voltage swing is available for the quasi-isothermal phases, increasing the heat pumped and thus HP system performance.

III. CONCLUSION

A power converter approach for electrocaloric applications enables significantly lower charging losses than LC resonant circuits previously reported in literature. Electrical field (voltage) variation is also possible and allows to use offset voltages (reducing the reactive power required) and to realize a thermodynamic cycle such as the Carnot-cycle which is more efficient than the previously typically implemented Brayton cycle. Using low leakage GaN transistors maintains a high performance also at very low thermal cycle frequencies with significant standby phases, and also enables high power density and efficiency of the electrical charging circuit as part of future heat pump systems. Application of the power converter approach to an actual heat pump prototype for demonstration purposes enables to experimentally investigate these promising system approaches for high performance electrocaloric heat pump system prototypes. The work contributes to accelerate the transition to future emission-free solid state heat pumps based on the electrocaloric effect.

ACKNOWLEDGMENT

This work was supported by the Fraunhofer Society in the Fraunhofer lighthouse project "ElKaWe - Electrocaloric Heat Pumps" (www.ElKaWe.org).

REFERENCES

[1] J. Shi, D. Han, Z. Li, L. Yang, S.-G. Lu, Z. Zhong, J. Chen, Q. M. Zhang, and X. Qian, "Electrocaloric Cooling Materials and Devices for Zero-Global-Warming-Potential, High-Efficiency Refrigeration," *Joule*, vol. 3, no. 5, pp. 1200–1225, 2019. [Online]. Available: http://dx.doi.org/10.1016/j.joule.2019.03.021

[2] S. Moench, R. Reiner, P. Waltereit, C. Molin, S. Gebhardt, D. Bach, R. Binninger, and K. Bartholome, "Enhancing Electrocaloric Heat Pump Performance by Over 99% Efficient Power Converters and Offset Fields," *IEEE Access*, vol. 10, pp. 46 571–46 588, 2022. [Online]. Available: https://doi.org/10.1109/ACCESS.2022.3170451

[3] E. Defay, R. Faye, G. Despesse, H. Strozyk, D. Sette, S. Crossley, X. Moya, and N. D. Mathur, "Enhanced electrocaloric efficiency via energy recovery," *Nature Communications*, vol. 9, no. 1, 2018. [Online]. Available: http://dx.doi.org/10.1038/s41467-018-04027-9

[4] D. E. Schwartz, "Thermodynamic Cycles and Electrical Charge Recovery in High-Efficiency Electrocaloric Cooling Systems," *International Journal of Refrigeration*, 2021. [Online]. Available: https://doi.org/10.1016/j.ijrefrig.2021.02.003

[5] Y. Meng, Z. Zhang, H. Wu, R. Wu, J. Wu, H. Wang, and Q. Pei, "A cascade electrocaloric cooling device for large temperature lift," *Nature Energy*, vol. 5, no. 12, pp. 996–1002, 2020. [Online]. Available: https://doi.org/10.1038/s41560-020-00715-3

[6] M. Almanza, T. Martinez, M. Petit, Y. Civet, Y. Perriard, and M. LoBue, "Adaptation of a Solid-State Marx Modulator for Electroactive Polymer," *IEEE Transactions on Power Electronics*, vol. 37, no. 11, pp. 13 014–13 021, 2022. [Online]. Available: https://doi.org/10.1109/TPEL.2022.3183437

[7] S. Moench, K. Mansour, R. Reiner, M. Basler, P. Waltereit, R. Quay, C. Molin, S. Gebhardt, D. Bach, R. Binninger, and K. Bartholomé, "A GaN-based DC-DC Converter with Zero Voltage Switching and Hysteretic Current Control for 99% Efficient Bidirectional Charging of Electrocaloric Capacitive Loads," *PCIM Europe 2022; International Exhibition and Conference for Power Electronics, Intelligent Motion, Renewable Energy and Energy Management*, 2022. [Online]. Available: https://ieeexplore.ieee.org/document/9862274

[8] D. Campolo, M. Sitti, and R. S. Fearing, "Efficient charge recovery method for driving piezoelectric actuators with quasi-square waves," *IEEE Transactions on Ultrasonics, Ferroelectrics and Frequency Control*, vol. 50, no. 3, pp. 237–244, 2003. [Online]. Available: https://doi.org/10.1109/tuffc.2003.1193617

[9] T. Zhang, X.-S. Qian, H. Gu, Y. Hou, and Q. M. Zhang, "An electrocaloric refrigerator with direct solid to solid regeneration," *Applied Physics Letters*, vol. 110, no. 24, p. 243503, 2017. [Online]. Available: https://doi.org/10.1063/1.4986508

[10] Y. Wang, Z. Zhang, T. Usui, M. Benedict, S. Hirose, J. Lee, J. Kalb, and D. Schwartz, "A high-performance solid-state electrocaloric cooling system," *Science*, vol. 370, no. 6512, pp. 129–133, 2020. [Online]. Available: http://dx.doi.org/10.1126/science.aba2648

[11] Y. Meng, J. Pu, and Q. Pei, "Electrocaloric cooling over high device temperature span," *Joule*, vol. 5, no. 4, pp. 780–793, 2021. [Online]. Available: https://doi.org/10.1016/j.joule.2020.12.018

[12] H. Johra, K. Filonenko, P. K. Heiselberg, S. Dall'Olio, K. Engelbrecht, and C. Bahl, "Cascading implementation of a magnetocaloric heat pump for building space heating applications," *10th International Conference on System Simulation in Buildings, Liege, Belgium*, 2018.

[13] T. Hess, L. M. Maier, P. Corhan, O. Schäfer-Welsen, J. Wöllenstein, and K. Bartholomé, "Modelling cascaded caloric refrigeration systems that are based on thermal diodes or switches," *International Journal of Refrigeration*, vol. 103, pp. 215–222, 2019. [Online]. Available: https://doi.org/10.1016/j.ijrefrig.2019.04.013

[14] D. Sette, A. Asseman, M. Gérard, H. Strozyk, R. Faye, and E. Defay, "Electrocaloric cooler combining ceramic multi-layer capacitors and fluid," *APL Materials*, vol. 4, no. 9, p. 091101, 2016. [Online]. Available: https://doi.org/10.1063/1.4961954

[15] P. J. Tipping and H. F. Gleeson, "Ferroelectric Smectic Liquid Crystals as Electrocaloric Materials," *Crystals*, vol. 12, no. 6, p. 809, 2022. [Online]. Available: https://doi.org/10.3390/cryst12060809

[16] U. Plaznik, M. Vrabelj, Z. Kutnjak, B. Malič, A. Poredoš, and A. Kitanovski, "Electrocaloric cooling: The importance of electric-energy recovery and heat regeneration," *EPL (Europhysics Letters)*, vol. 111, no. 5, p. 57009, 2015. [Online]. Available: https://doi.org/10.1209/0295-5075/111/57009

[17] U. Plaznik, A. Kitanovski, B. Rožič, B. Malič, H. Uršič, S. Drnovšek, J. Cilenšek, M. Vrabelj, A. Poredoš, and Z. Kutnjak, "Bulk relaxor ferroelectric ceramics as a working body for an electrocaloric cooling device," *Applied Physics Letters*, vol. 106, no. 4, p. 043903, 2015. [Online]. Available: https://doi.org/10.1063/1.4907258

[18] C. Molin, P. Neumeister, H. Neubert, and S. E. Gebhardt, "Multilayer Ceramics for Electrocaloric Cooling Applications," *Energy Technology*, vol. 6, no. 8, pp. 1543–1552, 2018. [Online]. Available: https://doi.org/10.1002/ente.201800127

[19] Y. Nouchokgwe, P. Lheritier, C.-H. Hong, A. Torelló, R. Faye, W. Jo, C. R. H. Bahl, and E. Defay, "Giant electrocaloric materials energy efficiency in highly ordered lead scandium tantalate," *Nature Communications*, vol. 12, no. 1, p. 3298, 2021. [Online]. Available: https://doi.org/10.1038/s41467-021-23354-y

[20] B. Neese, B. Chu, S.-G. Lu, Y. Wang, E. Furman, and Q. M. Zhang, "Large Electrocaloric Effect in Ferroelectric Polymers Near Room Temperature," *Science*, vol. 321, no. 5890, pp. 821–823, 2008. [Online]. Available: http://dx.doi.org/10.1126/science.1159655

[21] Y. Jia and Y. Sungtaek Ju, "A solid-state refrigerator based on the electrocaloric effect," *Applied Physics Letters*, vol. 100, no. 24, p. 242901, 2012. [Online]. Available: https://doi.org/10.1063/1.4729038

[22] H. Gu, X. Qian, X. Li, B. Craven, W. Zhu, A. Cheng, S. C. Yao, and Q. M. Zhang, "A chip scale electrocaloric effect based cooling device," *Applied Physics Letters*, vol. 102, no. 12, p. 122904, 2013. [Online]. Available: https://doi.org/10.1063/1.4799283

[23] Y. D. Wang, S. J. Smullin, M. J. Sheridan, Q. Wang, C. Eldershaw, and D. E. Schwartz, "A heat-switch-based electrocaloric cooler," *Applied Physics Letters*, vol. 107, no. 13, p. 134103, 2015. [Online]. Available: http://dx.doi.org/10.1063/1.4932164

[24] A. Bradeško, L. Fulanović, M. Vrabelj, A. Matavž, M. Otoničar, J. Koruza, B. Malič, and T. Rojac, "Multifunctional Cantilevers as Working Elements in Solid-State Cooling Devices," *Actuators*, vol. 10, no. 3, p. 58, 2021. [Online]. Available: https://doi.org/10.3390/act10030058

[25] R. Ma, Z. Zhang, K. Tong, D. Huber, R. Kornbluh, Y. S. Ju, and Q. Pei, "Highly efficient electrocaloric cooling with electrostatic actuation," *Science*, vol. 357, no. 6356, pp. 1130–1134, 2017. [Online]. Available: http://dx.doi.org/10.1126/science.aan5980

[26] T. Correia and Q. Zhang, Eds., *Electrocaloric Materials: New Generation of Coolers*. Berlin, Heidelberg: Springer Berlin Heidelberg, 2014.

[27] R. Faye, T. Usui, A. Torello, B. Dkhil, X. Moya, N. D. Mathur, S. Hirose, and E. Defay, "Heat flow in electrocaloric multilayer capacitors," *Journal of Alloys and Compounds*, vol. 834, p. 155042, 2020. [Online]. Available: https://doi.org/10.1016/j.jallcom.2020.155042

[28] S. Crossley, J. R. McGinnigle, S. Kar-Narayan, and N. D. Mathur, "Finite-element optimisation of electrocaloric multilayer capacitors," *Applied Physics Letters*, vol. 104, no. 8, p. 082909, 2014. [Online]. Available: https://doi.org/10.1063/1.4866256

[29] L. M. Maier, P. Corhan, A. Barcza, H. A. Vieyra, C. Vogel, J. D. Koenig, O. Schäfer-Welsen, J. Wöllenstein, and K. Bartholomé, "Active magnetocaloric heat pipes provide enhanced specific power of caloric refrigeration," *Communications Physics*, vol. 3, no. 1, 2020. [Online]. Available: http://dx.doi.org/10.1038/s42005-020-00450-x

[30] S. Crossley, B. Nair, R. W. Whatmore, X. Moya, and N. D. Mathur, "Electrocaloric Cooling Cycles in Lead Scandium Tantalate with True Regeneration via Field Variation," *Physical Review X*, vol. 9, no. 4, 2019. [Online]. Available: https://doi.org/10.1103/PhysRevX.9.041002

Design Considerations of a GaN-based Three-Level Traction Inverter for Electric Vehicles

Subhransu Satpathy
FREEDM Systems Center
NC State University
Raleigh, USA
ssatpat2@ncsu.edu

Partha Pratim Das
FREEDM Systems Center
NC State University
Raleigh, USA
ppdas@ncsu.edu

Subhashish Bhattacharya
FREEDM Systems Center
NC State University
Raleigh, USA
sbhatta4@ncsu.edu

Victor Veliadis
FREEDM Systems Center
NC State University
Raleigh, USA
jvveliad@ncsu.edu

Abstract—This paper demonstrates a GaN-based three-level inverter design for an electric vehicle (EV) traction application. The basic power block is a three-level active neutral point clamped (3L-ANPC) converter operating at 800 V DC bus voltage with the use of two paralleled 650 V, 60 A GaN E-HEMTs. The paper focuses on switching mode selection for better thermal performance of high-power 3L-ANPC inverters. The paper presents design details and guidelines for power layout structure and gate driver. The key parasitic inductances of the designed power block are extracted using ANSYS Q3D. A PLECS simulation model using the device loss parameters shows approximate power delivering capability of 41 kW for the designed forced air-cooled three-phase inverter. Experimental double pulse test results are presented with low voltage overshoot to verify the low inductance power loop design. Experimental results of the three-phase 3L-ANPC inverter operation at 70 kHz switching frequency are presented for delivered output power of 5.1 kW at 550 V DC.

Index Terms—GaN, Three-level Active Neutral Point Clamped (3L-ANPC), traction inverter, and electric vehicle.

I. INTRODUCTION

A high DC bus voltage (V_{DC}) of 800 V will benefit the EV fast chargers and also enable the design of faster electric motors with lower copper loss [1]. The use of traditional two-level inverters for a higher DC bus voltage will result in higher losses with the use of 1200 V rated switching devices [2]. Moreover, the high voltage dv/dt will cause more stress across the turn-turn and coil-coil insulation of the motor. This will require expensive insulation solutions for traction motors [3]. Typical approaches for dv/dt mitigation at inverter output include the use of LCR filters which add to the overall loss and result in a significant inductor voltage drop [4]. A simpler solution by slower switching of power devices for dv/dt stress mitigation leads to under-utilization of wide bandgap devices [5]. This has led to increased interest in three-level converter topologies, which can be realized with 600 V to 650 V rated power devices. The lower voltage switching enables higher efficiency, lower electromagnetic interference (EMI), and lesser motor loss [6]. Further, the reduced turn-turn voltage stress enables high dv/dt operation with no additional LCR filter. This paper selects a three-level active neutral point clamped (3L-ANPC) converter topology as the basic power design block. The 3L-ANPC topology offers the advantage of an active selection of zero

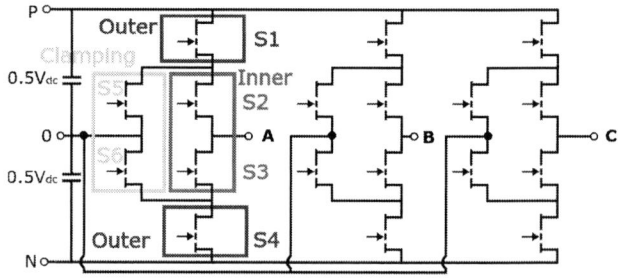

Fig. 1: Three phase 3L-ANPC topology showing the different switches: inner, outer and clamping.

states, allowing for smaller commutation loop inductance and better loss distribution compared to 3L-diode neutral point clamped (3L-DNPC) topology [7]. Also, 3L-ANPC offers better fault-tolerant operation for a single device short/open failure condition for a dual battery fed traction inverter [8].

Enhancement mode GaN HEMTs (GaN E-HEMTs) with lower switching charge (Q_g), zero reverse recovery loss, and smaller output capacitance produce significantly lower switching losses compared to similar rated SiC and Si power devices. A datasheet-based comparative evaluation chart of commercially available power semiconductor devices with different available ratings from various manufacturers is presented in Fig.2. A 650 V, 60 A GaN E-HEMT (GS66516B) has been used in this paper for the design of the power block. GaN E-HEMT manufacturers typically chose Si substrate for cost-benefit even though Si has significantly lower thermal conductivity compared to SiC ($k_{Si} = 1.5$ W/cm K and $k_{SiC} = 4.9$ W/cm K) [9]. This leads to the degradation in efficiency of the GaN E-HEMTs at higher temperatures, particularly due to significantly high conduction losses. The temperature-dependent switching loss and on-state resistance ($R_{ds,on}$) characteristics of the selected device are shown in Fig.5a and Fig.5b respectively. Hence, careful attention needs to be given to the thermal design of the power block for high-power inverters. In this regard, this paper presents the preferred switching mode, power layout, and gate drive design approach.

978-1-6654-8901-0/22 $31.00 © 2022 IEEE

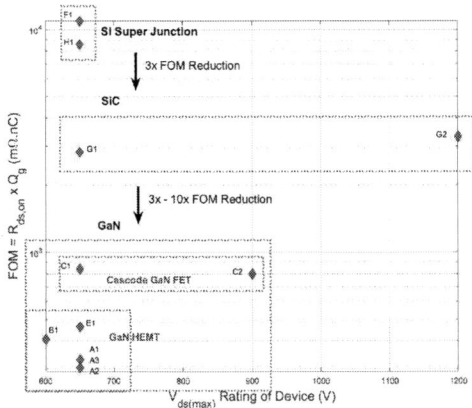

Fig. 2: FOM comparison of commercially available power semiconductor devices.

II. 3L-ANPC POWER BLOCK DESIGN

A. Switching Mode Selection

3L-ANPC topology enables several possible switching modes of operation to realize the P, 0, and N states. The P, 0 and N states correspond to the switching pole voltages (V_{AO}) $+0.5V_{dc}$, 0 and $-0.5V_{dc}$ respectively as shown in Fig.1. However, there are several switching possibilities to realize them. The use of the clamping switch produces two switching possibilities each for P and N states termed as (P_C, $P_{C'}$) and (N_C, $N_{C'}$). There are three switching possibilities for zero states termed as 0_+, 0_-, and 0_F. The switching states are shown in Table.I.

TABLE I: Switching States of P, 0 and N states

	S_1	S_2	S_3	S_4	S_5	S_6
P_C	1	1	0	0	0	1
$P_{C'}$	1	1	0	0	0	0
0_+	0	1	0	0	1	1
0_F	0	1	1	0	1	1
0_-	0	0	1	0	1	1
$N_{C'}$	0	0	1	1	0	0
N_C	0	0	1	1	1	0

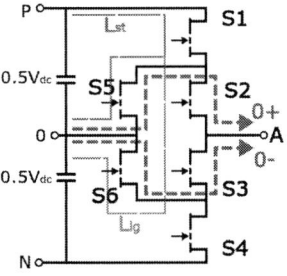

Fig. 3: Realization of different switching states in 3L-ANPC

The switching mode of 3L-ANPC influences the selection of devices and planning of power layout structure. Therefore it is considered the first step of the design process. It is preferred to select P_C and N_C for realizing P and N states to ensure that the steady-state voltages across the off devices are clamped to half DC bus. It is also preferred to avoid the power loop commutation at full load current using the higher inductance path, i.e., the longer loop L_{lg} shown in Fig.3. Therefore the two switching modes of operation compared here for power block design are - (1) short-loop and (2) full mode switching [10].

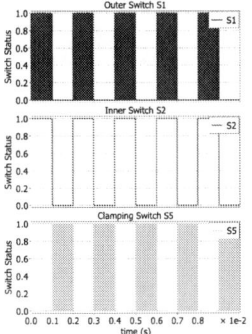 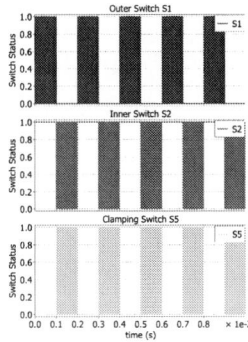

(a) Short-loop switching mode. (b) Full switching mode

Fig. 4: High frequency and line frequency switching of outer, clamping and inner switches in 3L-ANPC for short-loop and full switching mode.

Switching between $P_C \leftrightarrow 0+$ and $N_C \leftrightarrow 0-$ states results in power loop commutation through short loop path only with lower inductance L_{st} shown in Fig.3. This is termed short-loop switching. Outer and clamping switches operate at switching frequency while the inner device operates at line frequency for this switching mode, as shown in Fig.4. Therefore to lower the overall system cost, Si-based MOSFETs are often selected over wide bandgap (WBG) devices for inner devices in hybrid 3L-ANPC converter designs [11]. However, the inner devices that conduct during both on and off state result in significantly high conduction losses. Moreover, the switching implementation of the short loop requires an additional decision logic to select between $0_+/0_-$ based on the previous state P_C/N_C.

For high-power GaN-based 3L-ANPC design, the conduction loss dominates the overall inverter loss. The inner and clamping device conduction losses are significantly reduced using full mode switching [10]. The 0_+ and 0_- paths shown in Fig.3 are operated in parallel for 0_F mode. Hence it reduces the zero state conduction loss to half. However, full mode switching operation requires all the devices to operate at switching frequency, as shown in Fig.4. Therefore GaN E-HEMTs are used for all the switches in 3L-ANPC operation in this work. The clamping and inner devices operate synchronously in parallel for the full mode, sharing half current. This keeps the overall switching losses in a full mode similar to the short-loop mode. The total loss is reduced with better distribution across the devices. Further, the switching logic implementation is simplified because of a single zero state 0_F. The top inner switch S_2 and bottom clamping switch S_6

978-1-6654-8901-0/22 $31.00 © 2022 IEEE   193

share a common gate signal. Analogously, S_3 and S_5 share a common gate signal.

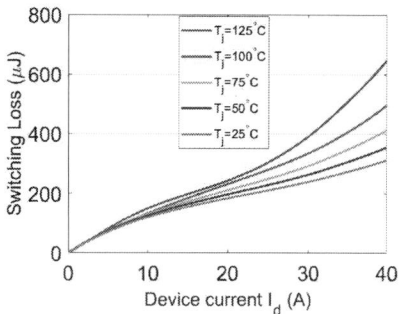

(a) Switching loss ($E_{on} + E_{off}$) vs junction temperature (Gate drive resistance: $R_{on}/R_{off} = 12\,\Omega\,/\,4\,\Omega$)

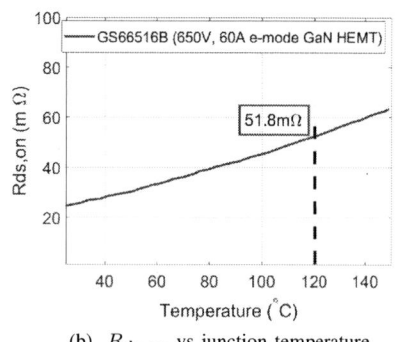

(b) $R_{ds,on}$ vs junction temperature

Fig. 5: Switching and conduction loss variation of selected GaN device (GS66516B) with junction temperature.

A comparative evaluation of short-loop and full switching mode is done using PLECS thermal simulation. The simulation requires a thermal model of the packaged device with temperature dependent turn-on loss (E_{on}), turn-off loss (E_{off}) and on-state resistance $R_{ds,on}$. Towards that, the junction-case Cauer RC network model of the packaged GaN E-HEMT is obtained from the device manufacturer datasheet [12]. For determination of E_{on} and E_{off}, the LTSpice model provided by the device manufacturer is used. The parasitic inductances input to the LTSpice simulation are extracted using ANSYS Q3D for the designed 3L-ANPC power block. The obtained switching losses and $R_{ds,on}$ of the selected device are shown in Fig.5a and Fig.5b respectively. The details of the case-heatsink thermal design are discussed in Section.II-B. The electrical and thermal parameters used for the simulation are tabulated in Table.II and Table.III respectively.

TABLE II: Electrical Parameters selected for simulation

Parameter	Value	Description
V_{dc}	800 V	DC bus voltage
I_{pk}	30 A	Peak line-line current
m	0.3 - 1	Modulation index
pf	0.5 - 1	Power factor
f_{ref}	1 kHz	Fundamental frequency of 3L-ANPC
f_{sw}	50 kHz	Switching frequency

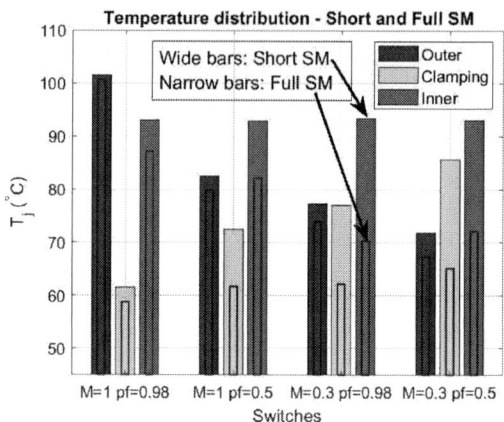

Fig. 6: Temperature rise of outer, clamping and inner switches during 3L-ANPC operation with short and full switching modes.

The thermal stress distribution across outer, inner, and clamping switches strongly depends on the operating modulation index (m) and power factor (pf). As we move to low m operation, the increased zero state period results in high conduction loss for inner and clamping switches. Similarly, as we move to low pf operation, the clamping switches get more often hard-switched, resulting in higher switching losses across them. The inner devices conduct during both P/N and 0 states. This makes inner device conduction losses always high irrespective of m and pf. These trends of thermal stress variation are observed in junction temperature (T_j) for outer, inner, and clamping devices in Fig.6. The thermal stress is determined for short-loop and full switching mode operation considering an ambient temperature of $T_{amb} = 45\,°C$ and a peak current of 30 A through each GaN E-HEMT. Fig.6 shows a major reduction in T_j of inner and clamping switches achieved particularly at lower m using the full switching mode. Therefore under the worst case operating condition of $m = 1$ and unity power factor, the three-phase inverter rating is estimated at $P_{max} = 41\,kW$. This estimation if for $T_{j(max)} = 125\,°C$ considering $T_{amb} = 45\,°C$ with two parallel GaN E-HEMTs delivering 34 A peak current per device at $V_{dc} = 800\,V$.

B. Thermal Design

The previous section highlighted the selection of switching mode that enables lower heating of GaN E-HEMTs for higher efficiency. This section focuses on the thermal design approaches for a GaN-based high-power converter design. Using top-cooled devices with top-side heat sink mounting typically prevents the placement of key surface mount technology (SMT) components like decoupling capacitors and gate drive circuits close to the device. This problem is often resolved by the design of multilayer FR4 PCBs. Multilayer PCBs require thermal vias or pedestal heat sinks as cooling solution. A custom pedestal heat sink increases the design complexity and is not discussed in this work. Multilayer PCBs are often

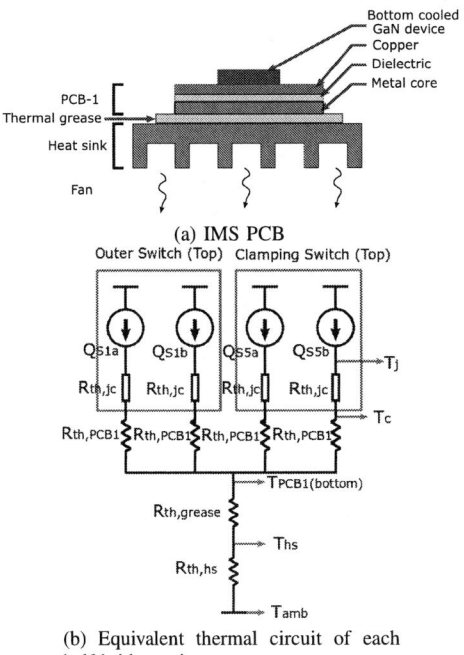

(a) IMS PCB

(b) Equivalent thermal circuit of each half-bridge unit.

Fig. 7: Thermal network for the IMS PCB based 3L-ANPC. S_{1a} and S_{1b} represent the the two parallel GaN E-HEMTs which form the S_1 switch.

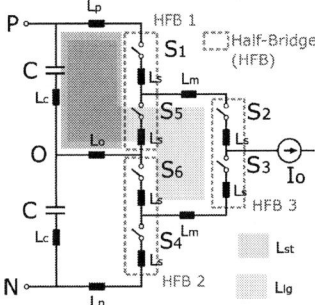

Fig. 8: Parasitic inductances in 3L-ANPC power block

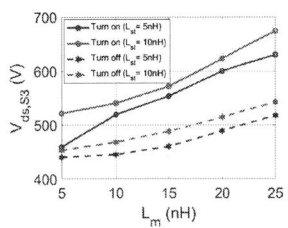
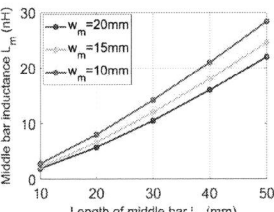

(a) Maximum inner device (S_3) voltage during $0 \rightarrow P$ switching transient for full mode switching with short loop inductance (L_{st}) and middle bus bar inductance (L_m).

(b) Design margin for deciding length and width of the middle bus bar.

Fig. 9: Impact of parasitic inductances on design of 3L-ANPC inverter.

used to create vertical flux cancellation paths to enable ultra-low power loop inductance [13]. However, the thermal vias needed for heat dissipation in multilayer PCBs result in high case to heat sink thermal resistance ($R_{th(c-hs)}$). Hence such design approaches are limited to lower power applications. The presented design approach uses bottom-cooled devices with insulated metal core substrate (IMS) PCBs (PCB-1 shown in Fig.10), which offer low $R_{th(c-hs)}$ for GaN-based high power block design [14]. This limits the power layout design to a single-layer PCB, as shown in Fig.10a. The different layers involved in heat extraction from the bottom-cooled GaN E-HEMT are shown in Fig.7a. The resulting equivalent thermal circuit is shown in Fig.7b, and resistances are presented in Table.III. The thermal conductivity (k), dielectric constant (ϵ_r), and thickness of the insulation material in IMS are the key parameters of selection. The presented design considers a low-cost di-electric material with $k = 3\,\mathrm{W/mK}$, $\epsilon_r = 4.8$ and thickness $t = 125\,\mu\mathrm{m}$. This results in estimated PCB thermal resistance of $R_{th,PCB1} = 2.92\,^\circ\mathrm{C/W}$. The estimated device-to-ground plane stray capacitance is $C_p = 34\,\mathrm{pF}$.

TABLE III: Thermal Parameters selected for simulation

Parameter	Value	Description
$R_{th,jc}$	0.27 °C/W	Junction-case thermal resistance
$R_{th,PCB1}$	2.92 °C/W	IMS PCB thermal resistance
$R_{th,grease}$	0.1 °C/W	Interface paste thermal resistance
$R_{th,hs}$	0.3 °C/W	Heat sink thermal resistance
T_{amb}	45 °C	Ambient temperature

C. Power Loop Design

The fast commutation of switches in a GaN-based 3L-ANPC requires careful attention to parasitic inductances to avoid device failure due to overvoltage. The key parasitic inductances in the 3L-ANPC are shown in Fig.8. The P, N and 0 bus bar inductances are denoted as L_p, L_n and L_o respectively. The source inductance is represented as L_s and the middle bus bar inductance is shown as L_m. The power loop design shown in Fig.8 is designed with an objective to realize a minimum short loop inductance L_{st}. ANSYS Q3D simulation is used to determine the estimated $L_{st} = L_c + L_p + L_o + 2L_s = 6.3\,\mathrm{nH}$. L_{st} forms the inductance in the commutation path between outer and clamping switches. The 3L-ANPC power block is realized as three half-bridge power blocks HFB-1, HFB-2 and HFB-3 shown in Fig.8. The required vertical flux cancellation path for enabling lower L_{st} is realized on the top side through PCB-2 by use of low profile headers, as shown in Fig.10a. PCB-2 contains high voltage decoupling capacitors with high self-resonant frequency ($> 10\,\mathrm{MHz}$) which serve as nearest charge reservoirs during switching transition to minimize overshoot. It should be noted that the provision for device current probing in the prototype design further adds to the power loop inductance.

The previous discussion has shown that full switching mode is preferred for high-power 3L-ANPC. In case of full mode the short loop (L_{st}) and long loop (L_{lg}) operate in parallel resulting in multi-loop commutation [15]. This can result in

978-1-6654-8901-0/22 $31.00 © 2022 IEEE   195

a high voltage overshoot at S_3 during hard-switched turn-on of S_1 for $0 \rightarrow P$ operation in full mode. However, turning off of clamping switch minimizes the overvoltage [15] during the more severe turn-on transient. During the turn-off operation, the switches commutate half of the total current and therefore have a lower overshoot despite a higher inductance. This modified form of full-mode operation is used in the LTSPice simulation results used for analysis. A detailed discussion of the modified full mode operation is not considered under the scope of this paper. The maximum voltage seen across switch S_3 is determined for the two critical hard-switched cases in $0 \rightarrow P$ operation - (1) turn-on transient for switch S_1 and (2) synchronous turn-off switching operation of $S5 - S3$. The modified full mode switching operation is used and the results are presented in Fig.9. The allowable L_m for a given design L_{st} can be determined based on acceptable overvoltage stress. A single layer IMS PCB does not allow the flux cancellation path design for L_m. Therefore the length and width of the PCB trace need to be carefully selected for the L_m path. Fig.9b provides a look-up reference for deciding the length and width of the PCB trace to ensure that the design margin for L_m is ensured. Inductance is determined from the PCB trace dimensions for a selected $4\,\mathrm{oz}$ copper thickness. For the designed, 3L-ANPC power block, Q3D simulation results estimate $L_m = 21\,\mathrm{nH}$.

D. Gate Drive Design

The complete gate drive consists of a totem pole stage and power supply stage formed by PCB stack assembly of PCB-2 and PCB-3 as shown in Fig.11. PCB-2 ensures a short and symmetrical gate drive loop for the two parallel connected GaN devices. Distributed gate and source resistors $R_{g,S1a} = R_{g,S1b} = 1\,\Omega$ are used to avoid gate ringing between parallel devices. Commercially available gate driver ICs with $I_{source} = 3.6\,\mathrm{A}$ and $I_{sink} = 8\,\mathrm{A}$ are selected to enable fast switching of two parallel GaN devices. The bipolar gate drive bias voltage of $6\,\mathrm{V}$ and $-4\,\mathrm{V}$ is generated using isolated DC-DC converters with very low $C_{iso} = 4\,\mathrm{pF}$. The gate discharge resistance and the anti-series connected zener diodes for protection are placed very close to the gate-source pads of the GaN E-HEMT. This ensures a safe gate-source voltage during switching transients.

III. Experimental Results

The designed GaN-based 3L-ANPC phase leg is shown in Fig.10. The complete gate drive stage with the decoupling capacitors is stacked on top of the power layout stage. The switching performance is evaluated using a double pulse test for the full switching mode. Preliminary experimental results of continuous three-phase inverter operation are also presented.

A. Double Pulse Test

The double pulse test (DPT) results with a single GaN E-HEMT per switch for full switching mode operation are shown in Fig.12. The gate resistances selected for the DPT are $R_{on} = 10\,\Omega$, $R_{off} = 2\,\Omega$, and $R_{g,S1a} = R_{s,S1a} =$

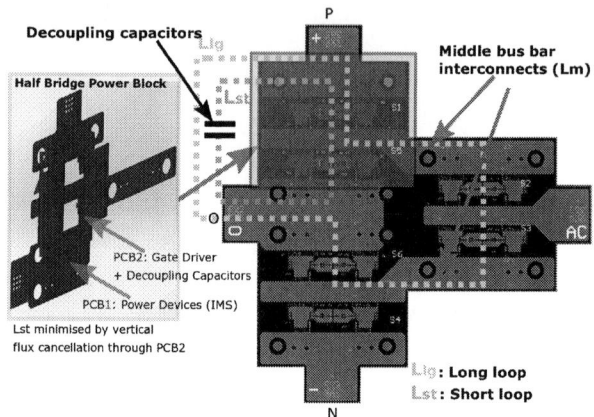

(a) Power layout structure of 3L-ANPC power block with two parallel $650\,\mathrm{V}$, $60\,\mathrm{A}$ GaN E-HEMTs.

(b) Hardware image

Fig. 10: 3L-ANPC phase leg with integrated gate drive stage.

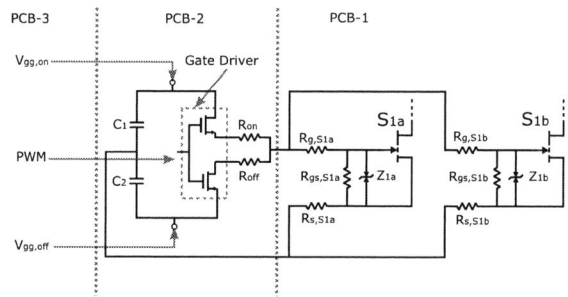

Fig. 11: Gate driver circuit.

$1\,\Omega$. The test results show low overvoltage during turn-on and turn-off switching transient. The maximum voltage seen at S_3 during hard-switched turn-on of S_1 is $V_{ds,S3(max)} = 415\,\mathrm{V}$. The maximum voltage seen at S_3 during the synchronous turn-off of $S_3 - S_5$ is $V_{ds,S3(max)} = 446\,\mathrm{V}$. The turn-on switching loss of S_1 is $310\,\mu\mathrm{J}$ for $di/dt = 4\,\mathrm{A/ns}$ and $dv/dt = 12\,\mathrm{V/ns}$ at $V_{dc} = 800\,\mathrm{V}$ and device current $I_d = 25\,\mathrm{A}$. Switches S_3 and S_5 operate synchronously during turn-off with a turn-off loss of $21\,\mu\mathrm{J}$ across each device.

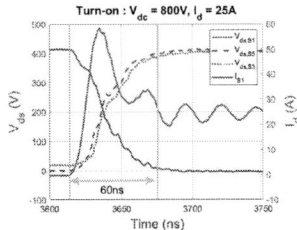 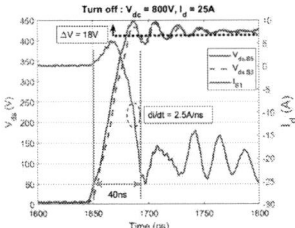

(a) Hard-switched turn-on of S_1 during $0 \rightarrow P$ transition. This results in overvoltage at S_3 and S_5 due to multi-loop commutation.

(b) Hard-switched turn-off of S_3 and S_5 during $0 \rightarrow P$ transition.

Fig. 12: Double pulse test (DPT) result for hard-switched $0 \rightarrow P$ operation with full switching mode operation.

The experimental results show noise-free safe switching operation of GaN E-HEMTs for 3L-ANPC. The short loop inductance can be determined from the turn-off switching transient of S_5 as $L_{st} \approx \Delta V/(di/dt) = 7.2\,\text{nH}$. This verifies the low short loop inductance of the design. The placement of voltage probing points close to drain-source terminals of GaN E-HEMTs is one of the major hindrances in capturing the actual device overvoltage waveforms.

B. Three-Level Inverter Test with RL load

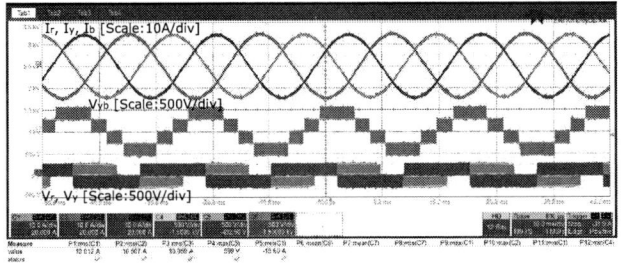

Fig. 13: 3-phase 3L-ANPC test results with RL load ($R = 16.4\,\Omega$ and $L = 1\,\text{mH}$).

The experimental results of the three-phase 3L-ANPC inverter operation are shown in Fig.13. The results demonstrate a continuous operation of the inverter at switching frequency $f_{sw} = 70\,\text{kHz}$ delivering power $P_o = 5.1\,\text{kW}$. The DC bus voltage is at $V_{dc} = 550V$, and the fundamental frequency of the generated AC waveform is $f_{ref} = 60\,\text{Hz}$.

IV. Conclusion

The key design features of a GaN-based 3L inverter for EV traction application are presented for 800 V DC bus voltage. The simulation models estimate the power-delivering ability of 41 kW for the designed inverter with two parallel GaN E-HEMTs. Switching mode selection is key in lowering the junction temperature rise in GaN devices for 3L-ANPC operation. An IMS-based power layout design of the 3L-ANPC phase leg further improves the thermal performance to enable high efficiency from GaN E-HEMTs. The design guidelines and the effect of key parasitic inductances in the power layout structure

are highlighted. The double pulse test results of the designed 3L-ANPC phase leg demonstrate low voltage overshoots at 800 V DC bus and device current of 25 A. Preliminary three-phase inverter operation is demonstrated at 5.1 kW for 70 kHz switching frequency at 550 V DC.

References

[1] I. Aghabali, J. Bauman, P. J. Kollmeyer, Y. Wang, B. Bilgin, and A. Emadi, "800-v electric vehicle powertrains: Review and analysis of benefits, challenges, and future trends," *IEEE Transactions on Transportation Electrification*, vol. 7, no. 3, pp. 927–948, 2021.

[2] S. Satpathy, S. Bhattacharya, and V. Veliadis, "Comprehensive loss analysis of two-level and three-level inverter for electric vehicle using drive cycle models," in *IECON 2020 The 46th Annual Conference of the IEEE Industrial Electronics Society*, 2020, pp. 2017–2024.

[3] X. Huang, B. Jiang, M. N. Alhallak, S. Nategh, Y. Liu, and A. Boglietti, "Experimental evaluation of conductor insulations used in e-mobility traction motors," in *2021 IEEE Workshop on Electrical Machines Design, Control and Diagnosis (WEMDCD)*, 2021, pp. 249–253.

[4] J. He, C. Li, A. Jassal, N. Thiagarajan, Y. Zhang, S. Prabhakaran, C. Feliz, J. E. Graham, and X. Kang, "Multi-domain design optimization of dv/dt filter for sic-based three-phase inverters in high-frequency motor-drive applications," *IEEE Transactions on Industry Applications*, vol. 55, no. 5, pp. 5214–5222, 2019.

[5] H. Kim, A. Anurag, S. Acharya, and S. Bhattacharya, "Analytical study of sic mosfet based inverter output dv/dt mitigation and loss comparison with a passive dv/dt filter for high frequency motor drive applications," *IEEE Access*, vol. 9, pp. 15 228–15 238, 2021.

[6] R. Teichmann and S. Bernet, "A comparison of three-level converters versus two-level converters for low-voltage drives, traction, and utility applications," *IEEE Transactions on Industry Applications*, vol. 41, no. 3, pp. 855–865, 2005.

[7] T. Bruckner, S. Bernet, and H. Guldner, "The active npc converter and its loss-balancing control," *IEEE Transactions on Industrial Electronics*, vol. 52, no. 3, pp. 855–868, 2005.

[8] A. Kersten, K. Oberdieck, A. Bubert, M. Neubert, E. A. Grunditz, T. Thiringer, and R. W. De Doncker, "Fault detection and localization for limp home functionality of three-level npc inverters with connected neutral point for electric vehicles," *IEEE Transactions on Transportation Electrification*, vol. 5, no. 2, pp. 416–432, 2019.

[9] S. Song, *Reliability of GaN-on-Si High-Electron-Mobility Transistors for Power Electronics Application*. PhD Thesis, Aalborg University, 2018.

[10] Y. Jiao and F. C. Lee, "New modulation scheme for three-level active neutral-point-clamped converter with loss and stress reduction," *IEEE Transactions on Industrial Electronics*, vol. 62, no. 9, pp. 5468–5479, 2015.

[11] M. Najjar, A. Kouchaki, J. Nielsen, R. Lazar, and M. Nymand, "Design procedure and efficiency analysis of a 99.3hybrid gan/si active neutral point clamped converter," *IEEE Transactions on Power Electronics*, pp. 1–1, 2021.

[12] *Bottom-side cooled 650V E-mode GaN transistor*, GaN Systems. [Online]. Available: https://gansystems.com/wp-content/uploads/2021/12/GS66516B-DS-Rev-211025.pdf

[13] E. Gurpinar, F. Iannuzzo, Y. Yang, A. Castellazzi, and F. Blaabjerg, "Design of low-inductance switching power cell for gan hemt based inverter," *IEEE Transactions on Industry Applications*, vol. 54, no. 2, pp. 1592–1601, 2018.

[14] "High power ims evaluation platform," *GaN Systems User's Guide*, 2017.

[15] H. Gui, R. Chen, Z. Zhang, J. Niu, J. Ren, B. Liu, L. M. Tolbert, F. F. Wang, D. Costinett, B. J. Blalock, and B. B. Choi, "Modeling and mitigation of multiloops related device overvoltage in three-level active neutral point clamped converter," *IEEE Transactions on Power Electronics*, vol. 35, no. 8, pp. 7947–7959, 2020.

978-1-6654-8901-0/22 $31.00 © 2022 IEEE

Novel High-Voltage-Gain High-Frequency Non-isolated Three-port DC-DC Converter with Zero Input Current Ripple and Soft Switching Capability

Zahra Saadatizadeh
Department of Electrical Engineering
University of Arkansas
Fayetteville, Arkansas, United States
zahras @uark.edu

Pedram Chavoshipour Heris
Department of Electrical Engineering
University of Arkansas
Fayetteville, Arkansas, United States
pedramc@uark.edu

Alan Mantooth
Department of Electrical Engineering
University of Arkansas
Fayetteville, Arkansas, United States
mantooth @uark.edu

Abstract—In this paper, a novel structure of three-port with zero voltage switching (ZVS) dc-dc converter is proposed. Three-port dc-dc converters are widely applicable in hybrid energy generating systems to provide substantial power to sensitive loads. One of the ports of the proposed converter is interfaced with a storage such as a battery which makes the converter capable of supplying the load even if the main renewable energy supply of the proposed converter (e.g. photovoltaic cells) is not able to generate power. The proposed converter operates in two operations of charging the battery port and the mode when the battery discharges to the load. In addition, it has ZVS operation for all utilized switches and can eliminate the input current ripple. The ZVS of the switches is provided by the leakage inductances of the used coupled inductors. Also, the voltage control of each of the terminals is achieved by three active switches. In this paper, the proposed topology is analyzed theoretically for all operating modes, also, the voltage and current equations of all components are calculated. Furthermore, the required soft switching and zero input currents ripple conditions are analyzed. Finally, the accuracy performance of the proposed converter and the analytical results is verified by the PSCAD/EMTDC simulation and experimental results for 1MHz switching frequency operation. Applying high frequency and using GaN Fets leads to achieve a more compact prototype of the proposed topology.

Keywords—*Three-port, non-isolated DC-DC converter, high-voltage-gain, high frequency, input current ripple cancellation, zero voltage switching*

I. INTRODUCTION

Generating power from photovoltaic (PV) cells has become one of the important clean energy sources of distributing energy. The main constraint about this source of energy is the great dependence of it to the environmental changes. To overcome this deficiency and provide a stable source of energy for the loads, especially sensitive loads, solar energy sources should be interfaced with energy storage such as a battery [1]-[4]. As a result, a three-port dc-dc converter (TPC) would be effective to achieve the above-mentioned goal. The general perception of such a converter can be shown in Fig. 1.

In design level, TPC can be designed through several SISO-based converters but in that case, there would be high number of components which leads to low efficiencies [5].

Typically, TPCs can be classified into three types; nonisolated [6]-[12], completely isolated [16]-[19], and partially isolated [20]-[25]. TPCs that have nonisolated structures have simple structures, therefore, they have utilized a smaller number of components in their structures and have low voltage stress on their active components. In [6], the nonisolated TPC suffers from pulsating currents which makes

the converter to have struggles with achieving maximum power point tracking of the PV cells as input voltage sources. The nonisolated TPC presented in [7] suffer from low conversion ratio and its gain is dependent on duty cycle. So, in order to achieve higher levels of voltages and powers, high duty cycles have to be applied to its active switches which would lead to severe spikes and switching losses.In [8], to avoid adopting extremely high duty cycles, there is a diode-capacitor cell used, however, the number of components has been increased. The presented completely nonisolated converter is presented in [9]. This converter has simple structure and compact, also, it uses leakage inductances for achieving soft switching capability but its input current is pulsating with the switching frequency. A nonisolated high step up TPC is presented in [10]. This converter provides high voltage conversion ratio through the implemented coupled inductors but since the winding of the coupled inductor affects the input voltage sources, these power generators would experience high input current ripples. In TPC presented in [11], the input current ripples from all of the ports are cancelled but on the other hand the converter has become complicated with multi-winding coupled inductors and the switches suffer from high voltage stress. Nonisolated TPC presented in [12] has a simple structure with low number of components but it suffers from high voltage stress on its switches and non-common grounding of its input voltage sources. Having high voltage gain is important characteristic for multiport converters to be applied in PV applications [13]-[15]. Two high voltage gain multi-port converters which have utilized diode capacitor circuits to increase voltage gain, have been presented in [13] and [14]. Other way for increasing voltage gain is utilizing coupled inductor cells in the power circuit of multiport dc-dc converters [15]. Another type of TPCs are completely isolated ones which usually have utilized one or more transformers with two or even more windings. The main merit of these converters is isolation of the source sides with the customer side which makes these converters safe, and the voltage regulation of these converters are simple. On the other hand, these converters have typically high weight, high volume and high voltage stress on their switches. [16]-[19].

The main idea comes from a nonisolated topology presented in Ref [12], which is shown in Fig. 2 (a). There are major improvements including the soft-switching capability of the switches, zero input current ripple, and higher conversion ratio of the converter which makes it suitable for many applications. Also, the size, weight, and operating power density of the proposed converter are improved by applying ZVS to the switches and integrating all three windings to one ferrite core, and overall having two coupled inductors in the power circuit which have decreased the

978-1-6654-8901-0/22 $31.00 © 2022 IEEE

number of passive components significantly. From Fig. 2, the derivation process of the proposed converter is illustrated. The proposed converter in [26] has low input current ripple, but the converter in [27] have high input current ripple. The proposed converter in [27] is able to achieve higher voltage gains through the coupled inductor in its topology.

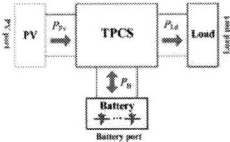

Fig. 1. General configuration of a three-port DC-DC converter

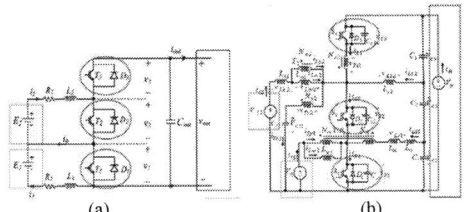

Fig. 2. The derivation process of the proposed converter; (a) The converter presented in [12], (b) The proposed TPC.

II. OPERATING PRINCIPLES OF THE PROPOSED CONVERTER

Fig. 3 shows the proposed non-isolated TPC and its equivalent circuits for both the charging and discharging modes of the battery. The proposed converter includes three switches, two coupled inductors, four capacitors and two inductors. In this figure, the diodes of D_1, D_2 and D_3 and the capacitors of C_{S1}, C_{S2} and C_{S3} are the internal diodes and the parasitic capacitors of the S_1, S_2 and S_3 switches, respectively. In this converter the capacitors of C_1, C_2 and C_3 are large enough that the voltages across them can be assumed as the constant voltages of V_{C1}, V_{C2} and V_{C3} that are equal to $V_{\ell 1}$, $V_{\ell 2} - V_{\ell 1}$ and $V_H - V_{\ell 2}$, respectively. The transformers turns ratio are considered the same values as $n_1 = n_2 = n$.

In general, for both charging and discharging operations of the proposed converter, there are six operating modes. In Fig. 4, the switching pattern is shown for both operations of the proposed converter. The equivalent circuits of the proposed converter for the six operating modes during a switching period is shown in Fig. 5. In this figure, sum of leakage inductors, L_k, and auxiliary inductors, L_r, is considered as L_s ($L_s = L_k + L_r$).

The voltage and current waveforms of the proposed converter for discharging and charging mode are shown in Fig. 6 (a) and 6(b), respectively.

A. Discharging mode (boost operation)

In discharging the battery operation, two voltage sources ($V_{\ell 1}, V_{\ell 2}$) are placed at low voltage side and one output load (R_H) is placed at high voltage side as the output port. $V_{\ell 1}$ is the voltage generated by the battery.

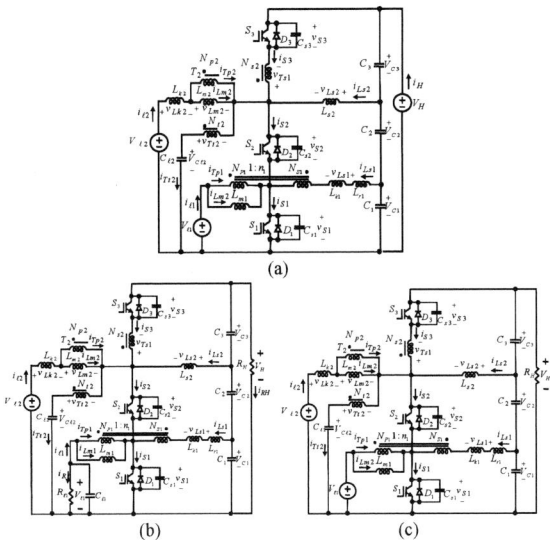

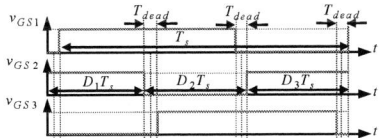

Fig. 3. Proposed TPC and its equivalent circuits; (a) proposed converter; (b) equivalent circuit in charging the battery (as buck-boost operation); (c) equivalent circuit in discharging the battery (as boost operation).

Fig. 4. Switching pattern of switches.

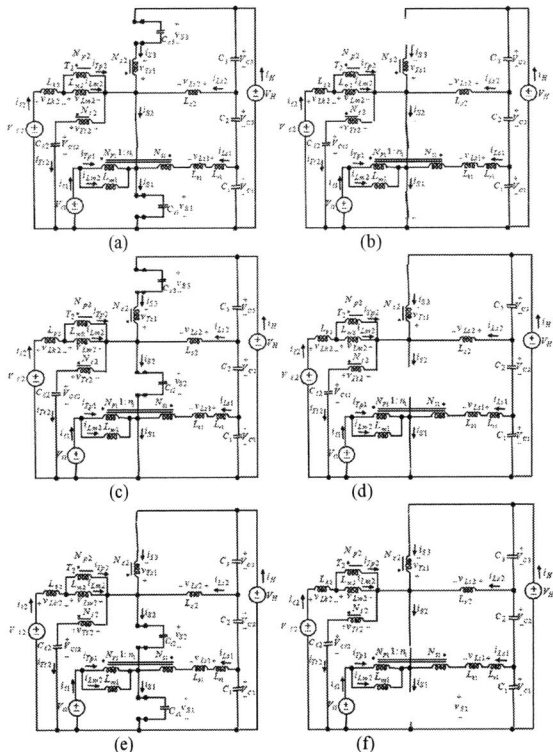

Fig. 5. Equivalent circuits of the proposed converter; (a) mode 1, (b) mode 2, (c) mode 3, (d) mode 4, (e) mode 5, (f) mode 6.

978-1-6654-8901-0/22 $31.00 © 2022 IEEE

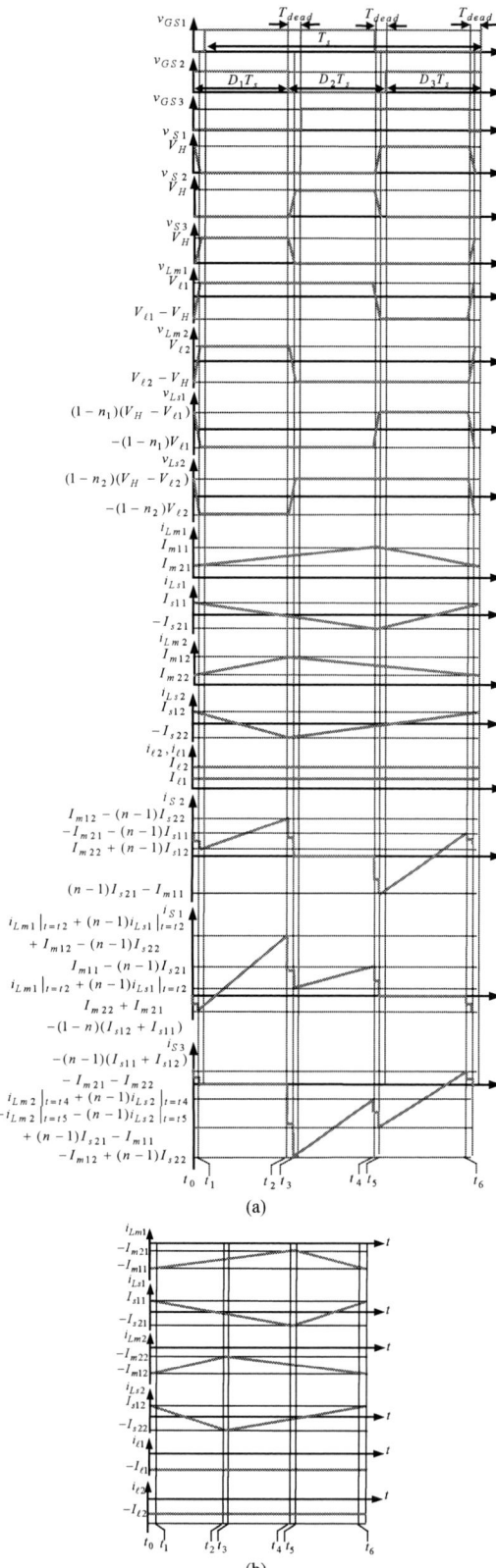

Fig. 6. Steady state theoretical voltage and current waveforms; (a) discharging mode; (b) charging mode.

The currents of inductances L_{m1}, L_{m2}, L_{s1} and L_{s2} have the initial values of I_{m21}, I_{m22}, I_{s11} and I_{s12}, respectively, at the moment $t = t_0$.

Mode 1 ($t_0 \le t < t_1$): The equivalent circuit of this mode is illustrated in Fig. 5(a). In this mode, the parasitic capacitors of C_1 and C_3 are discharged and charged, respectively. As a result, the following equations are obtained

$$i_{C3} + i_{Lk2} - i_{Ls2} - i_{C\ell2} + i_{Lm1} + n_1 i_{Ls1} - i_{Ls1} = i_{C1} \quad (1)$$

$$i_{Lk2} - i_{Lm2} - n_{s2} i_{C3} - n_{t2} i_{C\ell2} = 0 \quad (2)$$

The time interval of mode 1 is very short, as a result, the current values of the inductances are considered equal to their primary values. When the diode D_1, turns on this mode is finished and the ZVS turning on for switch S_1 is obtained. Therefore, capacitor voltage of v_{C1} is reduced from V_H to zero. The time interval of the first operating mode ($T_1 = t_1 - t_0$) is calculated as follows:

$$T_1 = \frac{(C_1 + C_3)(1 + n_{s2}/n_{t2})V_H}{I_{\ell2}(1/n_{t2} - 1) + I_{s12} - I_{m22}/n_{t2} - I_{m21} + (1 - n_1)I_{s11}} \quad (3)$$

Mode 2 ($t_1 \le t < t_2$): The equivalent circuit of this mode is illustrated in Fig. 5(b). The S_1 switch is turned on under ZVS operating condition. In this mode, the voltages of v_{Lm1}, v_{Ls1}, V_{C1}, v_{Lm2}, v_{Lk2}, v_{Ls2} and V_{C2} are equal to $V_{\ell1}$, $(n_1 - 1)V_{\ell1}$, $V_{\ell1}$, $V_{\ell2}/n_{t2}$, $(1 - 1/n_{t2})V_{\ell2}$, $V_{\ell2}$ and $V_{\ell2}$, respectively. The parameters I_{m21}, I_{s11}, I_{m22}, $i_{Lk2}|_{t=t1}$ and I_{s12}, are the primary values of i_{Lm1}, i_{Ls1}, i_{Lm2}, i_{Lk2} and i_{Ls2} currents respectively, the currents of inductors, switches and input currents at low voltage side are obtained as follows:

$$i_{Lm1} = I_{m21} + \frac{V_{\ell1}}{L_{m1}}(t - t_1) \quad (4)$$

$$i_{Lm2} = I_{m22} + \frac{V_{\ell2}}{n_{t2}L_{m2}}(t - t_1) \quad (5)$$

$$i_{Ls1} = \frac{(n_1 - 1)V_{\ell1}}{L_{s1}}(t - t_1) + I_{s11} \quad (6)$$

$$i_{Ls2} = \frac{V_{\ell2}}{L_{s2}}(t - t_1) + I_{s12} \quad (7)$$

$$i_{\ell1} = i_{Lm1} + n_1 i_{Ls1} = I_{m21} + n_1 I_{s11} + \frac{V_{\ell1}}{L_{m1}}(t - t_1) + \frac{n_1(n_1 - 1)V_{\ell1}}{L_{s1}}(t - t_1) \quad (8)$$

$$i_{\ell2} = i_{Lk2} = \frac{(1 - 1/n_{t2})V_{\ell2}}{L_k}(t - t_1) + i_{Lk2}|_{t=t1} \quad (9)$$

$$i_{S2} = i_{\ell2} - i_{C\ell2} - i_{Ls2} = i_{\ell2}\frac{i_{\ell2} - i_{Lm2}}{n_{t2}} - i_{Ls2} = \frac{(n_{t2} - 1)i_{\ell2} + i_{Lm2}}{n_{t2}} - i_{Ls2}$$
$$= \frac{I_{m22}}{n_{t2}} - I_{s12} + (1 - 1/n_{t2})i_{Lk2}|_{t=t1} + \frac{V_{\ell2}}{n_{t2}L_{m2}}(t - t_1) - \frac{V_{\ell2}}{L_{s2}}(t - t_1) + \frac{(1 - 1/n_{t2})^2 V_{\ell2}}{L_{k2}}(t - t_1) \quad (10)$$

$$i_{S1} = i_{\ell1} - i_{Ls1} + i_{S2} = I_{m21} + (n_1 - 1)I_{s11} + \frac{I_{m22}}{n_{t2}} - I_{s12} + (1 - 1/n_{t2})i_{Lk2}|_{t=t1}$$
$$+ \frac{V_{\ell1}}{L_{m1}}(t - t_1) + \frac{(n_1 - 1)^2 V_{\ell1}}{L_{s1}}(t - t_1) + \frac{V_{\ell2}}{n_{t2}L_{m2}}(t - t_1) - \frac{V_{\ell2}}{L_{s2}}(t - t_1) + \frac{(1 - 1/n_{t2})^2 V_{\ell2}}{L_{k2}}(t - t_1) \quad (11)$$

Mode 3 ($t_2 \le t < t_3$): The equivalent circuit of this mode is shown in Fig. 5(c). The time interval of the third operating mode ($T_3 = t_3 - t_2$) is calculated as follows:

$$T_3 = \frac{(1+n_{s2}/n_{t2}+C_2/C_3)C_3 V_H}{I_{m12}/n_{t2}-I_{s22}+(1-1/n_{t2})i_{Lk2}|_{t=t2}} \quad (12)$$

Mode 4 ($t_3 \leq t < t_4$): The equivalent circuit of this mode is illustrated in Fig. 5(d), where the switch S_3 is turned on in the ZVS condition. The voltages of v_{Lm2}, v_{Lk2} and v_{Ls2} are equal to $(V_{\ell2}-V_H)/[n_{t2}(1+n_{s2})]$, $(n_{t2}-1)(V_{\ell2}-V_H)/[n_{t2}(1+n_{s2})]$ and $-(V_{\ell2}-V_H)/(1+n_{s2})$, respectively. By considering the initial values of currents i_{Lm2}, i_{Lk2} and i_{Ls2} as I_{m12}, $i_{Lk2}|_{t=t3}$ and $-I_{s22}$, respectively, the following equations can be written

$$i_{Lm2} = \frac{(V_{\ell2}-V_H)/[n_{t2}(1+n_{s2})]}{L_{m2}}(t-t_3)+I_{m12} \quad (13)$$

$$i_{Ls2} = \frac{-(V_{\ell2}-V_H)/(1+n_{s2})}{L_{s2}}(t-t_3)-I_{s22} \quad (14)$$

$$i_{\ell2}=i_{Lk2} = \frac{(n_{t2}-1)(V_{\ell2}-V_H)/[n_{t2}(1+n_{s2})]}{L_{k2}}(t-t_3)+i_{Lk2}|_{t=t_3} \quad (15)$$

Mode 5 ($t_4 \leq t < t_5$): The equivalent circuit of this mode is illustrated in Fig. 5(e). The time interval of this mode ($T_5 = t_5 - t_4$) is calculated as follows:

$$T_5 = \frac{(C_1+C_2)V_H}{[I_{s21}(1-n_1)+I_{m11}]} \quad (16)$$

Mode 6 ($t_5 \leq t < t_6$): The equivalent circuit of this mode is illustrated in Fig. 5(f). The voltages v_{Lm1} and v_{Ls1} are calculated as $V_{\ell1}-V_H-n_{s2}(V_{\ell2}-V_H)/(1+n_{s2})$ and $(n_1-1)[V_{\ell1}-V_H-n_{s2}(V_{\ell2}-V_H)/(1+n_{s2})]$, respectively. By considering the primary values of inductors currents i_{Lm1} and i_{Ls1} equal to I_{m11} and I_{s12}, the following equations can be written:

$$i_{Lm1} = I_{m11}+\frac{V_{\ell1}-V_H-n_{s2}(V_{\ell2}-V_H)/(1+n_{s2})}{L_{m1}}(t-t_5) \quad (17)$$

$$i_{Ls1} = -I_{s21}+\frac{(n_1-1)[V_{\ell1}-V_H-n_{s2}(V_{\ell2}-V_H)/(1+n_{s2})]}{L_{s1}}(t-t_5) \quad (18)$$

$$i_{\ell1}=i_{Lm1}+n_1 i_{Ls1}=I_{m11}-n_1 I_{s21}+$$
$$\left[\frac{V_{\ell1}-V_H-n_{s2}(V_{\ell2}-V_H)/(1+n_{s2})}{L_{m1}}+\frac{n_1(n_1-1)[V_{\ell1}-V_H-n_{s2}(V_{\ell2}-V_H)/(1+n_{s2})]}{L_{s1}}\right](t-t_5)$$
$$(19)$$

$$i_{s2}=-[i_{Lm1}+(n_1-1)i_{Ls1}]=-I_{m11}+(n_1-1)I_{s21}-$$
$$\left[\frac{V_{\ell1}-V_H-n_{s2}(V_{\ell2}-V_H)/(1+n_{s2})}{L_{m1}}+\frac{(n_1-1)^2[V_{\ell1}-V_H-n_{s2}(V_{\ell2}-V_H)/(1+n_{s2})]}{L_{s1}}\right](t-t_5)$$
$$(20)$$

B. Charging mode (boost operation)

In charging mode operation, one voltage source of $V_{\ell2}$ is placed at low voltage side and two output loads $R_{\ell1}$ and R_H are placed at low and high voltage side, respectively. Charging of the battery is modelled as $R_{\ell1}$ in which $R_{\ell1}$ is in series with the battery. In this operation, the voltage waveforms across the components are same as discharging mode. Only, the direction of the currents passing through the inductors and switches are reversed with same magnitude as discharging mode at the same power.

III. STEADY STATE ANALYSIS

A. Voltage Gain Calculation

The voltage balance law for the inductors L_{m1} and L_{m2} is applied at the steady state and the following equations is resulted:

$$V_{\ell1}(D_1+D_2)T_s+\left[V_{\ell1}-V_H-n_{s2}\frac{V_{\ell2}-V_H}{1+n_{s2}}\right]D_3 T_s = 0 \quad (21)$$

$$V_{\ell2}D_1 T_s+\left[\frac{V_{\ell2}-V_H}{1+n_{s2}}\right](D_3+D_2)T_s = 0 \quad (22)$$

By simplifying (27) and (28) the voltage gains of $V_H/V_{\ell1}$ and $V_H/V_{\ell2}$ can be obtained as follows:

$$\frac{V_H}{V_{\ell1}} = \frac{1+n_{s2}D_1}{D_3} \quad (23)$$

$$\frac{V_H}{V_{\ell2}} = \frac{1+n_{s2}D_1}{D_2+D_3} \quad (24)$$

where, based on Fig. 4, the parameters of $D_1 T_s$, $D_2 T_s$ and $D_3 T_s$ are the interval times of switching and it is assumed that $D_1 = D_2 = D_3$.

B. Input Currents Ripple Elimination Condition

The required condition to eliminate the current ripple of input currents $i_{\ell1}$ and $i_{\ell2}$ can be obtained as following equations:

$$n_1(1-n_1)L_{m1} = L_{s1} \quad (25)$$

$$n_{t2}=1 \;\&\; L_{k2} \neq 0 \quad (26)$$

C. Auxiliary Inductors' Currents Calculation

The average value of i_{Ls1} and i_{Ls2} should be equal to zero in the steady state. As a result the following equations are obtained:

$$I_{s21}=I_{s11} = \frac{(1-n_1)V_{\ell1}(D_1+D_2)T_s}{2L_{s1}} \quad (27)$$

$$I_{s22}=I_{s12} = \frac{V_{\ell2}D_1 T_s}{2L_{s2}} \quad (28)$$

D. ZVS Conditions of Switches

At the beginning of the second operating mode, the current of switch S_1 (i_{s1}) is negative. Therefore, the internal diode of switch S_1 is conducting and ZVS can be achieved. As a result, the required condition for operating the switch S_1 under ZVS, is written as follows:

$$i_{S1}=i_{\ell1}-i_{Ls1}+i_{s2}=I_{m21}+(n_1-1)I_{s11}+\frac{I_{m22}}{n_{t2}}-I_{s12}+(1-1/n_{t2})i_{Lk2}|_{t=t1}$$

$$+\frac{V_{\ell1}}{L_{m1}}(t-t_1)+\frac{(n_1-1)^2 V_{\ell1}}{L_{s1}}(t-t_1)+\frac{V_{\ell2}}{n_{t2}L_{m2}}(t-t_1)-\frac{V_{\ell2}}{L_{s2}}(t-t_1)+\frac{(1-1/n_{t2})^2 V_{\ell2}}{L_{k2}}(t-t_1) \;<0$$

$$(29)$$

IV. PERFORMANCE COMPARISON

The characteristics of the proposed converter and some other conventional converters of the same type in [12], [25]-[27] are compared in Table I. The compared characteristics include voltage gain, normalized voltage stresses on the switches, components number (switch, diode, capacitor, coupled inductor, total number of components) the range of input currents ripple value, common-grounded capability of dc ports, ZVS operating capability for two operations of the proposed TPC. Considering Table I, the proposed converter comparing the compared converters, has some advantages of 1) higher voltage gain value between low and high voltage side, 2) higher total voltage gain, 3) low voltage stress on switches, 4) common-grounded feature of dc ports, 5) ZVS operating capability for switches, 6) zero input currents ripple at low voltage side and 7) bidirectional power flow at the same time.

TABLE I. COMPARISON OF HIGH VOLTAGE GAIN TPCs

DC-DC Converters	$G_{port\_1} = \frac{V_{o1}}{V_i}$	$G_{port\_2} = \frac{V_{o2}}{V_i}$	G_T	$\frac{\Sigma(V_S + V_D)_{max}}{V_{o\_max}}$	N_S	N_D	N_l	N_C	N_{Cl}	N_T	Δi_i	⏚	ZVS
[12]	$1/(1-D)$	$1/D$	$[1/(1-D)]+1/D$	3	3	-	2	4	2	11	Low	Yes	No
[25]	$2N/(1-D)$	$D/(1-D)$	$(2n+D)/(1-D)$	$2+(2/n)$	4	3	-	3	2	12	High	No	No
[26]	$1/(2-2D)$ $0.5<D<1$	$(1-D)/(2-2D)$ $0.5<D<1$	$(2-D)/(2-2D)$ $0.5<D<1$	3	4	2	2	3	-	11	Equal to zero	No	No
[27]	$1/(1-D)$	$1/(1-D)$	$2/(1-D)$	4	2	2	-	2	1	7	low	No	No
Proposed converter	$\frac{V_H}{V_{l1}} = \frac{1+n_{s2}D_1}{D_3}$	$\frac{V_H}{V_{l2}} = \frac{1+n_{s2}D_1}{D_2+D_3}$	$\frac{1+n_{s2}D_1}{D_3} + \frac{1+n_{s2}D_1}{1-D_3}$	3	3	-	1	4	2	10	Equal to zero	Yes	Yes

V. EXPEDRIMENTAL AND SIMULATION RESULTS

To reconfirm the theoretical analysis, the simulation and experimental results are given for both buck-boost operations (battery charging mode) and boost operation (battery discharging mode). This simulation is done in EMTDC/PSCAD software and is shown in Fig. 7. The values of different elements are summarized in Table II. The calculated theoretical values for the voltage gain, voltage stress on switches verify the simulation and experimental results.

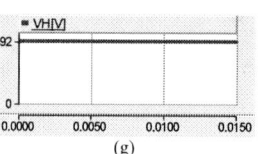

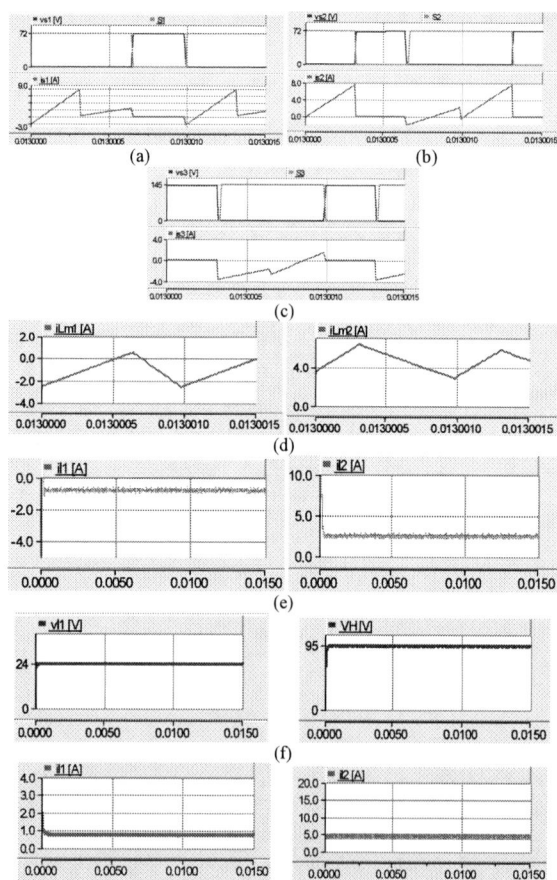

Fig. 7. Simulation results for buck-boost (charging mode of the battery) operation; (a) v_{GS1}, v_{S1} and i_{S1}; (b) v_{GS2}, v_{S2} and i_{S2}; (c) v_{GS3}, v_{S3} and i_{S3}; (d) i_{Lm1} and i_{Lm2}; (e) i_{l1} and i_{l2}; (f) V_{l1} and V_H; (g) i_{l1} and i_{l2} and V_H (discharging mode of the battery)

TABLE II. SIMULATION PARAMETERS

$L_{s2}=1\mu H$	$D_1=D_2=D_3=1/3$	$f_s=1MHz$
$L_{m1}=5\mu H$, $L_{m2}=1\mu H$	$P_{Battery}=20W$ $P_{output}=100W$-$200W$	
$L_{l1}=1.2\mu H$ $L_{l2}=0.8\mu H$	$C_1=C_2=C_3$ $=C_{l1}=C_{l2}=10\mu F$	$V_{l1}=24V$ $V_{l2}=48V$ $V_H=96V$
$R_H=100\Omega$ $R_{l1}=30\Omega$	$C_{s1}=C_{s2}=C_{s3}=2.5pF$	$n_{l1}=0.6$ $n_{s2}=n_{l2}=1$

TABLE III. CHARACTERISTICS OF THE PROPOSED TPC

Voltage conversion ratio;

$$\frac{V_H}{V_{l1}} = \frac{1+n_{s2}D_1}{D_3} \quad \frac{V_H}{V_{l2}} = \frac{1+n_{s2}D_1}{D_2+D_3}$$

$$V_H = [1+1(1/3)]/(1/3)\times 24 = 96$$

Voltage stresses of switches v_{S1}, v_{S2} and v_{S3};

$$v_{s3} = n_{s1}V_{l2} + V_H = 144V, \quad v_{s3} = n_{s1}V_{l2} + V_H = 144V$$

$$v_{s1} = v_{s2} = \frac{n_{s2}V_{l2} + V_H}{1+n_{s2}} = 72V$$

The circuit is implemented and operated for 200W for 1MHz frequency operation. The 3D model of the implemented circuit is shown in Fig. 8. The used components are listed in Table IV. The extracted experimental results are shown in Fig. 9. They have a reasonable match with the simulation and theoretical results.

Table IV Simulation parameters

Components	Specifications
Switches	GS66508B-MR
Gate drivers	LM5114AMF
Gate driver isolation cores	PC95ER9.5/5-Z Material: PC95
Ferrite Cores	P30/19-3F46 Material: 3F46

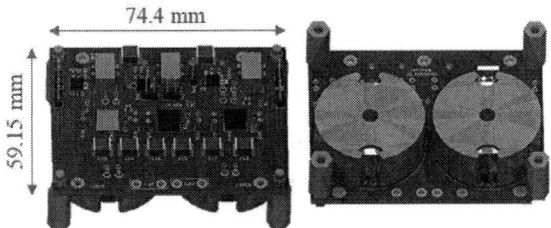

74.4 mm

59.15 mm

Fig. 8.   3D model of the implemented circuit.

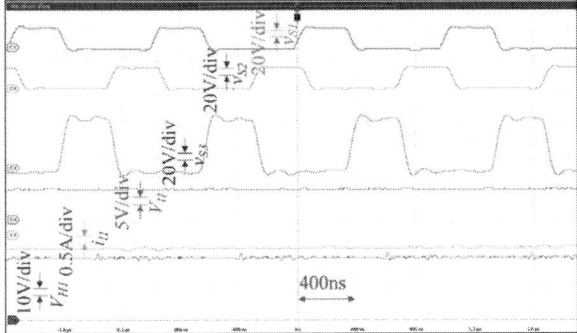

Fig. 9. Simulation results for boost operation; including v_{S1}, v_{S2}, v_{S3}, V_{c1}, i_{c1} and V_H (discharging mode of the battery).

Acknowledgment

Testing for this project was conducted at the National Center for Reliable Electric Power Transmission (NCREPT), the University of Arkansas' High-Power Test Facility.

References

[1] Z. Wang, Q. Luo, Y. Wei, D. Mou, X. Lu and P. Sun, "Topology analysis and review of three-port DC–DC converters," *IEEE Trans. Power Electron.,* vol. 35, no. 11, pp. 11783-11800, Nov. 2020.

[2] P. C. Heris, Z. Saadatizadeh and E. Babaei, "A New Two Input-Single Output High Voltage Gain Converter With Ripple-Free Input Currents and Reduced Voltage on Semiconductors," in IEEE Transactions on Power Electronics, vol. 34, no. 8, pp. 7693-7702, Aug. 2019.

[3] Z. Saadatizadeh, E. Babaei, F. Blaabjerg and C. Cecati, "Three-port high step-up and high step-down dc-dc converter with zero input current ripple," *IEEE Trans. Power Electronics.,* vol. 36, no. 2, pp. 1804-1813, Feb. 2021.

[4] S. H. Hosseini, Z. Saadatizadeh and P. C. Heris, "A New Multi-Input Multi-output DC-DC Bidirectional Converter with MPPT Control Applicable for PV," 2018 18th International Conference on Control, Automation and Systems (ICCAS), pp. 1272-1277, 2018.

[5] W. Jiang and B. Fahimi, "Multiport power electronic interface-concept, modelling, design," IEEE Trans. Power Electron., vol. 26, no. 7, pp. 1890–1900, Jul. 2011.

[6] B. Honarjoo, S. M. Madani, M. Niroomand, and E. Adib, "Non-isolated high step-up three-port converter with single magnetic element for photovoltaic systems," IET Power Electron., vol. 11, no. 13, pp. 2151–2160, Jun. 2018.

[7] G. Chen, Z. Jin, Y. Liu, Y. Hu, J. Zhang, and X. Qing, "Programmable topology derivation and analysis of integrated three-port DC–DC converters with reduced switches for low-cost applications," IEEE Trans. Ind. Electron., vol. 66, no. 9, pp. 6649–6660, Sep. 2019.

[8] B. Zhang, P.Wang, T. Bei, X. Li, Y. Che, and G.Wang, "Novel topology and control of a non-isolated three port DC-DC converter for PV-battery power system," in Proc. 20th Int. Conf. Elect. Mach. Syst., Sydney, NSW, Australia, pp. 1–6, 2017.

[9] R. Faraji, H. Farzanehfard, G. Kampitsis, M. Mattavelli, E. Matioli and M. Esteki, "Fully Soft-Switched High Step-Up Nonisolated Three-Port DC–DC Converter Using GaN HEMTs," IEEE Trans. Ind. Electron., vol. 67, no. 10, pp. 8371-8380, Oct. 2020.

[10] R. Faraji and H. Farzanehfard, "Soft-Switched Nonisolated High Step-Up Three-Port DC–DC Converter for Hybrid Energy Systems," IEEE Trans. Power Electron., vol. 33, no. 12, pp. 10101-10111, Dec. 2018.

[11] H. Zhu, D. Zhang, Q. Liu and Z. Zhou, "Three-Port DC/DC Converter With All Ports Current Ripple Cancellation Using Integrated Magnetic Technique," IEEE Trans. Power Electron., vol. 31, no. 3, pp. 2174-2186, March 2016.

[12] M. Marchesoni and C. Vacca, "New DC–DC Converter for Energy Storage System Interfacing in Fuel Cell Hybrid Electric Vehicles," IEEE Trans. Power Electron., vol. 22, no. 1, pp. 301-308, Jan. 2007.

[13] Z. Saadatizadeh, P. C. Heris and H. A. Mantooth, "Modular Expandable Multiinput Multioutput (MIMO) High Step-Up Transformerless DC–DC Converter," in IEEE Access, vol. 10, pp. 53124-53142, May 2022.

[14] Z. Saadatizadeh, P. C. Heris, E. Babaei and M. Sabahi, "A New Nonisolated Single-Input Three-Output High Voltage Gain Converter With Low Voltage Stresses on Switches and Diodes," in IEEE Transactions on Industrial Electronics, vol. 66, no. 6, pp. 4308-4318, June 2019, doi: 10.1109/TIE.2018.2864710.

[15] Z. Saadatizadeh, P. C. Heris, X. Liang and E. Babaei, "Expandable Non-Isolated Multi-Input Single-Output DC-DC Converter With High Voltage Gain and Zero-Ripple Input Currents," in IEEE Access, vol. 9, pp. 169193-169219, 2021, doi: 10.1109/ACCESS.2021.3137126.

[16] M. C. Mira, Z. Zhang, A. Knott, and M. A. E. Andersen, "Analysis, design, modeling, and control of an interleaved-boost full-bridge three-port converter for hybrid renewable energy systems," IEEE Trans. Power Electron., vol. 32, no. 2, pp. 1138–1155, Feb. 2017.

[17] S. Y. Kim, H. Song, and K. Nam, "Idling port isolation control of three-port bidirectional converter for EVs," IEEE Trans. Power Electron., vol. 27, no. 5, pp. 2495–2506, May 2012.

[18] L. Wang, Z. Wang, and H. Li, "Asymmetrical duty cycle control and decoupled power flow design of a three-port bidirectional DC–DC converter for fuel cell vehicle application," IEEE Trans. Power Electron., vol. 27, no. 2, pp. 891–904, Feb. 2012.

[19] N. D. Dao, D. Lee and Q. D. Phan, "High-Efficiency SiC-Based Isolated Three-Port DC/DC Converters for Hybrid Charging Stations," IEEE Trans. Power Electron., vol. 35, no. 10, pp. 10455-10465, Oct. 2020.

[20] H. Wu, K. Sun, L. Zhu, and Y. Xing, "An interleaved half-bridge three port converter with enhanced power transfer capability using three-leg rectifier for renewable energy applications," IEEE J Emerg. Sel. Topics Power Electron., vol. 4, no. 2, pp. 606–616, Jun. 2016.

[21] V. N. S. R. Jakka, A. Shukla, and G. D. Demetriades, "Dual-transformer based asymmetrical triple-port active bridge (DT-ATAB) isolated DC–DC converter," IEEE Trans. Ind. Electron., vol. 64, no. 6, pp. 4549–4560, Jun. 2017.

[22] Y. -E. Wu and I. -C. Chen, "Novel Integrated Three-Port Bidirectional DC/DC Converter for Energy Storage System," IEEE Access, vol. 7, pp. 104601-104612, 2019.

[23] J. Zeng, W. Qiao and L. Qu, "An Isolated Three-Port Bidirectional DC–DC Converter for Photovoltaic Systems With Energy Storage," IEEE Trans. Ind. Appl., vol. 51, no. 4, pp. 3493-3503, July-Aug. 2015.

[24] K. Wang, R. Zhu, C. Wei, F. Liu, X. Wu and M. Liserre, "Cascaded Multilevel Converter Topology for Large-Scale Photovoltaic System With Balanced Operation," IEEE Trans. Ind. Electron., vol. 66, no. 10, pp. 7694-7705, Oct. 2019.

[25] Y. Hu, W. Xiao, W. Cao, B. Ji, D. J. Morrow, "Three-Port DC-DC Converter for Stand-Alone Photovoltaic Systems," IEEE Trans. Power Electron., vol. 30, no. 6, pp. 3068-3076, June 2015.

[26] A. Ganjavi, H. Ghoreishy, and A. Ale Ahmad, "A novel single-input dual-output three-level dc-dc converter," IEEE Trans. Ind. Electron., vol. 65, no. 10, pp. 8101-8111, Oct. 2018.

[27] N. Nupur and S. Nath, "Minimizing Ripples of Inductor Currents in Coupled SIDO Boost Converter by Shift of Gate Pulses," IEEE Trans. Power Electron., vol. 35, no. 2, pp. 1217-1226, Feb. 2020.

978-1-6654-8901-0/22 $31.00 © 2022 IEEE

Comparison of Thermally Optimized SMD Packages for 100 V GaN HEMTs in 300 kHz Buck Converter High Current Applications

Dominik Koch*, Ankit Sharma†, Till Huesgen† and Ingmar Kallfass*

*Institute of Robust Power Semiconductor Systems, University of Stuttgart, Germany (dominik.koch@ilh.uni-stuttgart.de)
†Electronics Integration Lab, University of Applied Science Kempten, Germany (till.huesgen@hs-kempten.de)

Abstract—In this work an approach for a direct experimental comparison of the application-oriented performance between two high current GaN DC/DC converters based on $7\,m\Omega$, 100 V GaN HEMTs in a commercial off-the-shelf top-cooled package and $5\,m\Omega$, 100 V GaN HEMTs embedded in a thermally-optimized single-chip package is given. The two packaging versions are compared by a maximally identical implementation of the power and gate loops in a 48 V, 300 kHz buck converter with two parallel GaN HEMTs. In the single-chip package the die is directly mounted on a $12\times6\,mm^2$ copper-heat spreader offering a significantly lower thermal resistance to heatsink ($1.69\,K\,W^{-1}$) in comparison to the commercial off-the-shelf version ($2.45\,K\,W^{-1}$). For a direct benchmark, the COTS-based converter has an identical power-loop to the SCP-version with a novel gate-drive concept. Both versions have an efficiency of 96.8 % at 65 A output current (output power: 1.5 kW), while the commercial off-the-shelf version has a better efficiency for lower currents, due to its better hard-switching Figure-of-Merits and therefore lower switching losses and reaches an output current of up to 80 A (output power: 1.75 kW). A detailed analytical loss breakdown for the different transistors in dependence of the temperature and output current is given to proof the measured current point, where the efficiencies of both converters are identical, since the higher switching losses of the SCP version are compensated by lower conduction losses at higher currents and temperatures compared to the smaller COTS transistor. Finally, an outlook on further improvements for reaching higher output currents and potential converters for a more fair comparison of different thermally optimized SMD packages are given.

Index Terms—gallium nitride, high electron mobility transistor, DC/DC converter, high current application, parallelization, performance evaluation, loss breakdown

I. INTRODUCTION

To further enhance the output current of e.g. mild-hybrid converters, transistors are parallelized to achieve a higher current rating [1], [2]. Since Gallium Nitride (GaN) has intrinsic advantages it is suitable for high current operation in hard-switching applications like in 48 V buck converters. To further increase the output power, either improved thermal concepts on system level [3]–[5] or thermally optimized SMD packages [6], [7] are necessary to allow higher losses and high temperature operation, maintaining however low parasitic inductances. Especially, thermally optimized single-chip package (SCP), where the GaN die is directly sintered to a metal substrate have a decisive advantage of significantly improved heat spreading in comparison to commercial off-the-shelf (COTS) chip-scale GaN packages (either top- or bottom-side cooled) [6]. But,

since the improved heat spreading requires the SCP to be significantly larger than the COTS-version (cmp. Fig. 1), the power-loop, common source and gate-loop inductance can be affected negatively. To directly compare these two packaging concepts, in this work a $7\,m\Omega$, 100 V GaN HEMT in a COTS top-side cooled package [8] and a $5\,m\Omega$, 100 V GaN HEMT in a SCP package [6], [7] are investigated. The SCP has an integrated resistance thermometer (RTD) trace for package temperature sensing making it suitable for integration into an intelligent power module. Both HEMTs are compared in a hard-switched 300 kHz, 48 V to 24 V buck converter shown in Fig. 2. First, the common features and differences of the demonstrators are described. Then, system metrics such as efficiency and output power, as well as the switching behavior and the temperature rise are analyzed and compared. In the final chapter, the analytical derivation of the semiconductor losses is briefly described and used to explain the measurements.

II. HIGH CURRENT, 48 V DC/DC CONVERTER DESIGN AND SETUP BASED ON THE COTS AND SCP GAN HEMTs

For the sake of comparison, the setup of the converter PCB is as similar as possible for a fair comparison. Both demonstrators are designed on 4-layer $70\,\mu m$ copper thickness PCB with an 1 mm thick copper inlay, which carries the high currents (details in [9]). Fig. 1 is showing the true to scale drawing of the arrangement of the HEMTs on the backside

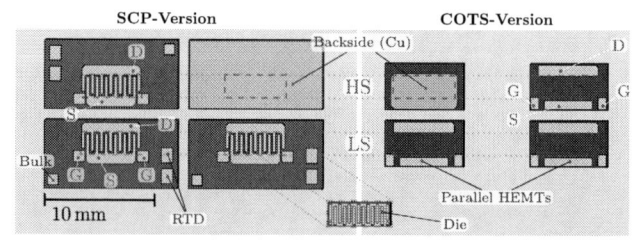

Fig. 1. True to scale comparison of the 2 x 2 half-bridge configuration of the GAN SCP- and the COTS-version. For the sake of comparison the distances between the dies is identical for both versions and defined by the SCP-version. The backside (bulk) potential is internally connected for the COTS-version and can be accessed at the topside of the SCP-version. Furthermore, the thermal pads on the backside (SCP: $12\times6\,mm^2$, COTS: $5.95\times3.1\,mm^2$) for the heat dissipation are shown. The size of the die, which has an identical size in both versions, is indicated in between both versions.

of the converter. The arrangement is done in a way, that the individual dies inside the different packages have the same spacing to each other. Since the SCP is significantly larger ($12 \times 6\,\text{mm}^2$ compared to $5.95 \times 3\,\text{mm}^2$) than the COTS-package, the SCP is defining the constraints, resulting in a non-ideal common source inductance. While the bulk of the SCP-version is accessible as a pad on the transistor, the COTS-version is closing the bulk loop internally. This bulk connection is also limiting the performance of the SCP-version to 65 A output current and explained in detail in [9]. Further details on the fabrication process of the SCP are given in [7], [9]. In Fig. 2 the top-side of the designed converter (version with SCP GaN HEMTs) is shown. On the top-side the gate driver concept with dedicated gate-boosters for each transistor [10], the gate-supply and PWM input as well as the temperature read-out circuit and the current sensor [9] is located. Furthermore, a $220\,\mu\text{F}$ ceramic DC-link and the power terminals for V_{BUS}, GND and the switch-node SW are indicated. The DC-link capacitors closing the power loop are located on the top-side, closing the vertical power loop (cmp. [9]). Since the top-side of the demonstrators are nearly identical, the SCP Version is not shown (cmp. [9]). The,

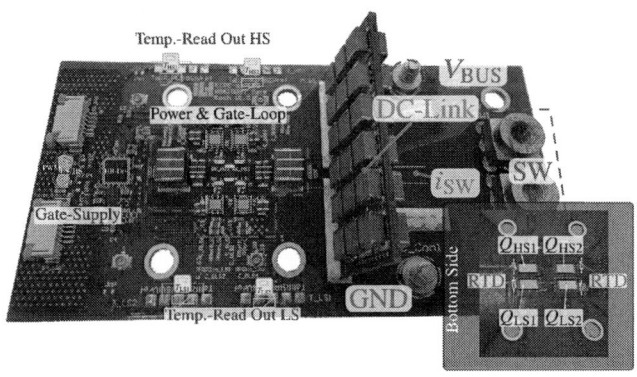

Fig. 2. Topside view of the DC/DC converter with the paralleled COTS GaN transistors in a half-bridge configuration at the backside (blue box). The topside contains the power- and the gate-loop, the gate-driver and supply, a temperature read out for either the integrated resistance thermometer (RTD) or an external RTD and the power terminals. Furthermore, a galvanically isolated phase current sensor[9] can be placed on the topside. The DC/DC converter is designed on a 4-layer PCB with 1 mm copper inlays for providing the current capability. The topside of SCP-based converter is identical. The backside containing the 2 x 2 GaN transistors in a half-bridge configuration is again shown in Fig. 1 and Fig. 5.

by a full-wave S-parameter simulation extracted, power loop inductances without components are $\approx 1.7\,\text{nH}$ for the SCP-version [9] and approximately $2.1\,\text{nH}$ for the COTS version shown in Fig. 2. The difference of about 23 % is due to the SCP is a true chip-scale package (distance between the centers of the source and drain pads of $2.95\,\text{mm}$), while the drain and source pin distance of the COTS package is about 10 % longer. The converter is mounted on a heatsink with forced convection. As thermal interface material (TIM) a high-performance silicone gap-filler with $0.5\,\text{mm}$ thickness and a thermal conductivity κ of $13\,\text{W/mK}$ is used. Further transistor and system level parameters are shown in Table I.

III. Demonstrator Measurements and System Metrics Evaluation for a 300 kHz, 48 V to 24 V DC/DC converter operation up to 80 A

Both converters described in the previous chapter are evaluated in 48 V input to 24 V output, 300 kHz DC/DC operation with an identical setup and an external air-coil. Fig. 3 is showing the comparison of the measured low- and high-side drain source voltages $v_{\text{ds,LS,HS}}$ and the switch-node current i_{sw} for both versions at an output current of 60 A. In the

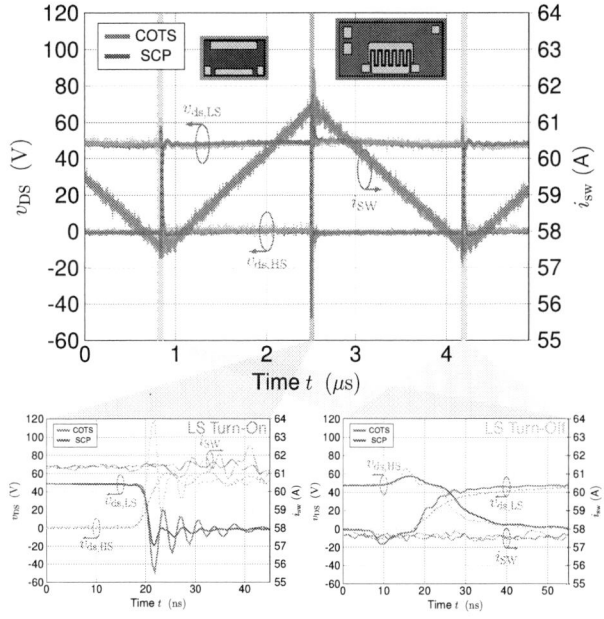

Fig. 3. Comparison of the low- and high-side drain source voltages $v_{\text{ds,LS,HS}}$ and the switch-node current i_{sw} between the nearly identical COTS-based version with the $7\,\text{m}\Omega$ GaN transistor (red) and the SCP-version ($5\,\text{m}\Omega$ GaN transistor, blue) over approximately one and a half period at 60 A output current in the mentioned DC/DC operation (top plot). In the bottom plots the more detailed comparison of the low side turn-on (bottom left) and the low side turn-off (bottom right) at this bias point is given.

top plot, showing approximately one and a half periods, it is directly visible, that the COTS-version has a significantly higher overshoot at the low-side (LS) turn-on. In the bottom left plot in Fig. 3 the magnified turn-on waveforms are showing an overshoot for the COTS-version (red) up to nearly 120 V, which is twice as large as the overshoot of the SCP-version ($\approx 60\,\text{V}$). Since the COTS-HEMT is smaller than the SCP-HEMT ($7\,\text{m}\Omega$ compared to $5\,\text{m}\Omega$) the parasitic transistor capacitances (cmp. Table I) are in average 25 % lower, resulting in nearly twice as fast switching transients (1 ns compared to 1.7 ns) with the identical gate-loop and driver circuitry. By taking the higher power-loop inductance into account, the overshoot is occurring due to the spacing and converter constraints set by the larger SCP version. The bottom right plot in Fig. 3 is showing the low-side turn-off, where, except of some small reverse conduction peaks between 10 ns to 20 ns, no over- or undershoots are visible. Again, a slightly faster switching transient can be observed

978-1-6654-8901-0/22 $31.00 © 2022 IEEE

TABLE I
PARAMETERS OF THE COTS AND SCP GaN TRANSISTORS AND IDENTICAL SYSTEM LEVEL PARAMETER FOR BOTH DEMONSTRATORS.

Type	$R_{\mathrm{DS,on}}$ (25 °C)	$C_{\mathrm{OSS}}(V_{\mathrm{ds}}=0\,\mathrm{V})$	$C_{\mathrm{OSS}}(V_{\mathrm{ds}}=48\,\mathrm{V})$	Q_{GD}	C_{RSS}	$R_{\mathrm{G,int}}$	$R_{\mathrm{th,j-hs}}$	Size
	5 mΩ	1020 pF	370 pF	4.8 nC	190 pF	0.52 Ω	1.69 K W^{-1} [11]	12×6 mm²
	7 mΩ	690 pF	280 pF	1.5 nC	130 pF	0.64 Ω	2.45 K W^{-1} [11]	7.1×4.1 mm²

System								
V_{BUS}	V_{out}	f_{sw}	V_{drv}	$R_{\mathrm{G,on/off}}$	$R_{\mathrm{PU/PD}}$	L_{out}	$\mathrm{DCR}_{\mathrm{L,out}}$	κ_{TIM}
48 V	24 V	300 kHz	+5 V/−6 V	19 Ω/1 Ω	2 Ω/0.2 Ω	15 µH	11 mΩ	13 W/mK

for the COTS-version (red). Although the overshoot of the COTS-version is twice as large as for the SCP-based converter, the efficiencies of both demonstrators are nearly identical. Fig. 4 is showing the efficiency η over the output current I_{out} for the converter with the COTS GaN HEMT (red) and the SCP GaN HEMT (blue). The SCP-version is limited to an output current of 65 A, which corresponds to an output power of above 1.5 kW at 24 V output voltage, due to a weakly designed bulk-via [7], [9]. Both systems have a nearly identical peak efficiency of approximately 99 % at 10 A and an overall efficiency of above 96.8 % for output currents up to 65 A. For currents between 20 to 65 A, the COTS-version has a 0.4 to 0.5 % higher efficiency than the SCP-version. At 65 A the COTS- and the SCP-version have an identical efficiency of 96.8 %, which is analyzed in detail in the next section. At a maximum current of 80 A, which is the limit due to the high overshoots at the low-side turn-on, the DC/DC converter based on the COTS GaN HEMTs has an efficiency of 95.3 %. The similar performance of both converters is also visible

in combination with the faster switching transients is resulting in higher ripple currents at the COTS converter, which are resulting in slightly higher temperatures. In the small top pictures, the backside of both converters is shown, which again illustrates the unnecessary parasitic inductance, which exists due to the comparability to the SCP version.

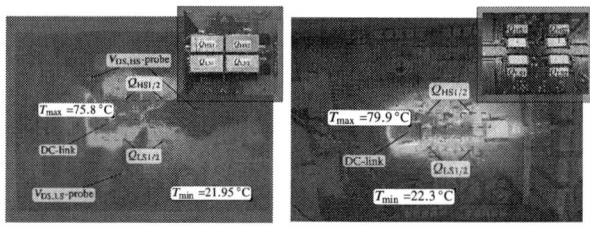

Fig. 5. Comparison of the infrared image for the COTS and SCP [9] version at an output current of 60 A in the 48 V input to 24 V output, 300 kHz DC/DC converter operation. The maximum temperature is occurring at the DC-link capacitor and is similar for both versions (COTS: 79.9 °C, SCP: 75.9 °C). The red and blue rectangle is indicating the position outline of the transistors on the backside. In the small top right pictures the GaN transistors on the backside of the converter PCBs are shown.

IV. ANALYTICAL LOSS BREAKDOWN OF COTS AND SCP GaN HEMTS IN HIGH-CURRENT AND TEMPERATURE DC/DC OPERATION

As mentioned in the previous section, the SCP-based version has a 0.4 to 0.5 % lower efficiency than the COTS-version in the output current range between 20 A and 65 A, although the SCP GaN HEMTs have a lower on-resistance of 5 mΩ making it more suitable for high current operation. Using on the one hand the hard-switching Figure-of-Merit (FOM), which takes into account the on-resistance $R_{\mathrm{ds,on}}$ and the output capacitance C_{OSS} [12]

$$\mathrm{FOM}_{\mathrm{cHS}} = {}^{1}/R_{\mathrm{ds,on}} \cdot C_{\mathrm{OSS}}, \tag{1}$$

and on the other hand the FOM which considers the gate to drain charge Q_{GD} and the charge Q_{GS2} necessary for the gate-voltage rise from the threshold voltage $V_{\mathrm{GS,th}}$ to the plateau voltage V_{PL} [13]

$$\mathrm{FOM}_{\mathrm{HS}} = R_{\mathrm{ds,on}} \cdot (Q_{\mathrm{GD}} + Q_{\mathrm{GS2}}), \tag{2}$$

the corresponding FOMs for both GaN HEMTs are depicted in Table II. Since the COTS device is ≈6 % better in the

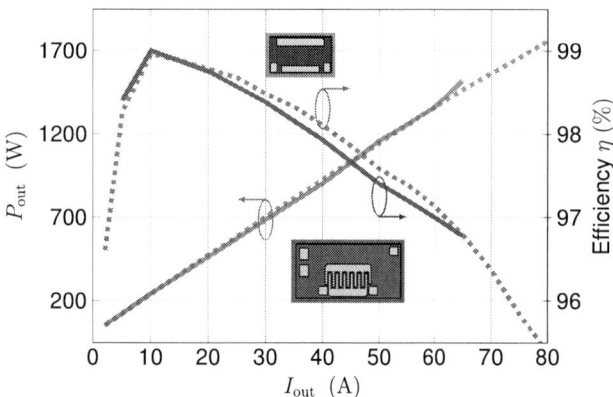

Fig. 4. Efficiency comparison η and output power P_{out} for the COTS-version (dashed) and the SCP-version (solid) in a 48 V input to ≈24 V output, 300 kHz DC/DC conversion in dependence of the output current I_{out}. The losses in the output coil were calculated analytically and subtracted. The duty cycle and thus the output voltage varies minimally between the two versions at higher currents.

in the infrared images (IR) shown in Fig. 5. Again, at an identical operation point of 60 A output current, the maximum temperature points are outermost DC-link capacitors with a maximum temperature of 79.9 °C for the COTS converter and 75.9 °C for the SCP-converter. The high voltage overshoots

978-1-6654-8901-0/22 $31.00 © 2022 IEEE

TABLE II
HARD-SWITCHING FOM FOR THE COTS AND SCP GAN HEMTS.

Type	FOM_{cHS}[1]	FOM_{HS}[2]
COTS	207 GHz	28.1 pC Ω
SCP	196 GHz	37.1 pC Ω

[1] the higher the better [2] the lower the better

FOM_{cHS} and 25 % better in FOM_{HS} than the SCP, an overall better switching performance of the COTS GaN HEMT is justified. To further proof the measurements and the theoretical assumptions due to the calculated FOM, an analytical loss estimation is performed, which is obtained from [14]–[16]:

By using the non-linear capacitances of the transistors, the voltage-dependent transconductance g, the plateau voltage V_{PL} and the external and internal gate-drive circuitry, the switching losses can be calculated for the turn-on and turn-off of the COTS and SCP GaN HEMTs. Furthermore, it is assumed, that the output current is shared equally by both paralleled transistors. The middle right plot in Fig. 6 is depicting the turn-on energy E_{ON}, respectively power P_{ON} and turn-off energy E_{OFF}, respectively power P_{OFF} for the COTS Q_{COTS} and SCP Q_{SCP} HEMTs in dependence of the drain source current i_{ds}. While the turn-off losses are comparably small for both Q_{COTS} and Q_{SCP}, the turn-on losses for the SCP are twice the turn-on losses for the COTS GaN for e.g. 20 A. Considering now the temperature dependent conduction losses P_{cond} in the middle left part of Fig. 6, the conduction losses are lower than the switching losses for any current at 25 °C and are becoming dominant ($P_{\text{cond}} \geq P_{\text{sw}}$) at either higher temperatures for both COTS and SCP GaN or higher currents for the COTS.

By adding up the individual loss components (gate-charge, output capacitance and reverse conduction losses are considered, but are comparably small and will therefore not be described in detail) the single transistor losses for either the high-side transistor Q_{HS} or the low-side transistor Q_{LS} are illustrated in the top plot in Fig. 6. For the sake of readability only the transistor losses at 25 °C are shown. While the low-side rectifier HEMT is mainly experiencing conduction losses, the COTS GaN HEMT has higher overall losses due to the higher on-resistance of 7 mΩ, the high-side HEMT is mainly dissipating switching losses. Here it is clearly visible, that the absolute difference between the smaller COTS GaN HEMT and the SCP GaN HEMT as HS transistor is larger than the difference for using them as a LS switch, resulting in overall lower transistor losses.

To obtain the overall system level losses two times the temperature-dependent GaN HEMT losses for each HS and LS from the top plot in Fig. 6 and the losses in the output coil are added and depicted in the bottom plot of Fig. 6. The solid lines are representing the losses at 25 °C, while the dotted lines are indicating the losses at 150 °C over the output current I_{out}. While the losses P_{dis} for the SCP-based converter are always higher than for the COTS-based converter at a temperature of 25 °C, there is a crossing of both loss curves at a temperature

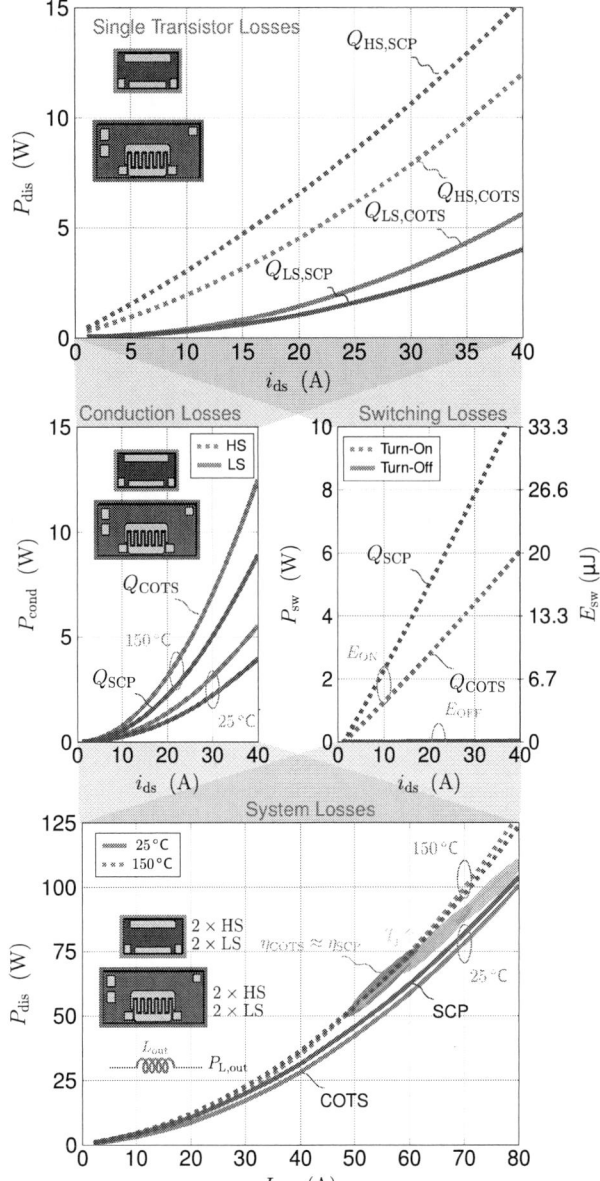

Fig. 6. Analytical estimation [14] of the transistor and main system losses in dependence of the output current for the COTS- and SCP-based versions with the operation constraints described in Table I. The top plot is showing the overall transistor losses (conduction and switching losses) for the SCP transistor $Q_{\text{LS/HS,SCP}}$ and the COTS transistor $Q_{\text{LS/HS,COTS}}$ in dependence of the switched current i_{ds}. The switched current is derived from the output current I_{out} by assuming an identical current sharing between the both parallel transistors. In the middle plots, again for both versions, the temperature-dependent conduction (left) and switching losses (right) for 25 °C and 150 °C are shown. The turn-off losses are significantly smaller than the turn-on losses and there is nearly no temperature dependence. In the bottom plot the overall system losses (two high-side and two low-side transistors and the losses in the output air coil) for 25 °C and 150 °C in dependence of the output current I_{out} are shown, highlighting the temperature and current dependent break-even point, where the higher switching losses of the SCP version are compensated by lower conduction losses at higher currents and temperatures in comparison to the smaller transistor of the COTS version.

of $150\,^{\circ}\mathrm{C}$ and in a current range between 40 to 50 A, where the efficiency η_{COTS} of the COTS converter is identical to the efficiency η_{SCP} of the SCP converter. Since both the switching losses and especially the conduction losses exhibit a nonlinear behavior over the temperature, the better switching behavior of the COTS-based converter is counterbalanced by the higher conduction losses compared to the SCP-based converter above a certain current value (as already mentioned, the conduction losses become more dominant than the switching losses). This intersection point is strongly temperature-dependent, shifts with higher temperatures closer to small currents and is indicated by the oval areas in bottom plot of Fig. 6. Considering the approximate temperatures from Fig. 5, the theoretical intersection point is between 60 A and 70 A and thus fits very well with the experimentally determined intersection point of the efficiencies of the two converters from Fig. 4. In addition, the analytically determined losses (e.g. $P_{\mathrm{dis}} \approx 74\,\mathrm{W}$ at 60 A output current and $150\,^{\circ}\mathrm{C}$) fit very well with the measured losses (whole system losses $P_{\mathrm{dis}} = 77.3\,\mathrm{W}$ at 60 A output current), whereby the analytical calculation resembles a worst-case estimate, since capacitance and ohmic losses are not considered in the analytical estimation.

Conclusion and Future Work

In this work a direct experimental comparison of the application-oriented performance between two high current GaN DC/DC converters based on $7\,\mathrm{m}\Omega$, 100 V GaN HEMTs in a COTS package and $5\,\mathrm{m}\Omega$, 100 V GaN HEMTs embedded in an SCP is given. Both converters are compared in a 48 V, 300 kHz DC/DC operation with two parallel HEMTs. To assure a fair comparison an identical implementation of the power and gate loops is assured. Both versions have an efficiency of 96.8 % at 65 A output current (P_{out}: 1.5 kW), while the COTS-version has a better efficiency for lower currents, due to lower switching losses and reaches an output current of up to 80 A (P_{out}: 1.75 kW). The COTS-based converter is offering better switching FOMs, but has higher overshoots due to the identical gate-drive and the non-optimal power loop, which was adapted to the SCP-based converter due to the required comparability. An analytical loss breakdown of conduction and switching losses for the COTS and SCP GaN HEMTs in dependence of the temperature and the output current is given to proof the measured $\approx 65\,\mathrm{A}$ output current as intersection point, where the efficiencies of both converters are identical. For future work, an SCP- and a COTS-based converter with the same die will be investigated to determine the influence of e.g. the package and the thermal performance without differences in the dynamic electrical behavior. Furthermore, the loss breakdown analysis will be improved by implementing more measurement data.

Acknowledgment

This work has been supported in the frame of the ECPE Joint Research Programme "Packaging Technology of GaN LV msPEBB Phase 2".

References

[1] J. Burkard and J. Biela, "Paralleling gan switches for low voltage high current half-bridges," in *2019 IEEE Energy Conversion Congress and Exposition (ECCE)*, 2019, pp. 3245–3252.

[2] P. P. Das, S. Satpathy, S. S. Shah, S. Bhattacharya, and V. Veliadis, "Paralleling of four 650v/60a gan hemts for high power traction drive applications," in *2021 IEEE Energy Conversion Congress and Exposition (ECCE)*, 2021, pp. 5269–5276.

[3] R. Letor, F. Scrimizzi, G. Longo, F. Iucolano, and M. Moschetti, "Compact design of dc/dc converter with new sti2gan solution," in *2020 AEIT International Conference of Electrical and Electronic Technologies for Automotive (AEIT AUTOMOTIVE)*, 2020, pp. 1–5.

[4] Reusch D., D. Gilham, Y. Su, and F. C. Lee, "Gallium nitride based 3d integrated non-isolated point of load module," in *2012 Twenty-Seventh Annual IEEE Applied Power Electronics Conference and Exposition (APEC)*, 2012, pp. 38–45.

[5] Wang K., B. Li, H. Zhu, Z. Yu, L. Wang, and X. Yang, "A double-sided cooling 650v/30a gan power module with low parasitic inductance," in *2020 IEEE Applied Power Electronics Conference and Exposition (APEC)*, 2020, pp. 2772–2776.

[6] J. Weimer, A. B. Sharma, T. Huesgen, D. Koch, and I. Kallfass, "Thermal study on leadframe dimensioning for high power dissipation and low inductance commutation cells," in *PCIM Europe digital days 2020; International Exhibition and Conference for Power Electronics, Intelligent Motion, Renewable Energy and Energy Management*, 2020, pp. 1–8.

[7] A. Sharma, J. Weimer, D. Koch, I. Kallfass, and T.Huesgen, "Asymmetric packages for optimal performance of gan-hemt using pcb fabrication technology," in *PCIM Europe digital days 2021; International Exhibition and Conference for Power Electronics, Intelligent Motion, Renewable Energy and Energy Management*, 2021.

[8] GaN Systems Inc., "Gn002 thermal design for ganpx packaged devices_r.201127," 2020, last access on 09/14/2022. [Online]. Available: https://gansystems.com/wp-content/uploads/2021/03/GN002-Thermal-Design-for-GaNPX-Packaged-Devices-Rev.201127-1.pdf

[9] D. Koch, A. Sharma, J. Weimer, M. Weiser, T. Huesgen, and I. Kallfass, "A 48 v, 300 khz, high current dc/dc-converter based on paralleled, asymmetrical & thermally optimized pcb embedded gan packages with integrated temperature sensor," in *2021 33rd International Symposium on Power Semiconductor Devices and ICs (ISPSD)*, 2021.

[10] D. Koch, J. Weimer, M. Weiser, J. Hueckelheim, and I. Kallfass, "Gate driver concept for parallel operation of low-voltage high-current gan power transistors for mild-hybrid applications," in *Applied Power Electronics Conference (APEC) 2021*, 2021.

[11] M. Ghebreslassie, A. B. Sharma, and T. Huesgen, "Thermal impedance evaluation of optimized pcb-based gan hemt single-chipprepackage using vgs method," in *Elektronische Baugruppen und Leiterplatten EBL 2022: DVS-Berichte, Band: 375*, 2022, pp. 43 – 48.

[12] M. Guacci, J. Azurza Anderson, K. L. Pally, D. Bortis, J. W. Kolar, M. J. Kasper, J. Sanchez, and G. Deboy, "Experimental characterization of silicon and gallium nitride 200 v power semiconductors for modular/multilevel converters using advanced measurement techniques," *IEEE Journal of Emerging and Selected Topics in Power Electronics*, vol. 8, no. 3, pp. 2238–2254, 2020.

[13] S. Biswas and D. Reusch, "Evaluation of gan based multilevel converters," in *2018 IEEE 6th Workshop on Wide Bandgap Power Devices and Applications (WiPDA)*, 2018, pp. 212–217.

[14] A. Lidow, D. Reusch, J. Strydom, J. Glaser, and M. de Rooij, *GaN transistors for efficient power conversion*, 3rd ed. Hoboken, NJ: John Wiley & Sons Inc, 2019.

[15] M. Moradpour and G. Gatto, "Efficiency analysis of two dc-dc universal converters for electric vehicles: Single-phase paralleled gan and two-phase sic-gan-based," in *2019 21st European Conference on Power Electronics and Applications (EPE '19 ECCE Europe)*, 2019, pp. P.1–P.7.

[16] N. Perera, A. Jafari, R. Soleimanzadeh, N. Bollier, S. G. Abeyratne, and E. Matioli, "Hard-switching losses in power fets: The role of output capacitance," *IEEE Transactions on Power Electronics*, vol. 37, no. 7, pp. 7604–7616, 2022.

978-1-6654-8901-0/22 $31.00 © 2022 IEEE

Design of High Power Converter with Single Low R_{on} Discrete SiC Device

Zibo Chen
Semiconductor Power Electronics Center (SPEC)
The University of Texas at Austin
Austin, Texas, USA
zibochen@utexas.edu

Chen Chen
Semiconductor Power Electronics Center (SPEC)
The University of Texas at Austin
Austin, Texas, USA
chenchen@utexas.edu

Qingyun Huang
Department of Electrical Engineering and Computer Science
University of Missouri
Columbia, Missouri, USA
qh7gp@missouri.edu

Alex Q. Huang
Semiconductor Power Electronics Center (SPEC)
The University of Texas at Austin
Austin, Texas, USA
aqhuang@utexas.edu

Abstract— **With the development of SiC technology, the on-resistance of discrete devices drops rapidly and can potentially replace the expensive power modules in high power converters. Unlike parallel discrete devices or a power module with built-in parallel dies, using a single device does not need to slow down the gate driver speed hence can achieve lower switching loss. However, there are still many challenges that need meticulous design considerations. This paper uses the 1200V TO-247-4 package device as an example to provide detailed guidance on device selection, loop optimization, insulation, thermal management, and mounting. A DC-DC converter design example is given with a very low loop inductance and excellent thermal performance. The air-cooled hardware demonstrates the hardware limitation is around 40kW in an 800V to 470V buck operation. Normalized to the R_{on} of the two devices used, this demonstrates an extremely high power density figure of merit of PDFOM=87kW/cm².**

Keywords—SiC, discrete SiC, TO-247, loop inductance, thermal management, figure of merit, design optimization, PDFOM

I. INTRODUCTION

Power modules are widely used in high power converters because of their high current ratings and ease of use. A power module often packages several dies in parallel, interconnecting the die into different configurations, such as a half-bridge module, full-bridge module, three-phase module, TNPC module, or ANPC module. Working with the gate driver solutions provided by the power module manufacturer, the converter designers are only required to focus on the bus bar, controller, sensors, system cooling, and EMI. However, convenience also comes with a relatively high penalty on performance and cost.

The discrete device packaging cost is low due to its ability to be produced in a very automated manufacturing line. Unlike multiple parallel dies in the power module, a discrete device also has better die utilization since every single device can be monitored, and less margin is required. Design with a discrete device is also more flexible and thus can fit designs with various housing constraints. Eight-hundred-volt DC bus is widely used

in different applications such as PV inverters, EV motor drives, energy storage converters, etc., where a 1200V power semiconductor is in need. IGBT or SiC modules are often used to simplify the design while offering the desired current ratings. With the fast development of SiC power semiconductor technology, the die size of SiC devices increases rapidly, resulting in a lower per die on-resistance (less than 10 mΩ) and high current rating (above 100A) of SiC MOSFETs. The upper limit of this trend is limited by the discrete package size. For example, at 1200V voltage level, Infineon released a 7mΩ trench MOSFET [1], and UnitedSiC has an 8.6mΩ cascode FET product [2]. The ultra-low R_{on} is already in the range of a conventional power module, which makes it possible to build a high power converter with only a single discrete device.

There are three major standard discrete packages for 1200V power semiconductors, through hole TO-247, surface mount D2PAK, and chassis mount SOT-227. Fig.1 collects the active discrete SiC packaging distribution from a major US vendor; the TO-247 package accounts for 70% of the discrete devices. The TO-247 package is popular in high power design because of its lower thermal resistance, high power dissipation, and better flexibility in busbar and gate driving circuit design. However, the large packaging inductance influences the gate driving for

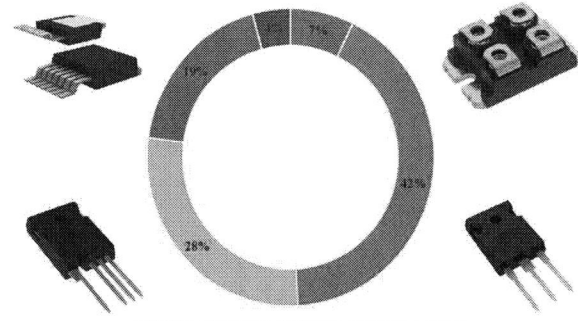

Fig. 1: Percentage of discrete SiC devices package (August 2022)

fast switching SiC devices. The new variation package TO-247-4 provides a Kelvin connection for the SiC device, which decouples the gate driving loop and the high power loop and accelerates the switching transient, thus reducing the loss.

This paper illustrates the considerations and hints for designing a high power converter using the TO-247-4 package SiC MOSFET. The benefits of TO-247-4 over TO-247-3 are discussed in Section II, and hints for device selection are given. Section III introduces the loop optimization and its experimental verification in Double Pulse Test (DPT). The insulation, thermal management, and mounting are discussed in Section IV. Finally, an example hardware is built and demonstrated to 40kW in 800V-470V buck operation.

II. 1200V LOW R_{ON} DEVICE SELECTION

A. TO-247-4 VS TO-247-3

The TO-247-3 package was introduced to the market in the 1980s, which fits well with the Si MOSFET and IGBT. SiC device has a much faster-switching speed than their Si counterparts. The superior performance of the SiC device pushes TO-247 packaging to its limitation on maximum temperature, current rating, and parasitic inductance. TO-247-4 package is developed to offer better driving reliability and lower switching loss for SiC MOSFET by providing a Kelvin connection. The parasitic inductance of the two packages is illustrated in Fig.2.

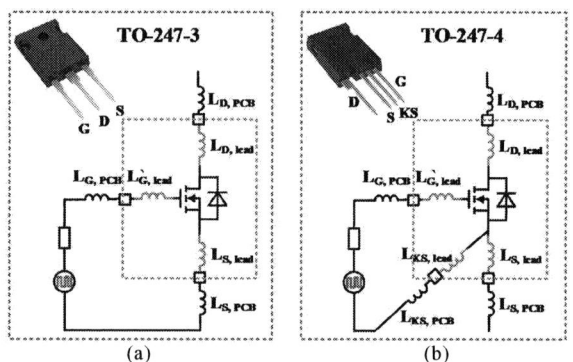

Fig. 2: Parasitic inductance. (a) TO-247-3. (b) TO-247-4

For the TO-247-3 package, the fast-changing current in the power loop will cause a voltage drop on the source side lead and PCB inductance, which is also part of the gate driving circuit, thus the actual gate to source voltage will not be the same as the gate driver chips. When the device turns off, the voltage across L_s will hold the gate for a longer time and slow down the current

drop; if the device turns on, the L_s voltage drop will pull down the V_{gs} increase and slow down the turn-on process. The prolonged switching process causes significant switching loss. So, it is common to slow down the device switching speed by using a large gate resistor R_g. On the other hand, the TO-247-4 package provides a Kelvin connection to decouple the gate driving loop from the power loop. The new structure can reduce the switching loss under the same test conditions; for example, the TO-247-4 IMZA120R007M1H reduces total switching loss by 57.8% over the TO-247-3 IMW120R007M1H SiC MOSFET, while E_{on} and E_{off} are reduced by 53.1% and 65.8%, respectively [1, 3]. Smaller R_g can also be used in the case of TO-247-4 since higher dI/dt will have less effect on the device performance. The lower limit of R_g, however, is related to the damping of the gate loop inductance shown in Fig.2.

B. Device Comparison

Recently more and more low R_{on} 1200V devices (<16mΩ) are becoming available, which is sufficient for many high power applications. However, selecting the best-fitting power semiconductor switch among many choices is still a question at the design stage. In the high power designs, the conduction loss usually dominates the full load so low on-resistance is desired. Using the device Figure of Merit (FOM) is a widely accepted method to evaluate different devices switching loss [4, 5]. In this way, the designer expects to use datasheet value to compare the gate driving speed, switching loss, reverse recovery loss, normalized cost, etc. Since conduction loss and switching loss are the major losses for high power applications, the HDFOM [4] can provide a quick comparison for hard-switching applications; a smaller number means a lower switching loss.

$$HDFOM = \sqrt{R_{on} \times Q_{gd}} \qquad (1)$$

Table I compares the characteristic of several 1200V low on-resistance SiC devices. The datasheet value from different manufacturers may be tested under different conditions, such as different gate voltage or conduction current; thus, the comparison may not be apple to apple. For example, for the on-resistance measurement, Wolfspeed tested it at 15V, Infineon tested it at 18V, and UnitedSiC tested it at 12V. Since the R_{on} is highly related to the gate voltage and current, one cannot directly compare the devices by the datasheet values. As is presented in [1], driving with 15V gate voltage has 27% more on-resistance than18V gate voltage. Although higher gate driving is attractive for low on-resistance, there still needs to be a trade-off between the gate driving loss, gate oxide degradation, and switching loss. The 175°C-25°C on-resistance ratio is another critical factor that

TABLE I - KEY CHARACTERISTICS COMPARISON OF 1200V LOW R_{on} SiC DEVICE.

Part number	Company	Structure	V_{gs} (V)	R_{on}(mΩ) 25°C	R_{max}(mΩ) 25°C	ΔR ratio	R_{on}(mΩ) 175°C	R_{175} /R_{25}	R_{on} Test Current (A)	R_{jc} (°C/W)	Q_{gd} (nC)	HDFOM
IMZA120R007M1H	Infineon	Trench	18	7	9.9	1.41	14	2	108	0.15	64	21.2
UF3SC120009K4S	United SiC	Cascode	12	8.6	11	1.28	18.2	2.12	100	0.15	40	18.5
G3R12MT12K	GeneSiC	Planar	18	10	13	1.3	16	1.6	100	0.2	112	33.5
IMZA120R014M1H	Infineon	Trench	18	14	18.4	1.31	27	1.93	54.3	0.25	32	21.2
NTH4L014N120M3P	Onsemi	Planar	18	14	20	1.43	29	2.07	74	0.17	98	37.0
C3M0016120K	Wolfspeed	Planar	15	16	22.3	1.39	28.8	1.8	75	0.27	61	31.2
UF3SC120016K3S	United SiC	Cascode	12	16	21	1.31	33	2.06	50	0.22	24	19.6

978-1-6654-8901-0/22 $31.00 © 2022 IEEE

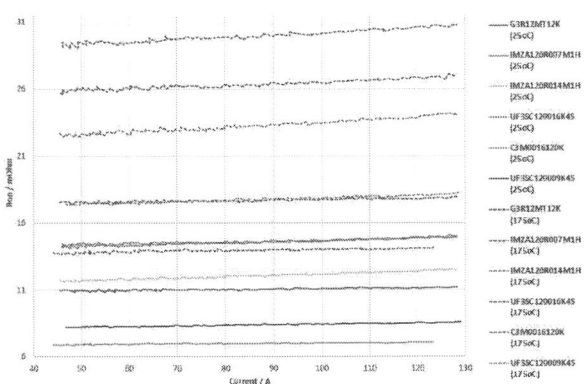

Fig. 3: R_{on} measurement of six SiC devices at 25°C (V_{gs}=19V)

requires attention, which involves the on-resistance change within the temperature range. Fig. 3 presents the R_{on} measurement result of six SiC devices using the measurement circuit introduced in [6]; the on-resistance is measured with the same 19V gate driving voltage and 25°C test condition. The estimated 175°C on-resistance for each device is plotted by the dash line which is the 25°C data multiplied the R_{175}/R_{25} ratio in Table 1. Thus, the resistance comparison can be revised for the same V_{gs} and taking in the temperature influence. The cascode SiC FET UF3SC120009K4S is selected in the following hardware design because of its superior performance.

III. LOOP INDUCTANCE OPTIMIZATION

A. Self-inductance of the TO-247 device

The TO-247 device typically has a die sinter on the metal frame (Copper plated with Nickel) and encapsulated with epoxy resin. The drain pad is directly connected to the package tab, and the gate and source require a bond wire to connect. SiC MOSFET has a much faster-switching speed than the Si counterpart. The dI/dt can reach above 6A/ns, which means 6V per nano Henry parasitic inductance. The self-inductance of the inductance of a SiC device is related to geometry and magnetic coupling. The TO-247 has relatively large parasitic inductance, but the magnetic coupling between each element is not significant and ignorable. The parasitic inductance mainly consists of the bond wire inductance, lead inductance, and PCB inductance once it's soldered. The typical parasitic inductance on the gate, drain, and source are 8.9nH, 4.3nH, and 7.5nH, respectively [7, 8]. The drain inductance is low because there is no bond wire, and the gate inductance is high because of a thin bond wire. The actual inductance varies by how much lead is

involved in the circuit. The longer the lead, the higher the inductance.

The TO-247 device self-inductance is extracted in Ansys Q3D, the lead is bent from the preset stand, and cut 4mm and 9mm from the stand, as is shown in Fig.4. The self-inductance is 11.9nH and 15.8nH, respectively. 5mm difference cause 3.9nH more inductance; thus, less lead should involve in the circuit as possible. It's worth mention bending the lead properly also needs attention to ensure it doesn't fail from the mechanical force; many vendors released a recommendation for their own devices to bend the lead from the prebuild stand [9].

B. Loop Optimization

Low commutation loop inductance is another crucial factor for the high voltage and high power design since dI/dt in SiC is very high and the loop inductance determines the overshoot voltage. As discussed in section II, TO-247 has higher internal inductance hence this is one of the key disadvantages of the TO-247 package. Minimizing additional inductance is very important. There are two loops requiring optimization. The first one is the power commutation loop which normally consists of the half bridge power semiconductor device and the decoupling capacitor, and the other is the gate driving loop completed by the gate driving circuit. A large power commutation loop inductance will cause a higher turn off voltage overshoot at a high current which can increase the turn off loss and kills the device. To reduce the overshoot voltage, either a low inductance loop is required, or use a higher external gate resistance to slow down the switching process, at the expense of higher switching loss. A high inductance gate driving loop will also require a larger gate resistor to damp the oscillation and provides long term reliability. The discrete packaging can directly access the gate and source PINs of each die thus having better design flexibility over the power modules in which the large gate resistor is built in to compensate unsymmetric gate loop of each die.

In a complete converter with a large DC link capacitor, two loops are actuallly formed. One is the power loop that defines the overshoot voltage and switching loss; the other is the capacitor loop between the low-value decoupling capacitor and the bulky DC link capacitor, which can induce undesired

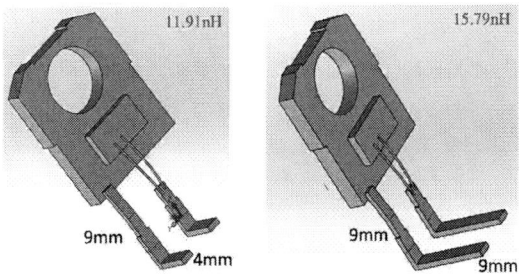

Fig. 4: Self-inductance extraction in Ansys Q3D

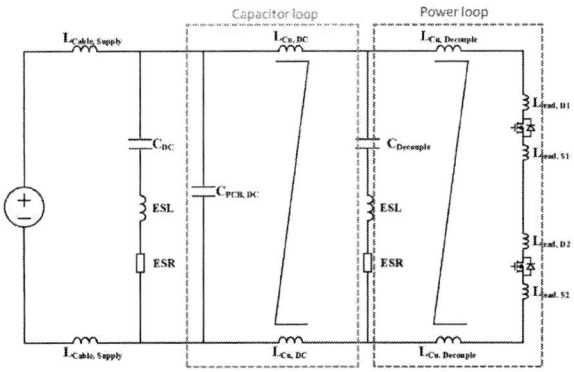

Fig. 5: Commutation loop inductance

978-1-6654-8901-0/22 $31.00 © 2022 IEEE

oscillation. The power loop inductance will separate into the device's internal inductance, mainly determined by the lead and bond wire, and the external inductance, the loop inductance on the PCB busbar. As is shown in Fig.5, the internal inductance has minimal magnetics coupling, so the value cannot change too much. Still, a well-designed PCB busbar can provide excellent magnetics cancelation to minimize the loop inductance, two neighborhood layers should be used for the forward and return paths [10, 11]. The PCB inductance can be estimate by Eq.2, where w is the copper width, l is the copper length, and e is the distance between two copper planes. There are many efforts to

$$L_{pcb} = \mu_0 \frac{e}{w} l \tag{2}$$

$$L_{pcb} = \mu_0 \frac{e}{w} l \left(\frac{1}{1 + e/w} + 0.024 \right) \tag{3}$$

$$f = \frac{0.35}{min(t_r, t_f)} \tag{4}$$

improve the calculation accuracy. When the w/e ratio is low Eq.3 [11] can be used. The parasitic inductance can be further extracted by Ansys Q3D as is discussed in [12], the simulation frequency (as in Eq.4) changes according to the device rise/fall time instead of the switching frequency as is presents in [13].

The decoupling capacitor is essential for the power semiconductor safe operation where 100 times C_{oss} value will be sufficient [14], and they are placed close to the device to minimize the power loop inductance. However, this is often limited by the physical constrain and layout of the PCB bus bar. A helpful hint is to remove the decoupling capacitor from the PCB bus bar and use a separate decoupling capacitor card so the theoretically lowest loop inductance can be achieved without obstructing the bus bar or gate driving circuit. This may increase the capacitor loop inductance and change the low-frequency oscillation between the two capacitors. Still, the power loop is fixed at its lowest value, providing more flexibility in the overall system design.

A design example is shown in Fig.6 and experimentally tested in a Double Pulse Test (DPT). Two test conditions are compared, the decoupling capacitor card is fixed at around 4mm from the stand, and the distance between the PCBs varies, so only the capacitor loop change. The first one places the decoupling capacitor card as close to the PCB bus bar as possible, so the capacitor loop inductance is minimized. The other one places the decoupling capacitor card as far from the bus bar as possible, so the capacitor loop inductance is at its highest value. The DC link voltage is 700V and the second turn off current is 118A, the turn-off overshoot voltage is around 985V for both conditions, and the power loop oscillation frequency is around 56MHz. The power loop inductance is extracted to be around 26nH by Eq.5; the C_{PCB} and C_{probe} are the parasitic inductance

$$L = \frac{1}{4\pi^2 f^2 (C_{oss} + C_{PCB} + C_{probe})} \tag{5}$$

$$L = \frac{1}{4\pi^2 f^2 (C_{Dec} + C_{PCB})} \tag{6}$$

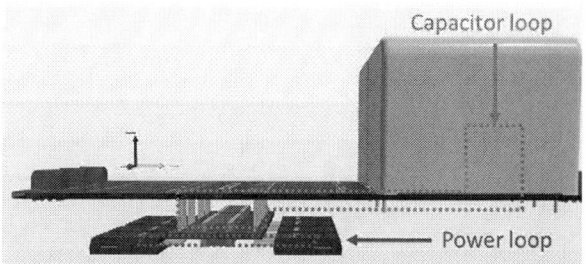

Fig. 6: Example for lowest loop inductance design

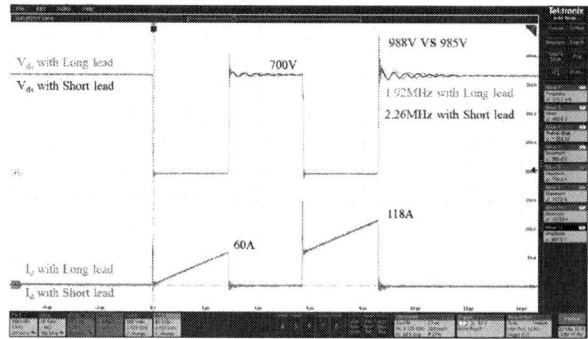

Fig. 7: DPT comparison for different capacitor loop

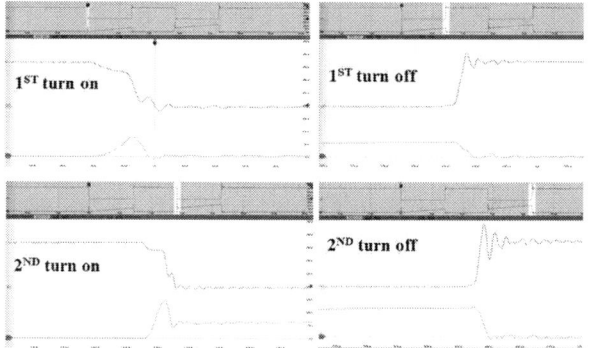

Fig. 8: Zoom in waveform for DPT comparison

induced by the PCB and probes. The capacitor loop inductance can be extracted from the 2.26MHz and 1.92MHz low-frequency oscillation by Eq.6, where C_{Dec} is the decoupling capacitor and the C_{PCB} is the parasitic capacitance from the PCB. The capacitor loop inductance is calculated as 24.3nH and 33.7nH respectively which sets the two boundary conditions for this kind of design. The DPT zoom-in waveform is presented in Fig.8, where the 60A turn on loss is around 1550µJ while the turn off loss is around 420µJ.

IV. INSULATION, THERMAL MANAGEMENT, AND MOUNTING

A. Mounting the SiC device in To-247 package

Mounting the TO-247 device looks effortless but needs to be careful. It also needs to tradeoff between the ease of assembly and thermal performance. A conventional method uses a screw to go through the predrilled mounting hole, but the tricky part is that the torque must follow the vendors' instructions. The screw will go through the device's encapsulated body and break the insulation layer, which can consequently cause insulation issues in high-voltage applications. Another risk is that the screw does not fit high vibration environments such as electric vehicle motor drive. In this case, the screw presses down on one side of the device, so the pressure on the interface material is uneven and causes a larger equivalent thermal resistance.

An alternative option is to use a clip to mount the device, such as U-clips, retaining clips, cam clips, saddle clips, and heatsink anchored clips. The clips usually sit above the die area and provide more even pressure than screws. However, a particular clip needs to fit with the heatsink. Using clips to mount the devices will add more workload to the production line. Some customized fasteners can be used for further mounting optimization. The third option is to directly solder or sinter the drain tab to an insulating substrate and mount it to the heatsink. Mechanical force stress would be on the device lead if no other supporting parts were applied. This approach itself is of higher cost but optimizes the overall system, and typically has lower thermal resistance than other approach.

B. Insulation and Thermal Management

Insulation and thermal are significant challenges for high voltage and high power converter design. TO-247 devices typically do not provide isolated thermal interface. Additional non conducting interface materials need to be used. Several common ones, including thermal pad, Al_2O_3 ceramic pad, AlN DBC, and Aluminum Core PCB [15], are experimentally tested. A side view of the thermal path of these solutions is illustrated in Fig.9.

Table I summarize the material properties: the insulation, thermal conductivity, and cost. DBC has the best thermal performance and excellent insulation, but the price is also high; Al_2O_3 ceramic pad also provides good thermal conductivity and insulation, but the assembly process is too complex. Aluminum Core PCB has similar thermal resistance as the Al_2O_3 pad, and the cost is very competitive. Still, the parasitic capacitance can be high because of the thin insulation layer and high E/d ratio. The thermal pad may not be desired in high voltage or harsh environments since reliability might be a concern.

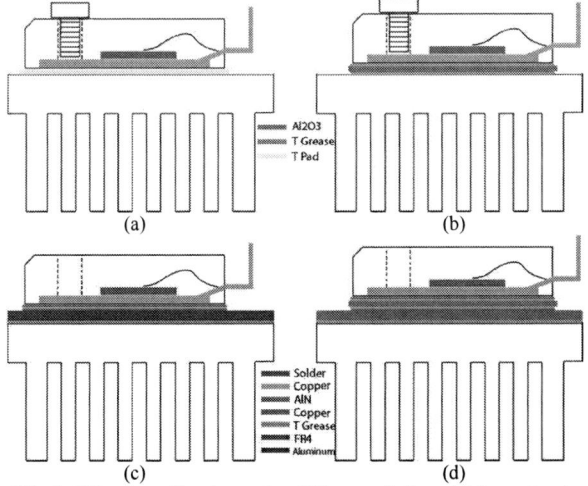

Fig. 9: Side view of implementing different solutions. (a) thermal pad. (b) ceramic pad. (c) MCPCB. (d) DBC. [15].

TABLE II - COMPARISON FOR DIFFERENT THERMAL INTERFACE

	Thermal Pad	Al_2O_3 Ceramic	Al based PCB	DBC
Dielectric Material	Silicone	94% Al_2O_3	FR4	AlN
Dielectric Constant	5	8.9	4.4	9
k (W/mK)	0.5-10	18-28	1-10	140-180
E/d Ratio	32.9	5	44	23.7
Volume Resistivity ($\Omega.cm$)	10^{10}	$>10^{14}$	$\geq10^{12}$	$>10^{14}$
Thermal Resistance	High	Medium	Medium	Low
Assembly Method	Screw/Clip	Screw/Clip	Solder	Solder
Cost	$	$$	$	$$$

V. EXPERIMENTAL RESULTS

A full bridge SiC converter is developed to verify the design concept and explore the limitation. The 3D model is shown in Fig.10 with four devices. One phase leg is first tested as a buck converter, the input voltage is 800V, the output voltage is 470V, and the switching frequency is 20kHz. The output capacitance is 120µF. Two 110µH inductors in series are used, each having three MAGNETICS Kool Mu MAX 0079097A7 cores in stack with 32 turns of 2790 strand AWG 40 wire. The inductor loss is calculated to be around 65W for each. The semiconductor loss is calculated by the UnitedSiC FET Jet Calculator [16]; as shown in Fig. 11, the high side device has a 141.3W loss, while the low side device has a 30.2W loss. The switching waveform

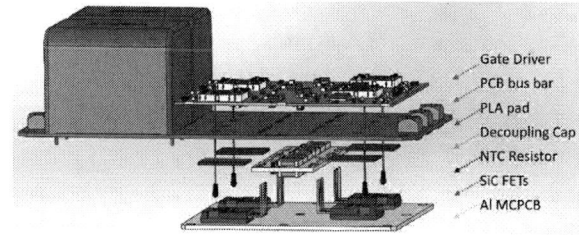

Fig. 10: 3D model of a full bridge SiC converter

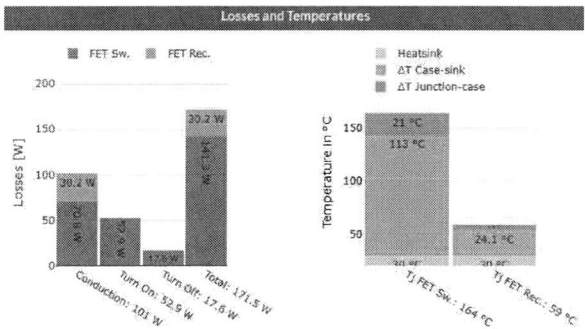

Fig. 11: Loss calculation by FET Jet Calculator [16]

and efficiency measurements are shown in Fig.12 and Fig.13. Fig.14 presents the thermal image at 40kW operation with thermal camera FLIR E60. The highest temperature observed is the high side device at 123.4°C. This test demonstrates that single discrete device can also achieve extremely high power level. The device utilization is very high and can be estimated by a new Power Density Figure of Merit.

$$PDFOM = \frac{P}{A} = \frac{P}{N \times \dfrac{R_{on\_sp}}{R_{on}}} = \frac{P \times R_{on}}{N \times R_{on\_sp}} \qquad (6)$$

Here A is the total chip area, R_{on} is the device R_{on}, N is the number of devices used, R_{on_sp} is the specific R_{on} of the technology. Using a R_{on_sp} for 1200V device as $2m\Omega$-cm^2, the buck converter achieved a PDFOM= 86kW/cm^2.

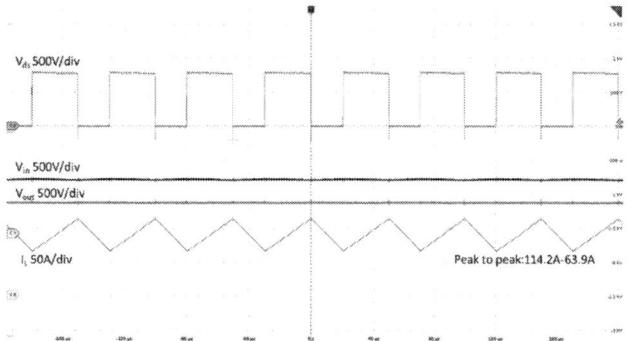

Fig. 12: Switching waveform of 40kW buck mode test

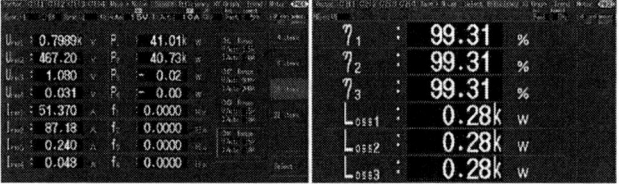

Fig. 13: Power analyzer measurement in 40kW buck mode test

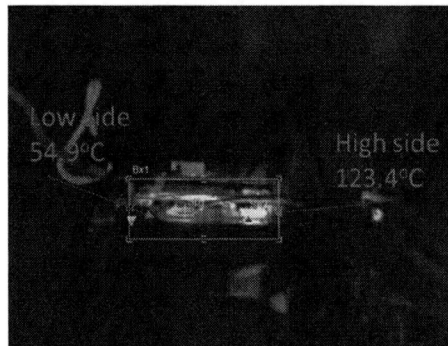

Fig. 14: Thermal image in 40kW buck mode test

VI. CONCLUSIONS

With the fast development of the SiC device, using a single discrete device can already construct a very powerful converter. This paper provides a fundamental design guidance on device selection, loop optimization, insulation, and thermal management. A DC-DC converter design example is given, which has perhaps the lowest loop inductance for a TO-247 device by utilizing a separate decoupling capacitor card. The hardware is tested at 40kW, 800V-470V buck mode, 141W loss on the high side device results in 123.4°C device temperature captured by the thermal camera.

ACKNOWLEDGMENT

This project was supported by Oak Ridge National Laboratory (ORNL) funded through the Department of Energy (DOE) - Office of Electricity's (OE), Transformer Resilience and Advanced Components (TRAC) program led by program manager Andre Pereira.

REFERENCES

[1] Infineon, "CoolSiC™ 1200 V SiC Trench MOSFET : Silicon Carbide MOSFET with .XT interconnection technology" IMZA120R007M1H Datasheet, Revised August, 2022.

[2] UnitedSiC, "1200V-8.6mW SiC FET" UF3SC120009K4S Datasheet, Revised December, 2019.

[3] Infineon, "CoolSiC™ 1200 V SiC Trench MOSFET : Silicon Carbide MOSFET" IMW120R007M1H Datasheet, Revised August, 2022.

[4] A. Q. Huang, "New unipolar switching power device figures of merit," in IEEE Electron Device Letters, vol. 25, no. 5, pp. 298-301, May 2004

[5] Q. Huang and A. Q. Huang, "Review of GaN totem-pole bridgeless PFC," in CPSS Transactions on Power Electronics and Applications, vol. 2, no. 3, pp. 187-196, Sept. 2017.

[6] Z. Chen, C. Chen, and A. Q. Huang, "Driver Integrated Online Rds-on Monitoring Method for SiC Power Converters," 2022 IEEE Energy Conversion Congress and Exposition (ECCE), 2022.

[7] K. Aikawa, T. Shiida, R. Matsumoto, K. Umetani and E. Hiraki, "Measurement of the common source inductance of typical switching device packages," 2017 IEEE 3rd International Future Energy Electronics Conference and ECCE Asia (IFEEC 2017 - ECCE Asia), 2017.

[8] T. Liu, T. T. Y. Wong and Z. J. Shen, "A New Characterization Technique for Extracting Parasitic Inductances of SiC Power MOSFETs in Discrete and Module Packages Based on Two-Port S-Parameters Measurement," in IEEE Transactions on Power Electronics, vol. 33, no. 11, pp. 9819-9833, Nov. 2018.

[9] UnitedSiC, "Through-Hole Lead Bending", Application Note UnitedSiC_AN0021, Revised June, 2019

[10] D. Reusch and J. Strydom, "Understanding the Effect of PCB Layout on Circuit Performance in a High-Frequency Gallium-Nitride-Based Point of Load Converter," in IEEE Transactions on Power Electronics, vol. 29, no. 4, pp. 2008-2015, April 2014.

[11] A. Letellier, M. R. Dubois, J. P. F. Trovão and H. Maher, "Calculation of Printed Circuit Board Power-Loop Stray Inductance in GaN or High di/dt Applications," in IEEE Transactions on Power Electronics, vol. 34, no. 1, pp. 612-623, Jan. 2019.

[12] I. Kovacevic-Badstübner, T. Ziemann, B. Kakarla and U. Grossner, "Highly accurate virtual dynamic characterization of discrete SiC power devices," 2017 29th International Symposium on Power Semiconductor Devices and IC's (ISPSD), 2017, pp. 383-386.

[13] Z. Zhang, B. Guo, F. F. Wang, E. A. Jones, L. M. Tolbert and B. J. Blalock, "Methodology for Wide Band-Gap Device Dynamic Characterization," in IEEE Transactions on Power Electronics, vol. 32, no. 12, pp. 9307-9318, Dec. 2017.

[14] Z. Chen, D. Boroyevich, P. Mattavelli and K. Ngo, "A frequency-domain study on the effect of DC-link decoupling capacitors," 2013 IEEE Energy Conversion Congress and Exposition, 2013, pp. 1886-1893.

[15] Z. Chen, C. Chen, and A. Q. Huang, "A 900V/4mΩ/80A Bidirectional SiC DC Solid State Contactor (SSC)," 2022 IEEE Energy Conversion Congress and Exposition (ECCE), 2022.

[16] UnitedSiC, "FET Jet Calculator", Available: http://https://unitedsic.com/fet-jet/. [Accessed: September 2022].

Comparative Investigation of Current-Source Inverters using SiC Discrete Devices and Power Modules

Feida Chen
Wisconsin Electric Machines and
Power Electronics Consortium
University of Wisconsin-Madison
Madison, USA
fchen89@wisc.edu

Sangwhee Lee
Wisconsin Electric Machines and
Power Electronics Consortium
University of Wisconsin-Madison
Madison, USA
sangwhee.lee@wisc.edu

Thomas M. Jahns
Wisconsin Electric Machines and
Power Electronics Consortium
University of Wisconsin-Madison
Madison, USA
jahns@engr.wisc.edu

Bulent Sarlioglu
Wisconsin Electric Machines and
Power Electronics Consortium
University of Wisconsin-Madison
Madison, USA
sarlioglu@wisc.edu

Abstract—The purpose of this paper is to investigate the impact of the SiC device packages on the commutation performance characteristics of current-source inverters (CSIs). The parasitic components in the CSI current commutation loop between the two-phase legs and output capacitors have a significant impact on the high-frequency performance of the SiC devices. To meet the elevated current requirements of high-power CSIs, it is often necessary to connect multiple discrete devices in parallel which increases the current commutation loop length. The selection of compact high-power SiC MOSFET and Schottky diode modules instead of discrete devices can be highly desirable to reduce the loop inductance and improve the system performance and power density. Two CSI benchtop prototype units, one with SiC discrete devices and the other with power modules, have been designed and tested, and the performances of the two CSIs are compared. The CSI with SiC power modules significantly reduces the inverter volume and lowers the parasitic inductance by 60% and the voltage ripple amplitude by 20% compared to the CSI unit with discrete devices.

Keywords—current commutation, current source inverter, parasitic inductance, SiC discrete devices, SiC power modules, wide bandgap

I. INTRODUCTION

Wide-bandgap (WBG) semiconductor devices using silicon carbide (SiC) and gallium nitride (GaN) have many attractive features, such as lower on-resistance, higher switching speed, and higher junction temperatures compared to conventional silicon (Si) devices [1-3]. The emergence of WBG power devices opens opportunities for the current source inverter (CSI) to provide a promising alternative drive configuration for EVs [4, 5], as shown in Fig. 1. The CSI replaces the dc-link capacitor of the voltage source inverter (VSI) with a dc-link inductor and three small output capacitors which provides more sinusoidal output voltage waveforms to the machine and lower common-mode EMI emissions compared to VSI [5]. By using WBG power switches, the CSI offers valuable performance advantages over the baseline VSI [6].

Financial support for this work is provided by the U.S. Department of Energy's Office of Energy Efficiency and Renewable Energy (EERE) under the Vehicle Technologies Office (VTO). Award Number DE-EE0008704.

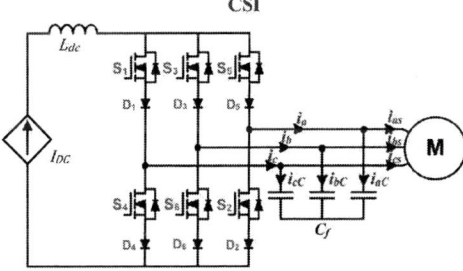

Fig. 1. CSI based motor drive system for traction applications.

A VSI switch commutation event involves the two switches and two diodes in the same phase leg. In contrast, the CSI current commutates between two phase legs and the output ac capacitors, as shown in Fig. 2. (d). The parasitic components in the commutation loop of a CSI have a significant impact on the high-frequency overshoot and resonance of the SiC switches. Han, *et al* [7] have investigated the influence of parasitic inductances on the performance of SiC MOSFETs for dc-dc converters from the loss and electromagnetic interference perspective. In [8], the commutation loop inductance for the SiC EV traction inverter is studied and analyzed. Torres, *et al.* [9] investigated the effect of power loop parasitic inductance on CSI switching transients. However, the comparative impact of SiC discrete devices and power modules on CSI switching performance has not received attention in the existing literature.

The objective of this paper is to investigate and compare the impact of discrete and module-based SiC switches on the parasitic inductance and resulting commutation performance of high-frequency CSIs. Section II explains the differences between the commutation loops and methods for CSIs and VSIs. In Section III, the impact of parasitic impedance in the commutation loop on SiC devices' switching behavior is investigated. In Section IV, two CSI benchtop units with 55 kW continuous power and 100 kW peak power ratings have been

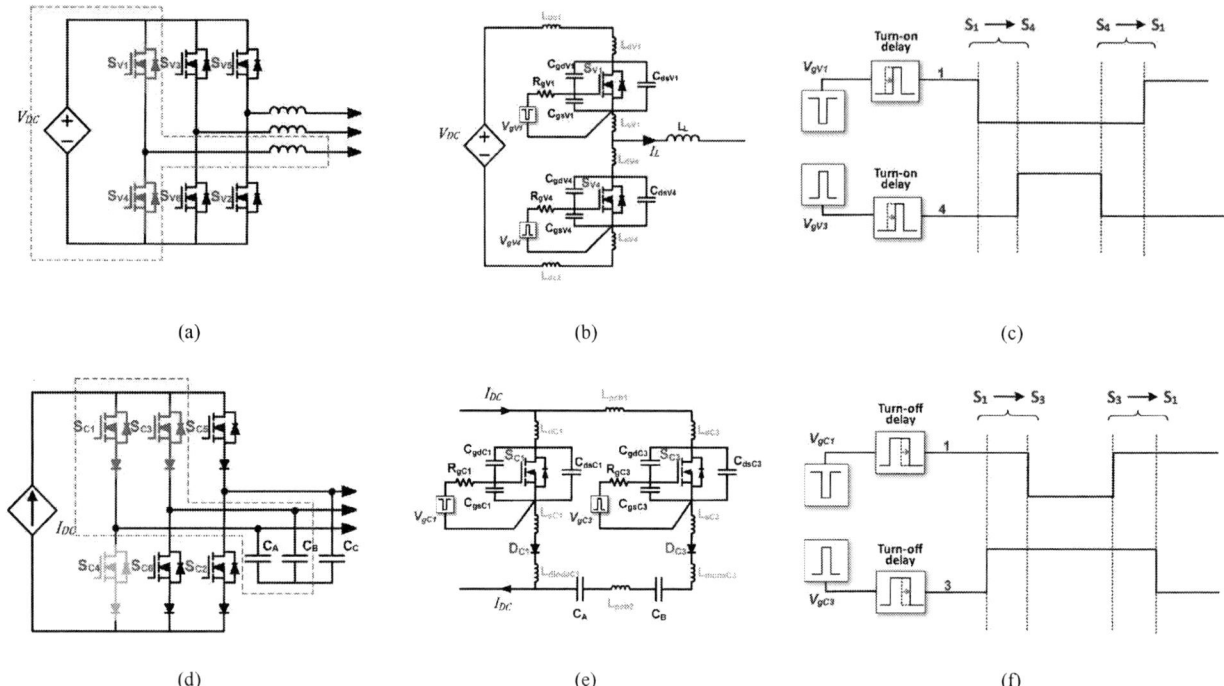

(a)   (b)   (c)

(d)   (e)   (f)

Fig. 2. (a) Basic 3-phase VSI topology showing commutation process between switches S_{V1} and S_{V4}; (b) VSI commutation loop with parasitic components; (c) VSI commutation gating sequence for S_{v1} and S_{v4} including deadtime periods; (d) Basic 3-phase CSI topology showing commutation process between switches S_{C1} and S_{C3} for commutation between a CSI active state and a zero state; (e) CSI commutation loop with parasitic components; (f) CSI commutation gating sequence for S_{C1} and S_{C3} including current overlap periods

designed and tested with SiC discrete devices and SiC modules, respectively. The parasitic inductance and performance of the two CSIs are compared and analyzed. Conclusions and future work are discussed in the last section.

II. COMPARISON OF COMMUTATION PROCESS BETWEEN VSI AND CSI

This section compares the commutation process between VSI and CSI using equivalent circuits and switching diagrams. Fig. 2 (a) and (d) show the typical operating conditions for VSI and CSI, respectively. It can be observed that the commutation process of VSI happens between the upper (S_{v1}) and lower switches (S_{v4}) of the same phase leg, while the commutation process of CSI happens between the S_{c1} and S_{c3}. Fig. 2 (b) and (e) show the typical commutation loop of VSI (S_{v1} and S_{v4}) and CSI (S_{c1} and S_{c3}), including the parasitic capacitances and inductances, respectively. The internal parasitic capacitances include gate-source capacitance C_{gs}, miller capacitance C_{gd} and drain-source capacitance C_{ds} of SiC MOSFETs. The parasitic inductances include drain parasitic inductance L_d, source parasitic inductance L_s, parasitic inductance on DC link L_{dc}, and parasitic inductance of the printed circuit board (PCB). R_g is the gate resistance of SiC MOSFETs which determines the switching speed. The subscript V represents VSI and C represents CSI. V_{DC} is the bus voltage, I_{DC} is the dc-link current, I_L is the load current; L_L is the load inductance. It can be observed the parasitic parameters in the CSI current commutation loop is different from VSI. Fig. 2 (c) and (f) show the commutation method for VSI and CSI, respectively. To

prevent short circuits of the DC link in VSIs, a dead time is necessary by adding a turn-on delay to both switches in the same phase leg. For CSIs, there must always be a current conduction path for the current source. To satisfy this constraint, it is a common practice to use an overlap time between an off-going and an on-coming switch connected to the same bus terminal, representing the dual of dead time in VSIs [9].

For VSI, the dc-link parasitic inductance has the biggest contribution to the commutation loop inductance. To minimize dc-link inductance, the dc-link capacitor should be very close to the phase legs. For CSI, the parasitic components in the CSI current commutation loop between the two phase legs and output capacitors have a significant impact on high-frequency SiC devices' performance. However, the parasitic inductance of the CSI commutation loop is not easy to reduce, being highly dependent on the device package and PCB layout.

III. IMPACT OF PARASITIC IMPEDANCE IN THE COMMUTATION LOOP ON SiC DEVICES SWITCHING BEHAVIOR

The resonance between parasitic inductance (L_s) in the current commutation loop and parasitic capacitance (C_J) of the switching device under switching events will cause voltage spikes in the output sinusoidal voltage waveforms of CSI. The resonance frequency between two parasitic components can be approximated as

$$f_{res} = \frac{1}{2\pi\sqrt{L_s C_J}} \qquad (1)$$

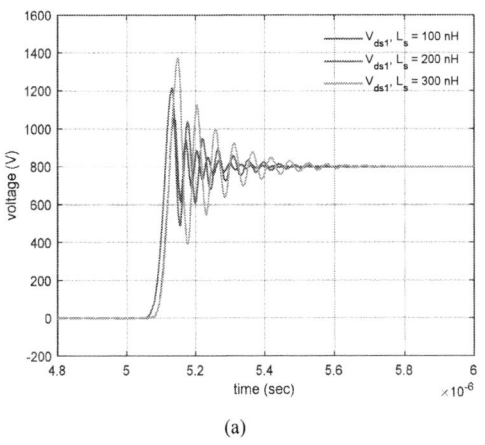
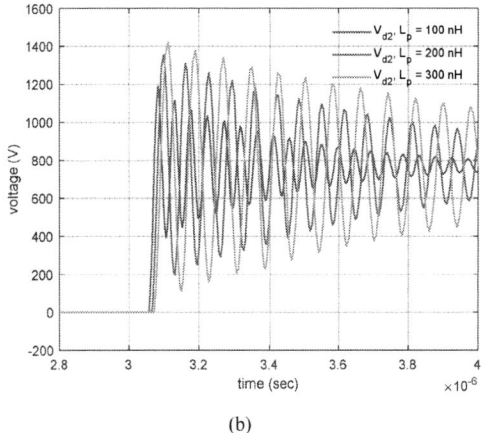

Fig. 3. Influence of parasitic inductance on device turn-off resonance. (a) SiC MOSFET V_{ds1}. (b) SiC Schottky diode V_{d2}

To reduce the influence of the switching oscillation caused by the parasitic inductance and parasitic capacitance in the current commutation loop, it is necessary to reduce the value of those parasitic components. The parasitic capacitance of a switching device is determined by the geometry of a switching device and cannot be changed easily. A practical way to reduce the switching oscillation is to minimize the parasitic inductance in the current commutation loop.

The influence of reducing the parasitic inductance L_s on device's turn-off voltage is demonstrated with LTSpice simulation and the results are shown in Fig. 3 using a CSI double-pulse circuit setup [10]. Based on the simulation result, it is identified that reducing the parasitic inductance L_s can reduce the turn-off oscillation significantly for both SiC MOSFET and SiC Schottky diode. It should also be noted that the voltage overshoot tends to be higher at the turn-off transient of SiC Schottky diode compared to that of SiC MOSFET.

IV. COMPARATIVE INVESTIGATION OF THE CSIS WITH DISCRETE AND MODULE SiC DEVICES

To meet the high current requirements of high-power CSIs, it is often necessary to connect multiple discrete devices in parallel, but the current commutation loop length will be increased. The selection of compact high-power SiC MOSFET and Schottky diode modules instead of multiple discrete devices can be highly desirable to reduce the loop inductance and improve the system performance and power density. In this section, two CSI benchtop units with 55 kW continuous power and 100 kW peak power ratings have been designed and tested with SiC discrete devices and SiC modules, respectively, as shown in Fig. 4. The parasitic inductance and performance of the two CSIs are compared and analyzed.

(a)

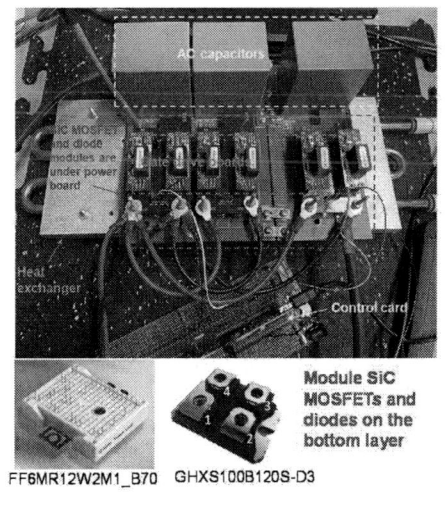

(b)

Fig. 4. (a) View of CSI unit with SiC discrete devices; (b) View of CSI unit with SiC power modules

A. Development of three-Phase CSIs based on SiC Discrete Devices and Power Modules

Fig. 4 (a) shows the fabricated CSI unit with SiC discrete devices. To meet the current rating requirements, 3 discrete SiC FETs (UF3SC120009K4S from UnitedSiC) are in parallel and 4 discrete SiC Schottky diodes (GC50MPS12-247 from GeneSiC) are in parallel. All the devices are on the bottom side of the PCB and mounted on a liquid heat exchanger.

Fig. 4 (b) shows the fabricated CSI unit with SiC power modules. A compact SiC MOSFET half-bridge module FF6MR12W2M1_B70 from Infineon that has comparable volume and electrical characteristics to multiple discrete devices is chosen as an alternative to UF3SC120009K4S discrete SiC FET with TO-247-4 package. Also, a SiC Schottky diode module GHXS100B120S-D3 from SemiQ in the SOT-227 package is selected as an alternative to the discrete SiC Schottky diode GD50SMPS12H. The 4-pass Aavid Thermalloy Hi-Contact™ cold plates (heat exchanger) are used to cool the power devices.

B. Parasitic Inductance Comparison

In this section, the parasitic inductance of the PCB layout of the module-based CSI is compared to the discrete-based CSI. The PCB layout with discrete devices is shown in Fig. 5 (a) and

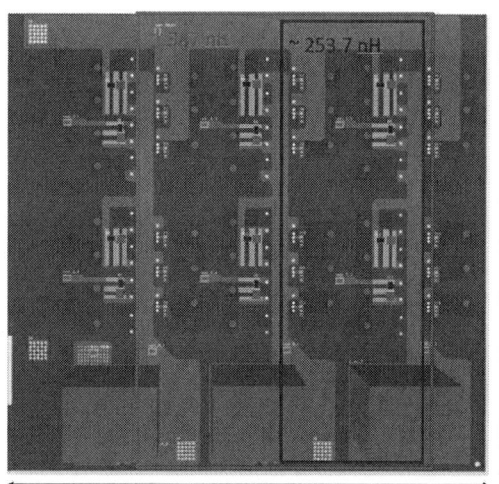

(a) Parasitic inductance values of CSI with SiC discrete devices

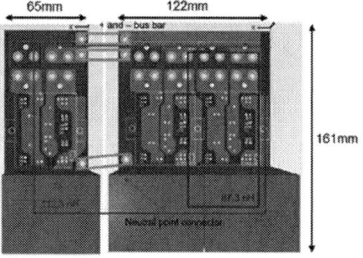

(b) Parasitic inductance values of CSI with SiC power modules

Fig. 5. Parasitic inductance comparison of CSIs with SiC discrete devices and SiC power modules

TABLE I.   COMPARISON OF ESTIMATED PARASITIC INDUCTANCE IN THE CURRENT COMMUTATION LOOP WITH DIFFERENT SiC MOSFET LAYOUT CONFIGURATIONS

Inductance	Discrete (UF3SC120009K4S)	Module (FF6MR12W2M1_B70)
Device	15~25 nH/device	8 nH/module
PCB trace	253.7~387 nH	92.7~112.3 nH
Total	347~480 nH	163.3~188.3 nH

that with modules is shown in Fig. 5 (b). The discrete devices are bent so that the metallic plate of discrete devices can have contact with the heat exchanger. This cooling scheme leads to the increased area in PCB layout resulting in increased length of the current commutation loop. On the other hand, modules can be directly mounted on the liquid-cooled heat exchanger and the length of the current commutation loop can be minimized.

The PCB layouts of the two CSIs with SiC discrete devices and modules are analyzed, and their parasitic inductances have been extracted using ANSYS Q3D. The total trace inductance of the current commutation loop in the PCB layout ranges from 253.7 nH to 387 nH using discrete devices, compared to a range using modules of 87.3 nH to 112.3 nH, which is approx. three times smaller. In the case of discrete SiC MOSFETs, the parasitic inductance of each package is approx. 15 to 25 nH, while the SiC MOSFET module is only about 8 nH. The total parasitic inductance of the CSI with SiC discrete devices is about 347 to 480 nH, while that of the CSI design with modules ranges from 163.3 nH to 188.3 nH. The estimated parasitic inductances in the current commutation loop with different SiC MOSFET layout configurations are summarized in TABLE I. As a result, the CSI unit with SiC power modules significantly lowered parasitic inductance by 60% compared to the unit with SiC discrete devices.

C. Experiment Results Comparison

Both CSIs have been tested using the same testbed and 3-phase RL passive load to evaluate the unit performance at 50 A peak phase current. The CSIs are controlled using a TMS320F28379D (Texas Instruments) DSP controller board. The circuit parameters and CSI's operation conditions are summarized in TABLE II, where L_{dc} is the dc-link inductance, C is the CSI's output ac filter capacitance, R is the load resistance, L is the load inductance, f_{sw} is the CSI's switching frequency, and f_1 is the modulated current's fundamental frequency. In addition, the phase current and line-line voltage are measured by high accuracy and high bandwidth voltage probes (Lecroy HVD3106A 1.5kV High Voltage Differential Probe) and current probes (LeCroy CP150 Current Probe,

TABLE II.   CIRCUIT PARAMETERS FOR CSI OPERATION

L_{dc}	C	R	L	f_{sw}	f_1
306 µH	22.5 µF	3.1 Ω	150 µH	25 kHz	200 Hz

LeCroy CP031 Current Probe) combined with a high bandwidth oscilloscope (LeCroy MDA810A 1 GHz, 2.5 GS/s, 8-channel).

The impact of the parasitic inductance in the commutation loop of CSIs with SiC discrete devices and modules can be observed in the experimental results shown in Fig. 5. Both CSIs exhibit very sinusoidal current, but the voltage waveform of the CSI with SiC discrete devices has 20% higher peak-to-peak voltage ripple (48.2% of peak fundamental l-l voltage) compared to the CSI with SiC power modules (27.8%). The voltage ripples are caused by the switching oscillation between the parasitic inductance in the current commutation loop and the parasitic capacitance of the switching device under switching events. The FFT analysis results comparison between CSI with discrete devices and power modules are shown in Fig. 6. It can be observed that the THD of the CSI with module devices (4.94%) is 1.15% lower than that with discrete devices (6.09%). The performance comparison between CSI with SiC discrete devices and power modules is summarized in TABLE III. The voltage ripples and THD of CSI with module SiC devices are significantly reduced compared to CSI with discrete devices due to the much lower parasitic inductance in the current commutation loops.

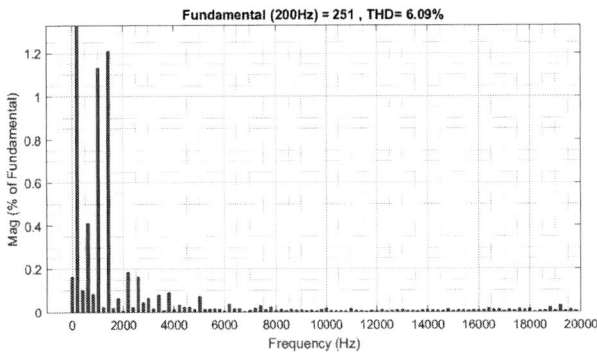

(a)

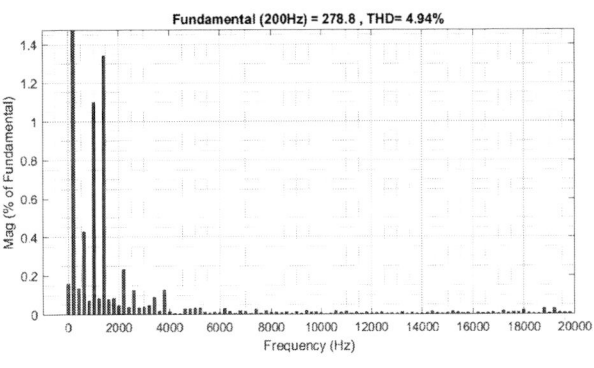

(b)

Fig. 6. FFT analysis comparison: (a) CSI with SiC discrete devices; and (b) CSI with SiC power modules.

TABLE III.   PERFORMANCE COMPARISON BETWEEN CSIs WITH SiC DISCRETE DEVICES AND POWER MODULES

	CSI with discrete devices	CSI with module devices
PCB area	1,175 cm²	301 cm²
Total parasitic inductance in the longest commutation loop	480 nH	188 nH
Voltage ripple	48.2%	27.8%
Voltage waveform THD	6.09%	4.94%

V. CONCLUSION

This paper investigated the impact of the SiC device packages on the parasitic inductance and commutation performance characteristics of CSIs. The current commutation loop and process of CSI and VSI are compared, and the impact of the parasitic inductance of the current commutation loop on the CSI switching performance is analyzed. Two CSI benchtop units with 55 kW continuous power rating have been designed and tested with SiC discrete devices and SiC power modules, respectively. The CSI with SiC power modules significantly reduces the inverter volume and lowers the parasitic inductance by 60% compared to the unit with SiC discrete devices. Both CSIs exhibit very sinusoidal current, but the voltage waveform

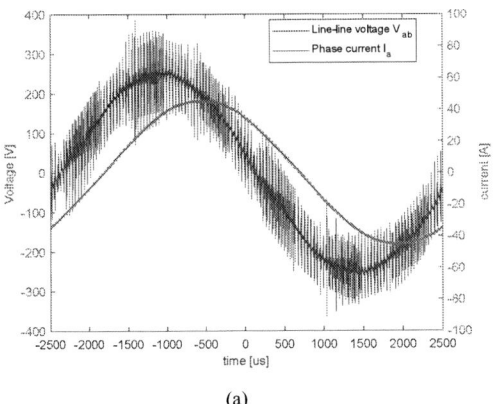

(a)

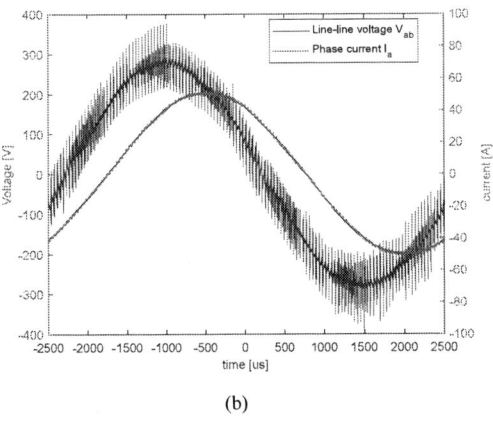

(b)

Fig. 5. Measured line-line voltage and phase current waveform comparisons for two CSI units during operation at 50 A pk current: (a) CSI with SiC discrete devices; and (b) CSI with SiC power modules.

of the CSI with SiC discrete devices has 20% higher peak-to-peak voltage ripple compared to the CSI with SiC modules due to the higher parasitic inductance in the current commutation loop.

ACKNOWLEDGMENT

This material is based on work supported by the U.S. Department of Energy's Office of Energy Efficiency and Renewable Energy (EERE) under the Vehicle Technologies Office (VTO), Award Number DE-EE0008704. The authors also gratefully acknowledge the support of the Wisconsin Electric Machines and Power Electronics Consortium (WEMPEC).

REFERENCES

[1] A. K. Morya et al., "Wide bandgap devices in AC electric drives: opportunities and challenges," in *IEEE Transactions on Transportation Electrification*, vol. 5, no. 1, pp. 3-20, March 2019.

[2] E. A. Jones, F. F. Wang, and D. Costinett, "Review of commercial GaN power devices and GaN-based converter design challenges," in *IEEE Journal of Emerging and Selected Topics in Power Electronics*, vol. 4, no. 3, pp. 707-719, Sept. 2016.

[3] X. Ding, F. Chen, M. Du, H. Guo, and S Ren, "Effects of silicon carbide MOSFETs on the efficiency and power quality of a microgrid-connected inverter," in *Applied Energy*, 2016, 201: 270-283.

[4] F. Chen, H. Ding, S. Lee, W. Feng, T. M. Jahns, and B. Sarlioglu, "Current source inverter based large constant power speed ratio SPM machine drive for traction applications," in *Proc. 2020 IEEE Transportation Electrification Conference & Expo (ITEC)*, Chicago, IL, USA, 2020, pp. 216-221.

[5] F. Chen, W. Feng, H. Ding, S. Lee, T. M. Jahns and B. Sarlioglu, "Comprehensive efficiency analysis of current source inverter based SPM machine drive system for traction applications," in *Proc. 2020 IEEE Energy Conversion Congress and Exposition (ECCE)*, Detroit, MI, USA, 2020, pp. 3002-3009.

[6] H. Dai, T. M. Jahns, R. A. Torres, D. Han, and B. Sarlioglu, "Comparative evaluation of conducted common-mode EMI in voltage-source and current-source inverters using wide-bandgap switches," in *Proc. IEEE Transportation Electrification Conference and Expo (ITEC)*, Long Beach, CA, 2018, pp. 788-794.

[7] D. Han and B. Sarlioglu, "Comprehensive study of the performance of SiC MOSFET-based automotive DC-DC converter under the influence of parasitic inductance," in *IEEE Transactions on Industry Applications*, vol. 52, no. 6, pp. 5100-5111, Nov.-Dec. 2016.

[8] R. S. Krishna Moorthy et al., "Estimation, minimization, and validation of commutation loop inductance for a 135-kW SiC EV traction inverter," in *IEEE Journal of Emerging and Selected Topics in Power Electronics*, vol. 8, no. 1, pp. 286-297, March 2020.

[9] R. A. Torres, H. Dai, T. M. Jahns and B. Sarlioglu, "Operation and analysis of current-source inverters using dual-gate four-quadrant wide-bandgap power switches," in *Proc. IEEE Energy Conversion Congress and Exposition (ECCE)*, 2019, pp. 2353-2360.

[10] F. Chen, S. Lee, R. A. Torres, T. M. Jahns and B. Sarlioglu, "Performance Evaluation and Loss Modeling of WBG Devices based on a Novel Double-Pulse Test Method for Current Source Inverter," in *Proc. IEEE Transportation Electrification Conference & Expo (ITEC)*, 2021, pp. 219-224.

978-1-6654-8901-0/22 $31.00 © 2022 IEEE

A Medium-Voltage Transformer with Integrated Leakage Inductance for 10 kV SiC-Based Dual-Active-Bridge Converter

Zihan Gao[1], Haiguo Li[1], Fred Wang[1,2]

[1]Min H. Kao Department of Electrical Engineering and Computer Science, the University of Tennessee, Knoxville, TN, USA

[2]Oak Ridge National Laboratory, Oak Ridge, TN, USA

Email: {zgao15, hli96}@vols.utk.edu, fred.wang@utk.edu

Abstract— **Medium-voltage (MV) dual-active-bridge (DAB) converters have become an emerging technology thanks to high-voltage silicon carbide (SiC) devices and nanocrystalline magnetic materials. However, the need of phase-shift inductance and insulation requirements for the MV DAB may complicate the design of the transformer and the MV DAB converter, which can also induce higher loss and occupy more space. In this paper, the leakage integration and insulation techniques are discussed for a 6.7-kV/850-V DAB converter, meeting both the inductance and insulation requirements of the MV DAB converter. Ferrite cores with air gaps are inserted between the LV and MV windings without introducing high loss, and the MV winding is selectively shielded to avoid high parasitics and meet the insulation requirement. Test results have verified the effectiveness of this design.**

Keywords—Dual-active-bridge converters; leakage inductance; medium voltage transformer; high voltage insulation

I. INTRODUCTION

The medium-voltage (MV) dc/dc converters have been implemented widely where the high power and high voltage conversion are needed. Fast-switching and high-voltage wide band-gap semiconductor devices, as well as low loss and high saturation magnetic materials, have enabled more compact and efficient designs of the MV dc/dc transformers and converters, making the converters more efficient and power dense.

Over the years, there have been several successful MV transformer designs. In [1], hollowed shielded cable was utilized as the transformer winding, since the inner conductor and outer shielding can be seen as two sets of windings having a unity turns ratio, with a good insulation rating, and the center hollowed channel of the cable was used to provide coolant path for active winding cooling. In [2], an oil-type MV transformer was implemented in a 1.2 MVA shunting locomotive, similar to the conventional traction transformers, the transformer oil can serve as both the insulation material and the liquid coolant. The solid-state transformer (SST) is also a typical application, for microgrids, datacenter, or renewable energy generation. In [3, 4], a GaN/SiC-based converter for the datacenter was proposed, by increasing the switching frequency up to 500 kHz to achieve a high-power density. Reference [5] introduces a transformer with casted silicone and epoxy to improve the insulation performance, and the MV winding was coated with a shielding layer to confine the electric field in the dielectric material inside the MV winding, so that the low voltage winding does not need to be encapsulated and has better cooling. In [6], 10 kV SiC devices were used as the MV side switch, so that the single stage can be realized for 7 kV to 400 V conversion. In [7], an air-insulated transformer was introduced, having a very high power density as all the winding conductors can be directly cooled by the air flow. An air-core transformer has been designed in [8] to have light weight. As the air core cannot constrain the magnetic field, proper magnetic shielding is needed.

A. Leakage Integration

For dual-active-bridge converters, the transmitting power is determined by the phase-shift and the series inductance between the two bridges. The series inductance for DAB converters should be sufficient for stable power conversion and also realizing ZVS to reduce switching losses. It should also be noted that the series inductance may also need MV insulation, which also impacts the volume and power density of the converter design, and hence may need to be integrated into the transformer.

An important effect on nanocrystalline tape-wounded transformer is the eddy current in the lamination surfaces. As the lamination is mainly for eliminating the eddy current component in the magnetizing loop to reduce the core loss, it may not work on the leakage loop and may cause high eddy current loss [9]. Fig. 1 shows a thermal image during short circuit test of a core-type MV transformer, and the temperature distribution of the core is strongly uneven. Most of the heat has been concentrated on the outer surface of the core, where the eddy current is mainly concentrated. Similar phenomena have also been found in [10] and [11]. To tackle this issue, [10] introduced a ferrite shielding method to bypass the leakage flux so that the eddy current may not be generated in the lamination, but the effect is limited by test verification.

This work was supported by the Advanced Manufacturing Office (AMO), United States Department of Energy, under Award no. DE-EE0008410, and made use of the Engineering Research Center Shared Facilities supported by the Engineering Research Center Program of the National Science Foundation and DOE under NSF Award Number EEC-1041877 and the CURENT Industry Partnership Program.

978-1-6654-8901-0/22 $31.00 © 2022 IEEE

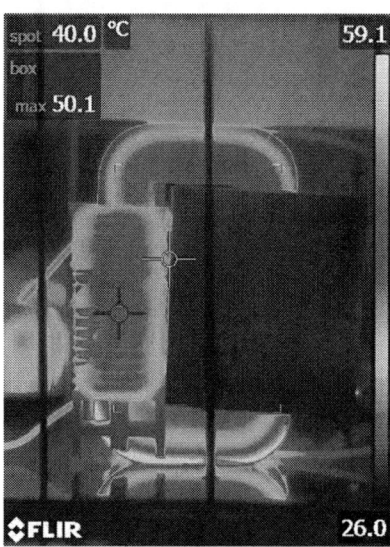

Fig. 1. Thermal image of a leakage inductance integrated core-type transformer with high eddy current loss.

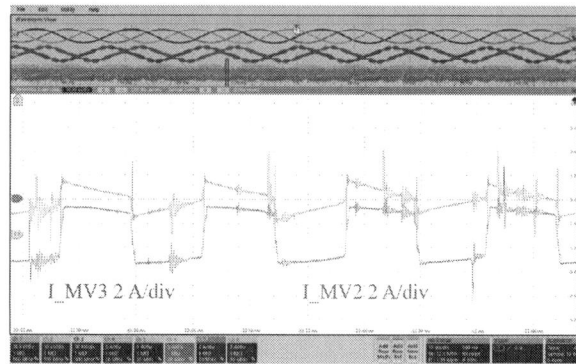

Fig. 2. Current spikes in transformers having high grounding capacitance.

Core type transformers can have high leakage inductance, and some of the leakage can be made further by central legs [11], and the adapted leakage layers can reduce the eddy current loss to a significantly low level. However, the structure of the dry-type MV winding has to fit into the core window with limited space, making the manufacturing process complicated.

B. Insulation and Parasitic Capacitance

The promising and environmentally friendly way of transformer insulation should be the dry-type insulation, which utilizes solid materials such as epoxy, silicone, or polyurethane as the insulation material, instead of using transformer oil. However, as the dielectric constants of the epoxy or silicone are usually above 3, while the transformer oil typically has a dielectric constant of 2.1-2.4, the parasitic capacitance of the dry-type transformer should have a larger parasitic capacitance [12, 13]. To confine the electric field of MV winding, and reduce common-mode electromagnetic interference, the MV winding may be shielded on the surface of the solid insulation, so that the LV windings and cores are not necessary to be encapsulated or cast, and therefore, the transformer cooling can be easier [5, 14]. However, due to the electric shielding on MV winding, the parasitic capacitance between the winding and ground also increases drastically [14]. In some applications, e.g., cascaded H-bridge (CHB) inverters, the common-mode voltage swing will be imposed on the transformer grounding capacitance, and causing significant switching losses on the CHB side devices. In Fig. 2, the transformer in dc/dc has high current spikes when the CHB devices switch and charge/discharge the parasitic grounding capacitance, causing significant switching losses. To reduce this effect, partial shielding can be chosen, so that only the critical surfaces are shielded while the other sides can remain unshielded [14].

C. Organization of the Paper

In this paper, a new leakage integration method having low eddy current loss will be introduced by using ferrite structures, and splitting nanocrystalline cores. In the meantime, dry-type transformer insulation consideration with partial shielding will be discussed. First, the leakage field of different types of transformer structures will be demonstrated, and the new structure having low eddy current loss will be introduced in Section II. Then, the dry-type insulation strategy will be discussed in Section III, with partial shielding. In Section IV, the transformer design based on the aforementioned method will be briefly discussed, and the test results will be shown. Finally, the conclusion will be drawn in Section V.

II. LOW LOSS LEAKAGE INTEGRATION

A. Leakge Integration with Ferrite Bridge

To begin with, the leakage inductance of the MV transformer should be clarified as the combination of the intrinsic leakage inductance that the transformer structure itself possesses without any external structures, and the external inductance that produced by structures which are artificially added to the transformer, i.e.

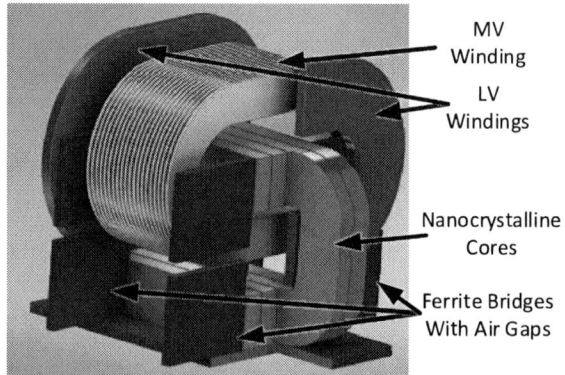

Fig. 3. Proposed transformer structure for leakage integration.

$$L_l = L_i + L_e \qquad (1)$$

Where L_l is the leakage inductance, L_i the intrinsic leakage inductance, L_e the external inductance. To find the transformer structure with low eddy current loss and high leakage integration capability, the intrinsic inductance should be minimized to have low leakage loss, while the external inductance should be maximized to a level that is desired for the transformer design.

One possible structure having low intrinsic leakage inductance is the shell-type transformer. However, the co-axial shell-type transformer is excluded, as the LV and MV winding placed together may compromise the transformer cooling and make the leakage integration complicated. As in shell-type transformers, the windings can be split or interleaved, a new structure can be found with low intrinsic leakage and can be externally integrated with ferrite structures. The new structure is shown in Fig. 3. As the LV winding is split into two sets, and interleaved with the MV winding, the intrinsic leakage flux should be lower compared to the core-type and non-interleaved conventional transformer so that the eddy current should be low.

However, thanks to the ferrite bridges added between the

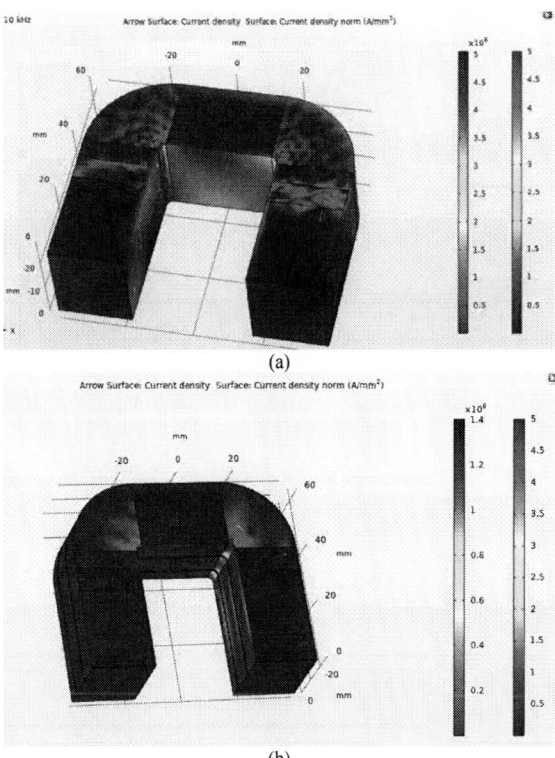

(b)

Fig. 5. Transformer eddy current simulation with proposed structure (a) one-piece transformer core, (b) split transformer cores.

LV and MV windings, high external leakage flux can be added and tuned by the air gaps. Even though the ferrite may introduce ferrite core loss when the transformer is loaded, as long as the flux density in the ferrite does not saturate, and the volume of the ferrite is relatively small, the overall loss when at short-circuit should be low compared to the other methods.

To verify the effectiveness of different structures, quick tests have been performed to test the losses and leakage inductance of the proposed structure. To make a fair comparison, the windings and core are set to the same as in Fig. 1. In Fig. 4 (a), as can be seen, without the ferrite bridge, the leakage inductance is around 150 µH, which is corresponding to L_i in (1), while after the ferrite bridge inserted, the leakage inductance increased to around 300 µH. That is, the external inductance of 150 µH has been added into the transformer. From Fig. 4 (b), the thermal image shows that unlike in Fig. 1 that the outer surface of the core was heated up, the surface of the inner window in Fig. 4 (b) was slightly heated, and the temperature rise is lower than in Fig. 1. The estimated eddy current loss is around 33 W, while in Fig. 1 it used to be 105 W.

B. Eddy Current Loss Reduction by Splitting Transformer Core

As can be seen in Fig. 4 (b), the shell-type transformer with interleaved winding still has some amount of eddy current loss on the inner surface of the transformer core, which may need to be further reduced. Note that the lamination reduces the

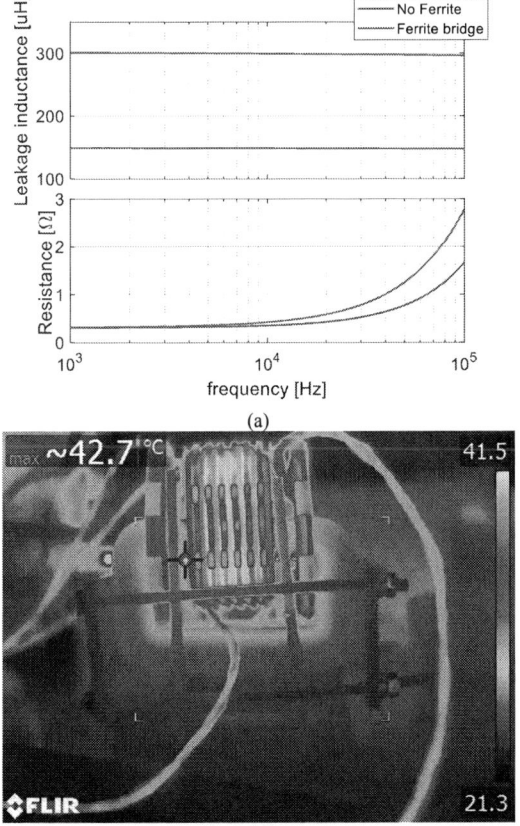

Fig. 4. Test results of the ferrite bridge (a) leakage inductance measurement, (b) thermal image during short-circuit test.

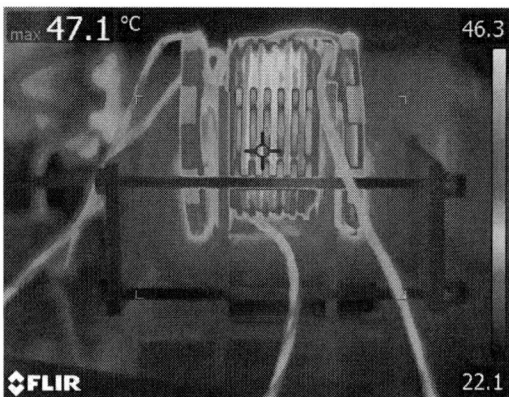

Fig. 6. Transformer short-circuit test with split core.

TABLE I
ELECTRIC PROPERTIES OF SELECTED MATERIALS

Part Number	Dielectric Constant (Relative)	Electric Field Strength (kV/mm)	Loss Tangent	Use
Dow TC-4605 HLV	4.4	24	0.08	Encapsulant
SOMOS PerForm	4	26.3	-	3D print winding bobbin/case
Loctite E-60NC	-	22	-	Epoxy adhesive
HV Putty	3.03	22	-	Electric field shaping

eddy current loss by cutting the solid area perpendicular to the magnetic field smaller, a similar strategy has also been adopted to reduce the gap losses for tape-wounded nanocrystalline cores [15, 16]. The same measure can also be taken to reduce the eddy current loss induced by the leakage inductance.

Therefore, simulation with split cut cores has been performed. Originally, the transformer core used is 50 mm in depth, and can be replaced with 5 thinner cores with 10 mm in depth, as it can be wounded easily with 10 mm wide nanocrystalline tape without any cutting process. The results are shown in Fig. 5. In Fig. 5 (a), the original one-piece transformer core has been used with 50 mm in depth, and in the inner window surface the current density of 2 A/mm^2 can be found with a wide range, and the loss was measured as 28 W. For Fig. 5 (b), the cores are cut into 5 smaller cores (mirrored part not shown). The area having eddy current is significantly shrank, only the edge regions have some current density around 2 A/mm^2, and the eddy current loss is measured as 9.4 W, which is 60% further reduced.

To verify the results, tests have also been done with replaced customized cores. From Fig. 6, the temperature of the eddy current surfaces has further decreased, and so did the total losses. For the proposed interleaved shell-type transformer, the total loss has been reduced to 111 W, and the estimated eddy current loss reduced to 9 W. From the thermal image it can be seen that the inner surface of the core now has no hot spot anymore. Therefore, the split core can further reduce the eddy current loss induced by the leakage flux.

III. INSULATION DESIGN

The insulation design of the MV dc/dc transformer can be challenging, as one weakest point found in the entire transformer may be damaged. Also, the material properties play a more important role, as not only the performance of each material matters, the coordination between different materials may also impact significantly on the overall insulation performance.

The insulation design starts with the material selection. Usually, the materials used do not only involve the insulating

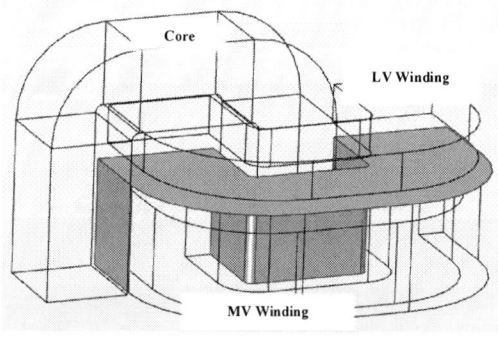

(a)

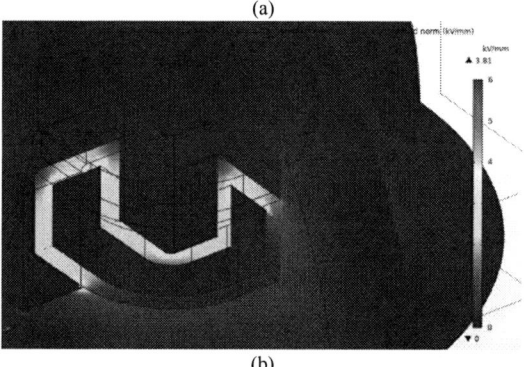

(b)

Fig. 7. Transformer insulation simulation with partial shielding (a) transformer structure, with shaded surface shielded, (b) electric field simulation results.

dielectric material, but also mechanical structural materials. Hence, careful study of the electrical properties of the mechanical parts is also required. To form an even electric field inside the MV winding encapsulation, the dielectric constants of the insulating material, and other structural materials, e.g., bobbin and case, adhesives, should be as close as possible. The insulation rating should also match, to avoid the electric breakdown in a significantly weak material. Table II shows the electric properties of the selected materials used in insulation design.

Similar to the partial shielding in [14], only the critical surfaces are close to the ground potential should be shielded, as the ground and high potential difference may impose a high electric field in the air, causing corona or flashover during operation. Hence, a partially shielded transformer has been

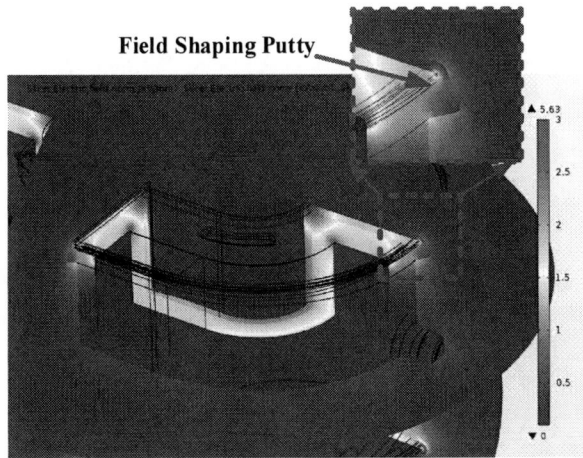

Field Shaping Putty

Fig. 8. Detailed insulation simulation with field shaping.

TABLE II
TRANSFORMER DESIGN PARAMETERS

Parameters	Values
Power Rating	16.7 kW
Voltage Rating	850/6.7 kV
Switching Frequency	10 kHz
Core	SC2062M1 (5x 10 mm depth cut)
Air Gap	2 x 0.2 mm
LV Winding	31 turn 259/AWG36 Litz Wire
MV Winding	244 turn 50/AWG36 Litz Wire
Ferrite Bridge	PLT64/50/4-3C95
Ferrite Bridge Air Gap	5 mm
Leakage Inductance	250 µH/15.5 mH
Insulation Rating (MV to ground)	13.4 kV peak
Power Density	8.5 kW/L

simulated in Fig. 7. The field distribution is only partially even because of the partial shielding, and the electric field extends to the air where the surface is not shielded. With partial shielding, the capacitance between the high potential winding and shielding ground is 102 pF, which is reduced by 50% compared to the fully shielded winding. The electric hot spot now is the edges of the shielding layer, which could be as high as 6 kV/mm and may cause a breakdown in the air, which can be reduced by applying silicone putty to form a rounded corner with low curvature in the real prototype. The detailed simulation with the silicone putty is shown in Fig. 8. By applying the silicone putty, the field strength can be reduced to 1.5 kV/mm on the edges of the shielding surfaces.

IV. TRANSFORMER TEST

A. Transformer design

With the proposed leakage integration and insulation, the MV DAB transformer has been designed for a 10 kV SiC-based DAB converter, serving as a front end of a CHB inverter. The design parameters of the transformer are shown in Table II. The photograph of the MV transformer with LV and MV power stages is shown in Fig. 9.

MV Power Stage

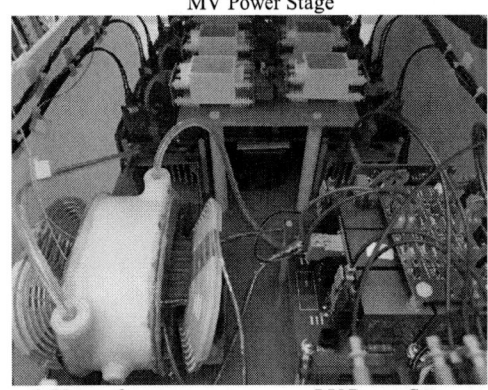

MV transformer   LV Power Stage

Fig. 9. Photograph of MV DAB transformer and DAB power stages

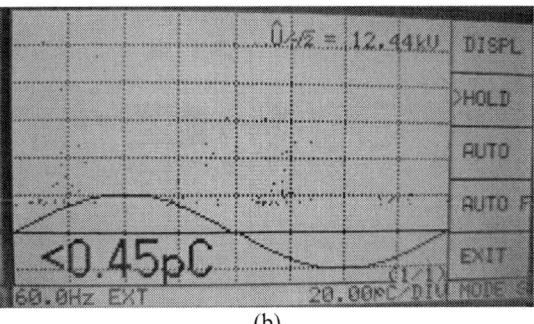

(b)

Fig. 10. PD test results of accumulative recording at 1.3 times of rated voltage for 3 minutes.

B. Transformer Test

After the transformer was built, several tests have been conducted to verify the design with leakage integration and insulation design. First, before the MV was applied, partial discharge (PD) test has been done to verify the winding and transformer insulation. The test followed IEEE standard C57.12.91-2020 that the MV winding was tested with a prestress of 1.8 times of rated voltage for 30 seconds, and then measured the PD level at 1.3 times of rated voltage for 3 minutes [17]. The test results are shown in Fig. 10. From the results, during prestress the PD level was around 45 pC, and most of the PD subsidized after the voltage reduced to 1.3 times of rated voltage, even though occasionally some PD or corona discharge appears. Most of the PD value was below 50 pC, meeting the requirement of IEEE Std C57.12.91.

After the PD test was finished, two transformers have been installed in the DAB converter and connected with the CHB converter forming a two-stage converter, tested from no-load to full load condition. The test waveforms are shown in Fig. 11. From the test waveforms, the transformer works as expected DAB transformers for both no-load and full-load conditions, and the waveforms of the two transformers agree with each other. From the test, the total loss of transformer is estimated as 119.52 W, achieving an efficiency of 99.3%, and the eddy current loss is estimated as 33 W.

978-1-6654-8901-0/22 $31.00 © 2022 IEEE

V. CONCLUSION

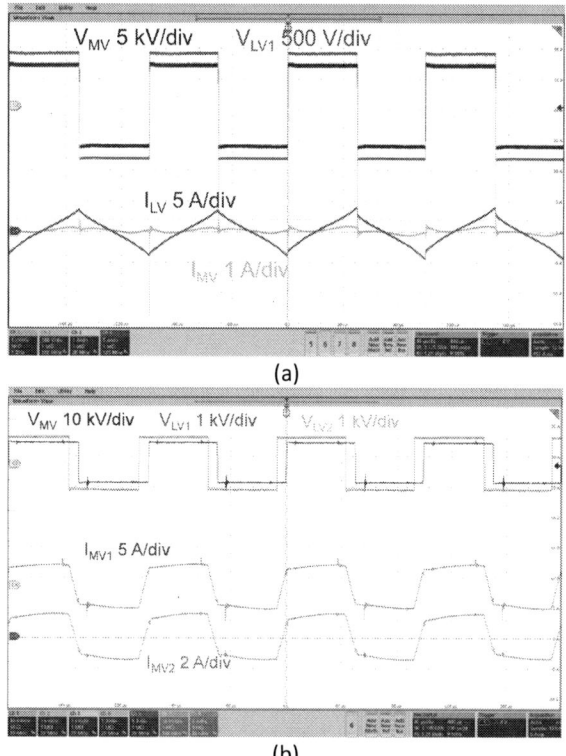

Fig. 11. Test results for MV DAB transformer/converter (a) no-load condition (b) full-load condition.

In this paper, a MV transformer design with low-loss leakage integration, as well as insulation design with field shaping has been introduced. Through the simulation and tests, the shell-type transformer with interleaved LV winding and ferrite structures can have better leakage integration capability and loss performance. The split transformer core can further reduce the eddy current loss induced by the leakage magnetic field, without sacrificing power density and other performances, but specification of manufacturer has to be followed when customizing the split cores. The insulation materials and insulation structure have been discussed to improve the insulation performance. Partial shielding is necessary for MV winding design because of reduced grounding capacitance, which may impact the switching losses of the converter having common-mode swings. Based on the discussions, a 850-V/6700-V 16.7-kW MV DAB transformer has been designed and tested. The MV winding has passed the PD standard in IEEE Std C57.12.91 up to 13.4 kV peak value, and the transformer achieved 99.3% power efficiency at full load condition, with the transformer having a power density of 8.5 kW/L.

ACKNOWLEDGMENT

The authors want to thank Powerex and Southern Company for providing help on this work.

REFERENCES

[1] L. Heinemann, "An actively cooled high power, high frequency transformer with high insulation capability," in *APEC. Seventeenth Annual IEEE Applied Power Electronics Conference and Exposition (Cat. No.02CH37335)*, 10-14 March 2002 2002, vol. 1, pp. 352-357 vol.1.

[2] C. Zhao *et al.*, "Power Electronic Traction Transformer—Medium Voltage Prototype," *IEEE Transactions on Industrial Electronics*, vol. 61, no. 7, pp. 3257-3268, 2014.

[3] S. Zhao, Q. Li, and F. C. Lee, "High frequency transformer design for modular power conversion from medium voltage AC to 400V DC," in *2017 IEEE Applied Power Electronics Conference and Exposition (APEC)*, 26-30 March 2017 2017, pp. 2894-2901.

[4] Z. Li, Y. H. Hsieh, Q. Li, F. C. Lee, and M. H. Ahmed, "High-Frequency Transformer Design with High-Voltage Insulation for Modular Power Conversion from Medium-Voltage AC to 400-V DC," in *2020 IEEE Energy Conversion Congress and Exposition (ECCE)*, 11-15 Oct. 2020 2020, pp. 5053-5060.

[5] Q. Chen, R. Raju, D. Dong, and M. Agamy, "High Frequency Transformer Insulation in Medium Voltage SiC enabled Air-cooled Solid-State Transformers," in *2018 IEEE Energy Conversion Congress and Exposition (ECCE)*, 23-27 Sept. 2018 2018, pp. 2436-2443.

[6] D. Rothmund, T. Guillod, D. Bortis, and J. W. Kolar, "99% Efficient 10 kV SiC-Based 7 kV/400 V DC Transformer for Future Data Centers," *IEEE Journal of Emerging and Selected Topics in Power Electronics*, vol. 7, no. 2, pp. 753-767, 2019.

[7] Z. Guo, R. Yu, W. Xu, X. Feng, and A. Q. Huang, "Design and Optimization of a 200-kW Medium-Frequency Transformer for Medium Voltage SiC PV Inverters," *IEEE Transactions on Power Electronics*, pp. 1-1, 2021.

[8] P. Czyz, T. Guillod, F. Krismer, J. Huber, and J. W. Kolar, "Design and Experimental Analysis of 166 kW Medium-Voltage Medium-Frequency Air-Core Transformer for 1:1-DCX Applications," *IEEE Journal of Emerging and Selected Topics in Power Electronics*, pp. 1-1, 2021.

[9] X. Li, W. Huang, B. Cui, and X. Jiang, "Inductance Characteristics of the High-Frequency Transformer in Dual Active Bridge Converters," in *2019 22nd International Conference on Electrical Machines and Systems (ICEMS)*, 11-14 Aug. 2019 2019, pp. 1-5.

[10] R. B. Beddingfield, S. Bhattacharya, and P. Ohodnicki, "Shielding of Leakage Flux Induced Losses in High Power, Medium Frequency Transformers," in *2019 IEEE Energy Conversion Congress and Exposition (ECCE)*, 29 Sept.-3 Oct. 2019 2019, pp. 4154-4161.

[11] B. Cougo and J. W. Kolar, "Integration of Leakage Inductance in Tape Wound Core Transformers for Dual Active Bridge Converters," in *2012 7th International Conference on Integrated Power Electronics Systems (CIPS)*, 6-8 March 2012 2012, pp. 1-6.

[12] H. Li, P. Yao, Z. Gao, and F. Wang, "Medium Voltage Converter Inductor Insulation Design Considering Grid Insulation Requirements," in *2021 IEEE Applied Power Electronics Conference and Exposition (APEC)*, 14-17 June 2021 2021, pp. 2120-2126.

[13] H. Li, P. Yao, Z. Gao, and F. Wang, "Medium Voltage Converter Inductor Insulation Design Considering Grid Requirements," *IEEE Journal of Emerging and Selected Topics in Power Electronics*, vol. 10, no. 2, pp. 2339-2350, 2022.

[14] T. Guillod, F. Krismer, and J. W. Kolar, "Electrical shielding of MV/MF transformers subjected to high dv/dt PWM voltages," in *2017 IEEE Applied Power Electronics Conference and Exposition (APEC)*, 26-30 March 2017 2017, pp. 2502-2510.

[15] Y. Wang, G. Calderon-Lopez, and A. J. Forsyth, "High-Frequency Gap Losses in Nanocrystalline Cores," *IEEE Transactions on Power Electronics*, vol. 32, no. 6, pp. 4683-4690, 2017.

[16] G. Calderon-Lopez, Y. Wang, and A. J. Forsyth, "Mitigation of Gap Losses in Nanocrystalline Tape-Wound Cores," *IEEE Transactions on Power Electronics*, vol. 34, no. 5, pp. 4656-4664, 2019.

[17] "IEEE Standard Test Code for Dry-Type Distribution and Power Transformers," *IEEE Std C57.12.91-2020 (Revision of IEEE Std C57.12.91-2011)*, pp. 1-102, 2021.

978-1-6654-8901-0/22 $31.00 © 2022 IEEE

Development of a 250°C 15kV Supercascode switch using SiC JFET technology

David E. Sanabria
Tetra Corporation
Albuquerque, United States
dsanabria@tetra-corporation.com

Randy Appert
Tetra Corporation
Albuquerque, United States
rappert@tetra-corporation.com

Steven G. E. Pronko
Tetra Corporation
Albuquerque, United States
s.pronko@tetra-corporation.com

Joshua Major
National Renewable Energy
Laboratory
Golden, United States
joshua.major@nrel.gov

Douglas DeVoto
National Renewable Energy
Laboratory
Golden, United States
douglas.devoto@nrel.gov

Karen Heinselman
National Renewable Energy
Laboratory
Golden, United States
karen.heinselman@nrel.gov

Jane .M. Lehr
Department of Electrical
and Computer Engineering
University of New Mexico
Albuquerque, United States
jmlehr@unm.edu

Nicolas Gonzalez
Department of Electrical
and Computer Engineering
University of New Mexico
Albuquerque, United States
negonzalezp2304@unm.edu

David S. Ginley
National Renewable Energy Laboratory
Golden, United States
david.ginley@nrel.gov

Abstract—**Tetra Corporation is developing a pulsed power drilling system for the geothermal industry. The system requires a high temperature switching technology to replace the existing technology based on Si thyristors with a temperature limit of 150°C. This work focuses on the development of such switch based on a Supercascode configuration using SiC JFET technology. The development presented contains thermal models, packaging design and processes, circuit simulations and prototypes of both the Supercascode switch and the high temperature packaged SiC JFETs.**

Keywords—Packaging, SiC, Supercascode, Cascaded Switch, JFET.

I. INTRODUCTION

There is an expanding need for high voltage and high temperature switch solutions for harsh environments in many applications. An example and the motivation for this work is geothermal drilling, Tetra Corporation is developing an electro-crushing drill for the geothermal industry named RePED 250. This drilling technology has the potential of increasing the geothermal deployment by significatively reducing the cost of drilling.

This drill is a pulsed power machine that requires electronic components such as high voltage switches capable of operation in high temperature environments found inside geothermal

TABLE I. REQUIREMENTS FOR THE RePED 250 SWITCH TECHNOLOGY.

Parameter	Minimum Value
Operating Voltage	10-15kV
RMS Current	100–150Arms
Peak Current	3,000 A peak current
Switch Pulsed Repetition Rate	300 Hz
Expected Lifetime	10^9 shots
Operating Temperature	250°C ambient

wells, which can reach about 250°C at 6.5km [1]. The requirements for the RePED 250 switches are listed in Table I.

High temperature environments limit the use of silicon-based electronics, however, SiC-based electronics such as SiC JFETs can operate at higher temperatures. Researchers have demonstrated operating temperatures up to 600°C in a controlled environment. However, a commercial switch product that meets the requirements needed for this application is not yet available. Most commercial products are rated to 175°C with a few exceptions rated to 225°C. A common limitation to bring the technology to the field is the packaging [2][3][4]. A high temperature packaging solution will be impactful in multiple applications and industries.

The work presented here explores materials and techniques to produce a high temperature packaging solution. The end goal is to build a 250°C JFET that later can be arranged in a 15kV Supercascode switch capable of operating at 250°C. This switch will provide a solution for our specific application and a viable alternative for other applications that require a high temperature high voltage switch.

The development of the switch encompasses numerous efforts that will be divided in two groups: High temperature packaging, and cascaded switch development. The packaging section shows the design process, including thermal models, materials, and fabrication techniques considered and used. The circuit design section shows the circuit topology, simulations, prototypes, and testing of Supercascode prototypes using commercially available devices with standard packaging. the integration of both efforts is ongoing work and results will be presented in a future publication.

The die to be packaged is manufactured by United Silicon Carbide (UF3N170006) and is a 1700V 5.7mΩ normally ON JFET. Selecting a JFET instead of a MOSFET is preferable since MOSFETs have a lower temperature capability. The

978-1-6654-8901-0/22 $31.00 © 2022 IEEE

oxide layer imposes a limit on the maximum operational temperature. JFETS lack this oxide layer, giving them the leading edge for high temperature applications [2].

II. HIGH TEMPERATURE PACKAGING

A. Materials

An illustration of a typical power module is shown in Fig. 1. The packaging encompasses all the elements shown that allow the interface between the die and the circuit and protect it from the environment. In high temperature applications one should pay special attention to the stresses caused by thermal expansion, the coefficient of thermal expansion (CTE) mismatch between all the materials needs to be minimized to reduce the strain caused.

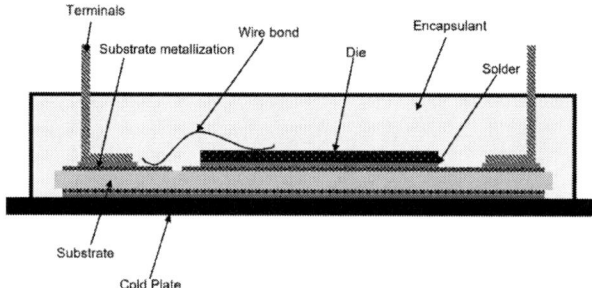

Fig. 1. Illustration of a typical power module internal layout.

1) Ceramic Substrate

One of the principal components in a SiC power module packaging is the ceramic substrate, which provides mechanical support, insulation for the thin semiconductor material and a thermal path for the heat to flow to the cold plate or base plate.

In high-temperature applications, the most common substrate material is Aluminum Nitride (AlN) due to the close match of the Coefficient of Thermal Expansion (CTE) with

TABLE II. THERMAL CYCLCING OF DIFFERENT SUBSTRATE TYPES [8].

Substrate type	0-350°C cycles	100-350°C	Failure mode
DBC (AlN,Al₂O₃)	12-13	50-75	Ceramic fracture
AMB (AlN)	10	100	Ceramic fracture
DBC (Si₃N₄)	78	1160	DBC bond failure
AMB (Cu,Si₃N₄)	145-640	>750*	Ceramic fracture
DBA (AlN)	>990*	>1380*	No failures to date
Cu TF (Al₂O₃)	116-128	>1380*	Adhesion failure
Ag TF on dielectric on Al baseplate	370-485	100*	Dielectric peeling
Ag TF on AlN	181 (270μm) 231 (210μm) 506* (140μm)	380*	Metal lifting
Ag TF on Zr-Al₂O₃	661*	380*	No failures to date
Cu TF on AlN	430*	50*	No failures to date
Cu TF on Zr-Al₂O₃	430*	50*	No failures to date

* Not yet failed (results to date)

SiC. For SiC the CTE is approximately $4.33 \cdot 10^{-6}$/K and for AlN is $4.5 \cdot 10^{-6}$/K. for this reason a direct bond copper (DBC) substrate made with AlN was chosen for this application.

Some researchers suggest that there are other substrate technologies that are more reliable, but the limited availability and the high cost of some of the materials discouraged us from using them. (Observe in Table I a list of substrates tested in [8]).

2) Wire Bonding

Research in high temperature electronics has shown that wire bonding accounts for 26% of the failures in SiC power modules. This is not a surprise given the stress caused by the high CTE mismatch between Al ($22 \cdot 10^{-6}$/K), the most common wire bonding material, and SiC ($4.33 \cdot 10^{-6}$/K) [6]. Observe an example of bond wire liftoff caused by bond and heel cracks in Fig. 2. Some manufacturers use ribbons instead of wires to improve the adhesion, but for high temperature applications wire bonds should be avoided.

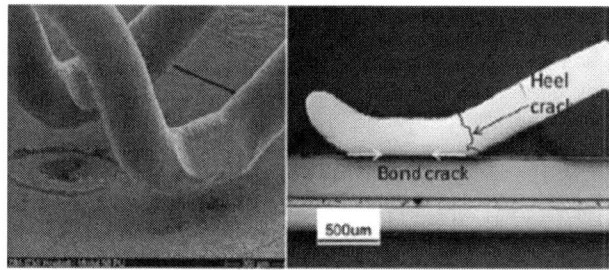

Fig. 2. Bond wire liftoff and cracks observed in SiC power modules [7].

3) Encapsulant

This area is of great interest. Common potting or encapsulant materials are either plastics or gels that do not perform well in temperatures above 200°C. we identified a few epoxies and ceramic encapsulants that can operate above 250°C and are listed in Table II. Due to the higher dielectric strength and thermal conductivity, the epoxy Duralco 865 by Cotronics was selected.

TABLE III. PROPERTIES OF ENCAPSSULANT MATERIALS CONSIDERED

Reference	Properties			
	Material	Temperature (°C)	Dielectric Strength (kV/mm)	Thermal Conduct (W/mK)
Duralco 862	Epoxy	316	19.7	0.6
Duralco 863	Epoxy	316	21.7	1.3
Duralco 865	Epoxy	260	27.6	2.9
Resbond 919	MgO ceramic	1538	10.6	0.6
Resbond 920	Al₂O₃ ceramic	1649	10.6	2.2

4) Die Attach and solder

Pb based solder has been extensively used in the industry, however the high temperature operation that SiC can achieve rules out most Pb solders available. The most promising die attachment technologies are Transient Liquid Phase (TLP), Silver sintering and electrically conductive adhesives. Researchers have found Silver sintering the most reliable for

978-1-6654-8901-0/22 $31.00 © 2022 IEEE

high temperature applications. The reason is likely the porous nature of the sintered bond, which helps alleviate some of the thermomechanical stress. Another characteristic of silver sintering is the low processing temperature. Observe in Fig. 3 the processing temperatures and the operating temperature of multiple solder, sintering, TLP and epoxies. Note that TLP and sintering has a low processing temperature compared to the operating temperature.

The Kyocera CT2700R7S pressureless silver sintering paste that was used to bond the die and the power contacts to the substrate, has a processing temperature of 200°C and once

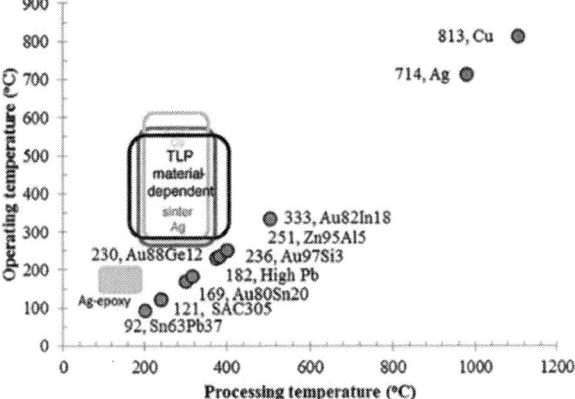

Fig. 3. Processing temperatures and operating temperatures for solder, sintering, TLP and Ag epoxy materials [10].

cured the operation temperature is greater than 300°C.

B. Final design and fabrication

The final design shown in Fig. 4 features a wire bondless power contact on top of the die that serves as electrical contact and as thermal path to improve the power dissipation.

Keep in mind that this design aims to test the high temperature materials and fabrication techniques and cost is not a design parameter we are following right now. Some of the parts can be simplified or miniaturized to save cost, but that will be the focus of future efforts.

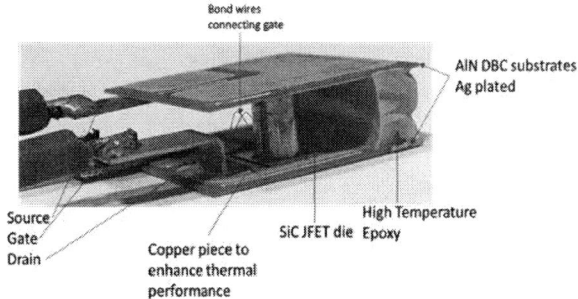

Fig. 4. Final design of the 250°C packaged JFET, the device is partially encapsulated to show the internals.

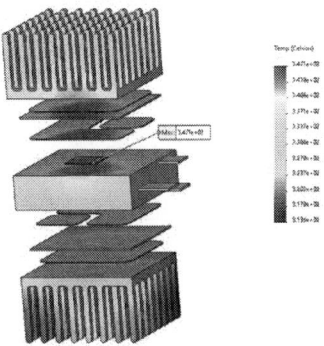

Fig.5. Thermal simulation of the prototype running at a 300Hz repetition rate at 250°C ambient air temperature.

A thermal model was generated to ensure that the switch could handle the power while operating at 250°C. Observe in Fig. 5 a thermal simulation of a worst-case scenario with a forced air-cooling system. The junction temperature reaches 347°C and the power dissipated by the switch is of around 50W.

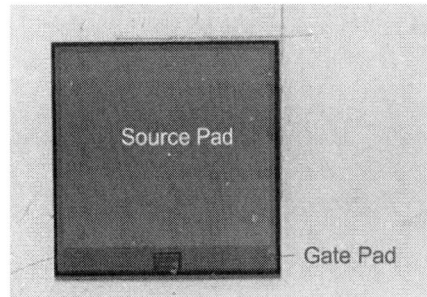

Fig. 6. Top side of the UFN170006 SiC JFET die from United Silicon Carbide.

The Gate Drain and Source leads are attached to the substrates using silver sintering as well as the cylindrical contact in top of the die connecting it to the top substrate and the source lead.

Fabrication begins with the preparation of the surfaces, to be able to use silver sintering, all copper pieces need to be plated with silver to form a strong bond with the sintered silver. Electroless plating is used to plate the copper in the substrates, the cylinder connector and all the leads.

The die also needs to be prepared including the source and gate pads found on top of the die as shown in Fig. 6, the metallization that the die has from the factory is an AlCu alloy that is compatible with wire bonding but not with silver sintering. An electron beam deposition technique is used to deposit a Nickel Silver stack. The Nickel is 10nm thick and the Silver is 250nm.

After plating all the surfaces, all the parts are joined using pressureless silver sintering. Once all the sintering is done the epoxy is poured inside. Vacuum degassing happens both before the pouring of the two-part mixture and after the pouring during the curing process. This reduces the number of voids and bubbles that can be trapped within the switch, decreasing the dielectric strength. Observe the finalized prototype in Fig. 7.

The prototypes prove to be valuable for the development of the fabrication techniques. Failure analysis of the prototypes allowed us to improve the fabrication techniques and the design of the switch. One of the most significant findings was the need to fine tune the amount of silver paste placed under the die; If too much paste was used the excess forms a meniscus around the die, as shown in Fig. 8, which effectively reduces the voltage holdoff leading to dielectric breakdown.

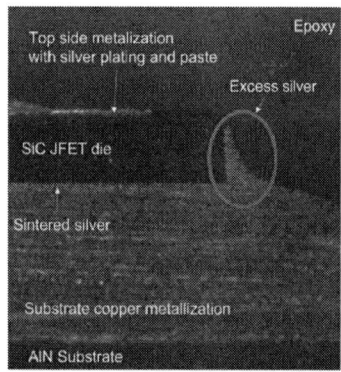

Fig. 7. fully encapsulated protoype after testing at 175°C.

Additional failures while testing have been recorded testing at 175°C. The devices are failing a high potential test at 1200V and the cause of the failure is being investigated. Electrostatic simulations show that the devices should be able to withstand 1700V. The hypothesis is that the epoxy reduces its dielectric strength with temperature quicker than anticipated. Further research is being conducted to find the cause of the failure.

Fig. 8. Excess silver forming a meniscus on the edge of the die. This is a cross section and magnification of one of the prototypes.

III. SUPERCASCODE SWITCH DEVELOPMENT

The RePED 250 system will require 15kV switches. Discrete JFETs can be arranged in a cascaded configuration to increase the voltage blocking, at least 10 stages will be needed to achieve the 15kV using the 1700V JFETs.

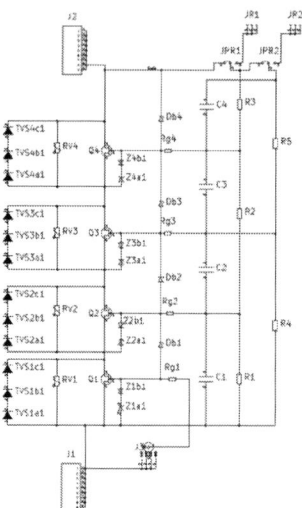

Fig. 9. Schematic of the first prototype, J1, J2, JR1 and JR2 are connectors to external circuitry. TVSXXX are tvs diodes, RV1-RV4 are MOVs, Db1-Db4 are biasing diodes, Z1xx-Z4xx are back to back zener diodes.

Initial exploration of different topologies in the literature such as the ones proposed in [11] by Li et al and in [12] by Garcia et al led to the first cascaded switch prototype, which is heavily based on the later. This prototype uses commercially available JFETs with a voltage rating of 1200V. The circuit diagram is shown in Fig. 9.

Both topologies are designed for current flowing in one direction so additional circuitry was added. This is to prevent the JFETS from operating in the third quadrant region and to prevent overvoltage. The modifications include Zener diodes and TVS-MOV pairs to protect the JFETs from overvoltage. All the modifications are necessary to operate the Supercascode in the RePED system reliably.

Note that the cascaded switch proposed by Garcia in [12] is composed of only JFETs; the first stage named Q1 in Fig. 9 is not a MOSFET like in a traditional Supercascode topology, it was replaced by a high voltage JFET, which transforms the Supercascode into a normally on cascaded switch. This modification comes with an advantage for high temperature applications.

Commercially available MOSFETS are limited to 150°C ratings with a small number of examples rated to 175°C. In almost every case, the MOSFET current ratings are insufficient to meet the goals of the program. Higher temperature, higher current capability MOSFETS are developing, but will be some years out before commercially available. This "barrier" has inspired exploration into a VJFET topology that does not require the MOSFET. The purpose of the MOSFET is to create a "Normally-Off" switch since the VJFET is a "Normally-On" device. To be in the Off state, the VJFET must have a negative bias signal applied to the gate-to-source terminals.

978-1-6654-8901-0/22 $31.00 © 2022 IEEE

In most pulsed power systems, high voltage is applied with a "command charge". High Voltage will not appear across the switches until this command is issued. This command can be inhibited until the negative gate bias is verified, ensuring that the VJFETs are in the Off state. Control logic of this nature is very common in pulsed power systems. Similar control logic can be implemented in power electronic systems such as inverters. If higher temperature MOSFETS become available, then a true normally-off switch assembly can be realized.

With the intention of simplifying the development of this cascaded switch, the first prototype has only four stages and it uses 1200V JFETs in a TO-247 package instead of the 1700V JFETs, which are expensive and sold as bare die.

Observe the fabricated prototype in Fig. 10, most of the components are surface mount to reduce parasitic inductances

Fig. 10. First cascaded switch prototype. External testing circuitry not displayed.

that affect the turn on and off dynamics.

The balancing resistors R1-R5 shown in Fig. 9 control the voltage sharing in steady state, the values were calculated considering the gate leakage current of the JFETs. A detailed explanation of the calculation process can be found in [12]. In this case the values are: R1=2MΩ, R2=4.7MΩ, R3=2.7MΩ, R4=4.42MΩ, R5=4.7MΩ.

The values of the balancing capacitors C1-C4 can be calculated mathematically, and the input capacitance of the JFETs plays a big role. However, in this case an optimization routine was used to find the values that would produce the best results. C1=3.9nF, C2=3.3nF, C3=3nF, C4=2.7nF. These

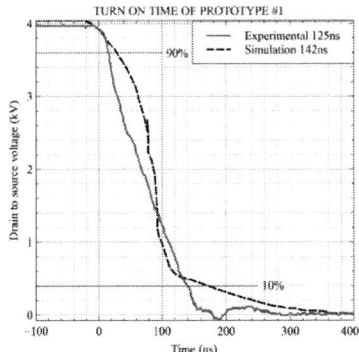

Fig. 11. Turn on time of the first prototype. Comparison between the circuit model and experimental data. Turn on time is measured based on the 10-90% drain to source voltage.

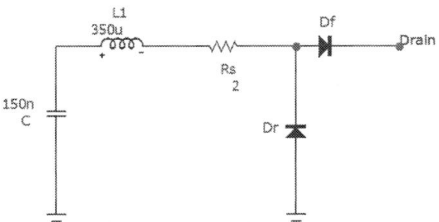

Fig. 12. RLC circuit tuned for 80A peak current and risetime of 7.5us.

capacitors in combination with the gate resistors control the switching speed and the synchronization of the JFET stack.

Testing on this prototype was successful, the result was a 4kV switch with a turn on time of 125ns 10-90%, which agrees with the simulation. Keep in mind that for our application, we do not require fast switching due to the nature of the circuit. Observe a comparison between the simulation and experimental data in Fig. 11.

The RePED drill has a bipolar pulse shape that Tetra has tuned to improve the drilling. It is possible to replicate the most important conditions that the switch will endure in the RePED 250 by tuning an RLC circuit. The tuning parameters come from the RePED system and are peak current, RMS current and risetime. With this approach, we have the flexibility of changing any of those parameters easily and independently.

Fig. 12 shows a diagram of the test platform used for this test and Fig.13 (a) shows the current through the switch for a low current test performed with the prototype of Fig.10. In this case Rs is the total equivalent resistance of the circuit and no additional resistance was added, C is 150nF and L is 350uH.

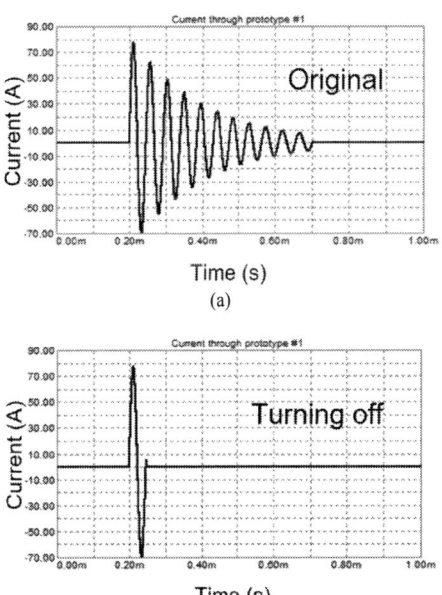

Fig. 13. Current through the prototype in 2 scenarios, (a) letting the switch on for 500us, and (b) turning off the switch after one full cycle at 44us.

Df and Dr are external diodes placed there to prevent the switch from experiencing reverse current which has proven to be harmful and to induce failures. This configuration also enables the ability to turn off the switch after the first cycle as shown in Fig. 13 (b). The turn off capability reduces the stress and power requirements in other components of the RePED system, increases reliability and potentially increases drilling speed.

Elevated temperatures also affect all the other components of the cascaded switch. For applications beyond 200°C ceramic capacitors almost the only alternative. Ceramic capacitors that can operate at high temperature lose about 40%-60% of their capacitance at 250°C [13][14]. Such variation affects the turn on and off time of the circuit and therefore increases losses. In our analysis, a decrease of 50% on the capacitance only translates to an increase of 11% in losses. This could be due to the long risetime of the circuit. Resistors decrease their power dissipation with temperature, the cascaded switch balancing resistors do not dissipate high power, each resistor does not exceed 0.5W.

IV. CONCLUSIONS

To design and build a high temperature package for high voltage applications is a difficult task, despite the advancements and discoveries of this work, a broader and more in-depth study of the materials needs to be done. One of the challenges for this type of application is the encapsulation material, it needs to combine good mechanical, chemical and electrical performance. When the temperature surpasses 200°C the number of commercial alternatives is greatly reduced.

The first cascaded prototype switch has produced promising results as a potential switching alternative for RePED 250. Further testing is required and will be done above ambient temperature. The work being done with the Supercascode prototypes will facilitate the integration of the discrete high temperature packaged JFETs into a cascaded switch that meets the requirements for RePED 250.

ACKNOWLEDGMENT

We would like to thank Dr. Peter Losee and United Silicon Carbide for all the support and for providing the necessary devices and expertise when was needed. We would also like to thank Dr. Christina DiMarino (Virginia Tech) for the shared knowledge and comments.

This work was authored in part by the National Renewable Energy Laboratory, operated by Alliance for Sustainable Energy, LLC, for the U.S. Department of Energy (DOE) under Contract No. DE-AC36-08GO28308. Funding was provided by the U.S. Department of Energy Advanced Research Projects Agency – Energy (ARPA-E). The views expressed in the article do not necessarily represent the views of the DOE or the U.S. Government. The U.S. Government retains and the

publisher, by accepting the article for publication, acknowledges that the U.S. Government retains a nonexclusive, paid-up, irrevocable, worldwide license to publish or reproduce the published form of this work, or allow others to do so, for U.S. Government purposes.

REFERENCES

[1] Tester, J. W., Anderson, B. J., Batchelor, A. S., Blackwell, D. D., DiPippo, R., Drake, E. M., M. N. (2006). The future of geothermal energy. Massachusetts Institute of Technology.

[2] Zetterling, CM. Integrated circuits in silicon carbide for high-temperature applications. MRS Bulletin 40, 431–438 (2015). https://doi.org/10.1557/mrs.2015.90.

[3] P. G. Neudeck, D. J. Spry, L. Chen, N. F. Prokop and M. J. Krasowski, "Demonstration of 4H-SiC Digital Integrated Circuits Above 800 °C," in IEEE Electron Device Letters, vol. 38, no. 8, pp. 1082-1085, Aug. 2017, doi: 10.1109/LED.2017.2719280.

[4] P. G. Neudeck et al., "Operational Testing of 4H-SiC JFET ICs for 60 Days Directly Exposed to Venus Surface Atmospheric Conditions," in IEEE Journal of the Electron Devices Society, vol. 7, pp. 100-110, 2019, doi: 10.1109/JEDS.2018.2882693.

[5] P. Mumby-Croft, D. Li, X. Dai, and G. Liu, "High-Performance Packaging Technology for Wide Bandgap Semiconductor Modules," in Disruptive Wide Bandgap Semiconductors, Related Technologies, and Their Applications, InTech, sep 2018.

[6] G. Bower, C. Rogan, J. Kozlowski, and M. Zugger, "SiC power electronics packaging prognostics," in IEEE Aerospace Conference Proceedings, 2008.

[7] E. Deng, Z. Zhao, Q. Xin, J. Zhang, and Y. Huang, "Analysis on the difference of the characteristic between high power IGBT modules and press pack IGBTs," 2017.

[8] Dean. P. Hamilton; Steve Riches; Michael Meisser; Liam Mills; Philip Mawby, "High temperature thermal cycling performance of DBA, AMB and thick film power module substrates — VDE Conference Publication — IEEE Xplore."

[9] A. A. Bajwa, E. M¨oller, and J. Wilde, "Die-attachment technologies for high-temperature applications of Si and SiC-based power devices," in Proceedings - Electronic Components and Technology Conference, vol. 2015-July, pp. 2168–2174, Institute of Electrical and Electronics Engineers Inc., jul 2015.

[10] K. S. Siow, Die-attach materials for high temperature applications in microelectronics packaging: Materials, processes, equipment, and reliability. Springer International Publishing, 2019. doi: 10.1007/978-3-319-99256-3.

[11] B. Gao, A. J. Morgan, Y. Xu, X. Zhao and D. C. Hopkins, "6.0kV, 100A, 175kHz super cascode power module for medium voltage, high power applications," 2018 IEEE Applied Power Electronics Conference and Exposition (APEC), 2018, pp. 1288-1293, doi: 10.1109/APEC.2018.8341182.

[12] L. A. Garcia Rodriguez, L. Gill, J. Mueller and J. Neely, "A High-Voltage Cascaded Solid-State DC Circuit Breaker Using Normally-ON SiC JFETs," 2021 IEEE 12th Energy Conversion Congress & Exposition - Asia (ECCE-Asia), 2021, pp. 1554-1561, doi: 10.1109/ECCE-Asia49820.2021.9479388.

[13] Randall Kirschman, "Part VII of High-Temperature Electronics," IEEE Press, 1999, pp.711-716, doi: 10.1109/9780470544884.ch92.

Next Generation of GaN Single-Board High Power Stages

Rahil Samani
Department of Electrical Engineering
University of Calgary
Calgary, Canada
rahil.ghatrehsamani@ucalgary.ca

Juncheng Lu
Applications Engineering
GaN Systems Inc.
Ottawa, Canada
llu@gansystems.com

Ignacio Galiano Zurbriggen
Department of Electrical Engineering
University of Calgary
Calgary, Canada
ignacio.galiano@ucalgary.ca

Abstract— **Insulated metal substrate (IMS) boards have been widely used in industry for their remarkable electrical and thermal performance. However, the IMS boards require a separate driver board that affects the power density. With Gallium Nitride enhancement-mode high electron mobility transistors (GaN E-HEMTs) being adopted in modern power converters, power density is becoming critical more than ever. This paper proposes the next generation of single-board IMS power stages (i.e., IMS4). In this board, the power board and the driver board are embedded into a single board. Furthermore, a CerStrate™ ceramic substrate single-board solution is proposed that maintains the exact electrical specifications as observed in IMS4 and significantly enhances the thermal performance. The superiority of these single-board solutions is verified through finite element analysis (FEA) simulations and experiments.**

Keywords— *Ceramic Substrate, Datacenter, GaN, Hybrid Multi-Layer IMS, Power Stage Module.*

I. INTRODUCTION

The power stage is one of the essential subsets in any power supply unit (PSU) as it contributes to 10-20% of the overall system size and cost. In addition, the power stage derives almost 50% of the total system losses [1], [2]. Therefore, it requires special design considerations. GaN Systems' reference designs take advantage of the insulated metal substrate (IMS) platform that can be used for designs with Gallium Nitride enhancement-mode high electron mobility transistors (GaN E-HEMTs) for high-power applications [3]. IMS boards consist of a copper connection layer, a dielectric layer and a metal base. The thickness of the dielectric layer ranges from 0.04 - 0.1 mm. The metal base is usually made of aluminum, which facilitates heat dissipation and improves system-level cooling. The IMS thermal conductivity varies from 3 to 7 W/m·K, depending on the thermal properties of the dielectric material [4]- [6].

The current high-power modules include a driver board and a separate half-bridge IMS board. Fig. 1 shows the history of three different versions of IMS boards developed at GaN Systems [7]-[9]. In particular, IMS 1 is designed for high power

applications (~3-12 kW), and IMS 2/3 is developed for mid-high power applications (~1-3 kW). The IMS2 and IMS3 boards are very similar, and their difference is in the thermal conductivity of the dielectric layer. Despite the benefits of the IMS boards, the current sandwich configuration of the driver board and the power board requires careful attention to parasitic inductance. In addition, with growing demands regarding high power density converters, it is required to develop single-stage boards that maintain the expected electrical and thermal features. In particular, it is required to propose solutions that have a low profile, ease of design and assembly, and acceptable switching performance. This paper proposes two candidates for this purpose:

A. Hybrid Multi-Layer IMS4 Board

The hybrid multi-layer IMS4 board features the above-mentioned design requirements. In particular, it has the power

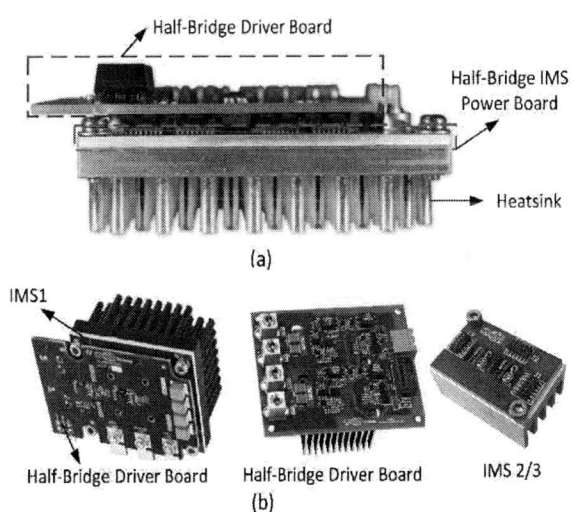

Fig. 1: Current generation of the power-stage IMS boards with separate half-bridge gate driver board and power board (a) side-view, (b) IMS1 and IMS2/3.

978-1-6654-8901-0/22 $31.00 © 2022 IEEE

stage in a single-layer format combined with the multi-layer driver board all in one single board. IMS4 provides an excellent all-around solution with an optimal electrical layout suitable proposed for most applications (e.g., onboard charger, traction inverter, datacenter). The suggested power range for this board is 1-12 kW.

B. Ceramic Substrate CerStrate™ Board

The CerStrate board has a similar idea of combing the driver board and the power board into one single board. However, it takes advantage of ceramic inlay substrate that is placed under GaN devices for better thermal conduction. The type of ceramic used is AlN, which features a very high thermal conductivity ceramic (i.e., 170 W/m.K). Therefore, the CerStrate board maintains IMS4 advantages but takes another step further by significantly enhancing the thermal performance. The suggested power range for this board is 3-22 kW.

II. TECHNOLOGY COMPARISON

Fig. 2 and Fig. 3 show the top views and the cross-sectional views of the next generation power stages presented in this paper. As shown in these figures, the driver stage is adjacent to the half-bridge power stage, while sharing a single board. In terms of thermal performance, the IMS4 board acts relatively similar to the IMS3 board. The bottleneck in the IMS board design is the dielectric layer. This dielectric layer is unavoidable as it ensures electrical isolation between the aluminum substrate and the copper traces. However, the dielectric layer has poor thermal conductivity. On the other hand, the CerStrate board does not have this limitation. Due to the presence of the ceramic layer right under the GaN devices, the CerStrate board offers outstanding thermal performance while maintaining comparable electrical performance.

In this section, finite-element analysis (FEA) is conducted to compare the thermal performance of the three different power stage solutions, which are: IMS3, IMS4, and CerStrate. Prior to the system-level comparison of these three technologies, the board-level comparisons are performed. To do so, we use ANSYS Icepak 2021 R1. The boundary conditions of the simulations are as follows:

1) Ambient temperature is set to 25 °C.

2) The model is in an open space.

3) The IMS temperature is fixed at the ambient temperature of 25 °C. This assumes an ideal heatsink being placed on the IMS board. This method is used to calculate the thermal resistance from the junction of the GaN device to the board.

4) In all simulations, power loss is considered such that it emulates a junction temperature around 100 °C for the GaN device. This is due to the thermal conductivity degradation of the die with respect to the temperature. In order to have a fair comparison, all die temperatures are studied at around 100 °C.

Fig. 4 demonstrates the thermal performance comparison of these three solutions. It is observed that the IMS3 and IMS4 act thermally similar. IMS4 has slightly better thermal performance due to the larger area of the board. Besides, we can

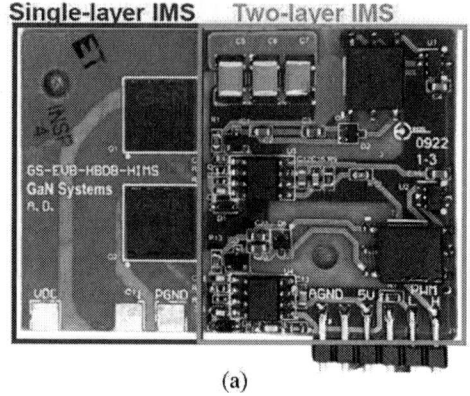

(a)

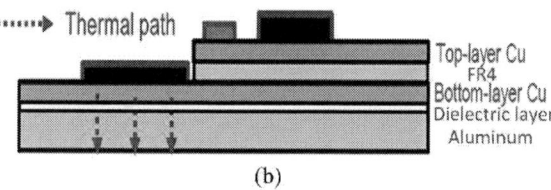

(b)

Fig. 2: Hybrid multi-layer IMS4 board (a) top view, (b) cross-sectional view.

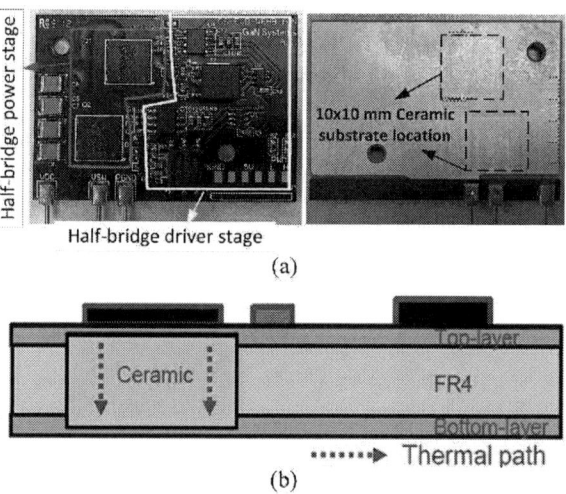

Fig. 3: CerStrate board (a) top view, (b) cross-sectional view.

see that the CerStrate board had a better heat distribution, as shown in the cross-sectional heat contours of the board. This is due to the excellent thermal conductivity of the AlN ceramic. Table 1 summarizes the junction to IMS thermal resistance of these three solutions. A figure of merit to understand the thermal performance of the proposed design is the junction to IMS thermal resistance, given by:

$$R_{th,J-IMS} = \frac{T_j - T_{IMS}}{P_{GaN}} \quad (1)$$

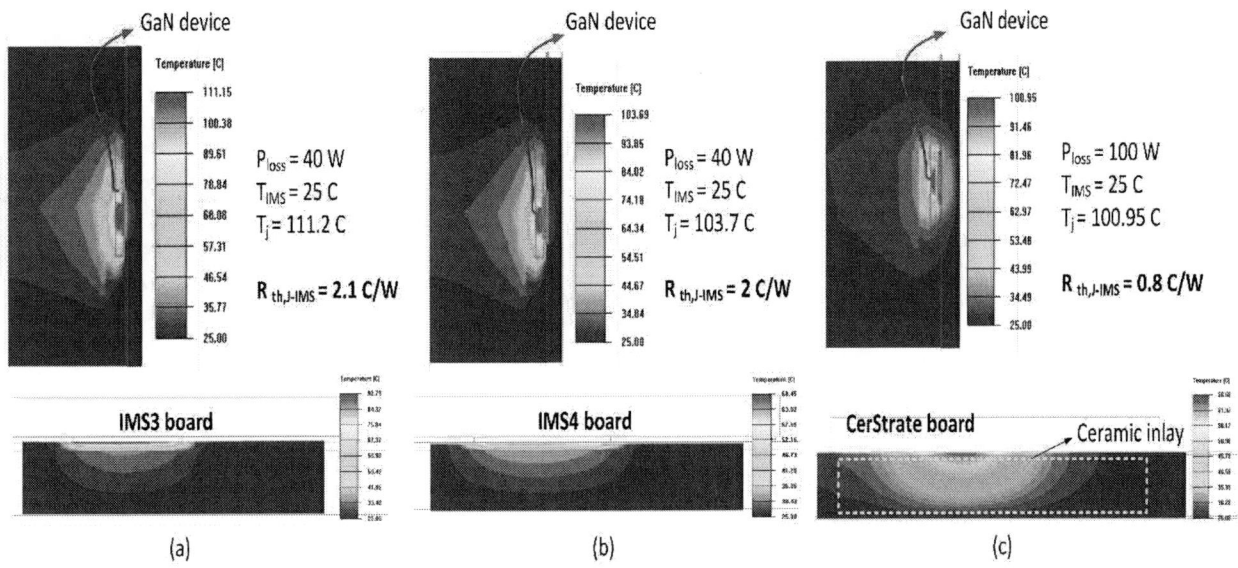

Fig. 4: Thermal comparison of the three technology of power stage boards: (a) IMS3, (b) IMS4, (c) CerStrate.

where $R_{th,J\text{-}IMS}$ (°C/W) is the thermal resistance from junction to board, T_j (°C) is the junction temperature, T_{IMS} (°C) is the board temperature, and P_{GaN} (W) is the total GaN power loss.

To better understand the heat flow trends in different boards, the heat flow equations can be used. The conductive heat flow can involve multiple directions and time-dependent terms, as follows:

$$\frac{\partial}{\partial x}\left(k\frac{\partial T}{\partial x}\right) + \frac{\partial}{\partial y}\left(k\frac{\partial T}{\partial y}\right) + \frac{\partial}{\partial z}\left(k\frac{\partial T}{\partial z}\right) + \dot{q} = 0 \quad (2)$$

where $\dot{q}$ is the heat flux on a surface (W/m²), k is the thermal conductivity (W/m.K), T is the temperature (°K).

For steady-state situations, we can simplify the problem into a one-dimensional system, where heat flows in one direction, and temperatures have reached the steady state (i.e., time-independent). Thus, the thermal resistance is given by:

$$R_{th} = \frac{l}{k.A} \quad (3)$$

where l is the length in the direction that heat flows, and A is the cross-sectional area. According to this equation, we can see how adding the ceramic layer improves the thermal resistance and how enlarging the area of the board or shrinking the thickness of layers helps with improving the thermal resistance.

Once the board-level performance of these three solutions is compared, we need to consider them in a real system where a heatsink is attached to the board with thermal grease. Fig. 5 depicts the simulation result of $R_{th,J\text{-}HS}$ for different grease thicknesses. According to [9], the grease thickness applied to the heatsink will most likely be around 30 μm. According to this figure, the grease thickness would be the bottleneck in the

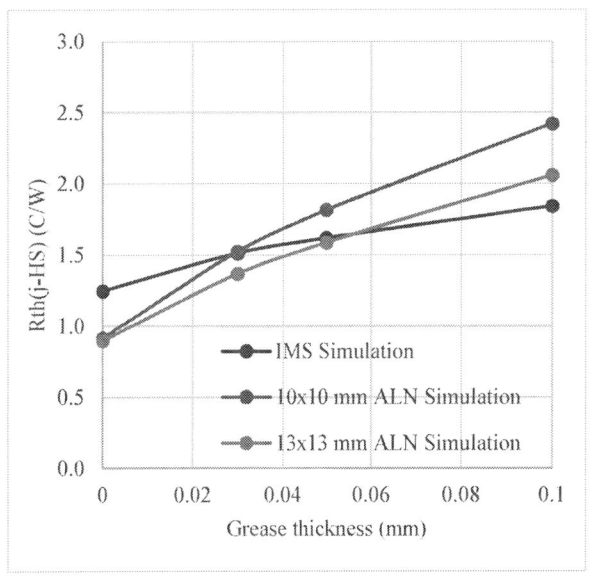

Fig. 5: FEA simulation results comparing $R_{th(j\text{-}HS)}$ vs grease thickness for IMS3, CerStrate with 10x10 mm AlN, and CerStrate with 13x13 mm AlN.

system-level design. Based on Eq. (3), it is desired to make sure that the grease thickness is always minimal. If the grease thickness cannot be appropriately controlled (i.e., sufficient force is not applied), then the CerStrate board may lose its thermal advantage over the IMS boards. This is due to the fact that the grease is a poor thermally conductive material. In the IMS board, heat has a more horizontal path to travel, whereas the CerStrate board is more meant to conduct heat in the vertical direction.

978-1-6654-8901-0/22 $31.00 © 2022 IEEE   235

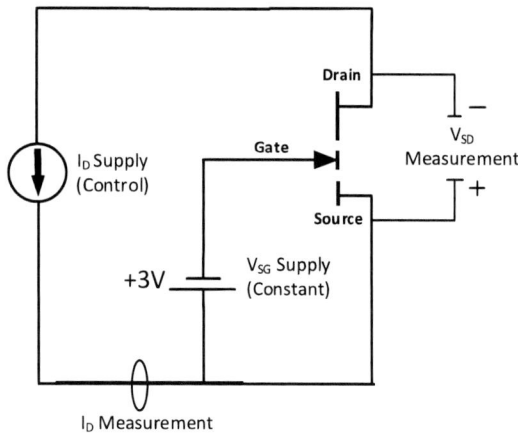

Fig. 6: Reverse conduction excitation circuit for the thermal experiments.

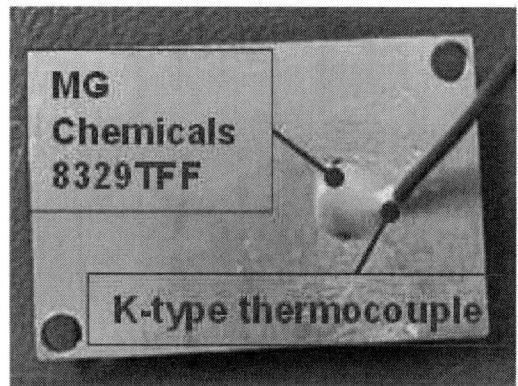

Fig. 7: Measuring IMS temperature using k-type thermocouples.

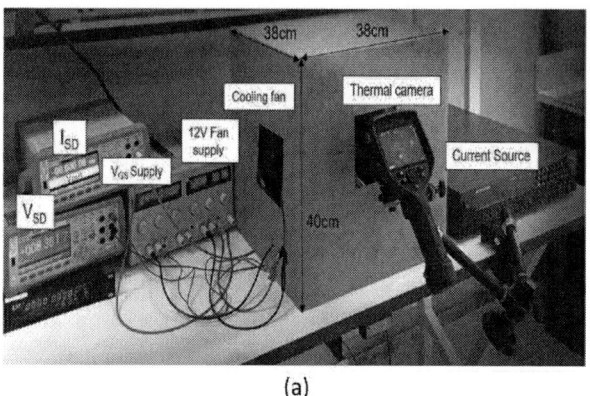

(a)

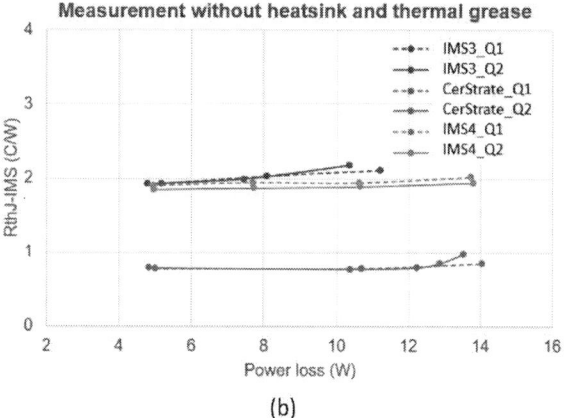

(b)

Fig. 8: (a) Thermal experimental platform and setup, and (b) experimental verification of Junction-to-IMS R_{th} without the heatsink and thermal grease.

Furthermore, it can be observed from Fig. 5 that the bigger the area of the AlN ceramic is, the lower the overall thermal resistance of the board would be. This can be adjusted with Eq. (3) as well.

III. EXPERIMENTAL RESULTS

To verify the performance of the proposed solutions, experimental tests are performed. In particular, two sets of experiments need to be conducted: 1) thermal experiments to verify the superiority and feasibility of the proposed solutions, and 2) electrical experiments to ensure acceptable switching performance.

A. Thermal Performance

Fig. 6 depicts the circuit diagram used to do the thermal test to measure the thermal resistance. In this circuitry, reverse conduction is considered to maximize the power loss at any given current. To compare the IMS3, IMS4, and CerStrate boards, we need to measure the junction to IMS thermal resistance $R_{th,J-IMS}$. To do so, k-type thermocouples are attached to the back of the board right under the GaN device. Fig. 7 shows the installation of the thermocouple using a thermally adhesive material. This board is then placed in the thermal chamber to be tested under forced air cooling. The reason to do forced air cooling is that the temperatures reach the steady state faster. Fig. 8(a) shows the platform used for the thermal experiment. The junction temperature of the GaN device is captured using the IR camera, and the IMS temperature is measured using the thermometer with k-type thermocouples. The results of the thermal test are shown in Fig. 8(b). It is experimentally observed that the IMS3 and IMS4 board are thermally comparable, and the CerStrate board has even more advantages compared to the IMS solutions. The measured $R_{th,J-IMS}$ are well aligned with the FEA simulations in Table 1.

Table 1: Thermal resistance of different power stage technologies

Board type	Simulated $R_{th,\ J-IMS}$ (°C/W)	Experimental $R_{th,\ J-IMS}$ (°C/W)
IMS3	2.1	2.1
IMS4	2	1.95
CerStrate	0.8	0.9

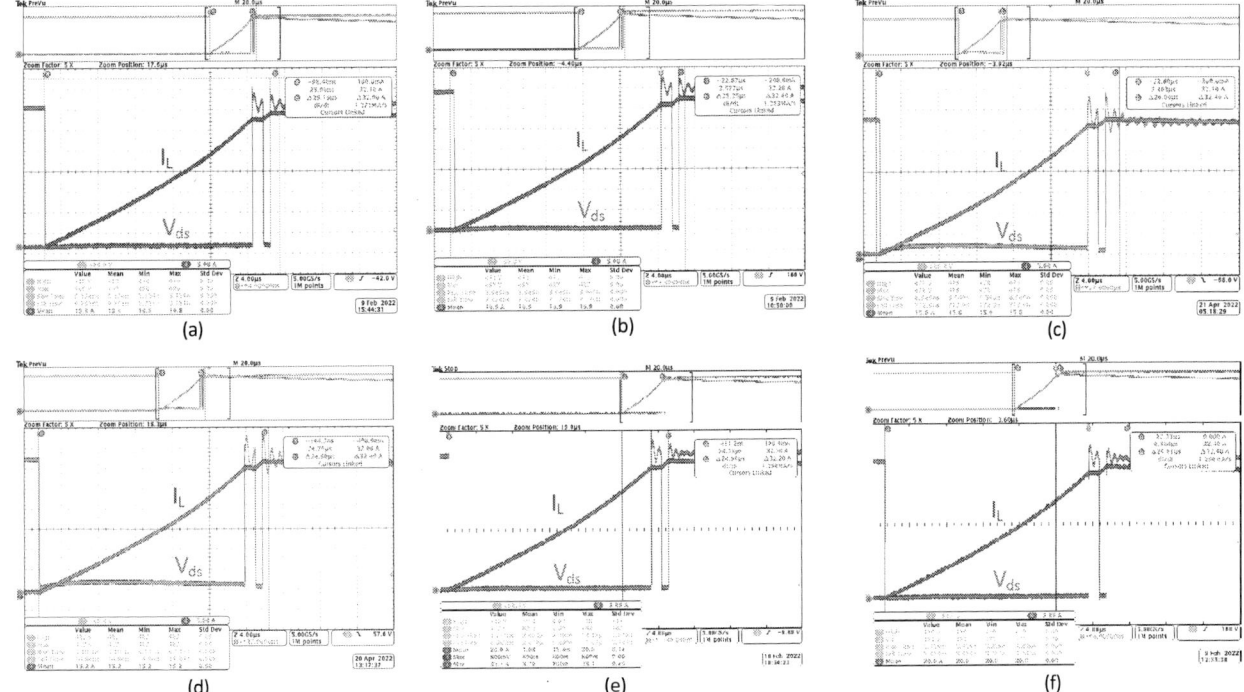

Fig.9: Double pulse test results on the three half-bridge power stage boards using GS-065-030-2-L from GaN Systems : (a) IMS3 high-side device, (b) IMS3 low-side device, (c) IMS4 high-side device, (d) IMS4 low0side device, € CerStrate board high-side device, and (f) CerStrate low-side device.

B. Electrical Performance

To verify the acceptable switching performance of the GaN devices on the proposed boards, double pulse tests (DPTs) are performed. Fig. 9 shows the DPT results of the high-side and low-side devices mounted on the half-bridge IMS3, IMS4, and CerStrate boards. Accordingly, it is shown that all three boards have comparable switching performance.

IV. CONCLUSION

In this paper, the next generation of GaN single-board power stage has been introduced. Formerly, different versions of IMS boards were employed in a sandwich configuration with their driver card mounted on top. To eliminate multiple board assemblies, lower profile, lower parasitics, and comparable thermal performance, two new solutions were proposed in this paper: IMS4 and CerStrate boards. These boards take advantage of combining both driver and power stage cards into one single board, which provides acceptable switching performance and enhances the thermal performance as well.

REFERENCES

[1] J. Frazor, "Power Solution Options for Data Center Applications", 2018. [Online]. Available: https://www.ti.com/lit/pdf/slyy155.

[2] Texas Instruments, "LMG5200 80-V, 10-A GaN Half-Bridge Power Stage," LMG5200 datasheet, Mar. 2015 [Revised Oct. 2018].

[3] J. L. Lu, D. Chen and L. Yushyna, "A high power-density and high efficiency insulated metal substrate based GaN HEMT power module," *2017 IEEE Energy Conversion Congress and Exposition (ECCE)*, 2017, pp. 3654-3658.

[4] U. Drofenik, A. Stupar and J. W. Kolar, "Analysis of Theoretical Limits of Forced-Air Cooling Using Advanced Composite Materials With High Thermal Conductivities," in IEEE Transactions on Components, Packaging and Manufacturing Technology, vol. 1, no. 4, pp. 528-535, April 2011.

[5] Drofenik, Uwe & Kolar, Johann & Member,. (2006). "A Thermal Model of a Forced-Cooled Heat Sink for Transient Temperature Calculations Employing a Circuit Simulator", IEEE Transactions on Industry Applications. 126. 10.1541/ieejias.126.841.

[6] GaN Systems, GN002 Application Note, "Thermal Design for GaNPX® Packaged Devices," [Online]. Available: https://gansystems.com/wp-content/uploads/2021/07/GN002_Thermal-Design-Guide-for-Top-Side-Cooled-GaNpx-T-Devices_Rev-210720.pdf.

[7] "GSP65R25HB-EVB Evaluation Board | GaN Systems", GaN Systems, 2020. [Online]. Available: https://gansystems.com/wp-content/uploads/2021/09/GSP65RXXHB-EVB_Technical-Manual_Rev_210914.pdf.

[8] "GS-EVB-IMS2-66504B-GS Evaluation Board | GaN Systems", GaN Systems, 2020. [Online]. Available: https://gansystems.com/wp-content/uploads/2020/11/GS-EVB-IMS2-XX-Technical-Manual-Rev-201023.pdf.

[9] "GS-EVB-IMS3-66516B-GS Evaluation Board | GaN Systems", GaN Systems, 2020. [Online]. Available: https://gansystems.com/wp-content/uploads/2020/12/GS-EVB-IMS3-665xxB-GS-Technical-Manual-Rev.-201207.pdf.

100V GaN for Highly Efficient 1kW Motor Drive Applications

Asantha Kempitiya, Hrach Amirkhanian, Srikanth Yerra and Kapil Kelkar

Infineon Technologies, El Segundo, California, USA.

Email: Asantha.Kempitiya@infineon.com

Abstract—In this work, Infineon's 100V 3mΩ CoolGaN™ is evaluated and compared with a 100V similar R_{dson} MOSFET using OptiMOS™5 silicon technology in a three-phase inverter for battery powered motor drive applications. In order to minimize conduction losses, 30ns dead time is achieved for GaN due to its lower gate and output charge in comparison to 60ns of dead time for silicon. A motor-generator setup is developed to evaluate a 48V drone motor up to ~1kW of inverter power. For a motor speed of 4000RPM, with a bus voltage of 48V and 100 kHz switching frequency, GaN shows higher efficiency across the entire load range. At 1/3rd load (~300W), 98.67% and at full load (~900W) 98.10% power efficiency values are achieved. At light load (~45W), a significant increase of ~5% is also observed in efficiency. This experimentally demonstrates that 100V CoolGaN™ is ideal for developing high frequency switching inverters with high power densities and motor efficiencies in the kilowatt power range.

Keywords— Gallium Nitride, silicon, MOSFETs, Optimos, motor drives, inverters, battery powered applications, drones, power tools, e-bikes, robotics.

I. INTRODUCTION

Size, weight and cost are key considerations for battery powered motor drive applications such as power tools, drones and electric bicycles. These applications can span power ranges from a fraction of a kilowatt to tens of kilowatts while their widespread adoption relies on having extended battery life and / or higher inverter power densities. The introduction of wide bandgap technologies such as GaN have enabled hard-switched power electronic topologies such as inverters to operate efficiently at higher frequencies and power levels in several kilowatts compared to their silicon predecessors [1]. Previous studies have demonstrated that higher switching frequency operation can also result in higher motor efficiency, lower motor torque ripple and faster control response [2]. In this work, a comprehensive investigation of GaN vs silicon performance is experimentally carried out to demonstrate the benefits of GaN power switches in motor drives operating at higher switching frequencies in the kilowatt power range.

II. METHODOLOGY

To experimentally investigate the value proposition of GaN for battery powered motor drive applications in this study, two inverters comprising of GaN and silicon MOSFETs are evaluated with respect to switching frequency and inverter output power. For a fair comparison in current capability and forward conduction performance, Infineon's 100V 3mΩ CoolGaN™ is evaluated and compared with a similar R_{dson} OptiMOS™5 silicon technology. Both GaN and silicon devices are packaged in top side cooling capable packages. An aluminum heatsink is then attached on top of the PCB to provide a cooling path for the power dissipated

from the top-side cooled power devices. A thermal pad with thermal conductivity of 12.5W/mK serves as the thermal interface material between the power device and the heatsink. While the evaluated motor drive system in this study is passively cooled by ambient air, depending on the specific application such as power tool, drone or e-bike, a similar motor drive system may have forced air flowing over the heatsink that will serve to further improve the thermal performance of the system.

The motor generator test setup that is developed for this study facilitates the measurement of inverter and motor efficiency so that the benefits of high switching frequency operation enabled by GaN can be investigated. System efficiency is also reported to assess the overall impact of using GaN over silicon. The key contribution of this work is that it provides a comprehensive investigation of GaN vs silicon and experimentally demonstrates that 100V CoolGaN™ is ideal for developing highly efficient motor drives with high switching frequencies in the kilowatt power range.

III. MOTOR GENERATOR TEST SETUP

A. Motor Drive Inverter Configuration

Fig 1. illustrates the block diagram of the test setup used to evaluate a three-phase inverter implemented with GaN and silicon. The GaN motor drive inverter [3] comprises of a low-profile design with six top-side cooled 3mΩ GaN power devices in PQFN 3x5 package mounted on a four-layer PCB as shown in the inset of Fig.1. The GaN switches are driven by Infineon's 1EDN7126G [4] dedicated GaN driver with 1.5 A peak source / sink current capability with active bootstrap clamp to avoid overcharging of the bootstrap capacitor. The silicon motor drive inverter comprises of six top side cooled silicon power MOSFETs in PQFN 5x6. Both motor drive boards use Infineon's TLI4971 in-phase hall effect current sensor for monitoring the inverter phase currents. In order to control the power switches of the motor drive inverter, Infineon's XMC 1400 drive card [5] equipped with the 32-bit XMC1402 micro-controller [6] is connected to the motor drive inverter as shown in the inset of Fig.1. A sensorless Field Oriented Control (FOC) algorithm is implemented in the micro-controller.

B. Drive and Load Motors

The motor drive inverter drives a 48V drone motor at a specific speed under speed control. A power analyzer is used to measure the inverter efficiency, while motor efficiency is measured by a torque and speed sensor coupled between the

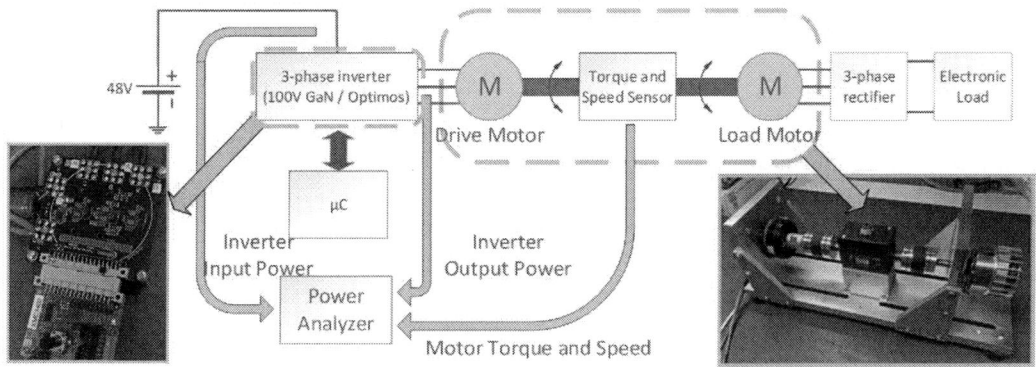

Fig. 1: Motor-generator test setup for evaluating GaN and silicon three phase inverters.

drone and load motors, as shown in the inset of Fig. 1. The load motor is connected to a three-phase rectifier terminated by an electronic load. The electronic load under current control will determine the required torque by the drone motor at a specific rotational speed.

IV. EXPERIMENTAL RESULTS

A. Switching waveforms

Figure 2 shows voltage and current waveforms for the GaN motor drive inverter. The phase current waveform shows an electrical frequency of ~682Hz which translates to a mechanical rotational speed of approximately 4000RPM for the drone motor consisting of 20 poles. The shape and amplitude of the current waveform in Fig. 2a show that sensorless FOC algorithm successfully regulates the motor speed under speed control for an inverter output power of 500W.

Various voltage probing methods from the utilization of passive probes with PCB connectors to active differential probes were investigated, and the use of single-ended passive probes directly soldered onto the PCB pads connected to the Kelvin gate, drain and source of the power device was chosen. Fig.2 shows switching waveforms for the GaN inverter using this methodology.

The conduction losses for GaN in a hard-switched inverter are strongly correlated to the effective dead time between the complementary switches of an inverter half-bridge. This is because the voltage drop of GaN during reverse conduction is much larger than that for a MOSFET body diode. As a result, it is important to avoid unusually large deadtimes in GaN-based motor drives to improve inverter efficiency. Since GaN has a lower gate and output charge, a micro-controller dead time setting of 30ns is used. On the other hand, the micro-controller dead time setting used for the silicon inverter is 60ns. The controller dead time ensures that an effective dead time at the gate of the complementary power switches of a half-bridge exists so that any shoot-through current in the phase leg of the inverter is prevented for an inverter operating at a specific power level.

B. Impact of switching frequency on phase current

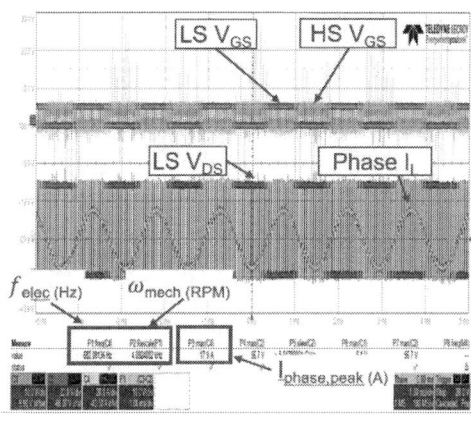

(a)

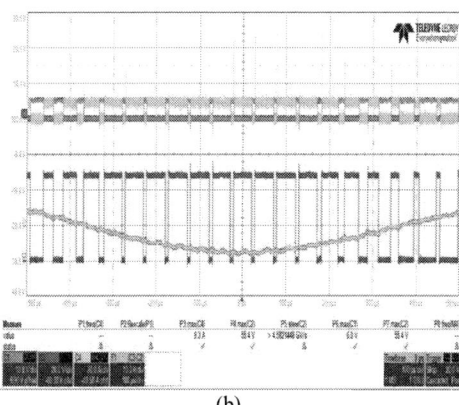

(b)

Fig. 2: Switching and phase current waveforms for three phase GaN motor drive inverter during operation.

978-1-6654-8901-0/22 $31.00 © 2022 IEEE

Fig. 4 shows the phase current waveforms for 20kHz and 100kHz switching frequencies for an inverter output power of 500W and 900W, respectively. These waveforms demonstrate that when witching frequency increases, the ripple on the phase current decreases for both inverter output power levels. This ripple current causes higher RMS currents to flow through the motor windings resulting in higher conduction losses. By allowing an inverter to drive a motor with a higher switching frequency also results in lower harmonic content in the phase current. This is ideal for generating smoother torque for motion-sensitive applications such as robotics and medical equipment.

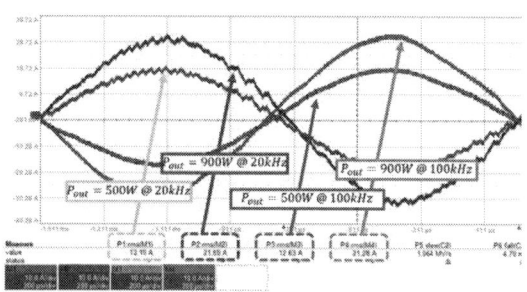

Fig. 3: Effect of switching frequency on inverter phase current.

C. Inverter Efficiency

Fig. 4(a) and 4(b) shows the inverter efficiency (primary y-axis on the left) as a function of inverter output power for switching frequencies of 20kHz and 100kHz, respectively. The secondary y-axis (on the right), shows the difference in inverter efficiency between GaN and silicon motor drive inverters. For a switching frequency of 20kHz, GaN shows higher peak efficiency of 99.03% at ~300W, while silicon shows higher efficiency of 98.53% at 900W. At 300W, GaN efficiency is 0.19% higher while at 900W silicon efficiency is only higher by 0.05%.

For a switching frequency of 100kHz, it shows that GaN has higher efficiency across the entire load range. GaN shows higher efficiency of 98.67% and 98.17% at 300W and 900W, respectively. GaN efficiency is 0.77% and 0.18% higher at 300W and 900W, respectively. At light loads (~45W), a massive ~4.52% efficiency increase is observed over silicon. This indicates that GaN would be advantageous in applications with load profiles consisting majority of light load operation such as E-bikes.

D. Motor Efficiency

Fig. 5 shows the motor efficiency (primary y-axis on the left) as a function of inverter output power when driven by the GaN motor drive inverter. The secondary y-axis (on the right), shows the difference in motor efficiency for 20kHz and 100kHz, respectively. The peak motor efficiency at 300W is approximately 85.61%. When increasing switching frequency from 20kHz to 100kHz, at light load (~45W) motor efficiency increases by ~2.56% and at 300W, motor efficiency increases by ~0.63%. For full load (~900W), motor efficiency increases by ~1.75%. As the inverter output power increases, the current through the motor also increase

resulting in increased losses and temperature on the windings of the motor causing motor efficiency to drop.

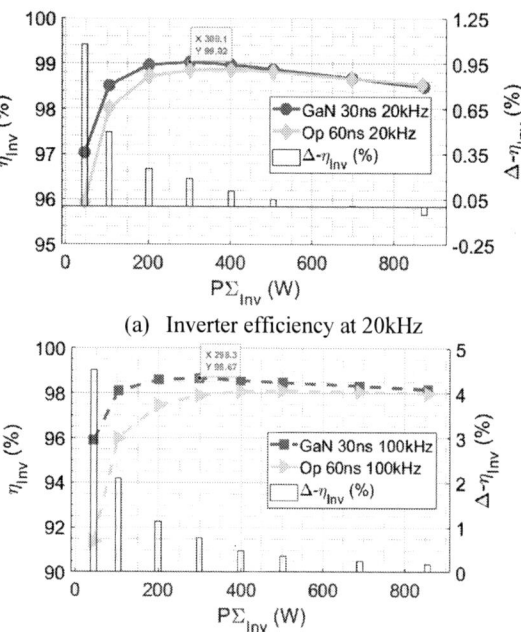

(a) Inverter efficiency at 20kHz

(b) Inverter efficiency at 100kHz

Fig. 4: Inverter efficiency comparison of 100V 3mΩ CoolGaN™ vs silicon OptiMOS™5 Technology.

E. System Efficiency

From the inverter and motor efficiency plots of Fig. 4 and 5, it is clear that when increasing switching frequency from 20kHz to 100kHz, the system efficiency is also increased. The measured system efficiency increases by approximately 1.85% at light load (45W) and 1.5% at full load (900W) for the GaN motor drive operating at 100kHz. This is because the decrease in inverter efficiency when operating at a higher switching frequency is compensated by a much higher increase in motor efficiency. The superior high frequency switching performance of GaN results in a smaller decrease in inverter efficiency when the switching frequency increases compared to silicon.

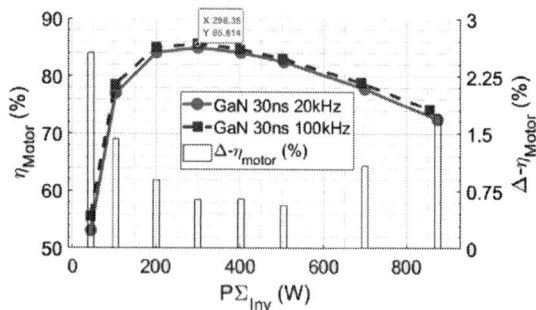

Fig. 5: Motor efficiency comparison with GaN motor drive inverter at 20kHz and 100kHz.

CONCLUSION

This work experimentally investigates the value proposition of GaN for battery powered motor drive applications. A comprehensive performance comparison of two inverters comprising of Infineon's 100V 3mΩ CoolGaN™ and OptiMOS™5 silicon MOSFET technology is carried out with respect to switching frequency and inverter output power. The motor generator test setup facilitates the measurement of inverter and motor efficiency as a function of inverter output power and switching frequency. The inverter, motor and system efficiency is presented and analyzed demonstrating that 100V CoolGaN™ is ideal for developing highly efficient motor drives operating at high switching frequencies in the kilowatt power range.

REFERENCES

[1] B. J. Baliga. "Power semiconductor device figure of merit for high frequency applications," IEEE electron Device Lett., Vol. 10, p. 455, Oct. 1989.

[2] H. Järvisalo, J. Korhonen, J. Honkanen and P. Silventoinen, "Considerations for a high-speed PMSM drive featuring a GaN-ANPC inverter," 2017 19th European Conference on Power Electronics and Applications (EPE'17 ECCE Europe), pp. P.1-P.6., 2017.

[3] M. Wattenberg, E. A. Jones and J. Sanchez, "A Low-Profile GaN-Based Integrated Motor Drive for 48V FOC Applications," PCIM Europe digital days 2021, 2021, pp. 1-8.

[4] Infineon Technologies, "EiceDRIVER™ 200 V high-side TDI gate driver IC optimized for CoolGaN™ SG HEMTs and Silicon MOSFETs", 1EDN71x6G datasheet, Oct. 2021.

[5] Infineon Technologies, "XMC1400 Drive Card", AP32370 Application Note, Dec. 2018

[6] Infineon Technologies, "XMC1402-Q064X0064 AA", XMC1400 AA-Step Datasheet, Nov. 2017

Compact Three-Level GaN Power Module Suitable for Active-Neutral-Point-Clamped (ANPC) Three-Level Converter

Ziwei Liang
Department of Electrical Engineering and Computer Scienec
University of Tenneessee
Knoxville, TN, USA
Knoxville, USA
zliang7@utk.edu

Liyan Zhu
Department of Electrical Engineering and Computer Scienec
University of Tenneessee
Knoxville, TN, USA
Knoxville, USA
liyan@utk.edu

Hua Bai
Department of Electrical Engineering and Computer Scienec
University of Tenneessee
Knoxville, TN, USA
Knoxville, USA
hbai2@utk.edu

Yue Sun
Department of Electrical Engineering and Computer Scienec
University of Tenneessee
Knoxville, TN, USA
Knoxville, USA
ysun79@vols.utk.edu

Abstract—**This paper presents a compact three-level (3L) Gallium Nitride (GaN) power module with low parasitic parameters. The power loop, gate loop and PCB layout are designed and optimized carefully to achieve: 1) low parasitic parameters; 2) good thermal performance; 3) compact size and high integration; 4) easy utilization on building 3L ANPC converter. To achieve higher power rating, for two high-frequency (HF) switch legs, three GaN Systems' GS665016T are paralleled and for the low-frequency pair, two switches are paralleled considering the much lower switching loss. The decoupling capacitors, driver circuits and auxiliary power supplies are all integrated into the module for the easy utilization and small parasitics. Double-pulse test (DPT) is applied to test the electrical characteristics and a bidirectional three-phase 3L ANPC converter is built using the proposed modules to verify the electrical performance and thermal performance.**

Keywords—*Three-level power module, GaN devices, integrated gate drive, decoupling cap*

I. INTRODUCTION

Compared with the Si devices, the wide-bandgap (WBG) GaN power high electron mobility transistors (HEMTs) can help to increase the power density significantly with the advantages of smaller size, lower $R_{ds,on}$ and faster switching speed[1]. Since the GaN HEMTs are very sensitive to the parasitic parameter and the heat dissipation, some two-level (2L) half-bridge GaN based power modules have been proposed to maximize the switching and thermal performance[2-3]. Reference [2] offers an double-sided cooling design for better thermal performance. Reference [3] integrate the decoupling cap, gate driver and the auxiliary power supply together to have a compact design and supply simpler interface for user.

So far, the maximum blocking voltage of mature GaN HEMTs which are introduced on the market is around 650V, such as 650V GaN transistor from GaN System. Such devices can only serve for two-level(2L) converters with DC voltage lower than 600V considering the enough marge of the Safe Operation Area (SOA). For any application with DC voltage over 600V, it is usually recommended to use 900-1200V SiC

MOSFETs [4]. To employ GaN HEMTs in such higher voltage application, one of solution is series two 650V GaN devices. However, such structure requires great simultaneity of two series devices to ensure the balanced voltage distribution. Another solution is to use the multi-level structure yielding the lower voltage stress of switches, lower current ripple and reduction of the filter size [4-7]. One of the most popular three-level topology for the Grid-connection high-power application is 3L-ANPC [6, 8].

Combining the good electrical characters of GaN HEMT and the benefits of 3L-ANPC, a compact 3L GaN based power module with integrated gate drive, auxiliary power supply and decoupling caps is proposed in this paper by using GaN Systems' 650V/60A device (GS66516T). The simple interface makes it convenient to be used to build a high-performance and compact 3L-ANPC converter.

In this paper, Section II describes the PCB design considerations, including the power loop design , gate loop design. Section III introduces the thermal simulation and parasitic simulation. Section IV shows the DPT and bidirectional power experiment results to verify the electrical and thermal performance of the proposed power module.

II. PCB DESIGN

A. Schematic of 3L-ANPC Phase Leg

The schematic of one 3L-ANPC phase leg is shown in Fig.1. It consists of six switches. Based on the modulation proposed in [9], switch S1-S4 will switch with high frequency (HF) and S5 and S6 will switch with grid frequency which means that the switching loss of S5 and S6 will be much lower than the switching loss of S1-S4.Terminal $+V_{dc}$, V_{dc_mid} and -V_{dc}, will be connected to the active , neutral point and negative DC bus. v_{sw} will be connected to the grid side.

To enhance the power capability, three GaN devices are paralleled for the high-frequency switch and two GaN devices are paralleled for the low-frequency switch considering the different loss dissipation. Paralleled devices can reduce the on-state resistance and share the switching current. The diagram of

978-1-6654-8901-0/22 $31.00 © 2022 IEEE

higher HF switch leg including the integrated gate driver, auxiliary power supply and decoupling cap is shown in Fig.2. The other two switch legs have the similar structure.

B. Integrated Gate Drive Circuit

Isolated gate driver and isolated auxiliary power supply are the key components for the gate driver circuit. Based on the gate drive circuit guidance from GaN System [10], each group of paralleled GaN devices will co-share 1) one-channel isolated gate driver, Si8271AB-ISR; 2) Turning on and off resistor, R_{on} and R_{off}; 3) Gate-source resistor R_{gs}; 4) Isolated power supply. Additionally, one gate resistor R_g and one source resistor R_s are added for each GaN HEMT for the paralleled GaN HEMTs [10]. To reduce the negative V_{gs} spike, R_{on} and R_{off} are optimized and set to 13Ω and 0Ω respectively. R_g and R_s are both 1Ω and R_{gs} is 4.7kΩ.

A negative V_{gs_off} can help to enhance the system stability in real application, -3V_{gs_off} is used for this design. To provide +6 and -3V for the isolated gate drivers, a 12V-9V isolated DC-DC module (PDS1-S12-S9-M-TR) and a following voltage divider is used as the auxiliary power supply. +6V and -3V is generated by connecting a 6V Zener diode and a 1kW resistor in series on the output of the isolated DC-DC module. The main components used in the integrated power module are summarized in TABLE I.

Gate loop design is important for GaN based power module since the GaN HEMTs are sensitive to the gate loop parasitic inductance. Multi-layer PCB is adopted in this design to realize flux canceling [11-12]. Additionally, to minimize the gate loop area and gate loop distance, gate drivers should be put close to the GaN devices. The turn-on gate loop of the paralleled GaN device S2 is shown in Fig. 3. In Fig.3, yellow line represents the shared gate loop and blue, green and black lines represents the gate loops for three GaN HEMTs S2-1, S2-2, S2-3 respectively. Solid line shows the gate current loop from gate driver to the GaN HEMTs, and the dashed line shows the gate current loop back to gate driver. From Fig.3, we can find that, by using the multi-layer PCB, gate current flows through top layer and returns through the internal layer 1, the vertical structure is formed, which can help to reduce the gate loop area and realize flux canceling. Therefore, low gate-loop parasitic inductance can be achieved.

C. Power loop design.

The parasitic inductance of the power loop has a lot of effects on the switching performance, such as voltage spike on gate drive loop, voltage overshoot on Vds and voltage ringing across decoupling cap [13]. Therefore, it's important to reduce the power loop inductance. Similar to the gate loop design, vertical structure is also adopted to the power loop by using multi-layer PCB. To increase the current capability and reduce the conducting loss, eight-layer PCB is employed in the power module design. Eight-layer PCB also enables the possibility of simple interface by putting all HV terminal to one side of the module, which can simplify the employment of such power module in 3L-ANPC converter design.

The top view and bottow view of the proposed GaN power module are shown in Fig. 4. From Fig.4, we can find that the

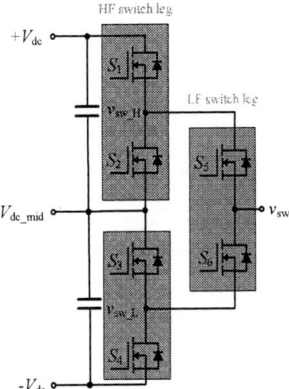

Fig. 1 Schematic of one 3L-ANPC phase leg.

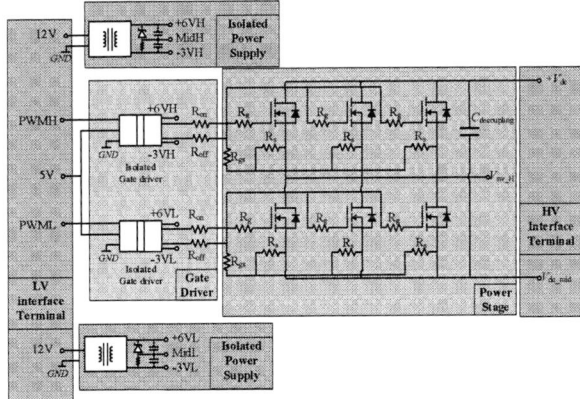

Fig. 2 Diagram of higher HF switch leg.

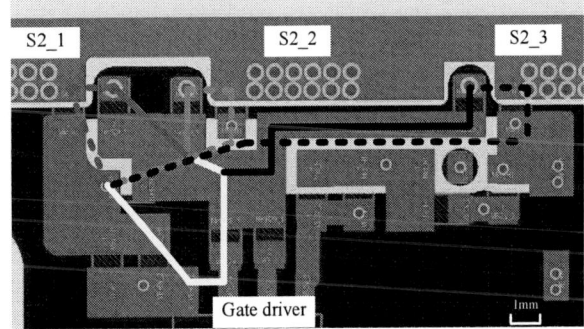

Fig. 3 Turn-on gate loop for paralleled GaN HEMTs.

TABLE I.   MAIN COMPONENTS

Component	Part Number	Value
Isolated gate driver	Si8271AB-ISR	--
Isolated power supply	PDS1-S12-S9-M-TR	--
Decoupling cap	C2220X124KDRAC 7800	2*4*0.12µF
GaN HEMT	GS66516T	--

978-1-6654-8901-0/22 $31.00 © 2022 IEEE   243

double-side layout is adopted in this module. All the GaN devices are assembled on the bottom layer, and the integrated gate driver, auxiliary power supply and the decoupling cap are assembled on the top layer. The power loops are on internal layers. Specifically, the power loops of two HF switch legs distribute symmetrically on the PCB and have the similar structure. The side view of the PCB and the power loop of one HF switch leg is shown in Fig.5 where the thickness of PCB is scaled up for demonstration purpose. Two groups of LF switches, S_5 and S_6 locate on the two edges of PCB as shown in Fig. 4.

To achieve better thermal performance, an air-cooling heat sink is customized and optimized to match the size of PCB as shown in Fig. 6.

III. SIMULATION VERIFICATION

A. Thermal Simulation

With the custom heat sink, the thermal simulation is conducted based on the assumption that the average loss of each GaN device is 6.5W. High-speed Fan, OD6038 - 12HHBXC01A, is used for cooling the module. A simplified thermal simulation model is built and imported into Ansys Icepack. In the simulation, airflow is set to 0.037 m³/s based on the fan's parameter and the ambient temperature is set to 70°C for potential EV on-board charger application. As shown in Fig.7, the simulation result shows that highest temperature is 95.3 °C, which means the maximum junction-to-case thermal resistance is 0.27 °C/W.

B. Parasitic Parameter Extraction

To analyze the parasitic parameters caused by the PCB layout, a simplified power loop is modeled in the ANSYS Q3D to extract the power loop parasitic inductance as shown in Fig. 8. The power loop inductance based on the Q3D analysis is 1.27nH. The comparison between the proposed power loop inductance and the power loop inductance in previous literatures is provided in TABLE II. This result verified that a small power loop inductance is introduced by such PCB layout, which can help to reduce the overshoot on drain-source voltage during turning-off process.

IV. EXPERIMENT VERIFICATION

A. DPT

DPT test is conducted separately for two HF switch legs to verify the HF switching performance. The test diagram is shown in Fig. 9. One 200μH inductor is used as the load inductor. From Fig. 9, we can find that the corresponding LF switch is also included in the test setup, and during the DPT, the LF switch will keep on state to keep consistent with the ANPC control schematic.

The DPT test results are shown in Fig. 10. Based on the test results, the overshoots on drain-source voltage at 350V/42A are 4.3% and 5% for higher and lower HF switch leg respectively. Similar overshoots verified that the symmetrical

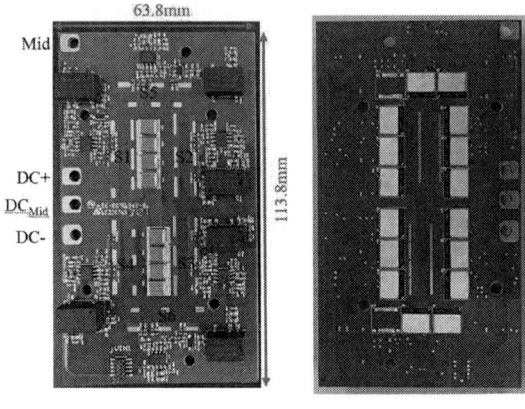

(a) Top View   (b) Bottom View

Fig. 4 Proposed GaN power module.

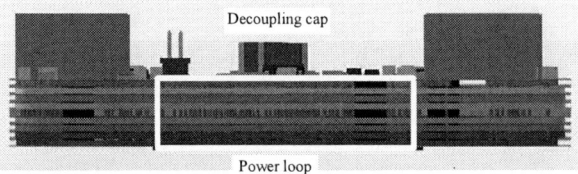

Fig. 5 Power loop of one HF switch leg.

Fig. 6 GaN module with installed custom heatsink.

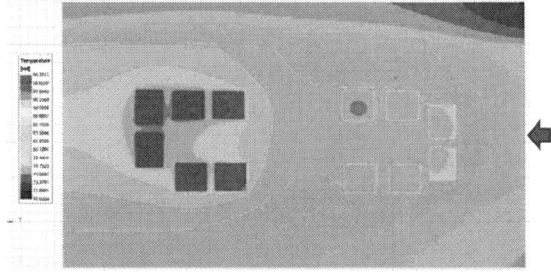

Fig. 7 Thermal simulation result @ Ploss_ave = 6.5W per GaN.

design of two HF switch legs. Small overshoots verified that low power loop parasitic inductance is achieved.

B. Conveter power test

To verify the electrical and thermal performance, a bidirectional three-phase 3L ANPC converter is built. The bidirectional power test results of the 3L converter is shown in

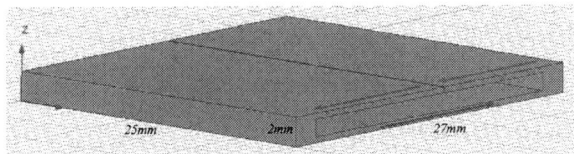

Fig. 8 Simulation model of power loop.

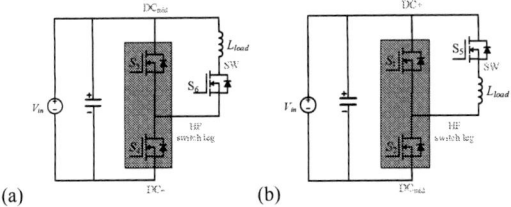

Fig. 9 Schematic of the DPT for the HF switch legs in the 3L modules (a) lower switch leg ; (b) higher switch leg.

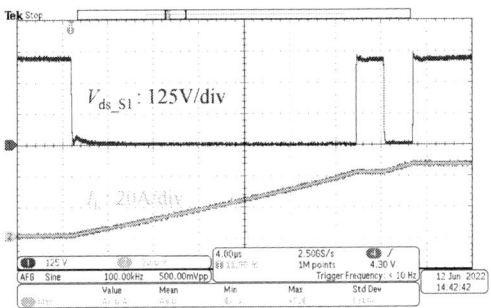

Fig. 10 DPT test waveform @ 350V/42A.

TABLE II.   COMPARISON OF POWER LOOP INDUCTANCE

Reference	Layout	Loop Inductance(nH)
[2]	Vertical	1.7-2.4
[3]	Vertical	0.77
[4]	Vertical	1.2
[14]	Lateral	5
This paper	Vertical	1.27

Fig. 11. Fig. 11(a) shows the 7.2kW PFC mode with three-phase 220V_{phase} AC input and 620V DC bus voltage output. Highest temperature is 46°C with around 20°C ambient temperature. Fig.11(b) shows the 8kW inverter mode test result with 600V DCbus voltage input and 180V_{phase} AC output. Only 5% overshoot voltage in 350V/42A DPT test and the bidirectional 3L converter-based power test with the maximum efficiency up to 99% verified the electrical and the thermal performance of the proposed GaN module.

V. CONCLUSION

This paper proposed an integrated 3L GaN power module to employ the 650V GaN deivces in the over-600V converters.

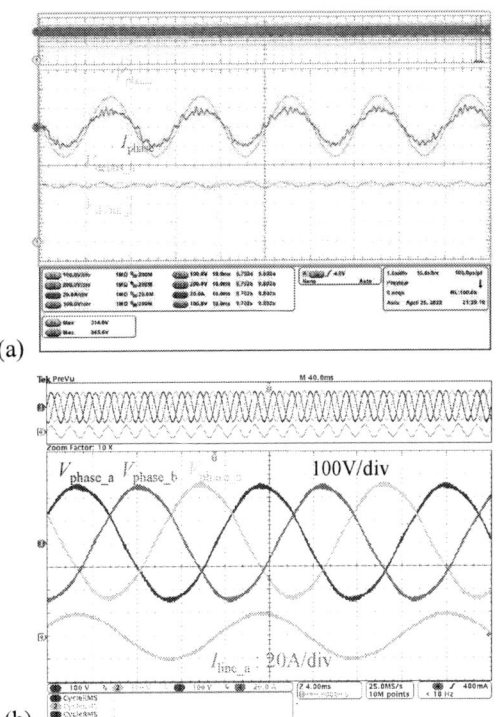

Fig.11 3L converter power test: (a) PFC mode @7.2kW; (b) inverter mode @8kW.

High-compactness is achieved by the integration of the isolated gate drivers, isolated auxiliary power supply and the decoupling cap together. Gate driver circuit is carefully designed to ensure the successful operation of paralleled GaN HEMTs. Low power-loop parasitic inductance and simple interface are achieved, which benefits from the optimized power loop design. The electrical and thermal performance has been verified by the DPT and bidirectional power test based on the three-phase 3L-ANPC converter, which is built by using the proposed GaN module. The experiment results show that the proposed GaN module could be a good candidate to employ the GaN HEMTs in the over 600V 3L converter design.

ACKNOWLEDGMENT

Authors would acknowledge the sponsorship and support from Magna Inc.

REFERENCES

[1] J. Millan, P. Godignon, X. Perpina, A. Perez-Tomas, and J. Rebollo, "A Survey of Wide Bandgap Power Semiconductor Devices," IEEE Trans. Power Electron., vol. 29, no. 5, pp. 2155-2163, May. 2014.

[2] K. Wang, B. Li, H. Zhu, Z. Yu, L. Wang, and X. Yang, "A Double-Sided Cooling 650V/30A GaN Power Module with Low Parasitic Inductance," in 2020 IEEE Applied Power Electronics Conference and Exposition (APEC), 2020, pp. 2772-2776.

978-1-6654-8901-0/22 $31.00 © 2022 IEEE

[3] Y. Yan, L. Zhu, J. Walden, Z. Liang, H. Bai and M. H. Kao, "Packaging A Top-cooled 650 V/150 A GaN Power Modules with Insulated Thermal Pads and Gate-Drive Circuit," 2021 IEEE Applied Power Electronics Conference and Exposition (APEC), 2021, pp. 2345-2350.

[4] W. Qian, J. Lu, H. Bai and S. Averitt, "Hard-Switching 650-V GaN HEMTs in an 800-V DC-Grid System With No-Diode-Clamping Active-Balancing Three-Level Topology," in IEEE Journal of Emerging and Selected Topics in Power Electronics, vol. 7, no. 2, pp. 1060-1070, June 2019.

[5] S. Kushwaha and M. T. Shah, "Bi-directional three-phase three-level neutral point clamped converter with capacitor voltage balancing scheme for unity power factor and low % THD," in Proc. IEEE 1st Int. Conf. Power Electron., Intell. Control Energy Syst. (ICPEICES), Delhi, India, Jul. 2016, pp. 1–6.

[6] M. Valente, F. Iannuzzo, Y. Yang and E. Gurpinar, "Performance Analysis of a Single-phase GaN-based 3L-ANPC Inverter for Photovoltaic Applications," 2018 IEEE 4th Southern Power Electronics Conference (SPEC), 2018, pp. 1-8.

[7] Y. Mei, X. Li, and Y. Qi, "A model predictive control method for threelevel bi-directional DC-DC converter in renewable generation system," in Proc. 18th Int. Conf. Elect. Machines Syst. (ICEMS), Oct. 2015, pp. 417–421.

[8] Z. Xia, Z. Liu and J. M. Guerrero, "Multi-Objective Optimal Model Predictive Control for Three-Level ANPC Grid-Connected Inverter," in IEEE Access, vol. 8, pp. 59590-59598, 2020.

[9] E. Gurpinar, D. De, A. Castellazzi, D. Barater, G. Buticchi, and G.Francheschini, "Performance analysis of SiC MOSFET based 3-level ANPC grid-connected inverter with novel modulation scheme," in Proc. of IEEE COMPEL, pp. 1-7, 2014.

[10] GaN Systems, "GN012 Application Note: Gate Driver Circuit Design with GaN E-HEMTs."

[11] E. Gurpinar, F. Iannuzzo, Y. Yang, A. Castellazzi, and F. J. I. T. o. I. A. Blaabjerg, "Design of low-inductance switching power cell for GaN HEMT based inverter," vol. 54, no. 2, pp. 1592-1601, 2017.

[12] D. Reusch and J. J. I. T. o. P. E. Strydom, "Understanding the effect of PCB layout on circuit performance in a high-frequency gallium-nitridebased point of load converter," vol. 29, no. 4, pp. 2008-2015, 2013.

[13] J. Lu, H. K. Bai, S. Averitt, D. Chen and J. Styles, "An E-mode GaN HEMTs based three-level bidirectional DC/DC converter used in Robert Bosch DC-grid system," 2015 IEEE 3rd Workshop on Wide Bandgap Power Devices and Applications (WiPDA), 2015.

[14] J. A. Brothers and T. Beechner, "GaN Module Design Recommendations Based on the Analysis of a Commercial 3-Phase GaN Module," 2019 IEEE Energy Conversion Congress and Exposition (ECCE), Baltimore, MD, USA, 2019, pp. 4109-4116.

978-1-6654-8901-0/22 $31.00 © 2022 IEEE

Design of High Current, High Power Density GaN Based Motor Drive for All Electric Aircraft Application

Waqar A. Khan, Armin Ebrahimian, S. Iman Hosseini S., Nathan Weise

Department of Electrical and Computer Engineering
Marquette University
Milwaukee, WI, USA
waqar.khan@marquette.edu, armin.ebrahimian@marquette.edu, iman.hosseini@marquette.edu, nathan.weise@marquette.edu

Abstract— **The past decade has seen noticeable advancements in the miniaturization of wide-band gap semiconductor technology making it feasible to be applied in demanding applications such as more-electric-aircrafts (MEA). Multiple high speed electric motors operating in parallel with current ratings in hundreds of amps are required to match the power delivery capability of conventional jet engines. As a result, the power electronics driving the motors require special attention to parallel multiple active switches in order to meet the high current requirements. This paper presents the design of an axially stator mounted 250 kW integrated modular motor drive for MEA applications. The design considerations presented include optimal component placement and layer stackup to minimize voltage overshoots during switch commutation, optimal gate drive layout to synchronize the current sharing of multiple GaN active switches, and the selection process of the on-board current sensor. Lastly, experimental test results are presented that verify the efficacy of the proposed design.**

Index Terms—**Integrated modular motor drives, MEA, GaN FETs, High Power Density Power Converters.**

I. INTRODUCTION

Integration of the electric machine and the drive system into one single structure, known as the concept of integrated modular motor drive (IMMD), has been thoroughly investigated in the past decade. Elimination of connecting cables between the electric machine and the drive system, increased power density, and modularity are some of the benefits of IMMDs [1]. These benefits make IMMDs a practical solution for aviation electrification where high power density and efficiency are essential requirements [2].

The emergence of wide band gap devices (WBGDs) has increased the practicality of designing power converters with high power density and efficiency [4]. Their intrinsic benefits like low on-state resistance and the capability to operate at high junction temperature make them suitable for high current/switching frequency applications [3]. For aviation applications, harsh operating conditions should be considered in

---

The information, data, or work presented herein was funded in part by the Advanced Research Projects Agency-Energy (ARPA-E), U.S. Department of Energy, under Award Number DE-AR0001352. The views and opinions of authors expressed herein do not necessarily state or reflect those of the United States Government or any agency thereof.

the design procedure including, vibrations, high environmental temperature, and EMI [5], [6]. From the point-of-view of power electronics in a drive system, ambient temperature (cooling system design), switching frequency, thermal dependency of the on-state resistance of the switch, and the number of parallel switches are the design variables, giving rise to a multivariate electrothermal problem.

The extremely fast switching speed of WBGDs during the turn on/off transition could lead to high values of dv/dt. As a result, the motor insulation could be adversely affected and in the worst case, high dv/dt could lead to insulation failure [7]. In the case of a non-optimal PCB design, the presence of parasitic elements will lead to severe overvoltages/overcurrents leading to detrimental performance and eventual failure of the active switches. Hence, component placement and in general PCB layout should be optimized to achieve a design with minimum parasitics. Compact component layout with proper placement of decoupling capacitors with respect to the active switches and optimal placement of signal/power nets within adjacent layers can maximize the mutual flux cancellation effects between them, leading to minimization of the parasitics [8], [9].

In [10], a GaN-based IMMD is designed and compared to its silicon (Si) counterpart. A more thorough study on the electrothermal design of the converter is presented in [11]. However, neither of the IMMD designs in the aforementioned references dealt with paralleling active switches as the power levels were relatively low.

This paper aims to bridge the gap by presenting the design of a high current paralleled switches GaN-based PCB busbar for a 250 kW axially stator mounted integrated modular motor drive. Design considerations presented in the paper include the optimal layer stack up of the busbar to minimize the common source, gate and power loop inductances, the layout of gate drive signal paths to drive multiple GaN switches in parallel, and the choice of onboard current sensor mainly driven by space constraints. Finally, experimental results are presented that verify the efficacy of the proposed design.

978-1-6654-8901-0/22 $31.00 © 2022 IEEE

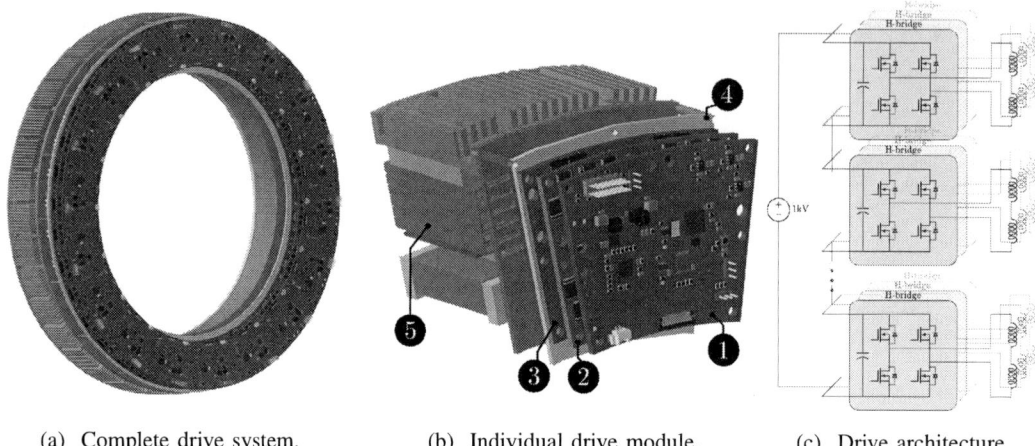

(a) Complete drive system.   (b) Individual drive module.   (c) Drive architecture.

Fig. 1: (a) Complete view of the axially stator iron-mounted fully integrated modular motor drive. (b) Individual drive module: 1) Communication board; 2) DC capacitor board; 3) GaN PCB; 4) GaN heatsink; 5) Additively manufactured motor coils; (c) Drive topology [3].

II. DRIVE TOPOLOGY

The structure of 3ϕ IMMD mounted directly on the stator of a surface permanent magnet motor (SPM) is shown in Fig. 1a with nominal system parameters at rated test conditions shown in Table I. The stator is comprised of a total of 36 coils (6 double coil pairs per phase), in which each double coil pair is driven by a GaN full-bridge converter thus leading to a total of 18 independent drive modules. Fig. 1b shows a zoomed in view of a single drive module comprising (from top to bottom) three individual PCBs, a cooling interface, and a motor segment. The PCB stack up consists of: 1) Local communication board (Top); 2) DC capacitor board (Middle); 3) GaN full-bridge converter (Bottom). The GaN FETs on the bottom PCB are oriented towards the direction of the heatsink which forms a part of the cooling jacket, lying between the PCBs and motor coils. It is to be noted that a singular cooling mechanism (cooling jacket) is used to extract the heat from both the motor coils and the active switches, simultaneously. The motor coils are directly connected to the full-bridge converter PCB, thereby eliminating the losses which would otherwise incur in the connection cables. The DC capacitor board stabilizes the DC voltage, whereas the communication board provides sensing, control, and PWMs to drive the local module.

The complete topology of the drive system is shown in Fig. 1c. Six individual drive modules are connected in series and thus share an equivalent motor's current per phase. Motor back emf per phase is thus the summation of emf produced by individual coils in any phase. Additionally, the distributed control regulates the DC capacitor voltages of individual modules actively and maintains them at $V_{DC}/6$ for stable operation. Since the drive topology shows symmetry and modularity, the subsequent sections will focus on the design of a single GaN full-bridge converter. In [3], a study was carried out using qualitative figure-of-merit (FOM) based on the switch's $R_{ds,on}$ and

TABLE I: Nominal System Parameters

Parameters	Value	Parameters	Value
V_{DC}	1 kV	Coil Resistance (R_s)	46 mΩ
I_{coil}	268.12 Arms	Coil Inductance (L_s)	137 μH
P_{rated}	250 kW	PM Flux (λ_{pm})	60.6 mWb
$f_{o,rated}$	1.25 kHz	Switching Frequency (f_{sw})	20 kHz
Power Factor ($\cos\phi$)	0.8	Target Efficiency (η)	$\geq$ 98%

output capacitance C_{oss} to determine the most optimal active switch for the application under consideration. Additionally, the paper addressed the number of active switches required to meet coil current (I_{coil}) requirement. Consequently, it was found that 8 EPC2034C GaN FETs per switch position satisfy the desired current and efficiency requirements for heatsink-ambient thermal resistance $R_{sa} < 0.2^o C/W$ [12].

III. DESIGN OF SINGLE PHASE FULL BRIDGE INVERTER MODULE

The front and back view of the 16-layer GaN full-bridge converter PCB is shown in Fig. 2. The PCB comprises of two physically symmetric half-bridges on each side. The board is 2.4 mm in thickness with 2 oz copper per layer. The specific power density is 26.13 kW/in^3 with a gravimetric power density of 3.713 kW/gm.

A. Layer Stackup

The layer stackup of the 16-layer full-bridge converter is shown in Fig. 3. The nets within the layers are arranged in a way to maximize the mutual flux cancellation between adjacent layers. There are three main contributors that degrade the performance of GaN FETs during switching: 1) Power loop inductance; 2) Gate loop inductance; 3) Common source inductance. The high-frequency power loop inductance L_{loop} is the inductance comprising parasitic inductance from the positive terminal of the HF decoupling capacitor, through

978-1-6654-8901-0/22 $31.00 © 2022 IEEE

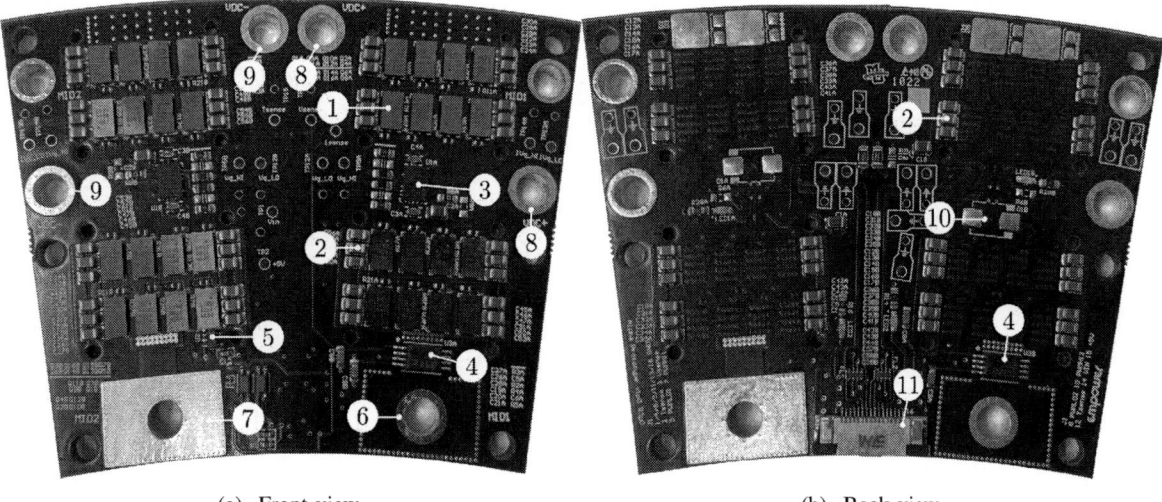

(a) Front view. (b) Back view.

Fig. 2: 16-Layer modular GaN full-bridge converter PCB. Legend: 1) EPC2034C; 2) Decoupling capacitors; 3) LMG1210 gate driver; 4) ACS37612 current sensor; 5) Temperature sensor; 6) Right half-bridge PHASE output; 7) Left half-bridge PHASE output; 8) $V_{DC}+$; 9) GND; 10) Bootstrap diode; 11) Connector.

the top and bottom device, ground plane, and back to the decoupling capacitor. This inductance adversely affects the switching speed as well as induces voltage overshoot during the commutation process through $L_{\text{loop}} di_{DS}/dt$, where i_{DS} is instantaneous drain-source current during the commutation process [13]. The optimal layout to minimize the power loop inductance was explained in [3] and is reiterated here for the sake of clarity. The high-side (HS) FET and the low-side (LS) FET are placed adjacent to each other (thus forming a power pole) on the 1st layer (top layer), while the HF decoupling capacitors are placed beside the drain of HS switch. The return path for the HF currents is through the 1st inner layer or the 2nd layer from the top. PCB vias connect the ground of the HF decoupling capacitors to the LS source terminal, where the dielectric thickness is kept as small as possible (dictated by the PCB manufacturer) to maximize mutual flux cancellation. As a result, the first layer consists of both the signal nets and $V_{DC}+$ with GND occupying layer 2 as shown in Fig. 3. In addition, the bottom layer or layer 16 is reserved for signal routing with the adjacent layer 15 reserved for GND.

The gate loop inductance is determined by the proximity of the gate driver relative to the position of the power poles as well as the mutual inductance between the go and return paths of the gate drive currents. To minimize this inductance, the gate driver must be located as close as possible to the switches, with the return path of the gate currents lying in the adjacent layer to the go path gate current layer.

The common source inductance L_{cs} is the inductance shared by the drain-source power current path and the gate drive loop [14]. Minimization of this inductance is critical as it adversely impacts the switching speed of the devices. As this inductance increases, the effective gate drive voltage and gate drive currents are effectively reduced, leading to longer

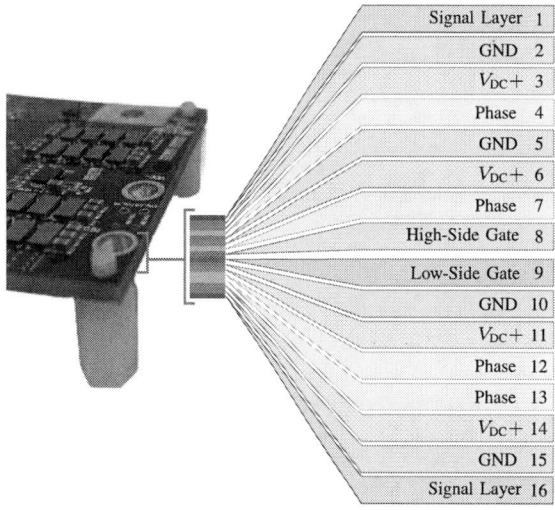

Signal Layer	1
GND	2
$V_{DC}+$	3
Phase	4
GND	5
$V_{DC}+$	6
Phase	7
High-Side Gate	8
Low-Side Gate	9
GND	10
$V_{DC}+$	11
Phase	12
Phase	13
$V_{DC}+$	14
GND	15
Signal Layer	16

Fig. 3: Layer stackup of the full-bridge converter.

switching transitions and thus increased switching losses. To minimize this inductance added by the PCB layout, it is essential to minimize the interaction between the HF power loop and the gate driver loop. This is accomplished in the following two ways: 1) The nets carrying the gate drive signals to FETs are placed away from Layers 1 and 2 (carrying HF power loop currents). In the design under consideration, Layer 8 and Layer 9 are chosen as the go paths for HS and LS gate currents with Layer 7 (PHASE) and Layer 10 (GND) nets forming the return paths for the gate currents, respectively; 2) The direction of the gate drive currents should be orthogonal to the main power loop currents. This is accomplished by orthogonal placement of gate drive resistances with respect to the decoupling capacitors.

978-1-6654-8901-0/22 $31.00 © 2022 IEEE

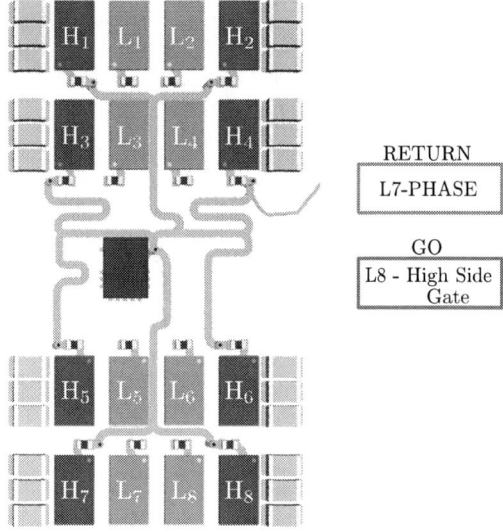

(a) High-side gate drive signal routing.

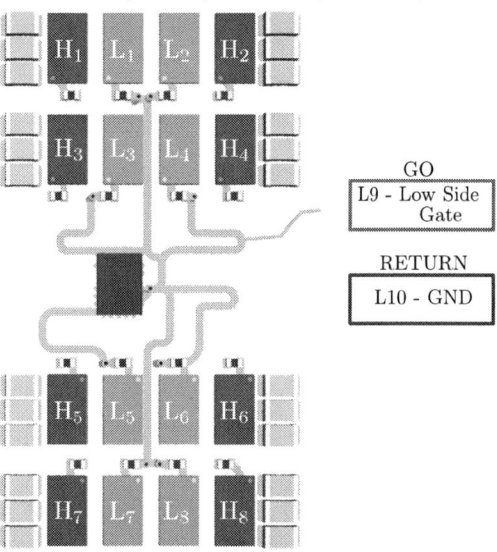

(b) Low-side gate drive signal routing.

Fig. 4: Layout of the gate drive signals. High-side FETs are indicated with purple, while low-side FETs are shown in green. LMG1210 gate driver is shown in blue.

Once the crucial HF power loops and gate drive loops are laid out, the remaining power nets can be distributed accordingly in the remaining unoccupied layers. Net $V_{DC}+$ should be adjacent to the GND net with the remaining layers dedicated to the PHASE output. Since a full-bridge per module is formed by two symmetric half-bridges, an isolation boundary is necessary in the PHASE layer to separate the two half-bridge PHASE outputs.

B. Gate Drive Signal Routing

As mentioned in the prior section, the distance of the gate driver with respect to the switches should be as small

and equal as possible to minimize the gate loop inductance. To that end, the most suitable layout geometry is a circle with the gate driver located at the origin and GaN switches equally distributed along the circumference. However, because of limited space on the PCB this cannot be accomplished. Therefore, the chosen geometry that best respects the space constraints is that of a rectangle with the gate driver located at its center on the top layer. The gate drive signal routes for HS and LS switches are then descended to their respective layers through vias as shown in Fig. 4a and Fig. 4b, respectively. The position of all HS and LS active switches are marked with H_m and L_m where m $\in \{1..8\}$. The go and return paths for the gate drive currents along with layer number are indicated on the right as well. A drawback of the chosen geometry is the unoptimal placement of the switches with respect to the gate driver. If the routing to individual gate drive resistances is performed by direct connection from the gate driver (thus leading to the minimum length of the gate drive nets), it can lead to uncoordinated turn-on and turn-off events during switching as the gate nets length will be unequal. This can lead to unequal current sharing among the GaN devices, and temperature hot spots resulting in device failure. To mitigate this issue, the nets are curved in such a way that the total net length for all active switches are approximately equal. For the case under study, the average length of the net from the HS common gate drive node to HS gate drive resistors is 17.458 mm and from LS common gate drive node to LS gate drive resistors is 14.434 mm, respectively. An additional practice to reduce gate loop inductance is to choose the gate signal track width as wide as permissible by the layout. In this design, a width of 20 mil provided acceptable results. Finally, to reduce the oscillations during turn-on and turn-off events, gate resistors of $R_G = 4.7 \ \Omega$ were used for all discrete FETs.

To determine the efficacy of the adopted layout practices, experimental results were carried out by measuring the gate drive signals at the gate node of FETs. The LS measurement was carried out by Tektronix TPP1000 1 GHz probes, while the HS measurements were carried out (with respect to PHASE output) using Tektronix TMDP0200 200 MHz differential probes at $V_{DC} = 0V$. The results for both the HS turn-on and turn-off and LS turn-on and turn-off are shown in Fig. 5. The figures exhibit a critically damped turn-on and turn-off transitions with gate signals of all 8 FETs well matched, lying on top of each other. If all the discrete switches were made from the same GaN wafer by the manufacturer, they will share a common gate threshold voltage and thus will exhibit simultaneous turn-on and turn-off using the presented layout guidelines.

C. Current Sensing

Current sensing is required for closed-loop operation and for the protection of the converter [15]. The following options were considered: 1) Direct resistive measurement; 2) Invasive hall effect sensing; 3) Non-invasive coreless hall effect sensing. Based on the current rating, the resistive measurement will require multiple current sense resistors to be connected

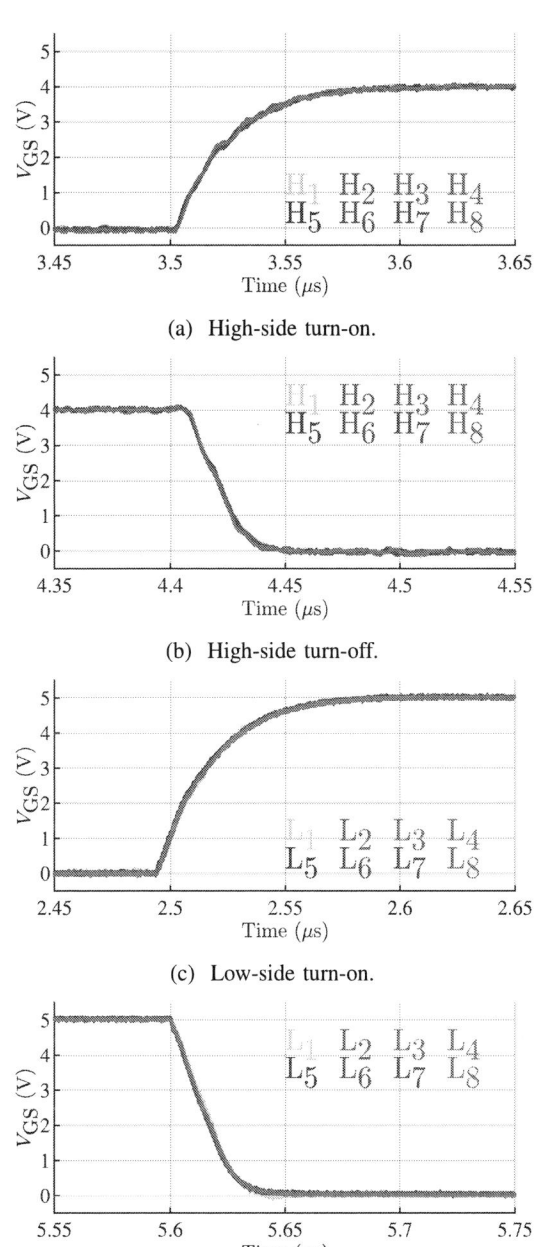

(a) High-side turn-on.

(b) High-side turn-off.

(c) Low-side turn-on.

(d) Low-side turn-off.

Fig. 5: Turn-on and turn-off waveforms for high-side and low-side switches at $V_{DC} = 0V$.

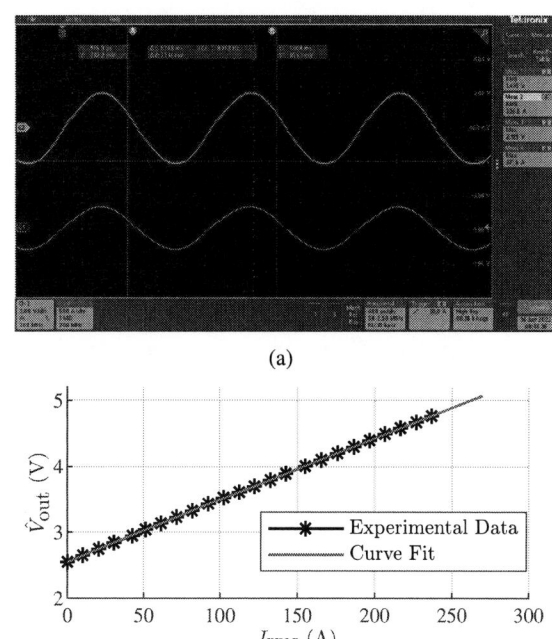

(a)

(b)

Fig. 6: (a) Experimental test results: ACS37612 output voltage (Ch2: 2 V/div), PHASE current (Ch4: 500 A/div) (b) Peak output voltage of current sensor against phase current.

board space available and therefore were rejected in favor of the subsequent technology.

Non-invasive coreless differential hall effect sensors such as ACS37612 are designed to be implemented in applications requiring measurement in hundreds of amps [17]. The sensors lack external field concentrator cores and thus are substantially smaller in size compared to the prior technology, favoring the application under study. Current flowing through a busbar or a PCB trace underneath the sensor generates a magnetic field that is sensed by linear Hall IC and converted into a proportional voltage, thus eliminating the need for additional signal conditioning circuitry. The sensitivity of the sensor is fixed but sensor gain is variable and depends on the PCB layout underneath the chip. A design tool provided by the manufacturer can be used to estimate the gain based on the track layout.

Experimental tests were conducted to verify the sensor gain by connecting the PHASE of a half-bridge to a California Instruments BP-30-1-480 AC source. The output of the current sensor in the form of voltage was measured by TPP1000 1Ghz probe, while the results were captured on Tektronix MDO3034 oscilloscope. Experimental waveforms at $I_{rms} = 226$ Arms are shown in Fig. 6a, while Fig. 6b shows a plot of peak sensor output voltage against injected PHASE current. The gain of the sensor remains linear throughout the test range and is determined to be 6.7 mV/A. The sensor output is saturated to its peak limit at a test current of $I_{rms} = 260$ Arms, thus limiting the operation to 97% of the rated coil current. This

in parallel. The sense resistors will incur additional loss and will require cooling. In addition, dedicated circuity will be required for signal conditioning and isolation. Invasive hall effect sensing can be accomplished by dedicated current sense ICs such as ACS773 [16]. These devices act as transducers converting input current into a proportional voltage through a fixed gain. The current flows into the device through bulky terminals (designed to handle high currents) which generates a magnetic field that the Hall IC converts into a voltage. The devices however are substantial in size relative to the limited

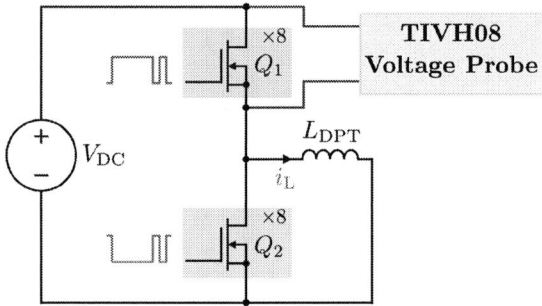

Fig. 7: Schematic of the double pulse test setup.

can be rectified in subsequent prototype versions by widening the current conduction path placed below the hall effect current sensor (equivalent to reducing the sensor gain).

IV. HARDWARE VERIFICATION

To verify the effectiveness of the proposed design, experimental tests were conducted to determine the voltage overshoot during the switch commutation process. The details of this experiment are provided in the subsequent section.

A. Double Pulse Tests Results

Power loop inductance results in the voltage overshoot on devices during switch commutation in a power pole. In [3], guidelines were provided to minimize the power loop inductance for the prototype under consideration. ANSYS Q3D software was used to analyze the stray inductance in the power loop which was found to be $L_{\text{Loop}} = 0.7$ nH.

To evaluate the overshoot via experimental tests, double pulse tests (DPT) were conducted on the prototype GaN PCB. The gating signals were generated from TI TMS320F28379D DSP. The schematic layout of the DPT is shown in Fig. 7. Due to the symmetric layout, the DPT test results of both the HBs are identical. The inductance of the test inductor is set to $L_{\text{DPT}} = 20.7$ μH. The waveforms were captured on Tektronix 1 GHz/6.25 GS/s MSO58 oscilloscope with Tektronix TIVH08 800 MHz optically isolated voltage probe for high-side switch drain-source voltage measurement and Tektronix TCP0150 20 MHz current probe for measurement of the test inductor current. The experimental waveforms captured at $V_{\text{DC}} = 130$ V and $i_{\text{L}} = 200$ A is shown in Fig. 8. The figure shows paralleled high-side switches drain-source voltage on the top and instantaneous inductor current on the bottom. The turn-off and turn-on waveforms of the high-side switches are shown in Fig. 9a and Fig. 9c with the zoomed in view shown in Fig. 9b and Fig. 9d, respectively. It can be seen that overshoot during the turn-off process is 140 V, only 10 V higher than the DC bus voltage. Similarly, the undershoot during the turn-on process is -10V as well. The results indicate that the voltage overshoot during the commutation process is $\approx 7.8\%$ above the test voltage.

B. Downscale Steady-State Test Results

This section discusses the downscale steady-state test results obtained by continuous testing of the converter by driving an

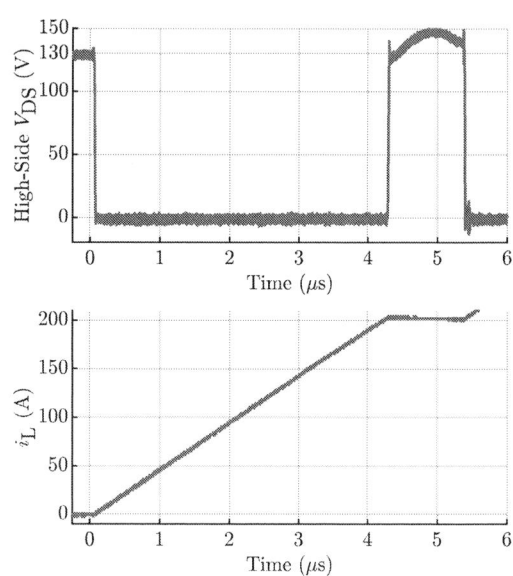

Fig. 8: Experimental double pulse test results at $V_{\text{DC}} = 130$ V, $i_{\text{L}} = 200$ A.

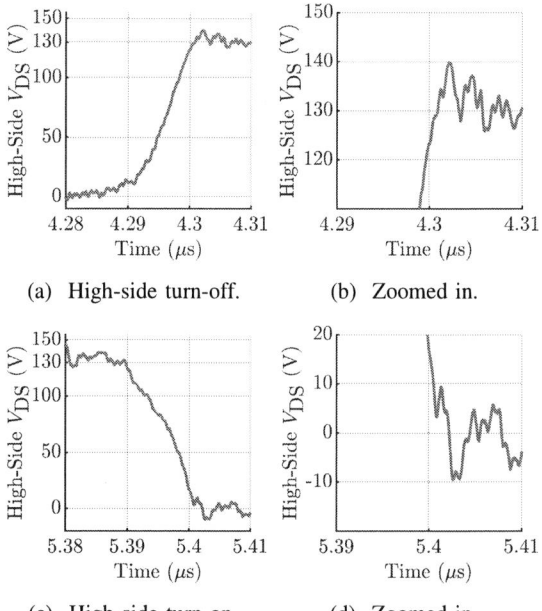

(a) High-side turn-off.

(b) Zoomed in.

(c) High-side turn-on.

(d) Zoomed in.

Fig. 9: Turn-off and turn-on waveforms for high-side switches at $V_{\text{DC}} = 130$V, $i_{\text{L}} = 200$ A.

RL load with $R_{\text{L}} = 1$ Ω and $L_{\text{L}} = 2.7$ mH at $V_{\text{DC}} = 110$ V, $f_{\text{o,rated}} = 1.25$ kHz and $f_{\text{sw}} = 20$ kHz. The modulation is carried out with unipolar PWM scheme and the test results for load voltage and load current are shown in Fig. 10. The maximum voltage during switch commutation is found to be 122 V, just 10% above the test voltage which is in accordance with results obtained in the prior section indicating the effectiveness of the adopted layout techniques. Tests are being scaled up in voltage and currents which will be presented in the future publications.

978-1-6654-8901-0/22 $31.00 © 2022 IEEE   252

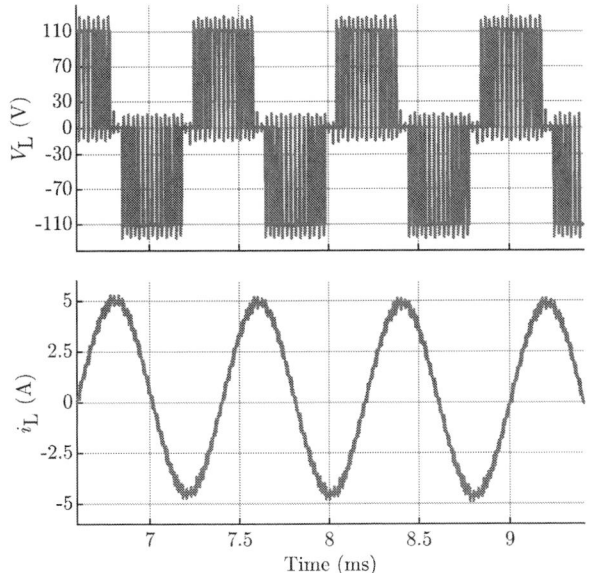

Fig. 10: Steady-state test results obtained at V_{DC} = 110 V, $\hat{i}_L$ = 5 A.

V. CONCLUSION

This paper has presented the design of a high current, high power density 16-layer GaN PCB busbar for an integrated modular motor drive intended to be used in a more-electric-aircraft application. The topology of the motor drive is modular, in which six individual GaN full bridge converters (modules) each driving a motor coil pair are connected in series to form a single phase. Independent operation of motor coils in series results in fault tolerance and division of DC-link voltage, thus allowing the use of low voltage high current GaN power transistors. Multiple GaN switches are connected in parallel to fulfill the IMMD high current requirements. This paper addresses the optimal component placement and layer stack up that aim to reduce the voltage overshoots during switch commutation, and presents gate drive layout guidelines to ensure equal current sharing among paralleled switches. Finally, the proposed PCB design is verified through experimental test results.

REFERENCES

[1] A. H. Mohamed, H. Vansompel, and P. Sergeant, "Electrothermal design of a discrete gan-based converter for integrated modular motor drives," *IEEE Journal of Emerging and Selected Topics in Power Electronics*, vol. 9, no. 5, pp. 5390–5406, 2021.

[4] S. Yin, K. J. Tseng, R. Simanjorang, Y. Liu, and J. Pou, "A 50-kw high-frequency and high-efficiency sic voltage source inverter for more electric aircraft," *IEEE Transactions on Industrial Electronics*, vol. 64, no. 11, pp. 9124–9134, 2017.

[2] W. Lee, S. Li, D. Han, B. Sarlioglu, T. A. Minav, and M. Pietola, "A review of integrated motor drive and wide-bandgap power electronics for high-performance electro-hydrostatic actuators," *IEEE Transactions on Transportation Electrification*, vol. 4, no. 3, pp. 684–693, 2018.

[3] A. Ebrahimian, W. A. Khan, S. Iman Hosseini S, and N. Weise, "Electrothermal design of a gan-based axially stator iron-mounted fully integrated modular motor drive," in *2022 IEEE Transportation Electrification Conference & Expo (ITEC)*, 2022, pp. 733–739.

[5] J. Wang, Y. Li, and Y. Han, "Integrated modular motor drive design with ¡roman¿gan¡/roman¿ power ¡roman¿fets¡/roman¿," *IEEE Transactions on Industry Applications*, vol. 51, no. 4, pp. 3198–3207, 2015.

[6] L. Verkroost, J. Van Damme, D. V. Bozalakov, F. De Belie, P. Sergeant, and H. Vansompel, "Simultaneous dc-link and stator current ripple reduction with interleaved carriers in multiphase controlled integrated modular motor drives," *IEEE Transactions on Industrial Electronics*, vol. 68, no. 7, pp. 5616–5625, 2021.

[7] R. A. Torres, H. Dai, W. Lee, T. M. Jahns, and B. Sarlioglu, "Development of current-source-inverter-based integrated motor drives using wide-bandgap power switches," in *2019 IEEE 15th Brazilian Power Electronics Conference and 5th IEEE Southern Power Electronics Conference (COBEP/SPEC)*, 2019, pp. 1–6.

[8] R. S. Krishna Moorthy, B. Aberg, M. Olimmah, L. Yang, D. Rahman, A. N. Lemmon, W. Yu, and I. Husain, "Estimation, minimization, and validation of commutation loop inductance for a 135-kw sic ev traction inverter," *IEEE Journal of Emerging and Selected Topics in Power Electronics*, vol. 8, no. 1, pp. 286–297, 2020.

[9] M. Abarzadeh, W. A. Khan, N. Weise, K. Al-Haddad, and A. M. EL-Refaie, "A new configuration of paralleled modular anpc multilevel converter controlled by an improved modulation method for 1 mhz, 1 mw ev charger," *IEEE Transactions on Industry Applications*, vol. 57, no. 3, pp. 3164–3178, 2021.

[10] Z. Gao, D. Jiang, W. Kong, C. Chen, H. Fang, C. Wang, D. Li, Y. Zhang, and R. Qu, "A gan-based integrated modular motor drive for open-winding permanent magnet synchronous motor application," in *2018 1st Workshop on Wide Bandgap Power Devices and Applications in Asia (WiPDA Asia)*, 2018, pp. 73–79.

[11] A. H. Mohamed, H. Vansompel, and P. Sergeant, "An integrated motor drive with enhanced power density using modular converter structure," in *2021 IEEE International Electric Machines & Drives Conference (IEMDC)*, 2021, pp. 1–6.

[12] (2020) Datasheet: Epc2034c. [Online]. Available: https://epc-co.com/epc/Portals/0/epc/documents/datasheets/EPC2034_datasheet.pdf

[13] D. Reusch and J. Strydom, "Understanding the effect of pcb layout on circuit performance in a high frequency gallium nitride based point of load converter," in *2013 Twenty-Eighth Annual IEEE Applied Power Electronics Conference and Exposition (APEC)*, 2013, pp. 649–655.

[14] D. Reusch, "Effectively paralleling gallium nitride transistors for high current and high frequency applications," EPC, Tech. Rep., 2016. [Online]. Available: https://epc-co.com/epc/Portals/0/epc/documents/application-notes/AN020%20Effectively%20Paralleling%20Enhancement%20Mode%20Gallium%20Nitride%20Transistors.pdf

[15] M. Abarzadeh, A. Ebrahimian, W. A. Khan, N. Weise, and K. Al-Haddad, "Systematic design of improved lead-lag direct power model predictive controller for multilevel active-front-end rectifier: A comparative study," in *IECON 2021 - 47th Annual Conference of the IEEE Industrial Electronics Society*, 2021, pp. 1–8.

[16] Datasheet: Acs773. [Online]. Available: https://www.allegromicro.com/en/Products/Sense/current-sensor-ics/Fifty-To-Two-Hundred-Amp-Integrated-Conductor-Sensor-ICs/ACS773

[17] Datasheet: Acs37612. [Online]. Available: https://www.allegromicro.com/en/products/sense/current-sensor-ics/sip-package-zero-to-thousand-amp-sensor-ics/acs37612

Design of Three-Level Flying Capacitor Totem Pole PFC in USB Type-C Power Delivery for Aircraft Applications

Tianyu Zhao, Rolando Burgos and Bo Wen
Center for Power Electronics and Systems
Virginia Polytechnic Institute and State University
Blacksburg, VA 24061, USA
Emial: tianyuz@vt.edu, roburgos@vt.edu, wenbo@vt.edu

Andrew McLean and Rodrigo Fernández Mattos
Collins Aerospace
Shirley, Solihull, United Kingdom
andrew.mclean@collins.com, rodrigo.fernandez-mattos@collins.com

Abstract— **This paper provides an optimization process of a 100-W three-level flying capacitor totem-pole PFC in USB Type-C Power Delivery (PD) for aircraft applications. The switching frequency and PFC inductor current ripple are varied to set the PFC inductance variation range and design the differential mode (DM) and common mode (CM) filter to meet the EMI requirement in the DO-160 standard. The design process for all the magnetic components is described with results at different switching frequencies and current ripples. The total volume and loss for all cases with both single-stage and two-stage filters are plotted and compared. Pareto fronts for the two cases are drawn to show the optimal design parameters for the application.**

Keywords— *power factor correction, flying capacitor three-level converter, EMI filter*

I. INTRODUCTION

To increase aircraft profitability and respond to environmental constraints, the concept of More Electric Aircraft (MEA) [1] was raised to replace conventional non-electrical systems such as pneumatic, hydraulic and mechanical with electrical systems in the aircraft. With the development of consumer electronics devices, the USB charging standard has evolved to the Type-C version. USB Type-C Power Delivery (PD) enforces 4 output voltage levels with a maximum power of up to 100W which dramatically reduces the charging time. To utilize this emerging charging standard in the MEA system, higher power density and efficiency are required for the charging system. Furthermore, compared with standard FCC and CISPR 22 EMI standards which are widely applied in consumer electronics, DO-160 has more than 20 dB lower requirement at the higher frequency range. This imposes challenges on the line filter design.

The flying capacitor (FC) multi-level topologies are promising in medium voltage applications like high-speed motor drives [2,3] and grid-connected converters. The advantage of shrinking the filter sizes makes it more suitable for aerospace applications. However, the high capacitance values and stored energy of the flying capacitors limit this topology to be applied in the high-power (e.g. MVA) application [4]. Meanwhile, under this power level, multilayer ceramic

capacitors (MLCCs) with high energy storage density [5] can be applied to compensate for the disadvantage.

In the normal two-level PFC converter, the switching frequency is increased to the MHz range to minimize the EMI filter size [6]. Soft-switching technique is necessary to maintain high efficiency but makes the control strategy of the PFC stage more complicated. Also, the soft-switching condition can be easily affected by the control delay time [7]. For the three-level topology, the effective frequency is doubled at the filter side to allow a lower switching frequency. Moreover, the conventional PFC control method can still be applied to the three-level topology. The increased voltage level also brings the benefit to use low-voltage devices. From the device Figure-of-Merit (FoM) aspects considering device conduction and switching losses, low-voltage devices have an advantage over high-voltage devices [8].

This paper provides the whole design process of a 100-W three-level FC totem-pole PFC which is the front-end of USB Type-C PD for aircraft applications. The detailed specification is shown in Table 1. To achieve high power density and efficiency based on the DO-160 requirement. The loss and volume of semiconductors, EMI filters, and PFC inductors are evaluated under different switching frequencies and current ripples.

II. OPERATION PRINCIPLE OF THREE-LEVEL FLYING CAPACITOR TOTEM POLE PFC

Compared to traditional totem pole PFC, this topology replaces the half-bridge high-frequency phase-leg with a three-level FC phase-leg as shown in Fig. 1. Because of the extra voltage level brought by the flying capacitors, a three-level voltage waveform can be seen in half of the line cycle. Owing

TABLE I. CONVERTER SPECIFICATION

Input	115 V rms, 360–800 Hz
Output	200 V
Power	100 W
EMI Standard	DO-160
Altitude	15 kft

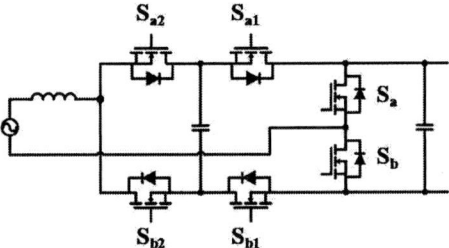

Fig. 1. Topology of three-level FC totem-pole PFC

to the line-frequency phase leg, a total five-level waveform can be seen at the ac nodes. Phase-shifted PWM (PS-PWM) method is applied to the three-level phase-leg which can naturally keep the flying capacitor voltage balanced. The Flying capacitor is charged and discharged in each switching cycle. Fig. 2 shows the flying capacitor charging and discharging condition when the input voltage V_{ac} is lower and higher than half of the bus voltage in the positive line cycle. The modulation waveform and switching node voltages are shown in Fig. 3.

III. CONVERTER OPTIMIZATION

To achieve high efficiency and power density, loss and volume of the semiconductor and EMI filter are optimized as these two parts share most of the whole converter. The switching frequency and PFC inductor current ripple are the two main variables that set the range for the inductance selection. Higher switching frequency leads to higher semiconductor loss, but it also requires smaller inductances so that the magnetic loss and size can be reduced. The PFC inductor current ripple leads to the trade-offs between the PFC inductance and the DM inductance. Since both inductors contribute to DM noise attenuation, a systematic optimization of the two inductors is needed. Therefore, the optimization process is summarized as follows. Firstly, the PFC inductance is selected under different switching frequencies and current ripples. Then the DM and CM filters are designed to match the EMI standard in DO-160. Both

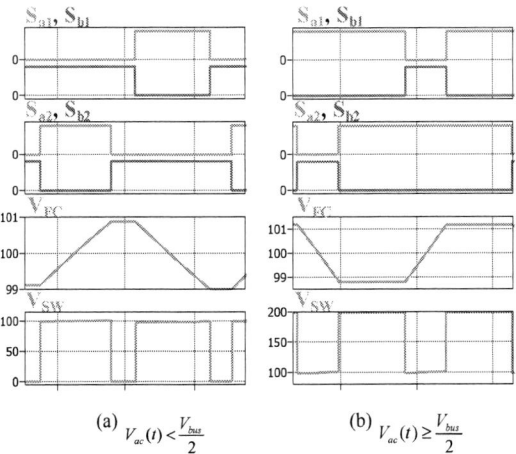

Fig. 2. Flying capacitor charging and discharging condition in one switching cycle

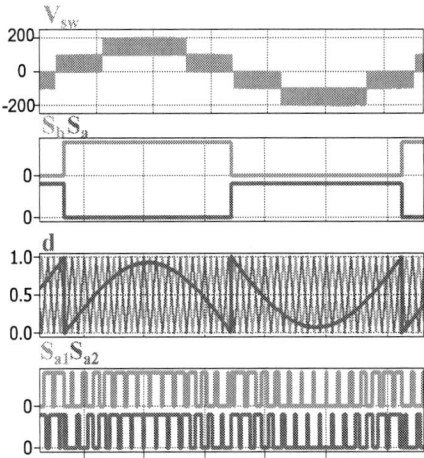

Fig. 3. Operation waveforms with PS-PWM for three-level FC totem-pole PFC.

single-stage and two-stage LC filters are considered. For each designed inductance, different losses and sizes can be obtained with different magnetic core selections. Finally, the optimum design can be observed from the total loss and volume.

A. PFC and DM Inductance Selection

Because of the unity power factor requirement, the duty ratio varies in a sine function which has little phase difference from the input voltage. The current ripple can be approximately calculated according to Equation (1)-(3).

$$V_{ac}(t) = V_{in}\sin(\omega t) \tag{1}$$

$$d(t) \approx \frac{V_{ac}(t)}{V_{bus}} \tag{2}$$

$$\Delta I_L = \begin{cases} \dfrac{V_{ac}(t)\cdot(1/2-d(t))}{L_{PFC}\cdot f_s} & V_{ac}(t) < \dfrac{V_{bus}}{2} \\[3mm] \dfrac{\left(V_{ac}(t)-\dfrac{V_{bus}}{2}\right)\cdot(1-d(t))}{L_{PFC}\cdot f_s} & V_{ac}(t) \ge \dfrac{V_{bus}}{2} \end{cases} \tag{3}$$

Fig. 4 Shows the PFC inductance varies with switching frequency under different current ripple percentages. The current ripple percentage is the ratio between the maximum current ripple and the inductor peak current. One of the advantages of using a multi-level phase-leg is shrinking the filter size by shifting the noise spectrum to a higher frequency range. For the DM noise of three-level FC topology, the noise peaks exist at the even times of switching frequencies. Due to the power factor requirement, the DM capacitor value is set to limit the reactive power below 5% of the total power at maximum load. Therefore, the DM capacitor is fixed at 60 nF. Based on the EMI standard, a specific PFC inductance will require a DM inductance as shown in Fig. 5. Because the EMI standard range is 150 kHz – 30 MHz, frequency below 75 kHz can avoid the standard effectively which shrinks the value a lot for the DM inductor. Compared with a single-stage filter, two-stage filter inductance decreases more with switching frequency.

978-1-6654-8901-0/22 $31.00 © 2022 IEEE

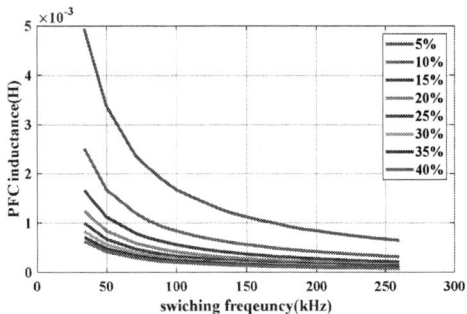

Fig. 4. PFC inductance variation as switching frequency under different current ripple percentage

B. CM Inductance Selection

The CM capacitor is fixed at 5.36 nF due to the limit of the ground leakage current [9]. All the parasitic capacitors to ground C_g are considered as Fig. 6 shows. According to [10], the equivalent circuit of three-level FC PFC including LISN can be obtained as shown in Fig. 7. The capacitance to ground C_g which decides the CM noise level is set as 2pF conservatively. The comparison of the noise from the equivalent circuit and time domain circuit simulation results are shown in Fig. 8. The simulated noise spectrum from the equivalent circuit matches well with the model at the dominant noise frequencies. Since the filter is designed based on these frequencies, the error at the high-frequency range doesn't affect the final results.

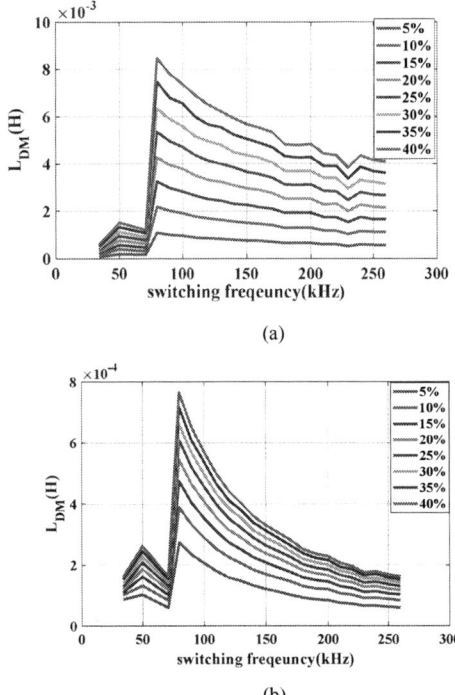

(a)

(b)

Fig. 5. (a) Single-stage filter DM inductance variation (b) Two-stage filter DM inductance variation under different switching frequencies and current ripple percentage.

C. Inductor design process

The inductor design begins with selecting the appropriate core size and materials. For the PFC inductor and DM inductor, the MPP core and Kool Mu core from Magnetics are selected due to their high inductance per turn A_L per volume. For the CM inductor, the nanocrystalline core is selected due to its high permeability. The toroidal core shape is selected for design simplicity. For the toroidal core, after selecting the appropriate core size and material according to the inductance and inductor current, the turns number N can be designed with the A_L value provided by the manufacturer. Then the winding gauge can be calculated based on the turns number and window area of the core. Here, the winding factor is assumed to be 40%. After these steps, the volume of the inductor is obtained.

The total inductor loss contains the winding loss and core loss. For the winding loss calculation, both the dc and ac losses are considered. The dc resistance is calculated according to the wire gauge and core geometry. For the ac resistance, both skin effect and proximity effect are evaluated according to [11]. The improved generalized Steinmetz equation (iGSE) [12] is adopted to calculate the core loss based on the current waveforms of the inductor from the simulation.

D. Flying Capacitor Selection

The capacitance value is decided by the capacitor charge and maximum current ripple limit because this voltage ripple is applied to the device bias. In the different periods of a line cycle,

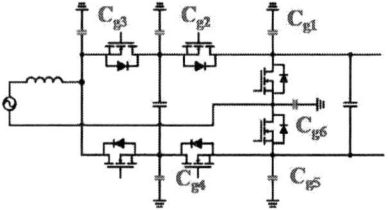

Fig. 6. Three-level FC totem-pole PFC and its CM parasitics.

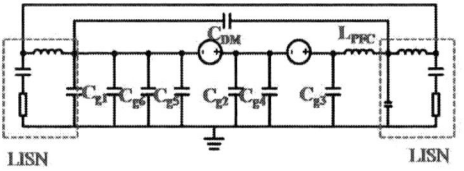

Fig.7. Three-level FC totem-pole PFC model for CM noise analysis

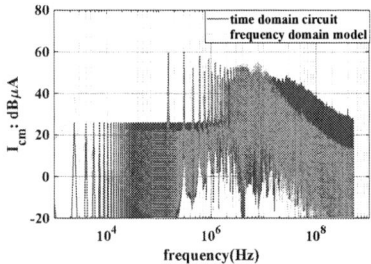

Fig. 8. CM noise comparison between time-domain circuit and frequency domain model analysis when switching frequency is 150 kHz.

the flying capacitor charge and discharge time are as Equation (4) shows.

$$C = \begin{cases} \dfrac{i_L(t) \cdot d(t)}{\Delta V \cdot f_s} & V_{ac}(t) < \dfrac{V_{bus}}{2} \\[2ex] \dfrac{i_L(t) \cdot (1-d(t))}{\Delta V \cdot f_s} & V_{ac}(t) \geq \dfrac{V_{bus}}{2} \end{cases} \qquad (4)$$

The flying capacitor voltage ripple calculation is also verified with the simulation waveform. Fig. 9 shows the voltage ripple ac component in the positive half of a line cycle and the calculation oscillation magnitude.

E. Optimization Results and analysis

Optimization results with single-stage and two-stage filters are shown in Fig. 10. Since the volume of semiconductors and capacitors are not affected by the switching frequency and current ripple, only the volume of the PFC inductor and EMI filter are plotted in the figure. For the designs with the single-stage filter, lower switching frequency results stand out. This indicates the advantage of the single-stage filter on the volume. For the two-stage filter case, better performance of high-frequency results shows the advantage on loss.

Comparing the Pareto front of the two cases in Fig. 10, the single-stage filter shows comparable performance compared

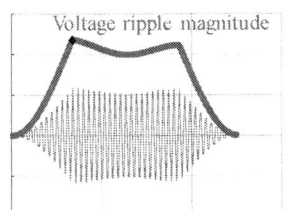

Fig. 9. Flying capacitor voltage ripple ac component and oscillation magnitude.

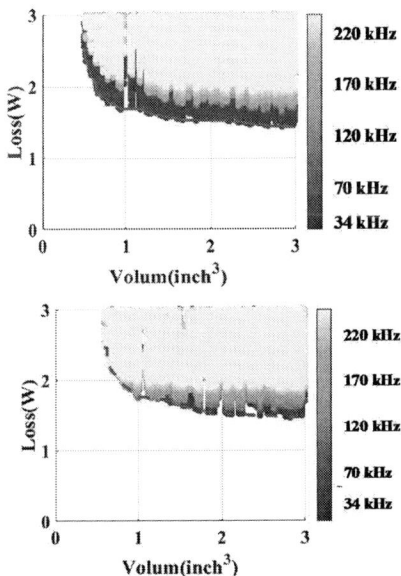

Fig. 10. Optimization results with (a) single-stage and (b)two-stage filters

with the two-stage case. However, the single-stage design is preferred since the switching frequency is much lower, which reduces thermal stress on the switching devices.

IV. CONCLUSION

The optimization process of a 100-W three-level FC totem-pole PFC in USB Type-C Power Delivery (PD) for aircraft applications was presented. By varying the switching frequency and PFC inductor current ripple, a trade-off between the total loss and volume was shown with the PFC inductor and EMI filter designed to meet the EMI standard in DO-160 requirement. In the final optimization results, the single-stage EMI filter shows comparable performance to the two-stage filter. With low thermal stress resulting from the low switching frequency, a prototype with the single-stage filter has been

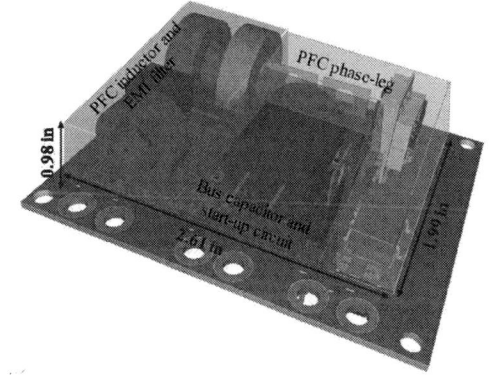

Fig. 11. Flying capacitor voltage ripple ac component and oscillation magnitude.

designed as Fig. 11 with a total converter volume as 5.1 in³. The efficiency and EMI performance will be verified with the experiment in the future.

REFERENCES

[1] Challenges Available: https://eng.umd.edu/~austin/ense622.d/lecture-resources/Boeing787-MoreElectricAircraft.pdf.

[2] J. Rodriguez, S. Bernet, B. Wu, J. O. Pontt and S. Kouro, "Multilevel Voltage-Source-Converter Topologies for Industrial Medium-Voltage Drives," in IEEE Transactions on Industrial Electronics, vol. 54, no. 6, pp. 2930-2945, Dec. 2007.

[3] N. Pallo, T. Foulkes, T. Modeer, S. Coday and R. Pilawa-Podgurski, "Power-dense multilevel inverter module using interleaved GaN-based phases for electric aircraft propulsion," 2018 IEEE Applied Power Electronics Conference and Exposition (APEC), San Antonio, TX, USA, 2018, pp. 1656-1661.

[4] D. Krug, S. Bernet, S. S. Fazel, K. Jalili and M. Malinowski, "Comparison of 2.3-kV Medium-Voltage Multilevel Converters for Industrial Medium-Voltage Drives," in IEEE Transactions on Industrial Electronics, vol. 54, no. 6, pp. 2979-2992, Dec. 2007.

[5] C. B. Barth, T. Foulkes, I. Moon, Y. Lei, S. Qin and R. C. N. Pilawa-Podgurski, "Experimental Evaluation of Capacitors for Power Buffering in Single-Phase Power Converters," in IEEE Transactions on Power Electronics, vol. 34, no. 8, pp. 7887-7899, Aug. 2019.

[6] Z. Liu, F. C. Lee, Q. Li and Y. Yang, "Design of GaN-based MHz totem-pole PFC rectifier," 2015 IEEE Energy Conversion Congress and Exposition (ECCE), Montreal, QC, Canada, 2015, pp. 682-688.

[7] Z. Huang, Z. Liu, Q. Li and F. C. Lee, "Microcontroller-based MHz totem-pole PFC with critical mode control," 2016 IEEE Energy Conversion Congress and Exposition (ECCE), Milwaukee, WI, USA, 2016, pp. 1-8.

[8] M. Guacci et al., "Experimental Characterization of Silicon and Gallium Nitride 200 V Power Semiconductors for Modular/Multi-Level Converters Using Advanced Measurement Techniques," in IEEE Journal of Emerging and Selected Topics in Power Electronics, vol. 8, no. 3, pp. 2238-2254, Sept. 2020.

[9] IEC Standards—Safety of Information Technology Equipment, IEC 60950, 1999

[10] S. Wang, P. Kong and F. C. Lee, "Common Mode Noise Reduction for Boost Converters Using General Balance Technique," in IEEE Transactions on Power Electronics, vol. 22, no. 4, pp. 1410-1416, July 2007.

[11] Kazimierczuk, M. K. , *High-Frequency Magnetic Components*, Germany: Wiley, 2013.FR

[12] K. Venkatachalam, C. R. Sullivan, T. Abdallah and H. Tacca, "Accurate prediction of ferrite core loss with nonsinusoidal waveforms using only Steinmetz parameters," 2002 IEEE Workshop on Computers in Power Electronics, 2002. Proceedings., Mayaguez, PR, USA, 2002, pp. 36-41.

978-1-6654-8901-0/22 $31.00 © 2022 IEEE

AUTHOR INDEX

Adina, Nihanth 77
Agamy, Mohammed 164
Agarwal, Anant K. 7, 22, 45, 49
Aghdam, Sima A. 164
Al-Jassim, Mowafak M. 49
Alkhalid, Khalid 77
Alsaif, Faisal 77
Amirkhanian, Hrach 238
Appert, Randy 227
Bach, Hoang L. 99
Bai, Hua K. 105
Bai, Hua 242
Baliga, B. Jayant 1
Bartholomé, Kilian 186
Basler, Michael 186
Beach, Bob 64
Bech, Michael M. 180
Bhattacharya, Subhashish 159, 192
Biswash, Rahul 111
Borjas, Rob 86
Burgos, Rolando 254
Cao, Joe 64
Castagna, Maria E. 31
Cerantonio, Viviana 35
Chen, Chen 209
Chen, Feida 215
Chen, Zibo 209
Chenetz, Steve 58
Cheng, Bing 105
Cheng, Qianyi 154
Chini, Alessandro 31
Chowdhury, Nadim 40
Cioni, Marcello 31
Cole, Zach 154
Cong, Yizhou 154
Dalal, Dipen N. 180
Das, Partha P. 192
Deboer, Skylar 54
Deshpande, Amol 154
Devoto, Douglas 227
Ebrahimian, Armin 247
Evans, Daniel 142
Ewanchuk, Jeffrey 86
Flicker, Jack 26
Fu, Pengyu 154
Gafford, James 142
Gao, Zihan 148, 221
Garg, Reenu 58

Gervasi, Leonardo 35
Giandalia, Marco 68
Ginley, David S. 227
Gonzalez, Nicolas 227
Goyal, Amit 154
Green, Jordan 64
Guo, Zhicheng 91
Heinselman, Karen 227
Heris, Pedram C. 111, 198
Hsu, Fu-Jen 176
Hu, Boxue 154
Huang, Alex Q. 17, 82, 91, 209
Huang, Anqi 99
Huang, Chih-Fang 176
Huang, Chih-Feng 176
Huang, Qingyun 209
Huang, Yang 105
Huesgen, Till 204
Hung, Hsiang-Ting 176
Iman, Hosseini S. S. 247
Isamotu, Mohamed F. 40
Isukapati, Sundar B. 122
Iucolano, Ferdinando 31, 35
Jahns, Thomas M. 215
Jang, Seung Y. 11, 54, 122, 127
Jank, Michael P. M. 99
Jin, Fanning 105
Jin, Feng 170
Jin, Michael 7, 45, 49
Kallfass, Ingmar 204
Kaplar, Robert 26
Kelkar, Kapil 238
Kempitiya, Asantha 238
Khan, Waqar A. 247
Kheirollahi, Reza 116
Kim, Dongyoung 11, 54
Kirkeby, Mathias 180
Koch, Dominik 204
Kuball, Martin 35
Kumar, Ashish 154
Lee, Sangwhee 215
Lehr, Jane M. 227
Li, Ang 95
Li, Haiguo 148, 221
Li, Qiang 170
Li, Xiao 86, 154
Li, Zheqing 170
Li, Ziqian 95

Liang, Ziwei	242
Lin, Guan-Wei	176
Lin, I-Chi	176
Lin, Lung-Sheng	176
Liu, Tianshi	7, 45, 49
Liu, Wen	95
Lu, Fei	116
Lu, Juncheng	233
Lu, Wu	22
Luo, Fang	132, 138
Luo, Xixi	82
Lynch, Justin	127
Ma, Dihao	86
Maddi, Hema L. R.	22, 45
Mahadik, Nadeemullah	49
Major, Joshua	227
Mancini, Stephen A.	11
Mansour, Kareem	186
Mantooth, Alan	111, 198
März, Martin	99
Mattos, Rodrigo F.	254
McLean, Andrew	254
Meyer, Dennis	58
Miccoli, Cristina	35
Moench, Stefan	186
Morgan, Adam J.	54, 127
Moschetti, Maurizio	31
Munk-Nielsen, Stig	180
Nayak, Suvendu	22
Nielsen, Morten R.	180
Niroula, John	40
Odekirk, Bruce	58
Oeder, Thorsten	73
Olejniczak, Kraig	154
Oliver, Stephen	68
Palacios, Tomás	40
Parkhideh, Babak	142
Pelletier, Daniel	154
Pfost, Martin	73
Pomeroy, James	35
Prabowo, Yos	159
Pronko, Steven G. E.	227
Qian, Jiashu	45, 49
Quay, Rüdiger	186
Rajput, Nitul S.	40
Rauh, Hubert	99
Reiner, Richard	186
Riar, Baljit	86
Ribarich, Tom	68
Roy, Chondon	142
Saadatizadeh, Zahra	111, 198
Salemi, Arash	7

Samani, Rahil	233
Sanabria, David E.	227
Sarlioglu, Bulent	215
Satpathy, Subhransu	159, 192
Schletz, Andreas	99
Schrock, Emily	26
Sengupta, Arijit	164
Sharma, Ankit	204
Sharma, Shrivatsal	159
Shen, Yi	95
Sheridan, David	7
Shi, Limeng	45, 49
Shi, Xiaodong	105
Shimbori, Atsushi	17
Sirat, Ali P.	142
Soto, Jake	49
Speer, Kevin	58
Stahlbush, Robert	49
Stecklein, Gordon	64
Sun, Yue	242
Sung, Woongje	11, 54, 122, 127
Talesara, Vishank	22
Tarmoom, Ehab	58
Teng, Yue	99
Ul-Hassan, Mustafeez	132, 138
Vaughan-Edmunds, Llew	68
Veliadis, Victor	192
Verzellesi, Giovanni	31
Waltereit, Patrick	186
Wang, Fred	148, 221
Wang, Jin	77, 86, 154
Wang, Ke	154
Weise, Nathan	247
Wen, Bo	254
White, Marvin H.	7, 45
Wong, Christopher	64
Wu, Hailing	86
Wu, Yuxuan	132, 138
Xia, Xin	105
Xie, Qingyun	40
Xu, Yibo	22
Yang, Ta-Yung	176
Yao, Pengfei	148
Yen, Cheng-Tyng	176
Yerra, Srikanth	238
Yuan, Mengyang	40
Yuan, Tianlong	170
Yun, Nick	127
Zagni, Nicolò	31
Zhang, Hua	116
Zhang, Xuning	58
Zhang, Yue	77, 86

Zhao, Hongbo 180
Zhao, Shuyan 116
Zhao, Tianyu 254
Zhu, Liyan 242
Zhu, Shengnan 7, 45
Zurbriggen, Ignacio G 233

IEEE
445 Hoes Lane
Piscataway, NJ 08854-4141

ISBN 978-1-6654-8901-0